WACHSTUMSHORMON UND WACHSTUMSSTÖRUNGEN

DAS CUSHING-SYNDROM

ELFTES SYMPOSION
DER DEUTSCHEN GESELLSCHAFT FÜR ENDOKRINOLOGIE
IN DÜSSELDORF VOM 5. BIS 7. MÄRZ 1964

SCHRIFTLEITUNG

PROFESSOR DR. ERICH KLEIN
2. MED. KLINIK UND POLIKLINIK DER MED. AKADEMIE DÜSSELDORF

MIT 130 ABBILDUNGEN

SPRINGER-VERLAG BERLIN HEIDELBERG GMBH

1965

Library of Congress Catalog Card Number 55—39230

ISBN 978-3-540-03413-1   ISBN 978-3-642-88683-6 (eBook)
DOI 10.1007/978-3-642-88683-6

Titel-Nr. 6777

Inhaltsverzeichnis

III. Freie Vorträge

Nebennierenrinde und ihre Hormone

Schwangerschaft

Hypophyse

Unspezifische Hormonwirkungen

Alphabetisches Verzeichnis der Vortragenden und Diskussionsredner

Bahner, F., Prof. Dr., Med. Universitäts-Poliklinik, Heidelberg, Hospitalstr. 3.

Bayer, J. M., Prof. Dr., Chir. Universitätsklinik, Bonn, Venusberg.

Berg, D., Dr.. I. Med. Universitätsklinik, Frankfurt/M., Ludwig-Rehn-Str. 14.

Berswordt-Wallrabe, Frau I. von, Dr., Universitäts-Frauenklinik, Göttingen.

Berswordt-Wallrabe, R. von, Dr., Hauptlaboratorium der Schering A.G., Berlin N. 65, Müllerstr. 170—172.

Bethge, H., Dr., II. Med. Klinik und Poliklinik der Med. Akademie, Düsseldorf.

Bettendorf, G., Doz. Dr., Universitäts-Frauenklinik, Hamburg-Eppendorf.

Bierich, J. R., Prof. Dr., Universitäts-Kinderklinik, Hamburg-Eppendorf.

Börner, W., Doz. Dr., Poliklinik der Universität Würzburg.

Bottermann, K., Dr., II. Med. Klinik der Universität München.

Breckwoldt, M., Dr., Universitäts-Frauenklinik, Hamburg-Eppendorf.

Breuer, H., Prof. Dr., Chemisches Laboratorium der Chir. Universitätsklinik, Bonn, Venusberg.

Buchholz, R., Prof. Dr., Frauenklinik der Universität Marburg/Lahn.

Cerasi, E., Dr., Department of Endocrinology and Metabolism, Karolinska sjukhuset, Stockholm, Schweden.

Conrads, R., Dr., I. Med. Universitätsklinik, Frankfurt/M., Ludwig-Rehn-Str. 14.

Crughs, R., Dr., Zentrallaboratorium des Bluttransfusionsdienstes vom Niederländischen Roten Kreuz, Amsterdam, Holland.

Czygan, P. J., Dr., Universitäts-Frauenklinik, Hamburg-Eppendorf.

Daweke, H., Dr,. II. Med. Klinik und Poliklinik, Düsseldorf.

De Ruyter, H. A., Dr., Zentrallaboratorium des Bluttransfusionsdienstes vom Niederländischen Roten Kreuz, Amsterdam, Holland.

Dhom, G., Prof. Dr., Pathologisches Institut der Universität Würzburg, Luitpoldkrankenhaus.

Ditschuneit, H., Doz. Dr., I. Med. Universitätsklinik, Frankfurt/M., Ludwig-Rehn-Str. 14.

Dohrmann, R. E., Doz. Dr., Med. Universitäts-Poliklinik, Bonn.

Domenico, A., Dr., Hauptlaboratorium der Schering A. G., Berlin N 65, Müllerstr. 170—172.

Eckstein, Frau M., Dr., II. Med. Klinik und Poliklinik der Med. Akademie, Düsseldorf.

Elert, R., Prof. Dr., Frauenklinik der Med. Akademie Düsseldorf.

Fehlings, H., Cand. med., II. Med. Klinik und Poliklinik der Med. Akademie Düsseldorf.

Fetzer, S., Dr., Anatomisches Institut der Universität Bonn, Nußallee 10.

Fischer, W., I. Med. Klinik der Universität München.

Gandar, R., Dr., Dept. Gynécol. Obstétr., Université de Strassbourg, Hôpital Civil (France).

Geyer, G., Doz. Dr., I. Med. Universitäts-Klinik, Wien IX, Garnisongasse 13.

Glaubitt, D., Dr., II. Med. Universitäts-Klinik und Poliklinik, Mainz, Langenbeckstr. 1.

Goslar, H, G., Priv.-Doz. Dr., Anatomisches Institut der Universität Bonn, Nußallee 10.

Griebner, D., Dr., I. Med. Universitäts-Klinik, Frankfurt/M., Ludwig-Rehn-Str. 14.

Gries, F. A., Dr., II. Med. Klinik und Poliklinik der Medizinischen Akademie Düsseldorf.

Hammerstein, J., Priv.-Doz. Dr., Universitäts-Frauenklinik, Freie Universität, Berlin, Pulsstr. 4—14.

Herlyn, U., Dr., Universitäts-Frauenklinik, Göttingen.

Herrmann, M., Dr., Anatomisches Institut der Universität Bonn, Nußallee 10.

Hillebrecht, J., Dr., Anatomisches Institut der Universität Bonn, Nußallee 10.

Höfer, R., II. Med. Universitäts-Klinik, Wien IX, Garnisongasse 13.

Hohlweg, W., Prof. Dr., Universitäts-Frauenklinik, Graz.

Hornstein, O., Prof. Dr., Hautklinik der Med. Akademie Düsseldorf.

Horster, F. A., Dr., II. Med. Klinik und Poliklinik der Med. Akademie, Düsseldorf.

Husmann, F., Dr., Neurologische Klinik der Universität Würzburg.

Ijzerman, G. L., Dr., N. V. Organon, Oss, Holland.

Jahnke, K., Prof. Dr., II. Med. Klinik und Poliklinik der Med. Akademie Düsseldorf.

Jöchle, W., Dr., Brunntal über München, Hofoldingerstr. 12.

Kaiser, E., Dr., Frauenklinik der Med. Akademie Düsseldorf.

Karl, H. J., Doz. Dr., I. Med. Klinik der Universität München.

Klein, E., Prof. Dr., II. Med. Klinik und Poliklinik der Med. Akademie Düsseldorf.

Kleinfelder, H., Prof. Dr., Med. Poliklinik der Universität Würzburg.

Klesper, R., Dr., Med. Universitäts-Poliklinik, Bonn.

Knorr, D., Priv.-Doz. Dr., Universitäts-Kinderklinik, München, Lindwurmstr. 4.

Knuppen, R., Dr., Chem. Laboratorium der Chir. Universitäts-Klinik Bonn, Venusberg.

Köhler, M., Dr., Med. Poliklinik der Universität Homburg (Saar).

Kopera, H., Dr., N. V. Organon, Oss, Holland.

Kopetz, K., Dr.. II. Med. Klinik der Universität München.

Kracht, J., Prof. Dr., Pathologisches Institut der Universität Hamburg.

Krüskemper, H. L., Prof. Dr., Med. Universitäts-Poliklinik, Bonn.

Kühnau, W., Dr., Wiesbaden, Wilhelmstr. 52.

Lanzani, P., Dr., Forschungslaboratorium der Farmavigor S. p. A., Sesto S. Giovanni, Mailand, Italien.

Laron, Z., Dr., Beilinson Medical Center, Petah Tikva, Israel.

Laschet, U., Frau Dr., Nervenklinik, Landeck/Pfalz.

Liebau, H., Dr., Med. Poliklinik der Universität Homburg (Saar).

Löffler, G., Dr., Med. Poliklinik der Universität Homburg (Saar).

Lommer, D., Dr., Med. Poliklinik der Universität Homburg (Saar).

Louven, B., Dr., Universitäts-Poliklinik, Bonn.

Luft, R., Prof. Dr., Department of Endocrinology and Metabolism, Karolinska sjukhuset, Stockholm, Schweden.

Macht, G., Dr., I. Med. Universitätsklinik, Frankfurt/M., Ludwig-Rehn-Str. 14.

Maingay, D., Dr., Zentrallaboratorium des Bluttransfusionsdienstes vom Niederländischen Roten Kreuz, Amsterdam, Holland.

Marguth, F., Doz. Dr., Max Planck-Institut für Hirnforschung, Köln.

Mascitelli-Coriandoli, E., Dr., Forschungslaboratorium der Farmavigor S. p. A., Sesto S. Giovanni, Mailand, Italien.

Melani, R., Dr., I. Med. Universitäts-Klinik, Frankfurt/M., Ludwig-Rehn-Str. 14.

Mentzel, H. E., Dr., I. Med. Universitäts-Klinik, Frankfurt/M., Ludwig-Rehn-Str. 14.

Mertz, D. P., Doz. Dr., Med. Poliklinik der Universität Freiburg/Br.

Morcos, R., Dr., I. Med. Universitäts-Klinik, Frankfurt/M., Ludwig-Rehn-Str. 14.

Morer-Fargas, F., Dr., II. Med. Universitäts-Klinik, Hamburg-Eppendorf.

Mucci, A., Dr., I. Med. Universtitäts-Klinik, Frankfurt/M., Ludwig-Rehn-Str. 14.

Nadjmi, M., Dr., Neurologische Klinik der Universität Würzburg.

Neumann, F., Dr., Hauptlaboratorium der Schering A. G., Berlin N 65, Müllerstr. 170—172.

Oberdisse, K., Prof. Dr., II. Med. Klinik und Poliklinik der Medizinischen Akademie Düsseldorf.

Oertel, G. W., Doz. Dr., Institut für Hygiene und Mikrobiologie der Universität Homburg (Saar).

Ogris, E., Dr., II. Med. Universitäts-Klinik, Wien IX, Garnisongasse 13.

Overzier, C., Prof. Dr., II. Med. Universitätsklinik und Poliklinik, Mainz.

Paulsen, E., Doz. Dr., Ferring AB, Malmö, Schweden.

Pfeiffer, E. F., Prof. Dr., Abteilung für Klinische Endokrinologie, I. Med. Universitäts-Klinik, Frankfurt/M., Ludwig-Rehn-Str. 14.

Prader, A., Prof. Dr., Universitäts-Kinderklinik, Zürich, Schweiz.

Quabbe, H. J., Dr., Policlinique Universitaire de Médécine, Genève, Schweiz.

Raith, L., Dr., I. Med. Klinik der Universität München.

Reinauer, H., Dr. med., Phys. Chem. Institut der Med. Akademie Düsseldorf.

Rentsch, F., Dr., Anatomisches Institut der Universität Bonn.

Rétiene, K., Dr., I. Med. Universitäts-Klinik, Frankfurt/M., Ludwig-Rehn-Str. 14.

Sartory, A., Dr., I. Med. Universitäts-Klinik, Frankfurt/M., Ludwig-Rehn-Str. 14.

Scriba, P. C., Dr., II. Med. Klinik der Universität München.

Scheithauer, W., Dr., Institut für experimentelle Endokrinologie, Humboldt-Universität, Berlin.

Schild, W., Doz. Dr., Frauenklinik, Med. Akademie Düsseldorf.

Schirren, C., Doz. Dr., Universitäts-Hautklinik, Hamburg-Eppendorf.

Schmidt-Elmendorff, H., Dr., Frauenklinik, Med. Akademie Düsseldorf.

Schneider, H., Dr., Med. Poliklinik der Universität Homburg/Saar.

Schumann, G., Dr., I. Med. Universtäts-Klinik, Frankfurt/M., Ludwig-Rehn-Str. 14.

Schwarz, G., Dr., Med. Universitäts-Poliklinik, Heidelberg.

Schwarz, K., Doz. Dr., II. Med. Klinik der Universität München.

Schweinitz, H. A. von, Dr., II. Med. Klinik und Poliklinik der Med. Akademie Düsseldorf.

Sereno, L., Dr., Istituto di Patologia Medica, Policlinico Umberto I, Rom.

Silló-Seidl, G., Dr., Frankfurt/M., Fürstenberger Str. 147.

Škrabalo, Z., Dr., Med. Fakultätsabteilung des Allgemeinen Krankenhauses, Zagreb, Jugoslawien.

Sorge, F., Dr., I. Med. Universitäts-Klinik, Frankfurt/M., Ludwig-Rehn-Str. 14.

Stark, G., Doz. Dr., Universitäts-Frauenklinik, Mainz.

Steeno, O., Dr., Med. Universitäts-Klinik, St. Rafael, Löwen/Belgien.

Steinacker, H. G., Dr., Med. Poliklinik der Universität Homburg (Saar).

Stöcker, E., Dr., Pathologisches Institut der Universität Würzburg, Luitpoldkrankenhaus.

Tamm, J., Doz. Dr., II. Med. Universitäts-Klinik, Hamburg-Eppendorf.

Teller, W., Dr., Universitäts-Kinderklinik, Marburg/Lahn, Deutschhausstr. 12.

Thomsen, J., Dr., Anatomisches Institut der Universität Bonn, Nußallee 10.

Tonutti, E., Prof. Dr., Anatomisches Institut der Universität Bonn, Nußallee 10.

Touber, J. L., Dr., Zentrallaboratotium des Bluttransfusionsdienstes vom Niederländischen Roten Kreuz, Amsterdam, Holland.

Tripp, R., Dr., I. Med. Universitäts-Klinik, Frankfurt/M., Ludwig-Rehn-Str. 14.

Ufer, J., Dr., Berlin-Grunewald, Königsallee 57.

Vilaclara, J., Prof. Dr., Hospital Santa Cruz y San Pablo, Barcelona.

Vogel, G., Dr., Pharmakologisches Laboratorium, Farbwerke Hoechst A.G., Frankfurt-Höchst.

Weinges, K. F., Doz. Dr., Med. Poliklinik der Universität Homburg/Saar.

Werner, E., Prof. Dr., Universitäts-Kinderklinik, Berlin 19, Heubnerweg 6.

Wolff, H. P., Prof. Dr., Med. Poliklinik der Universität Homburg (Saar).

Zahnd, G. R., Dr., Policlinique Universitaire de Médécine, Genève, Schweiz.

Zicha, L., Dr., Med. Klinik der Universität Erlangen, Krankenhausstr. 12.

Zimmermann, D., Dr., Patholog. Institut der Universität Hamburg.

Zimmermann, H., Doz. Dr., II. Med. Klinik und Poliklinik der Med. Akademie Düsseldorf.

Ferring AB., Malmö, Schweden

Über die Chemie des Somatotropin

Von

F. PAULSEN

Referat

Einleitung

Da LI 1960 eine Übersicht gegeben hat, die die Literatur bis 1959 behandelt, werde ich mich hier auf die seither erarbeiteten Fortschritte beschränken. Weiterhin werde ich mich auf rein chemische Arbeiten begrenzen und die wichtigen immuno-chemischen Arbeiten auslassen, da sie in weiteren Vorträgen behandelt werden.

Das begrenzte Ziel, ein Hormon mit Wirkung auf das menschliche Wachstum zu finden, ist wahrscheinlich Schuld an den langsamen Fortschritten der STH-Chemie. Hätte man sich z. B. früher und intensiver mit der Rolle der Evolution für die Veränderungen der STH-Struktur oder der Gewebereceptoren für das STH beschäftigt, wären vermutlich die Fortschritte größer gewesen.

Die Chemie des STH-Peptides ist, wie die der anderen Peptidhormone, zunächst aus einem allgemeinen, genetischen Gesichtspunkt bedeutsam: Wir wissen, daß genetische Anomalien zu Änderungen der Proteinstruktur (structural gene mutations) oder der Menge des synthetisierten Proteins (control gene mutations) führen können. Der Prototyp einer strukturellen Gen-Mutation ist die Anomalie des Hämoglobin-Moleküls, wie sie PAULINGs glänzende Arbeiten über die Sichelzellen-Anämie aufgezeigt haben.

Tabelle 1. *Molekulargewicht des STH*

	1960 Ultrazentrifuge bei pH 2,32	1963 Sephadex pH 7,4
Homo. . . .	27 000[1]	29 000[2]
Primaten . .	25 000[1]	25 000
Walfisch. . .	39 000[1]	
Schwein . . .	41 000	23 000
Rind	45 000	24 000
Schaf	47 000	23 000

[1] Nach LI, 1960.
[2] Nach ANDREWS u. FOLLEY, 1963.

Tabelle 2. *Aminosäurenzusammensetzung von STH*

	Lys	His	Arg	Asp	Thr	Ser	Glu	Pro	Gly	Ala	Cys	Val	Met	Ileu	Leu	Tyr	Phe	Try(NH₂)
Homo[1]	13	5	14	27	14	23	34	12	13	12	6	12	4	10	31	10	15	32
Rind[2]	23	7	33	24	25	25	51	13	22	29	8	14	7	14	54	12	25	3

[1] Nach DIXON u. LI, 1962.
[2] Nach PARCELL, 1961.

Tabelle 3. *Aminosäurensequenz von Teilen des STH-Peptids*

Art	NH$_2$-Ende	C-Ende
Homo	Phe-Pro-Thr-Leu-Asp-Leu	-Leu-Phe
Primat.	Phe-	Ala-Gly-Phe
Walfisch	Phe-	Leu-Ala-Phe
Rind	Phe-Thr-Ala	-Phe
	Ala-Phe-Ala	
Schaf	Phe-	Al-Leu-Phe
	Ala-	

Es liegt nahe anzunehmen, daß ähnliche Anomalien in der Synthese der Peptid-Hormone vorkommen können. Diese Anomalien würden bei Menschen ihren Ausdruck als klinische Syndrome finden, die entweder einen Mangel oder eine verkehrte Synthese des Hormons widerspiegeln. Das gilt auch — und vielleicht besonders — für das STH.

Auch aus rein biochemischen Gründen ist die Chemie des STH eine wichtige Aufgabe. Eine Reihe von Tatsachen sprechen dafür, daß wir es hier mit einem viel größeren Molekül zu tun haben als bei den anderen Hypophysenhormonen, und daß eine bessere Kenntnis dieses Moleküls zu einem besseren Verständnis der Endokrinologie der Hypophysenvorderlappenhormone überhaupt führen kann. Ich denke hierbei an die Frage, ob es sich wirklich um ein großes homogenes Molekül von einigen hundert Aminosäuren handelt, oder um einen Komplex bisher unbekannter Natur, der aus mehreren Teil-Peptiden mit verschiedener Wirkung in loser Weise zusammengesetzt ist.

1. Herstellungsmethoden nach 1960

Seit 1960 sind etwa ein Dutzend neuer Methoden zur Herstellung von STH veröffentlicht worden. Zusammenfassend läßt sich sagen, daß diese Methoden in der Regel einfacher und schonender sind als die älteren Methoden, daß aber die Fortschritte in bezug auf Reinheitsgrad oder Ausbeute unbedeutend sind. Die früheren Methoden zur Gewinnung von STH bestanden gewöhnlich in Alkoholfraktionierung oder Salzfraktionierung bei niedriger Temperatur, in Verbindung mit Ionen-Austausch-Chromatographie. Diese Methoden waren zeitraubend und erforderten eine große Anzahl von einzelnen Operationen. 1961 beschrieb Wallace eine einfache Methode zur Herstellung von ovinem STH, die darauf aufgebaut war, daß alles STH durch eine Säule mit DEAE-Cellulose hindurchgeht, ohne absorbiert zu werden, während fast alle anderen Hypophysenhormone an diese Säule absorbiert werden.

Eine ähnliche Methode wurde von Wallace und Ferguson auch für das humane STH beschrieben. Wie gewöhnlich bei humanen Hypophysen lagen die Ausbeuten wesentlich höher. Ja, sie konnten das 10fache der Ausbeuten aus ovinen Hypophysen erreichen. Eichungsresultate für dieses Präparat wurden jedoch nicht mitgeteilt. Elektrophoretisch war es nicht homogen. Seine Prolactin-Aktivität betrug 4,1 IE/mg.

Reisfeld u. Mitarb., die schon einige Jahre vorher eine Methode zur Isolierung von kristallinem STH beschrieben hatten, teilten 1962 eine vereinfachte Fraktionierungsmethode für Aceton-Trockenpulver von humanen Hypophysen mit, die ähnlich hohe Ausbeuten gab, nämlich etwa 25 g STH/kg frischer Drüsen. Auch dieses Präparat enthielt eine hohe Prolactin-Aktivität. Ein Vorteil dieser Methode ist, daß sie gestattet, gleichzeitig humanes TSH in guten Ausbeuten aus einer Seitenfraktion zu gewinnen.

PAPKOFF, LI und LIU beschrieben 1962 eine Methode zur Isolierung von porcinem STH, die u. a. eine Gegenstrom-Verteilung und Gel-Filtrierung an Sephadex G 50 umfaßte. Sie erhielten ein hochgereinigtes Produkt, das bei der Ultra-Zentrifugierung als homogene Substanz auftrat. Im gleichen Jahre publizierten LI, LIU und DIXON eine Methode zur Isolierung von humanem STH mit u. a. Sephadex Gel-Filterung. Die dabei erhaltene scheinbar homogene und reine Substanz wurde für die Bestimmung der Aminosäurenzusammensetzung sowie der N-terminalen Sequenz benutzt. Es ist eigentümlich, daß diese anscheinend reinen humanen STH-Präparate keine größere biologische Aktivität zu haben scheinen als die alten STH-Präparate. Die Ausbeuten mit dieser Methode liegen wesentlich niedriger, nämlich bei 1—2,5 g/kg frischer Drüsen.

TOUBER und MAINGAY zeigten 1963, daß das nach der ursprünglichen Methode von RABEN hergestellte humane STH Verunreinigungen enthält, die mit der immunologischen Bestimmung von humanem STH in Serum interferieren. Durch Gel-Filtrierung an Sephadex konnten sie diese Verunreinigungen zum größten Teil entfernen. Auch REISFELD u. Mitarb. verwendeten die Sephadex Gel-Filtrierung, um ihr früher beschriebenes STH-Präparat weiter zu reinigen. In dieser Arbeit ist zum erstenmal von einer wirklichen Erhöhung der Aktivität/mg die Rede. Die beste der Fraktionen dieser Verfasser enthielt eine Aktivität von 2,2—2,5 IU/mg, während in den vorhergehenden Arbeiten die Aktivität immer um 1 IE/mg lag (Tab. 4). Wenn man diese Erhöhung der Aktivität im Verlauf von 30 Jahren von einer IE/mg auf 2,5 IE/mg für STH mit der analogen Entwicklung für andere Hypophysenvorderlappenhormone, z. B. dem ACTH vergleicht, muß die Steigerung der Aktivität/mg STH als recht unbedeutend bezeichnet werden. Im Verlaufe von 10 Jahren ist die Aktivität des ACTH von 1 IE/mg auf etwa 200 IE/mg gebracht worden. Ein solcher Vergleich scheint mir die Ansicht zu bestärken, daß wir mit unseren Methoden zur Reinigung oder Eichung des STH von den rechten Wegen abgekommen sind.

Tabelle 4. *Reinheitsgrad von Hypophysenhormonen*

Hormone	Höchster Reinheitsgrad	Niedrigste aktive Dosis		Ausbeuten per kg frischer Drüsen	
	IE per mg	IE	μg	IE	N.A.D.
Serumgonadotropin	12500	250	20		
Choriongonadotropin . . .	12500	250	20		
Vasopressin	400	5	12,5	90000	18000
Oxytocin	400	4	10	90000	22500
Corticotropin.	250	5	20	120000	24000
Thyreotropin.	30	3	100	2000	700
Somatotropin, porc.. . . .	2,5	10	4000	4500	450
Somatotropin, hum.. . . .	1	2	2000	30000	15000

1963 haben MANCHESTER und WALLIS anscheinend reines bovines STH durch Säulenchromatographie in 3 Fraktionen aufteilen können, die alle 3 in vitro STH-Aktivität zeigten, aber ganz verschiedene Wachstumswirkungen.

Schließlich haben Wallace und Ferguson 1963 ihre Methode der Isolierung des STH mit Hilfe von DEAE-Cellulose-Säulen weiter entwickelt. Die Ausbeuten an einem biologisch reinen STH betrugen 1,9—2,2 g/kg frischer Drüsen, das aber bei der von diesen Verfassern beschriebenen Stärke-Gel-Elektrophorese immer noch 2 verschiedene Hauptkomponenten zeigte.

2. Identität

Es ist selbstverständlich, daß ein Hormon mit so mannigfaltigen und komplizierten Wirkungen wie das STH immer wieder zu Forschungsergebnissen führt, die für eine Dissoziation dieser Wirkungen sprechen. Wenn jetzt eine Reihe solcher Ergebnisse besprochen wird, dann mag daran erinnert sein, daß entsprechende frühere Mitteilungen bei kritischen Nachuntersuchungen nicht aufrecht erhalten werden konnten.

Huggins und Ottaway beschrieben 1961 eine Methode, nach der sie aus einem reinen bovinen STH-Präparat ein Peptid abgetrennt hatten, das eine langdauernde Hypoglykämie hervorrief. Dieser „insulinähnliche" Faktor war ein Peptid, das sich gut charakterisieren ließ. Es hatte einen isoelektrischen Punkt bei pH 6 und die Amino-Endgruppe war Methionin. Das Molekulargewicht lag zwischen 5000 und 10000. Dieses Peptid stimulierte die Aufnahme von Glucose in das Zwerchfell von Ratten und verursachte eine schwache, aber andauernde Glykämie. Wenn diese Ergebnisse bestätigt werden können, müßten im gleichen STH-Molekül eine Teilfraktion mit hypoglykämischer Wirkung und eine andere Teilfraktion mit hyperglykämischer Wirkung vorliegen. Die Versuche von Ottaway wurden mit kristallinem bovinen STH als Ausgangsmaterial gemacht.

1964 zeigte Li, daß die in vitro meßbare, lipolytische Wirkung von humanem STH eine Eigenschaft des STH-Moleküls selbst ist, daß aber die Wachstumswirkung und die lipolytische Wirkung an verschiedene Teile der STH-Peptidkette gebunden sind.

3. Beziehungen zwischen STH und Prolactin

Unter den verschiedenen Problemen der parallelen biologischen und chemischen Fraktionierung des STH, die in den letzten Jahren bearbeitet worden sind, verdient eines ohne jeden Zweifel das größte Interesse, nämlich das Verhältnis von STH zu Prolactin.

Das eigentliche Interesse für diese Frage entstand erst durch eine rein chemische Arbeit von Ferguson und Wallace 1961. Mit ihrer Methode der Zonen-Elektrophorese an Stärke-Gel fanden sie nämlich, daß Prolactin mindestens 4 Komponenten hat, die alle die Prolactin-Wirkung auf den Kropf der Taube haben. Sie fanden nun ebenfalls, daß 3 reine Präparate von humanem STH, die von 3 verschiedenen Laboratorien nach verschiedenen Methoden hergestellt waren, alle das gleiche Bild zeigten wie das Prolactin. Wenn sie die gleiche Methode anwandten, um das STH aus Hypophysen von Schafen, Schweinen und Walfischen zu fraktionieren, so bekamen sie ein anderes Bild. Menschliches STH jedoch und ovines Prolactin zeigten die gleiche Zusammensetzung. Als sie nun humanes STH auf Prolactin-Wirkung hin eichten, fanden sie auch eine gute Prolactin-Wirkung von etwa 2,8 IE/mg. Ferguson und Wallace zogen aus ihren Ergebnissen den Schluß, daß die Prolactin-Wirkung nicht einer Verunreinigung zu verdanken, sondern eine inhärente Funktion des STH-Moleküls selbst ist.

Es folgten eine ganze Reihe von Arbeiten, die diese Ergebnisse bestätigten und erweiterten: CHADWICK, FOLLEY und GEMZELL bestätigten kurz danach, daß humanes STH auch bei Säugetieren Prolactin-Wirkung hat.

KOVACIC wies 1962 nach, daß humanes STH eine luteotrope Wirkung von gleicher Größenordnung hat, wie die Prolactin-Wirkung, die FERGUSON und WALLACE beschrieben hatten.

Das aus klinischem und praktischem Gesichtspunkt Interessanteste ist aber, daß McGARRY und BECK 1962 zeigen konnten, daß ovines Prolactin an hypophysären Zwergen metabolische Wirkungen zeigten, die denen des humanen STH entsprachen.

Wir selbst haben gefunden, daß unser ovines Prolactin eine STH-Wirkung hat, die etwa 3 mal so hoch ist, wie die von unserem porcinem STH. Ebenso zeigten einige klinische Versuche in Schweden, daß ovines Prolactin beim Menschen gewisse Wachstumswirkungen hat. Leider haben die Mitteilungen von McGARRY und BECK nicht zu den breiten Nachuntersuchungen geführt, die wünschenswert sind.

Vielleicht kann man das Verhältnis von STH zu Prolactin so deuten, daß wir mit unseren Extraktionsmethoden für STH ein sehr großes Peptid mit breiter Wachstumswirkung aus den Hypophysen extrahieren, daß aber dann noch andere Peptide, die auch eine Wachstumswirkung haben, übrig bleiben und mit anderen Methoden, z. B. der Extraktionsmethode für Prolactin gewonnen werden können. Damit hätten wir denn auch eine Erklärung für die eigentümliche Tatsache, daß der STH-Gehalt aller untersuchten Hypophysen außer der des Menschen so ungewöhnlich niedrig ist, verglichen mit ihrem Gehalt an anderen Hormonen.

4. Homogenität

Es ist klar, daß diese interessanten Ergebnisse die chemische Arbeit an der Fraktionierung von STH stark stimuliert haben. Ich will nur einige der wichtigsten erwähnen.

MANCHESTER und WALLIS gelang es 1963, bovines STH mit Säulen-Chromatographie in 3 Fraktionen aufzuteilen, die sich in bezug auf den Grad ihrer Wachstumswirkung unterschieden. Den 3 Fraktionen entsprach auch eine Dissoziation zwischen der Glucoseaufnahme im Zwerchfell, dem Einbau von Aminosäuren in Eiweiß und der Freigabe von freien Fettsäuren nach der Methode von DOLE. Auch TOUBER und MAINGAY teilten 1963 ein reines Präparat von humanem STH in 3 Fraktionen auf.

5. Haltbarkeit

PAPKOFF zeigte 1962, daß ein kontrollierter enzymatischer Abbau von sog. reinem STH zu keiner meßbaren Verminderung der Aktivität führt. Auch LI hat gezeigt, daß eine partielle Hydrolyse mit Chymotrypsin und Pepsin keinen Aktivitätsverlust im Tibia-Test wie am Taubenkropf ergibt.

LEWIS hat 1962 gezeigt, daß der Hypophysenhinterlappen eine Proteinase enthält, die den enzymatischen Abbau von Rinder-STH beschleunigt, während eine solche Proteinase im Vorderlappen nicht vorzukommen scheint. Diese Proteinase im Hinterlappen verändert die Aminosäurensequenz am NH_2-Ende des STH-Moleküls. Damit hätten wir vielleicht noch eine Erklärung zu der schlechten

Wirkung der Rinder-STH-Präparate, die nämlich in der Regel aus ganzen Hypophysen hergestellt sind. Vielleicht könnte man ihre Aktivität erhöhen, wenn man sie aus isolierten Vorderlappen herstellte.

In alkalischer Lösung wird das STH langsam abgebaut, während es in saurer Lösung offenbar sehr haltbar ist. Wir selbst haben gelegentlich Oxycellulose-Filtrat, also eine Seitenfraktion von unserer ACTH-Herstellung aus Schweine-Hypophysen, aufgearbeitet, die volle STH-Aktivität zeigte, obwohl sie jahrelang bei Zimmertemperatur gelagert war. Entscheidend für die Haltbarkeit ist auch das pH bei der initialen Extraktionsmethode. Ein alkalisches pH bei der initialen Extraktion führt zu einem STH, das auch bei niedrigem pH, wie pH 3, instabil ist.

6. Artspezifität

Die wesentliche Aufgabe für die STH-Chemiker besteht m. E. gegenwärtig darin, den Kern, der den verschiedenen STH-Peptiden gemeinsam ist, festzustellen. Ich denke hierbei nicht nur an den Teil des humanen STH-Moleküls, den Li und seine Mitarbeiter als "core" bezeichnen, und der sowohl in humanem STH wie in porcinen STH und ovinem Prolactin vorkommt, sondern auch an die Fraktion, die Laron 1963 mit enzymatischem Abbau von ovinem STH mit Pepsin erhielt. Die von ihm gefundene Komponente (f), die nicht abgebautem, humanem STH ähnlich war, könnte vielleicht den Kern des ovinen STH darstellen, der mit dem des humanen STH identisch ist. Die Darstellung dieses Kerns und, wenn möglich, seine Synthese sollte jetzt das Ziel unserer Arbeit sein. Ebenso wichtig ist vielleicht auch ein feineres Studium der Unterschiede der STH-Präparate von verschiedenen Arten mit mehr differenzierten biologischen Methoden als den traditionellen Eichungen.

Ich will nur daran erinnern, daß nach Li (1962) 100 γ bovines STH keine Prolactin-Wirkung zeigten, während 0,5 γ ovines STH eine klare laktogene Wirkung hatten. Man darf nicht die Möglichkeit ausschließen, daß die angewandten Eichungsmethoden zu grob sind, und daß es mit feineren Eichungsmethoden möglich wäre, Unterschiede zwischen den verschiedenen STH-Präparaten festzustellen, die den Unterschieden in der klinischen Wirkung besser entsprechen.

7. Wie kann STH industriell hergestellt werden?

Solange der Kern der STH-Peptide nicht festgestellt ist, seine Aminosäuren-Sequenz nicht bestimmt ist und seine Synthese nicht in Reichweite liegt, ist das wichtigste Problem der STH-Forschung in der Arzneimittelindustrie das Problem der Menge von STH. Wir können nicht damit rechnen, daß eine Herstellung von humanen STH aus menschlichen Hypophysen in großem Ausmaß jemals durchgeführt werden kann. Ich glaube auch nicht, daß eine solche Einsammlung von Menschenhypophysen als Rohstoff für die Herstellung eines Arzneimittels mit den moralischen Grundbegriffen der meisten Menschen in Übereinstimmung gebracht werden kann. Wir müssen uns daher gegenwärtig nach anderen Wegen umsehen. Derjenige, der wohl am meisten verspricht, liegt in der Möglichkeit, ovines Prolactin in großer Masse herzustellen und wie humanes STH in der Therapie zu verwenden. Nach meiner Überzeugung ist dies der Weg, der z. Z. am schnellsten eine Lösung des Problems bedeuten kann.

In Analogie zu der Herstellung von gonadotropen Hormonen aus Urin, könnte man sich natürlich Menschenurin als Ausgangsmaterial für humanes STH denken. Der Gehalt von normalem Erwachsenenharn beträgt 17 γ für die Menge von 24 Std. Eine Extraktion von 1000 l Urin würde demgemäß 17 mg oder etwa 3 Einzeldosen ergeben. Menschenurin ist also für die Gewinnung von STH in größerer Menge nicht geeignet. Ein anderes denkbares Ausgangsmaterial wäre menschliches Serum. Auch hier liegt aber der Gehalt an STH viel zu niedrig als das es für die praktische Gewinnung von STH in Frage käme. Menschliches normales Blutserum enthält nämlich nur etwa 2,0—7,8 μg/l Serum, also noch weniger als Urin.

Es ist möglich, daß die Placenta ein besseres Ausgangsmaterial ist. JOSIMOVICH u. Mitarb. haben nämlich gezeigt, daß in der menschlichen Placenta ein laktogenes Peptid vorkommt, das immunologisch dem humanen STH sehr nahesteht.

Tabelle 5. *STH-Gehalt* von *Hypophysen verschiedener Arten*

| | Eine Hypophyse | | Ausbeuten pro kg frischer Drüsen in |
	wiegt mg	enthält STH mg	g
Walfisch	20 000	20	1,0
Schwein	250	1	4,0
Rind.	2 000	4	2,0
Schaf.	500		
STH-Extrakt.		1,25	2,5
Prolactin-Extrakt		45	90
Primaten.			24
Homo	500—800	5	30

Tabelle 5 zeigt eine Zusammenstellung von Zahlen, die die Ausbeute an STH von verschiedenen Hypophysen angeben. Danach liegen die Ausbeuten aus Menschenhypophysen in einer Klasse für sich. Sie liegen mehr als 10 mal so hoch wie bei den meisten anderen Hypophysen, auch wenn die gleiche Extraktionsmethode angewandt wird. Eine Erklärung für dieses äußerst eigentümliche Verhältnis würde wahrscheinlich die STH-Forschung einen großen Schritt vorwärtsbringen.

Wenn wir den Gehalt von ovinem Hypophysen an Prolactin in STH umrechnen, bekommen wir Ausbeuten, die nach unserer Ansicht eine Herstellung in genügender Menge möglich machen. Eine Voraussetzung hierfür ist aber auch, daß die experimentellen und klinischen Untersuchungen mit Prolactin als STH intensiver betrieben werden als bisher.

Literatur

CHADWICK, A., S. J. FOLLEY, and C. A. GEMZELL: Lancet **1961**, 241—243.
FERGUSON, K. A., and A. L. WALLACE: Nature (Lond.) **190**, 632 (1961).
HUGGINS, A. K., and J. H. OTTAWAY: J. Endocr. **23**, 193—207 (1961).
KOVACIC, N.: Nature (Lond.) **195**, 1210 (1962).
LARON, Z., A. YED-LEKACH, KOWADLO-SILBERGOLD, and S. ASSA: Acta endocr. (Kbh.) suppl. 89, 9 (1963).
LEWIS, U. J.: J. biol. Chem. **237**, 3141—3145 (1962).
LI, C. H.: First Int. Congr. Endocr., Copenhagen 1960.
— J..gen. Physiol. **45**, suppl. 169—175 (1962).
— W. LIU, and J. S. DIXON: Arch. Biochem. suppl. 1, 327—332 (1962).

Li, C. H., A. Tanaka, and B. T. Pickering: Acta endocr. (Kbh.) suppl. 90, 155—162 (1964).
McGarry, E. E., and J. Beck: Lancet 1962 II, 915.
Manchester, K. L., and M. Wallis: Nature (Lond.) 200, 888—889 (1963).
Papkoff, H., C. H. Li, and W. K. Liu: Arch. Biochem. 96, 216—225 (1962).
Reisfeld, R. A., B. G. Hallows, D. E. Williams, N. G. Brink, and S. L. Steelman: Nature (Lond.) 197, 1206—1207 (1963).
— U. J. Lewis, N. G. Brink, and S. L. Steelman: Endocrinology 71, 559—563 (1962).
Touber, J. L., and D. Maingay: Lancet 1963 I, 1403—1405.
Wallace, A. L. C.: Nature (Lond.) 190, 535—536 (1961).
—, and K. A. Ferguson: J. Endocr. 23, 285—290 (1961).
— — J. Endocr. 26, 259—263 (1963).

Department of Pediatrics (Sharon Hospital) and Rogoff Medical Research Institute, Department of Experimental Biology of the Tel Aviv University, and Kupat Holim, Beilinson Medical Center, Petah Tikva, Israel

Immunological Aspects of Anterior Pituitary Hormones with Special Emphasis on Growth Hormone

By

Z. LARON

With 6 figures

Referat

Immunology, until recently a discipline connected with bacteriology, has in recent years evolved as a very useful tool in endocrinology. This has been brought about by the discovery that many hormones, including those secreted by the anterior pituitary gland, are proteins (*1*), by developing methods for the purification of these hormones, and by the refinement of immunological techniques. Thus, the administration of adjuvants has been found to enhance antigenicity (*2*): formalinisation of erythrocytes (*3, 4*), their treatment with tannic acid (*5, 6*), or coupling with diazotised benzidine (*7*), advanced the very delicate hemagglutination methods; radioactive labelling of purified hormones (*8*) led to the development of radio-immunoassays.

Table 1. *Use of immunological methods in the study of anterior pituitary hormones*

Test the degree of purification
Purification from contaminants
Test species specificity
Find localization in tissues
Immunoassay (quantitative determination)
Test biological tolerance, allergy (presence or development of antibodies)

Table 1 summarises the main uses of immunological methods in the study of anterior pituitary hormones as developed during recent years. We shall first discuss in detail the application of these methods in the study of growth hormone, later summarising their application to the study of other pituitary hormones.

1. Growth hormone

Growth hormone is prepared from either pituitaries frozen immediately after removal at autopsy (*9, 10*) or pituitaries stored in acetone (*11, 12, 13*). We prepare GH according to the method of RABEN (*11*), i.e. from acetone dried powder.

a) Immunisation and antigenicity

The repeated injection of a protein into a rabbit or guinea pig, with an adjuvant to elicit maximum response or without it, will produce antibodies in the serum which should be capable of interacting in vitro and of counteracting the antigen in vivo.

We use rabbits, injecting the antigen which is mixed with whole Freund's adjuvant (3), twice intradermally into two foot pads. Two weeks after the administration of a subcutaneous booster, antiserum titers ranging between 1:1000 to 1:20,000 or more are obtained. According to our findings, and to those of other investigators (14), in rabbits human growth hormone (HGH) is a more potent antigen than either bovine or sheep (ovine) growth hormone (BGH and SGH). So far, antiserum to HGH has also been obtained in rats (15), while antiserum to BGH has been obtained in guinea pigs (16). Antibodies to BGH have also been found in man (17, 18), and recently the development of anti-HGH antibodies has been reported in man (19, 20). We shall come back to these findings later.

Porcine growth hormone (PGH) seems to be an even weaker antigen than BGH (21, 22). This may be due in part to the lesser degree of purification of the preparations tested.

As growth hormones from various sources have different molecular weights and chemical structures (23, 24), it is not surprising that their antigenic behaviour is not identical. However, it is interesting that BGH and SGH, which have according to present knowledge a higher molecular weight than HGH, possess less antigenic power than the latter.

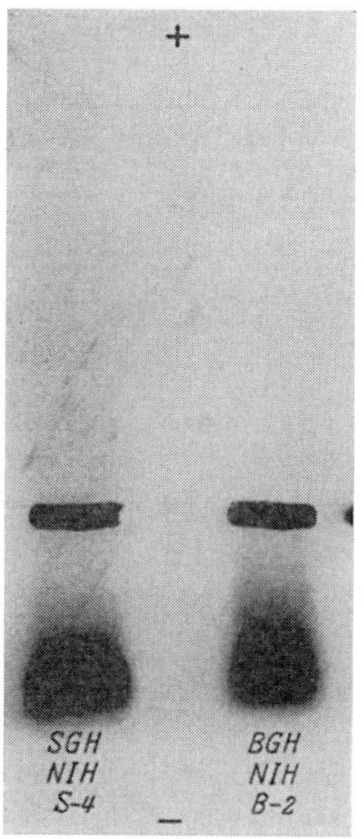

Fig. 1.

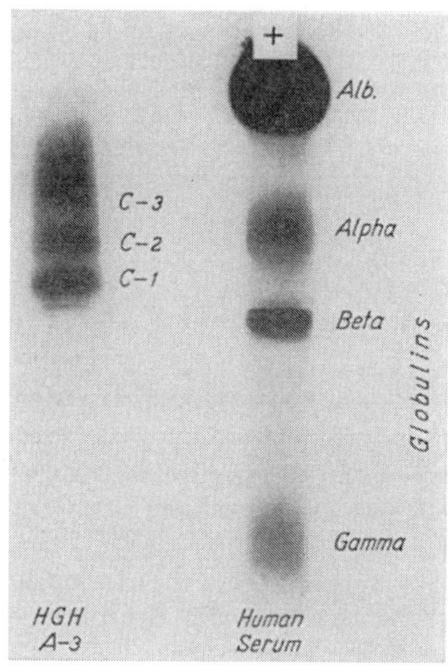

Fig. 2.

Fig. 1. Electrophoretic mobility in agar gel of bovine (BGH) and sheep growth hormone (SGH). Stained with amido-black

Fig. 2. Electrophoretic mobility in agar gel of human growth hormone (HGH) as compared to the mobility of human serum. "C" denotes electrophoretic component. Stained with amido-black

b) Electrophoresis

On agar gel electrophoresis, using a veronal buffer pH 8.2 (*25*), BGH and SGH appear as one component moving towards the cathode in an area between β- and γ-globulin (*26*) (Fig. 1). In this system, HGH appears as three anodal components corresponding to the area between α- and β-globulin (*27, 28*) (Fig. 2). The same number of components of HGH has been obtained by disc electrophoresis on polyacrylamide gels (*12*), but on starch gel electrophoresis at pH 8.0, four or more components of HGH are seen (*29, 30*). By the latter method, BGH, SGH and PGH also yield more than one component (*31*), but most SGH and PGH preparations are contaminated with other pituitary hormones, mainly prolactin. There is as yet no complete explanation for this electrophoretic heterogenicity of the GH preparations. The theories prevailing at present will be discussed later.

c) Precipitation and hemagglutination reactions

Immunological homogenicity has been well established as one of the criteria of the purity of proteins. The very sensitive techniques of immunology are very useful in detecting minute impurities in protein preparations.

When antiserum against HGH is tested by double diffusion in agar (Ouchterlony technique) it reacts with HGH and whole human pituitary extract by one precipitation line (Fig. 3) which proves the high degree of purification of the antigen.

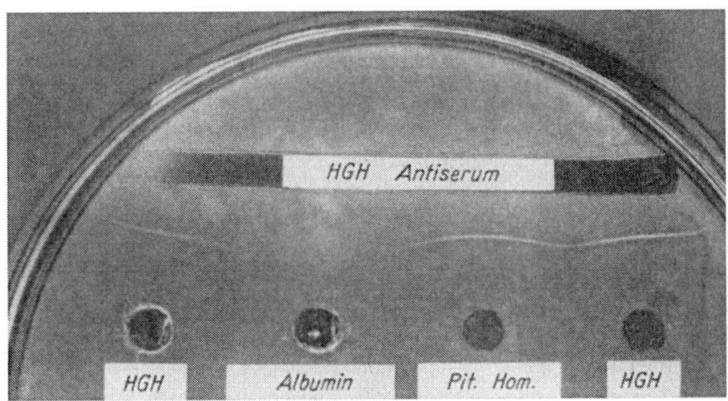

Fig. 3. Interaction of antiserum against HGH with HGH, homogenized human pituitary tissue and human albumin, as determined by the Ouchterlony agar gel double diffusion technique. Contact photo under water; without staining

The agar gel precipitation tests being relatively insensitive, only rarely we were able to see the serum albumin contamination in GH preparations (Fig. 4). The more sensitive hemagglutination tests (*27*) (Table 2) showed much more clearly the albumin and globulin contamination of HGH prepared from either fresh glands (Li's method) (*9*) or acetone powder (Raben's method) (*11*). We use formaline treated human RBC-Group 0 sensitized by tannic acid. It was possible to obtain more purified antiserum by absorption of the anti HGH antiserum with albumin which unfortunately also contained some globulin. Evidence of contamination of the GH preparations in use was also obtained by other investigators (*33, 34*).

We proceeded by eluting each electrophoretic component of HGH (*28*) individually. On agar gel double diffusion with antiserum to HGH, each formed one

Table 2. *Hemagglutination reactions of different HGH preparations and human plasma proteins with antisera to HGH or to human albumin*

Antisera \ Antigens	HGH A-2 1/6000	HGH A-3 1/6000	HGH Raben (Homo-6) 1/6000	Human plasma proteins			
				Albumin 1/1000	Cohn Fr. III 1/1000	Cohn Fr. IV 1/100	Gamma Globulin 1/100
Anti-HGH A-2 .	+	+	+	+	+	+	+
Anti-HGH A-3 .	+	+	+	+	+	+	+
Anti-HGH Li . .	+	+	+	+	+	+	+
Anti-HGH A-2 adsorbed with Hu. Albumin .	+	+	+	—	—	—	—
Anti-HGH A-3 adsorbed with Hu. Albumin .	+	+	+	—	—	—	—
Anti-HGH Li adsorbed with Hu. Albumin .	+	+	+	—	—[1]	—[1]	—[1]
Anti-Human Albumin . . .	+	+	+	+	+	+	+

[1] Sometimes these reactions were neither negative nor clearly positive.

precipitation line. All lines fused, indicating immunological identity (Fig. 5). When tested against HGH antiserum by the hemagglutination technique all fractions reacted, but tested against anti-human albumin serum, only the fastest moving component, 3, gave a positive hemagglutination reaction (Table 3). Next we

Table 3. *Hemagglutination reactions of electrophoretic components (C)[1] of human growth hormone (HGH) with antiserum to whole human growth hormone or to human albumin*

Antibody	Anti-HGH	Anti-human albumin
Antigen		
Area of application 0 .	+	—
HGH C 1	+	—
HGH C 2	+	—
HGH C 3	+	+

[1] For identification see Fig. 2.

Table 4. *Hemagglutination reactions of whole human growth hormone (HGH), its electrophoretic components (C)[1] and human albumin with antiserum to the individual components of human growth hormone (Anti-C)*

Antibody	Anti-C 1	Anti-C 2	Anti-C 3
Whole HGH	+	+	+
HGH C 1	+	+	+
HGH C 2	+	+	+
HGH C 3	+	+	+
Human albumin 1/1000 . .	—	—	+

[1] For identification see Fig. 2.

Fig. 4. Immunoelectrophoresis in agar gel of human albumin and its interaction with antiserum against HGH. Contact print under water without staining

immunised rabbits with the individual components of HGH and tested each antiserum with the various components. As seen in Table 4, each HGH component reacted with its own antiserum as well as with the antisera against the other components. Human albumin reacted only with the antiserum to component 3. It was thus demonstrated that the albumin contaminant of our HGH preparation is linked with only one electrophoretic component of HGH and can therefore be eliminated. Since this finding we have been immunising rabbits in our laboratory only with HGH electrophoretic component 1 so as to obtain antiserum against HGH free of anti-albumin antibodies.

The fact that all the electrophoretic components of HGH are immunologically identical, and the finding that all possess biological activity both in animals and in man (26, 31, 35) favours the idea that GH is originally homogenous and that these

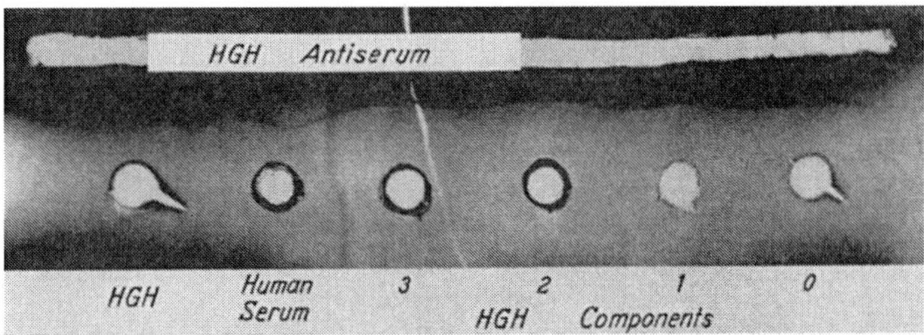

Fig. 5. Interaction of antiserum against whole HGH and its individual electrophoretic components as determined by the Ouchterlony agar gel double diffusion technique. (0 = origin). Stained with amido-black

components are induced by conditions after death, storage of the dissected gland, or by the process of hormone purification. REICHERT recently reported that purified GH preparations are contaminated with proteinases (36, 37) and LEWIS, using disc electrophoresis, was able to demonstrate changes in the electrophoretic components of HGH, BGH and sheep prolactin as the result of changes in pH, and temperature (38). Hot glacial acetic inactivates the hypophyseal proteinases, which may explain why growth hormone prepared by Raben's method shows a slower rate of degradation than GH prepared by other methods. Nevertheless, one cannot accept the presence of proteinases as the sole explanation of the presence of several electrophoretic components of HGH.

HGH has never been found to react with whole BGH, SGH or PGH, but HAYASHIDA and LI showed that HGH reacted with monkey GH, proving antigenic similarity between these two growth hormones (32). These immunological findings correspond to the biological species specificity of GH.

The bovine and sheep (ovine) growth hormone preparations available are apparently more homogenous on electrophoresis and are of lesser purity than human growth hormone when judged by the different immunological procedures.

On double diffusion in agar against their antisera, BGH preparations form between 2 to 3 precipitation lines (39). SGH reacts with BGH antiserum (32) and one or more of the precipitation lines fuse with the precipitation lines formed by BGH. Using the hemagglutination technique, we found that both BGH (NIH—GH—B$_7$) and SGH (NIH—GH—S 5) reacted not only with anti BGH serum but also with antiserum to bovine albumin and antiserum to sheep prolactin (Table 5). This heavy contamination of the BGH and SGH preparations with prolactin has also been demonstrated by bioassays (*31*).

Table 5. *Hemagglutination reactions of growth hormones of human (HGH), bovine (BGH), sheep (SGH) and porcine (PGH) origin with antisera against HGH, BGH, sheep prolactin and bovine albumin*

Antigens / Antisera	Anti-HGH A-5	Anti-BGH NIH B-7	Anti-bovine albumin Armour	Anti-sheep prolactin NIH-S-4
HGH, A-4; A-5	+	—	—	—
BGH; NIH-B-2,-6,-7	—	+	+	+
SGH; NIH-S-4,-5	—	+	+	+
PGH	—	+		+

d) Immunological behaviour of partially hydrolyzed growth hormone

After its secretion by the anterior pituitary, GH reaches the body organs via the blood stream. It is not known which tissues degrade the growth hormone molecule, but it is reasonable to assume that circulating GH also comprises GH modified by digestion. The relatively small quantities of GH excreted in the urine are further evidence of this assumption (*40*).

In the last few years immunoassays have been designed to measure the concentration of GH (*41, 42, 43, 44*). These methods will be described and discussed by Touber in the next paper (*45*), but the basic question can be asked now. What do we measure ? Is it the whole, biologically active hormone or a mixture of active hormone and its metabolites ? In order to clarify this question, we digested BGH

Table 6. *Hemagglutination reaction between pepsin digested human growth hormone (HGH) and antisera to HGH*

Antigen: HGH 1/100 % Digestion	Dilutions of antiserum					
	1/20	1/40	1/80	1/160	1/640	1/1280
0	++++	++++	++++	++++	++	++
33	++++	++++	++++	++++	++	—
50	++++	++++	++++	—	—	—
63	++++	++++	+++	—	—	—
70	++++	++	—	—	—	—
74	+	—	—	—	—	—

Table 7. *Hemagglutination reaction between pepsin digested bovine growth hormone (BGH) and antiserum to BGH*

Antigen: BGH 1/100 % Digestion	Dilution of antiserum				
	1/40	1/80	1/160	1/640	1/5120
0	++++	++++	++++	+++	++
31	++++	++++	++++	++	—
46	++++	+++	—	—	—
57	++++	—	—	—	—
63	+++	++	—	—	—

and HGH with pepsin and tested their immunogenicity at different degrees of hydrolysis so as to determine the relationship between the integrity of the GH molecule and its immunological potency. We found that peptic digestion split both growth hormones into several electrophoretic components, most of which moved towards the cathode (*39*). As illustrated in Tables 6 and 7, enzyme digestion progressively reduced the ability of both BGH and HGH to react with their respective antisera. Nevertheless, weak hemagglutination reactions were found even after 60 or 70% digestion. These findings demonstrate that the integrity of the growth hormone molecule does not seem necessary for immunological activity. The practical application of this is that in using immunological techniques for the detection of GH in biological fluids, even metabolites of GH would be detectable. The changing proportions between whole GH and its metabolites with reduced immunogenicity certainly influence the measurements by immunoassay, and possibly affect what is interpreted as the concentration of biologically active hormones.

We also studied the influence of heat on HGH, BGH and SGH and found that heating progressively reduced the immunological properties of these hormones (*46*). Of interest is the fact that HGH was more heat resistant than BGH or SGH (Tables 8 and 9). HGH retained precipitating properties when heated at 100°C up to 30 minutes and reacted in the hemagglutination test up to 60 minutes. BGH and SGH retained only weak hemagglutination properties when heated at 80°C for 15 minutes.

Table 8. *Hemagglutination reaction between heated human growth hormone (HGH) and antisera to HGH*

Antigen: HGH 1/100 Heated at 100° C	Dilution of antiserum				
	1/320	1/640	1/1280	1/2560	1/5120
0 minutes	+++	+++	+++	+++	+++
10 minutes	+++	+++	+++	—	—
20 minutes	+++	+++	+++	—	—
30 minutes	++	++	++	—	—
40 minutes	+++	+	—	—	—
50 minutes	++	++	—	—	—
60 minutes	—	—	—	—	—

Table 9. *Hemagglutination reaction between heated sheep growth hormone (SGH) and antiserum to bovine growth hormone (BGH)*

Antigen: SGH 1/100 Heating		Dilution of antiserum									
Temp. °C	Time in minutes	1/10	1/20	1/60	1/80	1/160	1/320	1/640	1/1280	1/2560	1/5120
0°	0'	++++	++++	++++	++++	++++	++++	+++	+++	+++	+++
60°	10'	++	+	+	+	+	+	+	+	—	—
60°	30'	+	+	+	+	+	+	+	—	—	—
70°	10'	+	+	+	+	+	—	—	—	—	—
70°	20'	+	+	+	+	—	—	—	—	—	—
75°	10'	+	+	—	—	—	—	—	—	—	—
80°	15'	+	—	—	—	—	—	—	—	—	—
80°	40'	—	—	—	—	—	—	—	—	—	—

e) The use of fluorescent antibodies

Using fluorescein-labelled rabbit anti-human growth hormone serum, Leznoff et al. (*47*) demonstrated green fluorescence in the cells of the human anterior pituitary. These authors concluded that the anti HGH antibody localized exclusively in the eosinophilic cells of normal pituitaries and in eosinophilic adenomata from patients with acromegaly. These findings, confirmed by Mene-ghelli and Scapinelli (*48*) and Grumbach (*49*), suggest that HGH is either produced or stored in the eosinophilic cells of the anterior pituitary. In another study using the same technique, Meneghelli and Scapinelli (*50*) demonstrated the presence of GH in the pituitaries of bovine fetuses as early as the 3rd month of pregnancy.

Studying human placental tissue at different gestational stages, Sciarra et al, (*51*) found that human placental tissue, as early as the 12th week of gestation. contains a protein immunologically related to GH, which is localized in the syncytial cytoplasm of the trophoblasts. Here it should also be mentioned that Josimovich et al. (*52*) isolated from human placentas a lactogenic fraction which, as these authors showed, is closely related immunologically to human pituitary growth hormone.

f) Antibodies against HGH in man

Prolonged treatment with HGH of children with pituitary insufficiency or growth retardation of undetermined origin often leads to a progressive reduction of the initially increased growth rate (*53, 54, 55, 56, 57*). The sera of some of these children were examined but no antibodies against HGH were found (*55, 57*). Trafford et al. (*58*) reported the development of a low titer of HGH antibodies ($^1/_{16}$ and $^1/_{128}$) in 2 out of 4 patients treated with HGH and mentioned that their growth continued at an increased rate despite the presence of the antibodies. The only high titer of anti-HGH antibodies (1:2000 to 1:5000) was found by Széky et al. (*59*) five months after initiation of HGH treatment in a $5^1/_2$ year old boy with familial growth retardation of undetermined origin. Interruption of HGH administration caused a decrease in the antiserum titer, which, however, rose again upon reinstitution of therapy. It is completely unclear why HGH should lead to the formation of autoantibodies and, if so, why only on rare occasions. The very modern concept of autoimmune disease, which has been proved beyond doubt for thyroid (*60*) and adrenal diseases (*61*), has been suggested by Conti et al. (*62*) who found HGH antibodies in three dwarfed patients who had never received any treatment and in two acromegalics treated by intrahypophyseal implantation of Y^{90}. Most interesting is another finding by the same investigators, in which anti HGH antibodies were found in 3 out of 10 patients who had been treated with bovine growth hormone, all of whom developed anti BGH antibodies. Considering the proven species specificity of GH and the repeated finding that BGH and SGH are biologically ineffective in man (*63*), it is very difficult to accept these findings. None the less, we too observed that one component of hydrolyzed bovine or sheep GH reacted with HGH antiserum (*39*). These findings are suggestive of common antigenic sites between HGH, BGH and possibly SGH.

The problem of antibodies against HGH is a very important one and more extensive studies on a larger scale using highly purified antigens are required in order to determine its extent and to understand its meaning and practical implications in diagnosis and treatment.

I shall now summarise certain important and interesting facts concerning the immunological properties of anterior pituitary hormones other than growth hormones.

2. Prolactin (Lactogenic hormone)

In man, separation of the lactogenic from the somatotrophic activity in the hormone preparations available has not been possible (*29, 64*); therefore one can

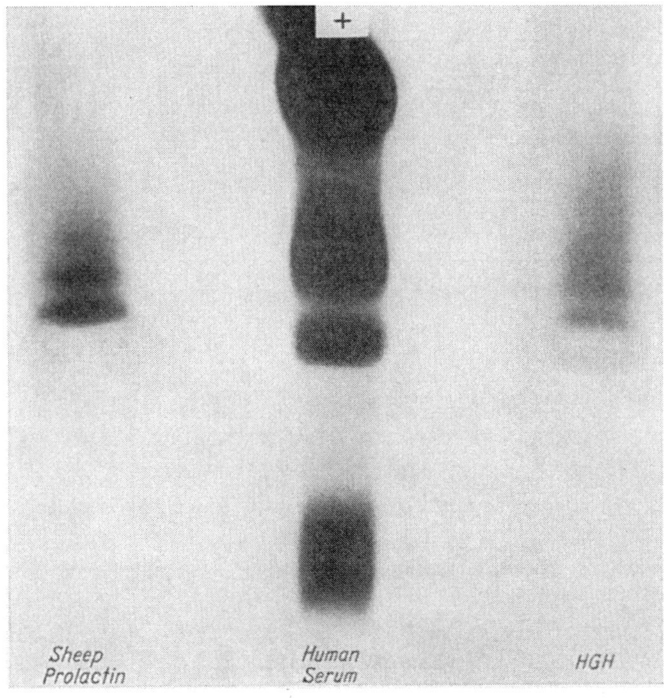

Fig. 6. Agar gel electrophoresis of sheep prolactin as compared to human growth hormone (HGH)

not be sure whether growth hormone and prolactin have separate identities in man, as would appear to be true in other species, such as cattle and sheep.

Sheep prolactin on agar gel electrophoresis presents as three anodal components with a mobility very similar to that of HGH (Fig. 6). However, neither whole sheep prolactin nor any of the individual electrophoretic fractions react immunologically with HGH (*65*). Sheep prolactin has been found to react with BGH both on agar gel double diffusion and in hemagglutinations tests, but this was explained as due to contamination with prolactin of the GH preparations (*65, 66*). These studies also demonstrated that the antigenic structures of sheep, bovine and goat prolactin are closely related.

3. Corticotrophin (ACTH)

Compared to growth hormone and prolactin, corticotrophin is a much weaker antigen. When tested with their homologous antisera only porcine ACTH showed a

                                    Z. Laron:

precipitation in the agar gel double diffusion test (*67*). Immunological similarity of human, bovine and porcine corticotrophin was shown by the fact that they not only reacted with their own antisera but also showed positive tests with the heterologous antisera (*67, 68*) in hemagglutination tests (Table 10). Lack of species specificity of ACTH is also evidenced by the efficiency of animal ACTH in man. It would seem that the anaphylactic reactions encountered when using these animal preparations in man (*69*) are due to species specific protein contaminants. We were not able to obtain antibodies by the immunisation of rabbits with the new synthetic β^{1-24} Corticotrophin (Ciba).

Table 10. *Hemagglutination reactions between ACTH of different species and their antisera*

Antigens	Antibodies					
	Anti human ACTH	Anti bovine ACTH	Anti porcine ACTH	Anti human GH	Anti sheep prolactin	Anti human albumin
Human pituitary homogenate	+	+	+	+		+
Human ACTH	+	—	—	—	—	+
Bovine ACTH	+	+	+[1]	—	+	—
Porcine ACTH	+	+	+[1]	—		—
Sheep ACTH	+	+	+[1]	—	+	+ ?
Sheep prolactin	+	+	+[1]	—	+	—
Human GH	—	—	—	+	—	±
Human albumin	+	—	—	—	—	+

[1] Positive precipitation tests (Ouchterlony).

4. Thyrotrophin (TSH)

Antisera have been prepared against both bovine and human thyrotrophin (*70, 71, 72*) but the purity of the antigens used, has been questioned. Recently Condliffe reported the purification of human TSH (*73*) and using this preparation Utiger et al. (*74*) showed that there was no immunological cross reaction between bovine and human TSH. Supporting evidence for this has been found earlier using cross inhibition tests (*75*). On the other hand, Oki found cross immunity between human and hog TSH (*76*). It is interesting to note the immunological species specificity of TSH, but lack of species specificity of its biological action as evidenced by the effectiveness of bovine TSH in man.

5. Follicle stimulating hormone (FSH)

The FSH preparations available are not yet completely purified and demonstrate on agar gel diffusion against their antiserum at least three antigenic components (*77, 78, 79, 80*). Using hemagglutination and hemagglutination inhibition tests no cross reaction could be demonstrated between bovine, sheep, pig or human pituitary FSH. Some cross reaction with LH has been found but this was assumed as contamination with either the hormone itself or the common presence of contaminating serum proteins. Human FSH antiserum reacted with human chorionic gonadotrophin (HCG) only when large amounts of HCG were used (*79, 80*).

As was the case with TSH, it would appear that failure to demonstrate cross reactions by immunological methods is not an absolute proof of biological species specificity.

6. Interstitial cell stimulating hormone (ICSH)

Investigations performed with sheep and human ICSH (*81, 82*) demonstrated that ICSH is a potent antigen but the preparations used were contaminated with serum proteins and to a lesser degree with GH. Tested by the agar gel double diffusion technique against sheep ICSH antiserum, human ICSH, HCG and PMSG (pregnant mare's serum gonadotrophin) gave no precipitation reaction. Using the same technique, HCG was found to cross react with antiserum to human ICSH.

We have seen from the present review that there is an increasing use of immunological methods in the study of the anterior pituitary hormones and that the degree of application of these methods is directly related to the degree of purification of the individual hormones.

Acknowledgement

I would like to express my gratitude for the great effort made by my collaborators, Mrs. SARA ASSA, Mrs. AVIVAH KOWADLO-SILBERGELD and Mrs. ARIANA YED in preparing this work. I also acknowledge the assistance of Mr. IOSEF SADOVNIK, of the Photography Laboratory of the Sharon Hospital.

References

1. LI, C. H., H. M. EVANS, and M. E. SIMPSON: J. biol. Chem. **159**, 353 (1945).

2. FREUND, J.: Amer. J. clin. Path. **21**, 645 (1951).

3. FLICK, J. A.: Proc. Soc. exp. Biol. (N. Y.) **68**, 448 (1948).

4. COLE, L. V., and V. R. FARRELL: J. exp. Med. **102**, 631 (1955).

5. BOYDEN, S. V.: J. exp. Med. **93**, 107 (1951).

6. STAVITSKY, A. B.: J. Immunol. **72**, 360 (1954).

7. —, and E. R. ARQUILLA: J. Immunol. **74**, 306 (1955).

8. HUNTER, W. M., and F. C. GREENWOOD: Nature (Lond.) **194**, 495 (1962).

9. LI, C. H.: J. biol. Chem. **211**, 555 (1954).

10. WALLACE, A. L. C., and K. A. FERGUSON: J. Endocr. **23**, 285 (1961).

11. RABEN, M. S.: Recent Progr. Hormone Res. **15**, 71 (1959).

12. REISFELD, R. A., U. J. LEWIS, N. G. BRINK, and S. L. STEELMAN: Endocrinology **71**, 559 (1962).

13. ELRICK, H., V. YEARWOOD-DRAYTON, Y. ARAI, F. LEAVER, and H. G. MORRIS: J. clin. Endocr. **23**, 694 (1963).

14. LI, C. H.: In G. E. W. WOLSTENHOLME, and M. P. CAMERON, Eds.: Ciba Foundation Colloquia on Endocrinology **14**, 20 (1962).

15. MOUDGAL, N. R., and C. H. LI: Endocrinology **68**, 704 (1961).

16. LI, C. H., H. PAPKOFF, and T. HAYASHIDA: Arch. Biochem. **85**, 97 (1959).

17. CONTI, C., L. SERENO, L. LUCHETTI, O. RECCHIA, and A. ISIDORI: Folia endocr. (Roma) **15**, 163 (1962).

18. LARON, Z., and S. ASSA: Manuscript in preparation.

19. SZÉKY, J., A. HÄSSIG u. A. PRADER: Helv. paediat. Acta **17**, 411 (1962).

20. TRAFFORD, J. A. P., D. A. LILLICRAP, and M. H. LESSOF: Lancet **1963** I, 1128.

21. PAPKOFF, H., N. R. MOUDGAL, and C. H. LI: Fed. Proc. **19**, 157 (1960).

22. LARON, Z., and S. ASSA: Unpublished data.

23. PAPKOFF, H., C. H. LI, and W. K. LIU: Arch. Biochem. **92**, 216 (1962).

24. LI, C. H.: Bull. N. Y. Acad. Med. **39**, 143 (1963).

25. GRABAR, P.: In D. GLICK, Ed.: Methods of Biochemical Analysis, vol. 7, p. 3. New York-London: Interscience Publ. Inc. 1959.

26. LARON, Z., S. ASSA, and J. MATOTH: Harefuah **64**, 1 (1963).

27. — — Acta endocr. **40**, 311 (1962).

28. — — Nature (Lond.) **194**, 491 (1962).

29. Ferguson, K. A., and A. L. C. Wallace: Nature (Lond.) **190**, 632 (1961).
30. Irie, M., and R. J. Barrett: Endocrinology **71**, 277 (1962).
31. Ferguson, K. A., and A. L. C. Wallace: Recent Progr. Hormone Res. **19**, 1 (1963).
32. Hayashida, T., and C. H. Li: Endocrinology **65**, 944 (1959).
33. Hirschfeld, I., C. A. Gemzell, and L. Wide: Nature (Lond.) **187**, 64 (1960).
34. Boucher, B. J.: Nature (Lond.) **188**, 1025 (1960).
35. Laron, Z., S. Assa, and R. Menache: J. clin. Endocr. **23**, 315 (1963).
36. Reichert, L. E. jr.: Biochim. biophys. Acta (Amst.) **50**, 191 (1961).
37. — Gen. and comp. Endocr. **2**, 283 (1962).
38. Lewis, U. J.: J. biol. Chem. **238**, 3330 (1963).
39. Laron, Z., A. Yed-Lekach, S. Assa, and A. Kowadlo-Silbergeld: Endocrinology, in press.
40. Geller, J., and A. Loh: J. clin. Endocr. **23**, 1107 (1963).
41. Read, C. H., and T. G. Bryan: Recent Progr. Hormone Res. **16**, 187 (1960).
42. Széky, J., L. Holländer, and A. Prader: Helv. paediat. Acta **16**, 691 (1961).
43. Greenwood, F. C., W. M. Hunter, and J. S. Glover: Biochem. J. **89**, 114 (1963).
44. Glick, S. M., J. Roth, R. S. Yalow, and S. A. Berson: Nature (Lond.) **199**, 784 (1963).
45. Touber, J. L.: Lecture at: 11. Symp. deutsch. Ges. f. Endokrinol. S. 22, Heidelberg, Springer, 1964.
46. Laron, Z., A. Yed-Lekach, S. Assa, and A. Kowadlo-Silbergeld: Acta endocr. in press.
47. Leznoff, A., J. Fishman, L. Goodfriend, E. McGerry, J. Beck, and B. Rose: Proc. Soc. exp. Biol. (N. Y.) **104**, 232 (1960).
48. Meneghelli, V., e E. Scapinelli: Boll. Soc. Ital. Biol. Spez. **37**, 277 (1961).
49. Grumbach, M. M.: In: Immunoassay of hormones. Ciba Foundation Colloquia on Endocrinology **14**, 373 (1962).
50. Meneghelli, V., and E. Scapinelli: Acta Anat. **51**, 198 (1962).
51. Sciarra, J. J., S. L. Kaplan, and M. M. Grumbach: Nature (Lond.) **199**, 1005 (1963).
52. Josimovich, J. B., B. L. Atwood, and D. A. Goss: Endocrinology **73**, 410 (1963).
53. Raben, M. S.: New Engl. J. Med. **266**, 82 (1962).
54. Aarskog, D.: Amer. J. Dis. Child. **105**, 368 (1963).
55. Prader, A.: Personal communication.
56. François, R.: Personal communication.
57. Laron, Z.: Unpublished results.
58. Trafford, J. A. P., D. A. Lillicrap, and M. H. Lessof: Lancet **1963 I**, 1128.
59. Széky, J., A. Hässig u. A. Prader: Helv. paediat. Acta **17**, 411 (1962).
60. Roitt, I. M., and D. Donisch: Lancet **1958 II**, 1028.
61. Blizzard, R. M., R. W. Chandler, M. A. Kyle, and W. Hung: Lancet **1962 II**, 901.
62. Conti, C., L. Sereno, L. Luchetti, O. Recchia, and A. Isidori: Folia endocr. (Roma) **15**, 163 (1962).
63. Bergenstal, D. M., and M. B. Lipsett: J. clin. Endocr. **20**, 1427 (1960).
64. Chadwick, A., S. J. Folley, and C. A. Gemzell: Lancet **1961 II**, 241
65. Laron, Z., and S. Assa: Nature (Lond.) **197**, 299 (1963).
66. Hayashida, T.: In G. E. W. Wolstenholme, and M. P. Cameron Eds.: Ciba Found. Colloquia Endocr. **14**, 338 (1962).
67. Laron, Z., B. Z. Arie, and S. Assa: in preparation.
68. McGarry, E. E., A. Ballantyne, and J. C. Beck: In G. E. W. Wolstenholme, and M. P. Cameron Eds.: Ciba Found. Colloquia Endocr. **14**, 273 (1962).
69. Kantor, Z., and Z. Laron: Ann. Pediat. **201**, 381 (1963).
70. Selenkow, H. A., F. M. Pascasio, and M. J. Cline: In G. E. W. Wolstenholme, and M. P. Cameron Eds. Ciba Found. Colloquia Endocr. **14**, 248 (1962).
71. Pascasio, F. M., and H. A. Selenkow: Endocrinology **71**, 254, (1962).
72. Reichlin, S., and R. L. Boshans: Proc. Soc. exp. Biol. (N. Y.) **111**, 139 (1962).
73. Condliffe, P. G.: Endocrinology **72**, 893 (1963).
74. Utiger, R. D., W. D. Odell, and P. G. Condliffe: Endocrinology **73**, 359 (1963).
75. Arquilla, E. R.: in discussion of ref. 70, p. 263.
76. Oki, K.: J. Jap. Soc. int. Med. **51**, 110 (1962).

77. SEGAL, S. J., K. A. LAWRENCE, M. PERLBACHS, and S. HAKIM: Gen. comp. Endocr. 1, 12 (1962).
78. WOLF, A.: Nature (Lond.) 198, 1308 (1963).
79. BUTT, W. R., A. C. CROOKE, F. J. CUNNINGHAM, and A. WOLF: J. Endocr. 25, 541 (1963).
80. McGARRY, E. E., and J. C. BECK: Fertil. and Steril. 14, 558 (1963).
81. MOUDGAL, N. R., and C. H. LI: Arch. Biochem. 95, 93 (1961).
82. — — Nature (Lond.) 191, 192 (1961).

Diskussion

W. TELLER (Marburg/Lahn):

Von PRADER und seiner Arbeitsgruppe ist gezeigt worden, daß es bei der Behandlung von hypophysären Zwergen mit menschlichem Wachstumshormon (HWH, Typ Raben) zum Auftreten von spezifischen Antikörpern kommen kann, die schließlich eine Therapieresistenz der Patienten nach sich ziehen. Diese Antikörper wurden mittels der passiven Hämagglutination nachgewiesen. Der Immunisierungsvorgang muß im Sinne einer Iso-Immunisation gedeutet werden.

In Ihrem Vortrag erwähnten Sie Autoren, die Anti-HWH-Antikörper auch schon *vor* der Behandlung mit HWH nachgewiesen haben. Diese mögliche Auto-Immunisierung wäre von besonderer Wichtigkeit bei der Beurteilung der Frage, ob eine HWH-Therapie bei einem hypophysären Zwerg erfolgversprechend sein wird oder nicht. Sollte es sich herausstellen, daß HWH-Autoantikörper in einem größeren Prozentsatz der Fälle von hypophysärem Minderwuchs gefunden werden, so erschiene es sinnvoll, *vor* Therapiebeginn alle Patienten auf das Vorhandensein von HWH-Antikörper zu testen, um eine Verschwendung des kostbaren Wachstumshormons möglichst zu vermeiden.

Haben Sie bei Ihren Untersuchungen Patienten mit einwandfreier Hypophysenvorderlappeninsuffizienz in bezug auf HWH-Autoantikörper vor der Behandlung getestet?

Ist die Antigenwirksamkeit der drei Fraktionen Ihrer Wachstumshormonpräparate annähernd gleich stark? Man könnte sich zumindest theoretisch vorstellen, daß bei denjenigen Patienten, die HWH-Antikörper bilden, ein Wachstumseffekt durch Zufuhr einer Unterraktion des HWH noch zu erzielen ist.

J. L. TOUBER (Amsterdam):

The HGH-Antibodies of Prof. PRADERs patient are 7^s-γ-Globulins as shown by immunoelectrophoresis with purified radioactive HGH. Before therapy no antibodies could be found. The binding capacity of these antibodies for HGH is very high and can explain the resistance to HGH-therapy.

D. KNORR (München):

STH von Anthropoiden ist beim Menschen auch wirksam. Ist bekannt, ob die STH-Präparate, auf deren Verabreichung sich bei Kindern hohe STH-Antikörpertiter entwickelten, nur aus menschlichen Hypophysen gewonnen waren, oder ob auch anthropoide Hypophysen mit verwertet waren?

Z. LARON:

Selbst unsere mit 24 mg HGH pro Woche behandelten Patienten entwickelten keine Antikörper. In Anbetracht der Tatsache, daß nur in gewissen Kliniken ausnahmsweise manche Kranken nach HGH Antikörper entwickelten, ist zu fragen, ob solche Befunde vielleicht von Besonderheiten des Präparates oder der Patienten abhängen.

L. SERENO (Rom):

We have no information on the effects of the presence of autoantibodies on calcium metabolism. The titer of those antibodies never resulted very high. Therefore their presence must be considered, in the sickness, as a secondary sign, and not as the most important etiogenic factor.

Aus dem Zentrallaboratorium des Bluttransfusionsdienstes vom Niederländischen Roten Kreuz
Amsterdam

Radioimmunologische Bestimmung von Wachstumshormon

Von

J. L. Touber, D. Maingay, H. A. De Ruyter und R. Croughs

Mit 1 Abbildung

Referat

In den letzten Jahren sind für die quantitative Bestimmung von Hormonen immunologische Methoden entwickelt worden. Für die Bestimmung des menschlichen Wachstumshormons im Serum sind bisher zwei immunologische Techniken brauchbar:

1. die passive Hämagglutinationsreaktion, zum erstenmal durch Read beschrieben, und

2. radio-immunologische Methoden, auf dem Prinzip der durch Berson und Yalow entwickelten Insulinbestimmung basierend.

Wir in Amsterdam haben beide Methoden angewendet. In diesem Vortrag will ich mich jedoch auf *die* Resultate beschränken, die mit der radioimmunologischen Bestimmung von Wachstumshormon erzielt wurden. Die Hämagglutinations-Hemmreaktion ist nach unseren Erfahrungen zwar für die Bestimmung von gereinigten Hormonpräparaten brauchbar, aber nicht für die Bestimmung von Wachstumshormon im Serum.

Zur Entwicklung einer immunologischen Hormonbestimmung muß man über ein möglichst reines Hormonpräparat und ein spezifisch dagegen gerichtetes Antiserum verfügen. Wir verwendeten ein nach der Methode von Raben durch Organon, Holland, bereitetes menschliches Wachstumshormonpräparat. Die biologische Aktivität dieses Präparates betrug 1,5 U.S.P.-Einheiten pro mg.

Durch Immunisation von Kaninchen mit diesem, in Freunds Adjuvans emulgiertem menschlichen Wachstumshormonpräparat, wurden die Antiseren gewonnen.

Alle gegen menschliches Wachstumshormon gerichteten Antiseren enthielten neben spezifischen Antikörpern auch Antikörper gegen menschliches Serumalbumin. Das Wachstumshormonpräparat ist also mit diesem Serumprotein verunreinigt. Wir konnten nachweisen, daß der biologische Effekt menschlichen Wachstumshormons, wie dies im sog. Tibiatest gemessen wird, durch das Antiserum quantitativ verhindert werden kann.

Prinzip einer radio-immunologischen Bestimmuung:

Einer kleinen Quantität radioaktiven Hormons (HGH) wird Antiserum in einer solchen Konzentration beigefügt, daß nach einer gewissen Inkubationsdauer ungefähr 60—70% des radioaktiven Hormons an Antikörper gebunden sind. Wird dem Inkubationsgemisch auch nicht-radioaktives Hormon zugesetzt, dann wird der Prozentsatz des radioaktiven, durch Antiserum gebundenen Hormons geringer, weil weniger Antikörper zur Verfügung stehen. Im Prinzip sind diese Methode und die Isotopen-Dilutionsmethode gleich. Der Prozentsatz des Antikörpergebundenen radioaktiven Hormons entspricht also der Quantität des im Inkubationsgemisch anwesenden, nicht radioaktiven Hormons.

Bei der Entwicklung dieser Methode wurden wir mit drei Problemen konfrontiert, nämlich:

1. die Bereitung von immunologisch unverändertem, radioaktivem Hormon mit einer hohen spezifischen Aktivität,

2. die Ausarbeitung einer Technik mit der das an Antikörper gebundene Hormon so gut möglich von dem freien Hormon geschieden werden konnte und

3. die Analyse störender Einflüsse, die teilweise durch die Unreinheit des verwendeten Hormonpräparates und teilweise durch den Einfluß von Serumproteinen verursacht wurden.

Die Bereitung von immunologisch unverändertem, radioaktivem Hormon mit einer hohen spezifischen Aktivität ist kein Problem mehr, seitdem wir uns zu diesem Zwecke der Jodierungsmethode von HUNTER und GREENWOOD bedienen. Die Methode ist schnell und erfordert nur wenig Hormon und radioaktives Jod, nämlich 5 μg Wachstumshormon und 2 mC radioaktives Jod. Man kann das Hormon markieren mit Jod[131], dessen Halbwertszeit 8 Tage beträgt, aber auch mit Jod[125], das eine Halbwertszeit von 60 Tagen hat. Chloramine-T dient als Oxydationsmittel. Die spezifische Aktivität des radioaktiven Hormons beträgt ungefähr 250 μC/μg.

Die Trennung von Antikörper-gebundenem und freiem Hormon ist viel schwieriger. Wir haben viele Trennungstechniken probiert: Elektrophorese auf Papier, auf Celluloseacetat, in Stärkegel und in Agargel, Chromatographie auf Papier und Immunpräcipitation mit Hilfe von gegen Kaninchen-γ-Globulin gerichteten Antiseren. Die Techniken, die wir am häufigsten anwenden, sind Elektrophorese in Agargel und Chromatographie auf Papier.

Das Autoradiogramm der Agargel-Elektrophorese von radioaktivem Wachstumshormon zeigte, daß ohne Zusatz von Antiserum sich die Radioaktivität hauptsächlich im Gebiet der α-Globuline befindet. Wenn dem Hormon jedoch ein wenig Antiserum zugesetzt wird, scheint ein Teil der Radioaktivität in das Gebiet der γ-Globuline zu geraten.

Wird das Gel nach der Elektrophorese in zwei Teile geteilt — wobei die Trennungslinie etwas weiter kathodenwärts als die Auftragestelle gelegt wird, selbstverständlich immer auf denselben Platz —, können das gebundene und das freie Hormon getrennt werden. Die zwei Gelteile werden separat gezählt. Die Verteilung der Radioaktivität kann auch mit einem automatischen "strip-counter" bestimmt werden. In diesem Falle muß man das Gel erst trocknen.

Zwei Faktoren haben aber eine genaue Trennung von Antikörper-gebundenem und freiem Hormon erschwert:

1. das verwendete Wachstumshormonpräparat enthält Unreinheiten, die in der radioimmunologischen Bestimmung störend wirken. Durch Gelfiltration über Sephadex G 200 können sie eliminiert werden.

2. Während der Inkubation im Serum von Mensch oder Tier wird ein Teil des so gereinigten radioaktiven Hormons an Serumproteine, und zwar an α_2-Makroglobuline gebunden. Dadurch verliert dieser Teil seine immunologische Reaktionsfähigkeit. Hält man den Serum- oder Plasmagehalt der Inkubationsmischung so niedrig wie möglich, dann kann diese sog. Inkubationsstörung des radioaktiven Wachstumshormons vermindert werden. Bei der radioimmunologischen Insulinbestimmung wird eine ähnliche Inkubationsstörung gefunden. Aus diesem Grunde bestimmten wir die untersuchten Seren in einer Konzentration von zehn Volumenprozent.

Genauso wie freies Insulin, wird auch freies Wachstumshormon an Papier adsorbiert. Das an Antikörper gebundene Hormon, aber auch der durch Inkubation veränderte Teil des Hormons wird mit den Serumproteinen mitchromatographiert. Die Inkubationsstörung kann aus — ohne Antiserum bereiteten — Kontrollinkubationsgemischen berechnet werden. Wenn dieser Faktor berechnet ist, kann nun der tatsächliche, an Antikörper gebundene Hormongehalt festgestellt werden.

Die Abb. 1 zeigt eine Standardkurve. Die Punkte stellen Doppelbestimmungen dar. Es erweist sich, daß sogar eine halbe Mikro-Einheit Wachstumshormon pro ml Inkubationsgemisch noch bestimmt werden kann, wenn man sehr kleine Konzentrationen von radioaktivem Wachstumshormon und Antiserum nimmt und während 4—6 Tagen bei +4°C

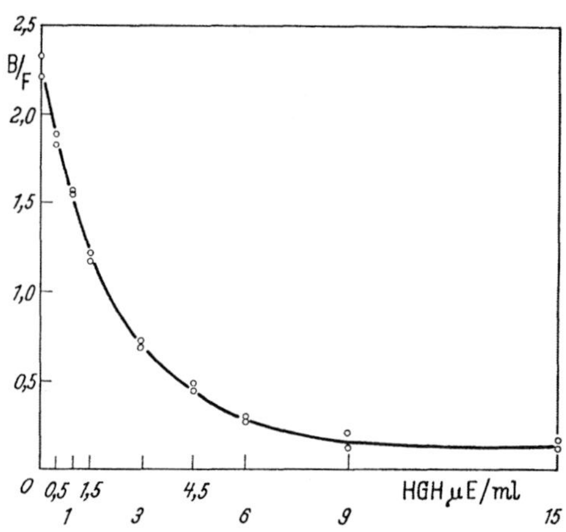

Abb. 1. Standardkurve der radioimmunologischen Bestimmung von menschlichem STH (HGH)

inkubiert. Das Antiserum wird z. B. in einer Verdünnung von 1 auf 200000 verwendet. Nachdem es bis heute niemandem gelungen ist, vollkommen reines Wachstumshormon zu bereiten, ziehen wir es vor, die Quantität von Wachstumshormon in biologischen Einheiten anzugeben. Die Angabe in Gewichtseinheiten ist also nicht reell. 0,5 μE unseres menschlichen Wachstumshormonpräparates betragen 0,33 μg.

Resultate

Bei einer Reihe von normalen Erwachsenen, Männern und Frauen, wurde an verschiedenen Tagen der Wachstumshormongehalt des nüchternen Plasma bestimmt. Die Werte schwankten zwischen 0 und 40 μE/ml. Nach Verabreichung von 100 g Glucose per os sank der Wachstumshormongehalt meistens; diese Senkung nach Glucose ist nur dann deutlich, wenn nüchtern der Wachstumshormongehalt des Plasmas hoch ist. So sank z. B. der HGH-Gehalt von 30 μE/ml auf 6μE nach einer halben Stunde, auf 3 μE nach einer Stunde und auf 0 μE nach zwei Stunden.

Von 9 Patienten mit Panhypopituitarismus oder hypophysärem Zwergwuchs hatten 8 keine nachweisbaren Mengen von Wachstumshormon im Serum. Bei einem Patienten, der nüchtern eine geringe Menge Wachstumshormon im Serum hatte, war nach 100 g Glucose keines mehr nachzuweisen. Auch bei Normalen kann in nüchternem Plasma oft kein Wachstumshormon gefunden werden. Dies ist also kein Grund für die Diagnose Panhypopituitarismus. Nach einer intramuskulären Injektion von menschlichem Wachstumshormon bei einem dieser Patienten, stieg der Wachstumshormongehalt im Plasma von 0 auf 14 μE/ml.

Bei 6 Patienten mit Akromegalie wurden mehr als 150 μE/ml Plasma gefunden. Wenn dieses Plasma nicht in einer Verdünnung von 1 auf 10, sondern von 1 auf 50 untersucht wurde, variierte das Wachstumshormonniveau zwischen 200 und ungefähr 750 μE/ml. Von den 8 Akromegalie-Patienten, die weniger als 100 μE/ml hatten, waren 5 mit Röntgenbestrahlung der Hypophyse behandelt worden. Einer dieser Patienten hatte 140 μE/ml *vor* und 54 μE/ml *nach* der Bestrahlung. Wir fanden bei einer jungen Frau mit unbehandelter aktiver Akromegalie, daß nach Verabreichung von 100 g Glucose der Plasma-Wachstumshormongehalt sank, und zwar von 480 auf 340 μE/ml.

Wir glauben, daß mit radioimmunologischen Techniken eine exakte quantitative Bestimmung von menschlichem Wachstumshormon möglich ist. Im allgemeinen sind unsere Resultate im Einklang mit den Ergebnissen von ROTH u. Mitarb. in New York und mit GREENWOOD in London.

Policlinique Universitaire de Médecine, Genève (Prof. E. Martin)

Beitrag zur Problematik der immunologischen Bestimmung von Wachstumshormon im Serum

Von

H.-J. Quabbe* und G. R. Zahnd

Die Anwendung der Hämagglutinationshemmungs-Methode zur immunologischen Bestimmung des Wachstumshormons im menschlichen Serum wurde zuerst von Read und Bryan (*1*) beschrieben. Das Prinzip besteht darin, daß von einer konstanten Menge Antikörper (AK) gegen menschliches Wachstumshormon (STH) um so mehr einer Reaktion mit sensibilisierten Hammelblutkörperchen (BK) entzogen wird, je mehr STH im Serum vorhanden ist. Wird in einer progressiven Verdünnungsreihe schließlich eine so geringe Konzentration an STH erreicht, daß der größte Teil des zugefügten AK freibleibt, so führt dessen Reaktion mit den sensibilisierten BK zur Agglutination derselben. Durch Vergleich des ersten Röhrchens mit voller Agglutination (Endpunkt der Hemmung) in einer Verdünnungsreihe bekannter Konzentration und der des zu testenden Serums erfolgt die Errechnung des gesuchten Wertes.

Die Bedeutung der mit dieser Technik gewonnenen Ergebnisse wurde bald durch die Erkenntnis erheblicher Unvollkommenheiten geschmälert. Neben der Anwesenheit eines unspezifischen Hemmfaktors im Serum standen dabei die schwankenden Eigenschaften der benutzten BK und Schwierigkeiten der klaren Bestimmung des Hemmungsendpunktes im Vordergrund. Weiter wurden diskutiert der Einfluß des zur Verdünnung benutzten Meerschweinchenserums, Heterogenität des Antiserums, die evtl. Freisetzung von STH von den sensibilisierten BK. Eine ausführliche Besprechung dieser Probleme fand auf dem Ciba Found. Coll. ''Immunoassay of Hormones'' (London 1961) statt.

Ähnliche Erfahrungen führten auch uns zu einigen methodischen Untersuchungen:

1. Das benutzte STH wurde von einem von uns (G. Z.) nach der Methode von Raben gewonnen und hat sich in vitro und in vivo als aktiv erwiesen (*2*). In der Immunodiffusion nach Ouchterlony gegen seinen AK zeigte sich eine einzige Linie; gegen menschliches Albumin wurde keine Präzipitation erzielt. Durch Stärkegel-Elektrophorese wurden vier Linien entwickelt, die mit den von anderen Autoren beschriebenen im wesentlichen übereinstimmen. Durch Injektion von nur 3 mg dieser Präparation konnte ein AK-Titer von 1/40000 erzielt werden (Kaninchen). Obwohl das Vorhandensein eines heterogenen Antigen/Antikörper-Systems trotz der zufriedenstellenden Ergebnisse von Immunodiffusion und Elektrophorese

---

* NATO-Forschungsstipendiat.

nicht ausgeschlossen ist, sollte doch in Anbetracht auch der niedrigen zur Immunisierung benutzten Antigenmenge sein evtl. Einfluß auf den Test auf das geringstmögliche beschränkt sein.

2. Um den Einfluß der wechselnden Eigenschaften frischer BK auszuschalten wurden diese einer Formol-Vorbehandlung unterworfen. Die nach der Methode von WEINBACH (3) behandelten BK wurden mit Merthiolat versetzt, zunächst bei $-20°C$ eingefroren und dann während ihrer Verwendung mehrere Wochen lang bei $+9°C$ aufbewahrt. Diese Zellen erwiesen sich bei der Bestimmung des Antiserum-Titers und im Agglutinationsbild als stabil. BK verschiedener Schafe wiesen nach gleicher Formolbehandlung geringe Unterschiede auf, jedoch blieben die einmal gefundenen Eigenschaften konstant.

Einer der Gründe für die Variationen der Ergebnisse bei Benutzung frischer BK scheint darin zu liegen, daß die Bestimmung des Antiserum-Titers mit ihnen eine erhebliche Schwankungsbreite besitzt und daher die im Test zur Anwendung kommende Antiserum-Verdünnung nicht optimal ist. Dieses Problem ist durch tägliche Titerbestimmungen unmittelbar vor dem Test nicht zu lösen, da keine Gewähr besteht, daß der gewonnene Wert dem „wahren" Titer näher ist als jeder andere. Bei Benutzung der formolbehandelten BK wurden diese Titerschwankungen nicht mehr gesehen und die Frage täglicher Titerbestimmungen wurde damit hinfällig.

Die Tatsache, daß mit formolbehandelten BK häufig der Hemmungsendpunkt genauer abzulesen ist, hängt u. a. damit zusammen, daß der mit diesen BK bestimmte Antiserum-Titer immer niedriger ist als mit frischen BK, daß also diese Zellen weniger leicht agglutinabel sind. Es muß daher eine gewisse Einbuße an Empfindlichkeit des Testes bei Verwendung dieser BK hingenommen werden.

3. Von Bedeutung für die Agglutinierbarkeit der BK und damit für die Empfindlichkeit des Testes ist auch die Menge des zur Sensibilisierung der BK benutzten STH. Werden höhere Mengen angewendet, so steigt die Gefahr einer späteren Freisetzung von STH und damit einer Verfälschung des Ergebnisses. Zu geringe Konzentrationen erzielen jedoch nur eine geringere Agglutinierbarkeit der BK und die Empfindlichkeit des Testes sinkt. So wurde z. B. der Antiserum-Titer um die Hälfte niedriger gefunden, wenn 100 μg anstelle von 250 μg zur Sensibilisierung benutzt wurden. Die optimale Menge STH zur Sensibilisierung sollte daher für jede benutzte Zellart, bzw. nach jeder Formolbehandlung erneut ermittelt werden. Die sensibilisierten BK müssen sofort benutzt werden und in den Kontrollen darf keinerlei Hemmung der Agglutination auftreten, wenn eine Freisetzung von STH von den BK weitestgehend ausgeschlossen werden soll.

4. Von verschiedenen Autoren wurde ein unspezifischer Einfluß höherer Proteinkonzentrationen auf die Agglutination vermutet und daher z. B. versucht, durch stärkere Verdünnung des Meerschweinchenserums diesen Faktor zu verringern. Ein Vergleich des Agglutinationsbildes und des Hemmungsendpunktes zeigte keinen erkennbaren Unterschied im Bereich der von uns geprüften Konzentrationen 1/200 und 1/400.

5. Die Vorbehandlung der Sera (Test-S., Anti-S., Meerschweinchen-S.) wird verschieden gehandhabt. Benutzung formolbehandelter BK macht zwar die Zerstörung des Komplementes unnötig, da diese Zellen auch in Gegenwart von Komplement nicht mehr hämolysieren. Die Frage der evtl. Ausschaltung eines

unspezifischen Hemmfaktors der Agglutination bleibt aber natürlich bestehen. Erhitzen der Sera 40 min lang auf 60°C, gefolgt von zweimaliger Absorption mit BK brachte keine Verbesserung der Ergebnisse, verglichen mit den ursprünglichen Bedingungen (Erhitzen während 30 min auf 56°C, gefolgt von einmaliger Absorption mit BK). Die Vorbehandlung des Antiserums in der angegebenen Weise verändert jedoch dessen Titer unregelmäßig in beiden Richtungen. Bei Verwendung formolbehandelter BK kann das Antiserum ohne Vorbehandlung benutzt werden, ohne daß das Agglutinationsbild oder die Testresultate hierdurch erkennbar beeinflußt werden.

6. Nach READ et al. soll das Agglutinationsbild dadurch zustande kommen, daß der AK an die Glasoberfläche adsorbiert wird und so imstande ist, nach der Reaktion mit den sensibilisierten BK letztere an der Sedimentation zu hindern. Wir benutzten Röhrchen mit gleichmäßig gerundetem Ende aus normalem und aus Pyrex-Glas. Wenn auch in beiden das Agglutinationsbild etwas verschieden ist, wurde doch kein Einfluß auf das Testergebnis gesehen. Trotzdem ist zu empfehlen, nur Röhrchen der gleichen Glassorte und selbstverständlich immer der gleichen Dimensionen zu verwenden.

Obwohl nach Berücksichtigung der diskutierten Faktoren und insbesondere durch Benutzung der formolbehandelten BK ein Teil der Unsicherheitsfaktoren des Testes verringert werden konnte, gelang es nicht, unsere Bestimmungsresultate wesentlich zu verbessern. Wegen ihrer Unzuverlässigkeit müssen wir daher die Methode als zur Klärung klinischer oder experimenteller Fragestellungen ungeeignet betrachten. Da die neuerdings entwickelten radioimmunologischen Dosierungsmethoden für STH — die bessere Ergebnisse versprechen — einen Teil der besprochen Schwierigkeiten mit dem Hämagglutinationshemmungs-Test gemeinsam haben (unspezifischer Hemmfaktor, Titer-Bestimmung und optimale Konzentration der Antiseren, Heterogenität des Antigen/Antikörper-Systems, Einfluß der Inkubationsbedingungen auf die Antigen/Antikörper-Reaktion), ist das weitere Studium dieser Probleme jedoch Voraussetzung für die Zuverlässigkeit auch dieser Bestimmungsmethoden.

Literatur

1. READ, C. H., and G. T. BRYAN: Recent Progr. Hormone Res. **16**, 187 (1960).
2. ZAHND, G. R.: Ann. Endocr. (Paris) **22**, 741 (1961).
3. WEINBACH, R.: Schweiz. Z. allg. Path. **21**, 1043 (1958).

Aus der Abteilung für Klinische Endokrinologie (Leiter: Prof. Dr. E. F. Pfeiffer) der
I. Med. Univ.-Klinik Frankfurt/Main (Direktor: Prof. Dr. F. Hoff)

Immunologische Untersuchungen und quantitativer Nachweis von Wachstumshormon beim Menschen

Von

F. Melani, R. Conrads, A. Sartory, F. Sorge, H. Ditschuneit u. E. F. Pfeiffer

Mit 2 Abbildungen

In den letzten Jahren ist es verschiedenen Arbeitsgruppen gelungen, STH aus Hypophysen von Menschen und Tieren nahezu rein darzustellen (*7, 9, 11*). Die noch vorhandenen Beimengungen bestehen vorwiegend aus Prolactin, das aber proteinchemisch vom STH nicht zu trennen ist und wahrscheinlich einer Wirk-Komponente des STH selbst entspricht (*5, 6, 7*). FSH, ACTH sowie TSH konnten dagegen vollständig eliminiert werden.

Im Ouchterlony-Plattentest beobachteten Li (*7*), Raben (*9*) und Grumbach (*5*) nur eine einzige Präcipitationsbande, die nach immunoelektrophoretischen Untersuchungen auf ein Protein des α 2-Bereiches zurückzuführen ist. Hayashida (*6*) berichtet dagegen, daß nach mehreren Tagen bei allen bisher zur Verfügung stehenden STH-Präparationen ein bis zwei zusätzliche Präcipitationsbanden auftreten, die möglicherweise auf Verunreinigungen zurückzuführen sind. Die evtl. vorhandenen Beimengungen sind aber derart gering, daß dadurch die Spezifität einer Messung von STH im Blut auf immunologischem Wege nicht wesentlich beeinträchtigt wird.

Mit den zur Verfügung stehenden STH-Präparationen lassen sich bei Kaninchen spezifische Antikörper erzeugen und damit ein System gewinnen, mit dem Wachstumshormon im Blut auf immunologischem Wege quantitativ bestimmt werden kann.

Die heute allgemein angewandten Bestimmungsverfahren beruhen entweder auf der Erythrocyten-Agglutination nach Boyden (*1*) oder aber auf dem von Yalow und Berson (*13*) für die immunologische Bestimmung von Insulin im Blut angegebenen Prinzip. Unsere eigenen immunologischen Studien mit menschlichem STH hatten zum Ziel, ein immunologisches Bestimmungsverfahren von STH im Blut nach dem Prinzip von Yalow und Berson auszuarbeiten.

Das von uns verwandte STH wurde von Reisfeld u. Mitarb. (*11*) dargestellt. Antikörper gewannen wir durch Immunisierung von Kaninchen mit komplettem Freundschem Adjuvans. Im Agargel-Diffusionstest nach Ouchterlony bildete sich bereits nach 24 Std eine deutliche Präcipitationsbande aus.

Der Titer des Antiserums im Erythrocyten-Agglutinationstest betrug bei Verwendung von Tannin-behandelten Hammelerythrocyten 1 : 12800. Nach ungefähr

4—6 Tagen beobachteten wir, wie auch Hayashida (6), eine zarte zweite Präcipitationsbande.

Bei papierelektrophoretischer Auftrennung des Antiserums zusammen mit einer kleinen Menge von 10 mμg/ml Jod^{131} markiertem STH fanden wir nur im γ-Globulinbereich einen scharfen Aktivitätsgipfel (Abb. 1). Diese Untersuchungen beweisen nicht nur das Vorhandensein von Antikörpern gegen STH, sondern liefern uns auch einen Beweis dafür, daß die Jodmarkierung des STH die antigenen

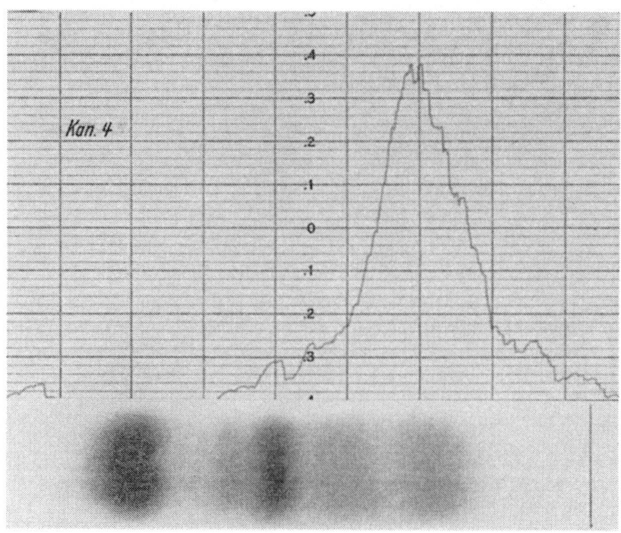

Abb. 1. Radiochromatogramm der Papierelektrophorese von Kaninchen-anti-human-STH-Serum nach Zusatz von 10 mμg/ml Jod^{131}-human-STH in vitro (spez. Aktivität 85 mC/mg)

Eigenschaften desselben nicht verändert hat. Die Markierung wurde nach Greenwood und Hunter (4) mit Chloramin T vorgenommen, und die Reinigung mit Hilfe einer Sephadex-Säule durchgeführt. Dabei erreichten wir eine spezifische Aktivität zwischen 60 und 140 mC/mg.

Bei der Behandlung von zwei hypophysären Zwergen mit 47,5 und 100 mg menschlichem Wachstumshormon über einen Zeitraum von 8 Wochen in dem einen Falle und 20 Wochen in dem anderen traten in dem Serum dieser Patienten ebenfalls Globuline mit bindenden Eigenschaften gegenüber menschlichem STH auf. Die Radiochromatogramme von den Papierelektrophoresen dieser beiden Kranken zeigen sowohl einen Radioaktivitätsgipfel im γ-Globulinbereich, als auch im Bereich der α 2-Globuline. Der Gipfel bei den γ-Globulinen ist wahrscheinlich auf Antikörper gegen das zur Behandlung verwandte menschliche STH zurückzuführen. Bei der Auftrennung von Normalseren wurden niemals Bindungen von STH im γ-Globulinbereich, sondern immer nur im Bereich von α 2-Globulinen und am Startpunkt beobachtet.

Für die Bestimmung von STH im Blut haben wir das von Yalow und Berson (13) für die Bestimmung von Insulin angegebene Prinzip benutzt, daß diese Autoren neuerdings auch für die Bestimmung von STH verwenden (3). Bei konstanter Menge von Jod^{131}-STH und konstanter Antiserumverdünnung wird die

Menge von antikörpergebundenem STH durch Zusatz steigender Mengen von nicht markiertem STH umgekehrt proportional beeinflußt. Die graphische Darstellung des Verhältnisses von gebundenem zu freiem STH ergibt den für Antigen-Antikörper-Reaktionen typischen hyperbolischen Kurvenverlauf (Abb. 2). Bei den meisten in der Literatur mitgeteilten Verfahren zur immunologischen STH-Bestimmung erfolgt die Trennung von freiem und gebundenem STH entweder mit der Papierelektrophorese oder durch Präcipitation mit Anti-γ-Globulin-Serum.

Wir benutzten zur Trennung von freiem und gebundenem STH den Ionenaustauscher Dowex 1 in der basischen Form (8). Das freie STH wird an den Ionenaustauscher gebunden, während das antikörpergebundene STH nach Zentrifugation im Überstand zurückbleibt.

Dieses Verfahren hat den Vorteil, daß auch STH-Präparationen mit niedriger spezifischer Aktivität verwandt werden können. In der dargestellten Eichkurve wurden 1 : 40000 verdünntes Kaninchen-Anti-STH-Serum und Jod¹³¹-STH in einer Menge von 0,25 mμg/ml verwandt. Zu diesem System wurden nicht markierte STH-Mengen zwischen 2 und 100 mμg/ml hinzugefügt, und die Proben bis zur Trennung von freiem und gebundenem STH mit Hilfe des Dowex 1 3 Tage bei +4° aufbewahrt.

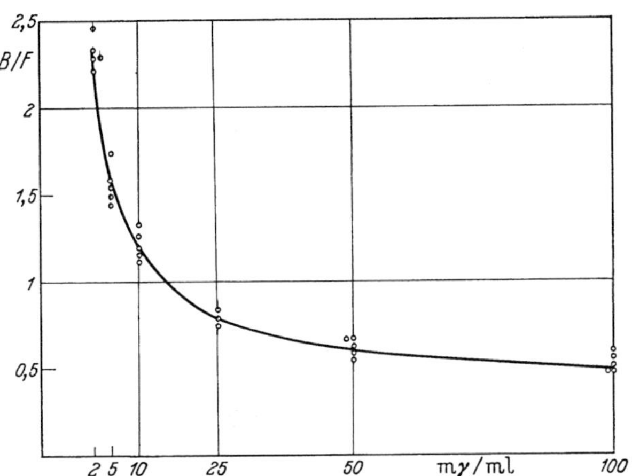

Abb. 2. Einfluß von steigenden Mengen von nichtmarkiertem STH auf das Verhältnis von antikörpergebundenem zu freiem Jod¹³¹-STH (B/F) Trennung von freiem und gebundenem STH mit Dowex 1 × 8. Jod¹³¹-STH 0,25 mμg/ml (spez. Aktiv. 9 mC/mg) Kaninchen-anti-STH-Serum 1:40000

Bei Zusatz von 100 mμg/ml STH ergibt sich ein Verhältnis von B/F von 0,53 und bei Zusatz von 2 mμg/ml ein Bindungsverhältnis von 2,26. Die größte Empfindlichkeit liegt bei diesem System zwischen 2 und 25 mμg/STH/ml.

Bei der Bestimmung des STH-Gehaltes von Serumproben stört die Bindung an α 2-Globuline. Dieser Störfaktor läßt sich aber durch starke Verdünnung des Serums weitgehend ausschalten. Unsere Serumproben wurden daher 10—20fach verdünnt und diese Verdünnung bei der späteren Berechnung berücksichtigt. Auch UTIGER u. Mitarb. (12) konnte mit der Doppel-Antikörper-Methode ebenso wie DOMINGUEZ u. Mitarb. (2) mit der Methode der Erythrocyten-Agglutination im Serum keine exakten quantitativen Messungen durchführen. Als Ursache werden von den Autoren unspezifische Serum-Plasmafaktoren angenommen, die möglicherweise mit der Bindung von STH an α 2-Globuline in Zusammenhang stehen. Bei 20facher Serumverdünnung fanden wir bei Stoffwechselgesunden Werte im Bereich zwischen 0 und 40 mμg/ml. Bei drei hypophysären Zwergen und zwei Patienten mit einem Sheehan-Syndrom war der STH-Spiegel dagegen, ebenso wie bei drei röntgenbestrahlten Akromegalen, nicht meßbar. Bei einer nicht röntgenbestrahlten, aktiven Akromegalie fand sich im Nüchternserum ein über dem Normalbereich liegender Wert.

Literatur

1. Boyden, S. V.: The absorption of proteins on erythrocytes treated with tannic acid and subsequent haemoagglutination by anti-protein sera. J. exp, Med. **93**, 107 (1951).
2. Dominguez, J. M., and O. H. Pearson: Immunologic measurement of growth hormone in human sera. J. clin. Endocr. **22**, 865 (1962).
3. Glick, S. M., J. Roth, R. S. Yalow and S. A. Berson: Hypoglicemia: A potent stimulus to secretion of growth hormone. Science **140**, 988 (1963).
4. Greenwood, F. C., and W. M. Hunter: The preparation of ^{131}J-labelled human growth hormone of high specific radioactivity. Biochem. J. **89**, 144 (1963).
5. Grumbach, M. M., and S. L. Kaplan: Immunochemical studies on human growth hormone: a consideration of the human growth hormone anti-human-growth-hormone system and its application to the assay. Ciba Found. Coll. Endocr. **14**, 63 (1962).
6. Hayashida, T.: Immunological studies with pituitary lactogenic hormone (Prolactin). Ciba Found. Coll. Endocr. **14**, 338 (1962).
7. Li, C. H., N. R. Moudgal, A. Trenkle, G. Bourdel, and K. Sadri: Some aspects of immunochemical methods for the characterization of protein hormones. Ciba Found. Coll. Endocr. **14**, 21 (1962).
8. Melani, F., H. Ditschuneit, H. H. Ditschuneit, R. Petzoldt e E. F. Pfeiffer: Metodo immunologico per la determonazione dell'insulina nel siero per mezzo di Insulina J^{131} e di anticorpi-anti-insulina. I. Simposio Internaz. sul Diabete Modena, Settembre 1963.
9. Raben, M. S.: Preparation of growth hormone from pituitaries of man and monkey. Science **125**, 883 (1957).
10. Read, C. H., and G. T. Bryan: The immunological assay of human growth hormone. Recent Progr. Hormone Res. **187**, 16 (1960).
11. Reisfeld, R. A., U. J. Lewis, N. G. Brink, and S. L. Steelman: Human growth hormone: Preparation from acetone powder. Endocrinology **71**, 559 (1962).
12. Utiger, R. D., M. L. Parker, and W. H. Daughaday: Studies of human growth hormone. I. A radio-immunoassay for human growth hormone. J. clin. Invest. **41**, 254 (1962).
13. Yalow, R. S., and S. A. Berson: Immuno-assay of endogenous plasma insulin in man. J. clin. Invest. **39**, 1157 (1960).

Department of Endocrinology and Metabolism, Karolinska sjukhuset, Stockholm, Schweden

Humanes Wachstumshormon und Diabetes

Von

R. Luft und E. Cerasi

Mit 10 Abbildungen

Referat

Das Wachstumshormon der Hypophyse ist seit mehreren Jahrzehnten in einen engen Zusammenhang mit dem Diabetes gestellt worden und wird als wesentlicher diabetogener Faktor der Hypophyse angesehen. Dies gilt für Tierversuche, und ich glaube, es ist nicht notwendig, die reichhaltige Literatur auf diesem Gebiet zu erwähnen. In der gegenwärtigen Diabetesforschung beim Menschen nimmt das Wachstumshormon eine periphere Stellung ein. Trotzdem sind wir der Meinung, daß es in diesem Zusammenhang wahrscheinlich eine wesentliche Rolle spielen wird.

Die allgemeine Wirkung des humanen Wachstumshormons (HGH) auf den Stoffwechsel wird von anderen Rednern in diesem Symposium behandelt. Ich möchte mich in diesem Zusammenhang nur bei einem Punkt aufhalten, nämlich dem Effekt des Hormons auf den Fettstoffwechsel. Unter Zufuhr von HGH an 11 nüchterne Versuchspersonen stieg die Menge der nicht veresterten Fettsäuren im Blut (FFA) als Zeichen einer gesteigerten Mobilisierung der Fettdepots an (Abb. 1). Die Zunahme der FFA bei Zufuhr von HHG kann man verhindern, wenn man gleichzeitig Glucose gibt. Dieser Effekt des HGH könnte auch für den Effekt des Hormons auf den Kohlenhydratumsatz von Bedeutung sein. Randle u. Mitarb. haben nämlich 1963 die Hypothese aufgestellt, daß eine Zunahme der FFA im Plasma die periphere Glucoseaufnahme und Glykolyse herabsetze, und so eine scheinbare Insulinresistenz hervorrufen kann.

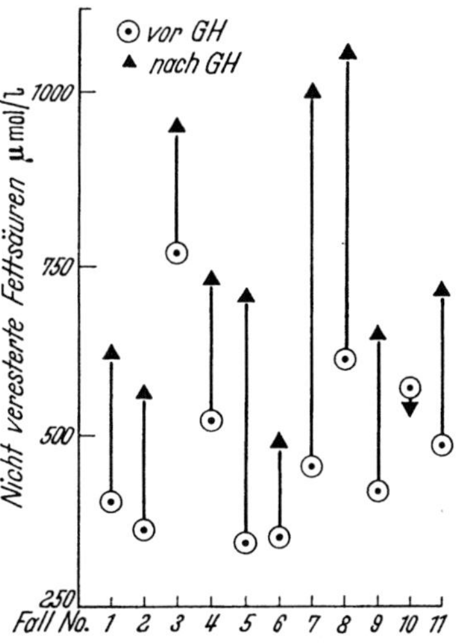

Abb. 1. Effekt von HGH (30 mg pro Tag während 3 Tage) auf nicht veresterte Fettsäuren im Blut bei 11 nüchternen gesunden Versuchspersonen (Ikkos, Luft, Gemzell u. Almvqist, 1962)

Ich werde zunächst drei Aspekte der Wirkung von HGH auf den Kohlenhydratstoffwechsel erörtern.

1. Können wir mit HGH auf den Kohlenhydratstoffwechsel von Nicht-Diabetikern einwirken?

2. Kann HGH die diabetische Situation bei Kranken mit Diabetes mellitus beeinflussen?

3. Kann man mit HGH bei Nicht-Diabetikern einen klinischen Diabetes hervorrufen, d. h. einen wirklich diabetogenen Effekt des Hormons zeigen?

Die Zufuhr von 30 mg HGH ruft bei nicht-diabetischen Versuchspersonen keine Hyperglykämie hervor, wenn auch der Nüchternblutzucker bei einigen von ihnen etwas ansteigt (Abb. 2). Dagegen konnten wir und andere nachweisen, daß das Wachstumshormon einen initialen, sehr kurz dauernden Abfall des Blutzuckers bewirkt. Diese Eigentümlichkeit ist nicht leicht verständlich, m. E. könnte es sein, daß das Wachstumshormon die Insulinabgabe der Pankreasinseln initial stimuliert.

Der Effekt des Wachstumshormons auf den Kohlenhydratstoffwechsel wird klarer, wenn wir dessen Wirkung auf die Glucosetoleranz bei 11 gesunden Personen, die täglich 30 mg des Hormons während dreier Tage erhielten, betrachten (Abb. 3). Wir bestimmten die Glucosetoleranz durch intravenöse Glucosebelastung und fanden eine definitive Senkung der Glucosetoleranz bei der Mehrzahl der Patienten. Ich möchte in diesem Zusammenhang auf die Fälle 3, 4 und 7 in dieser Gruppe aufmerksam machen, bei welchen die Geschwindigkeit der Glucoseabnahme die stärkste Senkung zeigte und eine wirklich diabetische Glucosetoleranz mit einem sog. diabetischen k-Wert erreichte.

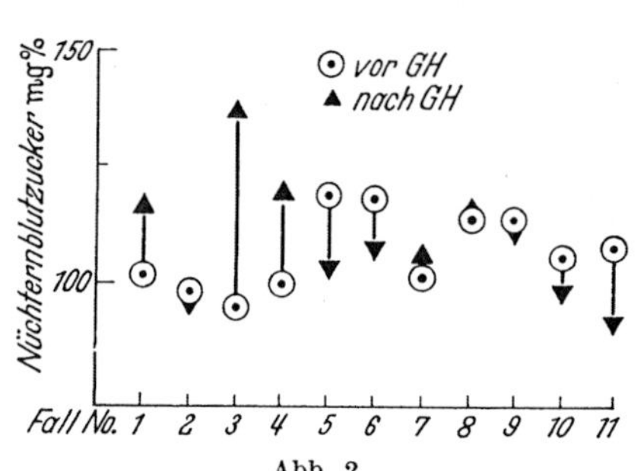

Abb. 2

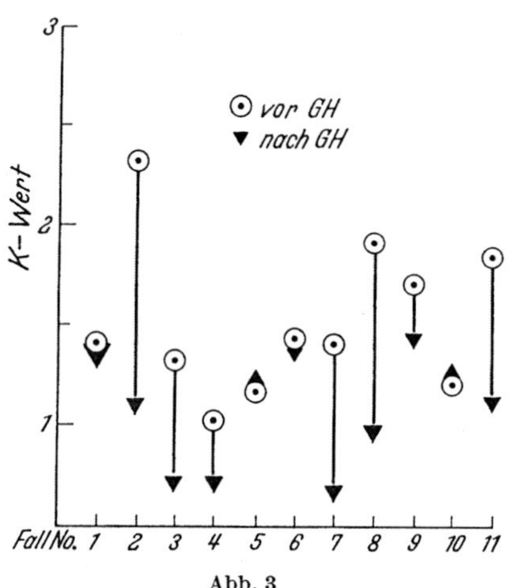

Abb. 3

Abb. 2. Effekt von HGH (30 mg pro Tag während 3 Tage) auf den Nüchternblutzucker bei 11 gesunden Versuchspersonen (Ikkos, Luft, Gemzell u. Almqvist, 1962)

Abb. 3. Effekt von HGH (30 mg pro Tag während 3 Tage) auf die Glucosetoleranz (k-Wert) von 11 gesunden Versuchspersonen (Ikkos, Luft, Gemzell u. Almqvist, 1962)

Die HGH-Dosen, die wir gaben, waren allerdings ziemlich groß (30 mg pro Tag). Kleinere Mengen von Wachstumshormon wirkten bedeutend weniger und nur in einzelnen Fällen auf die Glucosetoleranz. Außerdem möchte ich betonen, daß auch bei hypophysektomierten Nicht-Diabetikern, die Wachstumshormon erhielten, das gleiche Resultat erreicht wurde wie bei den gesunden Personen.

Den Mechanismus, durch welchen HGH die Kohlenhydrattoleranz bei gesunden Versuchspersonen ändert, kennen wir nicht.

Um ihn aufzuklären, sollte festgestellt werden, ob HGH Veränderungen in den Konzentrationen der intermediären Kohlenhydratmetaboliten im Blut induzieren konnte. Wir führten die Untersuchungen bei den schon genannten 11 Personen durch, die während dreier Tage 30 mg HGH täglich erhielten, und gaben ihnen eine einstündige Infusion von 15%iger Glucose vor und am letzten Tag der HGH-Zufuhr. Bestimmt wurden die Plasmakonzentrationen von Pyruvat, Lactat,

Citrat und α-Ketoglutarat (Abb. 4). Diese Bestimmungen geschahen unmittelbar vor der Glucoseinfusion, nach 60 min, also am Ende der Infusion, und nach 120 min, d. h. 2 Std nach Versuchsbeginn.

Wir können das leider etwas komplizierte Ergebnis so zusammenfassen, daß eine systematische Steigerung oder Veränderung der Nüchternkonzentrationen der genannten Metaboliten unter HGH nicht stattfand. Allerdings lag ein deutlicher Anstieg der Nüchternwerte von Pyruvat und Lactat in Fall 3 und 4 und von Citrat in Fall 3 und evtl. 7, also insgesamt in 3 von 11 Fällen vor. Außerdem sei

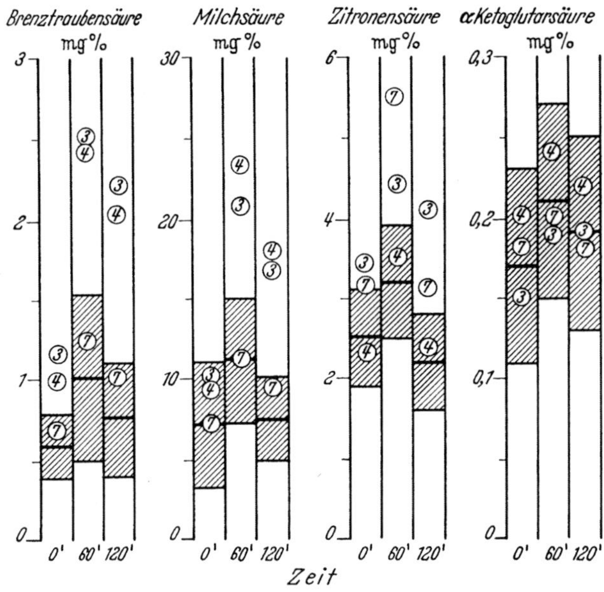

Abb. 4. Effekt von HGH (30 mg pro Tag während 3 Tage) auf die Konzentrationen von Kohlenhydratmetaboliten im Blut bei 11 gesunden Versuchspersonen (IKKOS, LUFT, GEMZELL u. ALMQVIST, 1962). Die schraffierten Teile sind die ± 2-Standarddeviationsgrenze für die Werte der gesamten Gruppe vor Wachstumshormonzufuhr. Die Kreise — die Ziffer innerhalb bezeichnet die Nummer des Falles — stellen diejenigen Fälle dar, bei denen HGH-Zufuhr eine über die Norm erhöhte Zunahme in einem der Metaboliten beobachtet wurde

betont, daß HGH unter Glucoseinfusion nur in diesen drei Fällen definitive Veränderungen in der Konzentration der Metaboliten erzeugte: eine Zunahme von Pyruvat und Lactat in Fall 3 und 4 und von Citrat in 3 und 7. Es mag hier darauf hingewiesen sein, daß es sich bei diesen drei Fällen um die gleichen handelte, bei denen die Glucosetoleranz nach HGH-Zufuhr am stärksten sank. Man könnte daraus die Schlußfolgerung ziehen, daß die Geschwindigkeit der Glucoseabnahme einer Verminderung, also einer Senkung der Glucosetoleranz bedarf, ehe wir Veränderungen in den genannten Metabolitenkonzentrationen unter Glucosezufuhr beobachten und messen können.

Wir können im Augenblick diese Resultate nicht anders deuten, als daß sie akute Effekte von hohen HGH-Dosen sind.

Ich komme nun zur *Frage* des Effektes von humanem Wachstumshormon auf diabetische Personen.

Die Abb. 5 zeigt einen solchen Versuch. Wir gaben einem früher hypophysektomierten erwachsenen Patienten mit juvenilem Diabetes 10 mg Wachstumshormon

und innerhalb weniger Stunden kam es zu einer Verschlechterung des diabetischen
Status mit einer schweren metabolischen Acidose. Wie sich später ergab, ist das
Resultat nicht das gleiche, wenn man hypophysektomierten erwachsenen Personen
mit sog. Erwachsenen-Diabetes 5—10 mg Wachstumshormon gibt. Man erzielt
auch hier eine Hyperglykämie, Glykosurie und vielleicht geringe Ketonurie, aber
keine signifikative Änderung des Säure-Basen-Haushaltes.

In diesem Zusammenhang soll erwähnt sein, daß wir mit 10 mg HGH täglich
kaum Veränderungen des Diabetes-Status bei nicht-hypophysektomierten Diabetikern vom sowohl erwachsenen als juvenilem Typ erhielten und auch nur mäßige
Veränderungen, als wir ihnen 30 mg HGH gaben.

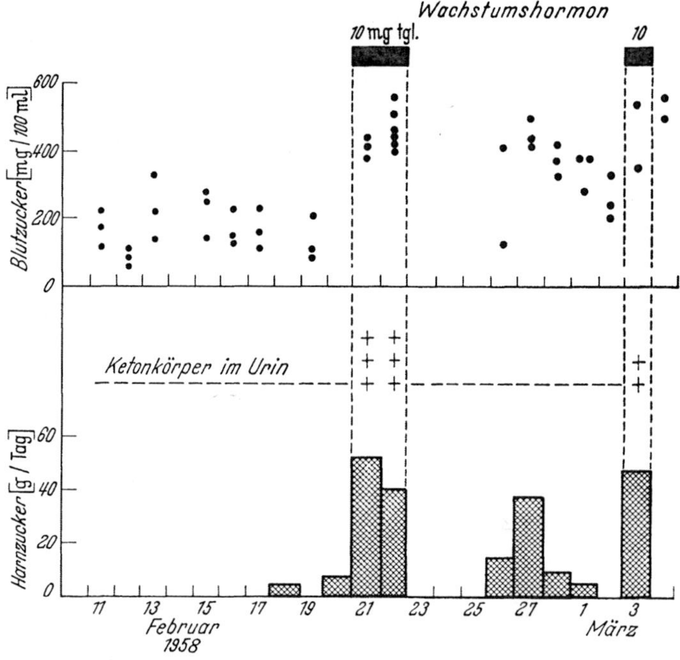

Abb. 5. Blutzucker, Zucker und Ketonkörper im Harn vor, während und nach Zufuhr von HGH zu einem
hypophysektomierten, juvenilen Diabetiker (Luft, Ikkos, Gemzell u. Olivecrona, 1958)

Aus diesen Studien können wir den Schluß ziehen, daß HGH den diabetischen
Status bei Patienten mit Diabetes mellitus verschlimmert, und daß der jugendliche
Diabetiker nach Hypophysektomie für das Hormon außerordentlich empfindlich ist.

Ich komme nun zur *Frage*, ob wir mit HGH einen klinischen Diabetes bei Nicht-
Diabetikern hervorrufen können. Nach einer üblichen Definition bezeichnen wir
einen Zustand mit herabgesetzter Glucosetoleranz aber im übrigen normalen Verhältnissen als latenten oder chemischen Diabetes. Im Hinblick darauf konnten wir
bei der Mehrzahl unserer Versuchspersonen eine diabetogene Wirkung des HGH
nachweisen.

Zu diesem Zweck verwendeten wir mit C^{14} markierte Glucose und gaben den
nicht-diabetischen Versuchspersonen 10—20 mg HGH täglich in 2—3 Tagen. Wir
führten die Versuche bei hypophysektomierten, nicht-diabetischen Personen aus,
weil wir annahmen, daß diese mehr als gesunde Versuchspersonen auf HGH

ansprechen würden. Ich möchte erwähnen, daß diese Patienten wegen Mamma-carcinom hypophysektomiert worden waren, daß sie sich aber in ausgezeichnetem Zustand bei Versuchsbeginn und während des Versuches befanden.

Die Abb. 6 zeigt, daß das Wachstumshormon eine Erhöhung des Nüchtern-blutzuckers bei diesen Patienten bewirkte, und daß alle eine Nüchternhyper-glykämie zeigten. Da es sich hier um Glucose im Plasma handelt, sind die angegebe-nen Werte erhöht. Die Hyperglykämie bei diesen Patienten war nach Nahrungs-aufnahme viel deutlicher unter Wachstumshormonzufuhr als vorher. Glykosurien

Fall Nr.	Nüchtern-Blutzucker mg-%	Harn-Zucker g/Tag	Ketonurie	Harn-Stickstoff g/Tag	Harn-Calcium mg/Tag
1	83 → 112				
2	72 → 109	0 → 11	0 → +++	9 → 3	300 → 550
3	82 → 153	0 → 86	0 → ++	14 → 7	250 → 600
4	48 → 125	0 → 7	—	10 → 6	300 → 450
5	67 → 120	0 → 4	—	8 → 4	80 → 250

Abb. 6. Nüchternblutzucker sowie Zucker, Ketonkörper, Stickstoff und Calcium im Harn nach Zufuhr von HGH, 10—20 mg pro Tag während dreier Tage, bei fünf hypophysektomierten, nicht-diabetischen Versuchspersonen (IKKOS u. LUFT, 1962)

verschiedenen Grades traten bei sämtlichen Fällen auf, und diese wie auch die Hyperglykämien verschwanden, als HGH abgesetzt wurde. In allen Fällen zeigte sich eine signifikative Stickstoffretention, und die Calciumausscheidung im Urin nahm unter Hormonzufuhr merkbar zu. Alle diese Veränderungen stimmen bis in Details mit den Stoffwechselveränderungen, die einen idiohypophysären Diabetes charakterisieren, überein; ein Zustand, der u. a. von HOUSSAY und YOUNG bei einer Reihe von Tieren mit bovinem Wachstumshormon demonstriert worden ist. Die Stickstoffretention zeigt zusammen mit der vermehrten Calciumausscheidung im Urin, daß die HGH-Präparate, die wir in diesen Versuchen verwendet haben, die erwartete Wachstumshormonaktivität besaßen.

Wir hielten es für bedeutungsvoll, den bei diesen Personen erzeugten Diabetes-status weiter aufzuklären. Am meisten interessierten uns etwaige Veränderungen in der peripheren Ausnutzung der Glucose. Dabei ergab sich, daß in allen Fällen unter Hormonzufuhr eine Nüchternhyperglykämie auftrat, und dem entsprach, daß die Hyperglykämie nach Nahrungsaufnahme viel stärker unter als vor Wachs-tumshormongabe war.

Ferner stieg der Glucosepool, den wir als Produkt von Nüchternblutzucker und Glucoseraum berechneten, bei sämtlichen Patienten unter HGH-Zufuhr an, wäh-rend der mit C^{14}-Glucose bestimmte Glucoseumsatz signifikativ sank. In diesen Fällen bestand eine nahe Parallele zwischen den Werten und Veränderungen in der Geschwindigkeit der Abnahme von gewöhnlicher Glucose und "turnover rate" von C^{14}-Glucose. Die sog. "turnover rate" der Glucose repräsentiert die Geschwindigkeit mit der die spezifische Aktivität beim Glucosepool durch Addition von nicht mar-kierter Glucose vermindert wurde. Wir nennen dies gewöhnlich die Glucosezufuhr ("glucose inflow"). Wenn wie in unseren Versuchen die Zuckerkonzentration und der Glucosepool während des C^{14}-Glucoseexperimentes konstant sind, beinhaltet die "turnover rate" der Glucose auch die Geschwindigkeit, mit welcher die Glucose

peripher verschwindet (Glucoseverbrauch oder "glucose outflow"). Deshalb können
die Werte für Glucosezufuhr und Glucoseverbrauch als identisch und als Produkt
von "turnover rate" und Glucosepool betrachtet werden.

Abb. 7 zeigt, daß Glucosezufuhr oder Glucoseverbrauch unter Wachstumshormon in einem Fall abnahmen, in den anderen vier Fällen aber unverändert
blieben.

Aus diesen Untersuchungen folgt, daß das humane Wachstumshormon bei
unseren Versuchen die absolute Menge Glucose, die peripher ausgenutzt wurde,
nicht änderte. Wir müssen dagegen berücksichtigen, daß Blutzuckerspiegel und
Glucosepool unter Wachstumshormonzufuhr stiegen. Es ist bekannt, daß die
periphere Ausnutzung von Glucose unter den gegebenen Verhältnissen dem Blutzuckerspiegel proportional ist. Wenn wir davon ausgehen, daß das Blutzuckerniveau unter Zufuhr von HGH erhöht war, so zeigen unsere Untersuchungen

Glucosezufuhr und Glucoseverbrauch in mg/min =
Glucose pool in g × turnover rate von Glucose in % pro min

	Vor HGH	Nach HGH
Fall 1	200	207⁻
Fall 2	207	169***
Fall 3	187	234⁻
Fall 4	139	151⁻
Fall 5	149	137⁻

Abb. 7. Effekt von HGH auf Glucosezufuhr und Glucoseverbrauch in den fünf Fällen von Abb. 6 (Ikkos u. Luft
1962)

folgendes: HGH vermindert den Glucoseverbrauch, also die periphere Ausnutzung
der Glucose, relativ zum erhöhten Blutzuckerspiegel.

Eine verminderte periphere Ausnutzung der Glucose ist charakteristisch für
den Diabetes mellitus. Außerdem haben Patienten mit Akromegalie, die oft eine
niedrige Glucosetoleranz aufweisen, auch eine verminderte Geschwindigkeit in der
Glucoseabnahme. Wir können daraus folgern, daß die Verminderung der peripheren
Ausnutzung der Glucose unter HGH im Vergleich zum Blutzuckerspiegel zeigt,
daß der temporäre diabetische Status, den wir mit dem Hormon beim Menschen
erzielten, was die Glucoseausnutzung betrifft, sich nicht von anderen Typen des
experimentellen Diabetes und vom klinischen Diabetes oder Diabetes bei Akromegalie unterscheidet.

Wir kommen nun zur wesentlichen Frage in diesem Zusammenhang: durch
welchen oder welche Mechanismen bewirkt das HGH eine verminderte periphere
Ausnutzung der Glucose. Ich kann hier nur einige Möglichkeiten skizzieren.

1. Die Zunahme von FFA im Plasma senkt die periphere Glucoseaufnahme.

2. Das HGH erzeugt eine verminderte Insulinabgabe des Pankreas.

3. HGH ruft eine Erhöhung von einem oder mehreren zirkulierenden Insulinantagonisten hervor.

Wir haben einige preliminäre Untersuchungen gemacht, um diese Aspekte zu
beleuchten. Ich habe bereits erwähnt, daß HGH die Menge der FFA im Plasma
erhöht und in diesem Zusammenhang auch früher berichtet, daß wir mit HGH eine
deutliche Insulinzunahme im Plasma von Nicht-Diabetikern, die früher hypophysektomiert worden waren, hervorriefen. Das Insulin wurde sowohl biologisch
als auch immuno-chemisch bestimmt. Die Abb. 8 zeigt die Plasmainsulinzunahme

bei zwei hypophysektomierten Patienten (1 und 2) und einem mit Panhypopituitarismus aus anderer Ursache. Wir registierten unter HGH eine Steigerung des Nüchterninsulins sowohl mit der immuno-chemischen als auch mit der biologischen Methode.

Fall	Therapie	k-Wert	Nüchtern-Blutzucker mg-%	Harnausscheidung mg/Tag				Insulin im Plasma (Nüchtern) μU/ml	
				Glucose	Ketonkörper	Stickstoff	Calcium	immunol.	biologisch
1	Kontrolle (6 Tage)	2,48	58 — 67	—	—	11,0—15,2	156 — 195	47 — 56	81 — 115
	HGH (3 Tage)	0,94	69 → 114	—	(+)	5,1— 9,1	211 → 290	69 → 99	137 → 193
2	Kontrolle (7 Tage)	1,28	60 — 66	—	—	12,1—12,7	123 — 148	29	270 — 280
	HGH (3 Tage)	0,84	88 → 138	—	+	5,2— 9,0	137 → 173	79	319
3	Kontrolle (7 Tage)	1,0	60 — 74	—	—	13,3—20,1	135 — 185	42 → 51	80
	HGH (4 Tage)	1,0	101 → 105	—	(+)	8,5—12,8	259 → 486	55 → 110	110

Abb. 8. Effekt von HGH auf Glucosetoleranz, Nüchternblutzucker und Insulin im Plasma (immunologisch und biologisch bestimmt) bei drei Versuchspersonen mit Panhypopituitarismus (LUFT u. CERASI, 1964)

In Abb. 9 sieht man die Veränderungen des Plasmainsulins nach i.v. Glucosezufuhr wie bei der üblichen i.v. Glucosebelastung. Bei dieser kommt es normalerweise zu keiner nennenswerten Steigerung des Plasmainsulins 30 min nach der

Plasmainsulin während eines i.v. Glucose-Toleranz-Testes (μU/ml)		Nüchternwert	30 min	k-Wert
Fall 1 . . .	Kontrolle	(29)	62	2,48
	HGH	79	147	0,94
Fall 2 . . .	Kontrolle	50	64	1,28
	HGH	99	185	0,84

Abb. 9. Effekt von HGH auf das Plasmainsulin während einer i.v. Glucosebelastung bei zwei Versuchspersonen mit Panhypopituitarismus (LUFT u. CERASI, 1964)

Zuckerzufuhr. Aus dem Bild geht hervor, daß eine deutliche Insulinzunahme während der HGH-Periode auftrat. Eine solche Reaktion findet man auch bei Altersdiabetikern und Prädiabetikern.

Die Abb. 10 demonstriert die Insulinwerte unter verlängerter Glucosezufuhr bei einem Patienten mit Panhypopituitarismus. Er war mit Cortison und Thyreoidea sicca substituiert. Vor der HGH-Zufuhr bewirkte die Glucoseinfusion eine mäßige Insulinsteigerung mit einem k-Wert von 1,69 nach Infusionsschluß. Unter der HGH-Periode erhielten wir mit derselben Glucosezufuhr ein bedeutend höheres Blutzuckerniveau. Der Insulinspiegel stieg wesentlich stärker als vor HGH an. Trotz der großen Menge von zirkulierendem Insulin war die Glucoseabnahme nach Infusionsende bedeutend langsamer mit einem k-Wert von 0,89.

Das HGH bewirkte somit bei Patienten mit Panhypopituitarismus eine Minderung der peripheren Glucoseaufnahme trotz einer erheblichen Steigerung der zirkulierenden Insulinmenge. Es bleibt jedoch noch zu zeigen, ob dieses Paradoxon durch die gleichzeitige Steigerung der FFA unter HGH (Randles Theorie), eine gesteigerte Produktion von spezifischen Insulinantagonisten oder durch andere Faktoren erklärt werden kann.

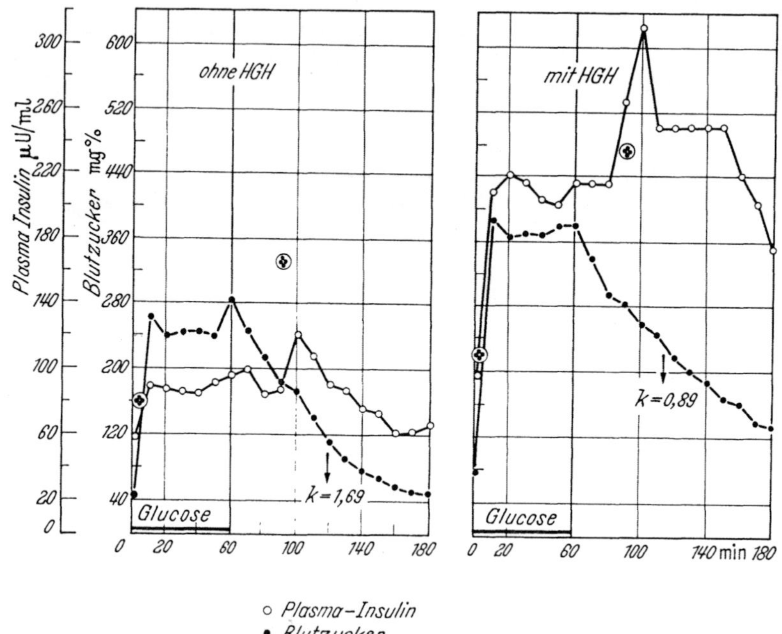

Abb. 10. Effekt von HGH auf das Plasmainsulin (immunologisch und biologisch bestimmt) während verlängerter Glucosezufuhr bei einem Patienten mit Panhypopituitarismus (Luft u. Cerasi, 1964)

Literatur

Cerasi, E., and R. Luft: Lancet **1963 II**, 1359.
Ikkos, D., u. R. Luft: Lancet **1960 I**, 897.
— — Acta endocr. (Kbh.) **39**, 567 (1962).
— —, C. A. Gemzell, and S. Almqvist: Acta endocr. (Kbh.) **39**, 547 (1962).
Karam, J. H., G. M. Grodsky, and P. H. Forsham: Diabetes **12**, 197 (1963).
Luft, R., and E. Cerasi: Ciba Found. Coll. Endocr. (1964, in press).
— D. Ikkos, C. A. Gemzell u. H. Olivecrona: Acta endocr. (Kbh.) **32**, 330 (1959).
Randle, P. J., P. B. Garland, C. N. Hales, and E. A. Newsholme: Lancet **1963 I**, 785
Vallance-Owen, J.: In: Advances in Metabolic Disorders, Vol. 1, R. Levine and R. Luft, Ed. New York-London: Academic Press 1964.

Diskussion

H. G. Goslar (Bonn):

In den letzten Jahren sind eine Reihe von Untersuchungen durchgeführt worden, welche sich mit der direkten Wirkung von Hormonen auf die Enzymsysteme befassen. So berichtete z. B. Schriefers auf dem Münchener Symposium 1961 über die Wirkung von Cortison auf die Glucose-6-Phosphatase der Leber. Der gesteigerte Gehalt an freien Fettsäuren im Serum läßt so eine Aktivierung der Gewebsesterasen durch das Hormon vermuten und der paradoxe Befund von Insulin- und Glucosegehalt des Blutes könnte auf Aktivitätsänderung u. a. der Leber-Glucose-6-Phosphatase hinweisen. Ich möchte den Herrn Vortragenden fragen, ob entsprechende Untersuchungen schon durchgeführt wurden bzw. in Angriff genommen sind.

R. Luft:

Ich habe vorläufige Ergebnisse, die ich jedoch nicht fixieren möchte.

Aus der Endokrinologischen Abteilung (Leiter: Prof. Dr. E. F. PFEIFFER) der I. Medizinischen
Universitätsklinik (Direktor: Prof. Dr. F. HOFF) Frankfurt am Main

Wachstumshormon und Insulinsekretion: Die Verhältnisse unter normalen und pathologischen Bedingungen

Von

E. F. PFEIFFER[1]

Mit 7 Abbildungen

Die Veränderungen des Stoffwechsels, die menschliches Wachstumshormon (STH hum.) bei Gesunden und Zuckerkranken auch bei nur kurzfristiger Gabe auslöst, hat LUFT (1964) beschrieben. Die Fähigkeit des Hormons, bei Dauerzufuhr einen permanenten (metahypophysären) Diabetes hervorzurufen, ist seit langem aus dem Tierexperiment bekannt [YOUNG (1937), CAMPBELL et al. (1950)].

Es soll meine Aufgabe sein, den Effekt des STH auf die Produktion und Sekretion von Pankreasinsulin zu untersuchen. Zu diesem Zweck haben wir

1. die Auswirkung der Hypophysektomie auf das Seruminsulin des Hundes sowie den Effekt der Substitutionsbehandlung mit STH auf seinen Insulinspiegel verfolgt,

2. die Insulinsekretion des partiell isolierten Hundepankreas nach STH im Kurzversuch direkt gemessen und mit dem Ausmaß der Stimulierung der Insulinsekretion durch Glucose und Sulfonylharnstoffe verglichen, sowie

3. die Veränderungen der Insulinaktivitäten und -konzentrationen im Blute gesunder und kranker Menschen nach menschlichem Wachstumshormon zu erfassen versucht.

1. Methodik

Zu den Tierversuchen wurden Hunde verschiedener Rassen im Gewicht von 10—15 kg verwandt.

Die Hypophyse wurde auf transbuccalem Weg entfernt und die Tiere mit DOCA i. m. (5 mg/Tag) substituiert. Das Pankreas wurde z. T. als einziges Organ, z. T. 2 Wochen nach der Hypophysektomie entfernt. Zu den Sekretionsstudien wurde das Hundepankreas durch Unterbindung aller venösen Abflüsse zu den Vv. portae, gastro-epiploicae und lienalis isoliert, aber in situ belassen, und die Insulinsekretion in dem aus der kanulierten V. pancreatico-duodenalis ausfließenden Blut gemessen. Die Ausflußgeschwindigkeit wurde auf 6—7 ml/min eingestellt und der Blutverlust durch Dauerinfusion von Hundeblut ersetzt. Dieses Verfahren

---

[1] Durchgeführt mit Unterstützung der Deutschen Forschungsgemeinschaft, Bad Godesberg.

erlaubte es, die gesamte Sekretion der Bauchspeicheldrüse zu erfassen, das Rezirkulieren von frisch sezerniertem Pankreasinsulin zu verhindern und trotzdem den Versuch über 3 Std auszudehnen.

Die Insulinaktivitäten und -konzentrationen wurden in den bei −28°C bis zum Untersuchungstermin aufbewahrten Seren von Menschen und Tieren bestimmt. Die Messung erfolgte einerseits mit Hilfe unserer Modifikation [Ditschuneit et al. (1962 a)] der biologischen Methode am isolierten Fettgewebe der Ratte [Martin et al. (1958), Renold et al. (1960)], andererseits nach einem immunchemischen Verfahren [Yalow und Berson (1960)]. Abweichend von derursprünglichen Technik trennten wir bei dem letzteren Verfahren jedoch das „gebundene" von dem „freien" Insulin durch Elektrophorese im Agargel oder durch Ausschüttelung in DOWEX 1 [Melani et al. (1963)].

Die Untersuchungen wurden z. T. bereits an anderer Stelle publiziert [Pfeiffer (1961/62), Pfeiffer et al. (1960/61 a−c), Müller-Ruchholtz et al. (1962), Ditschuneit et al. (1962 b), Sirek et al. (1963/64), Schöffling et al. (1963/64)], z. T. werden sie hier erstmals vorgetragen [Pfeiffer, Ditschuneit, Sartory et al. (1964)].

a) Hypophysektomie und Pankreatektomie und das Verhalten der Seruminsulinwirkung und des Blutzuckers nach STH

Beginnen wir mit dem Verhalten der Seruminsulinwirkung nach Hypophysektomie und dem Effekt der Dauerbehandlung mit STH auf den Insulinspiegel. Die

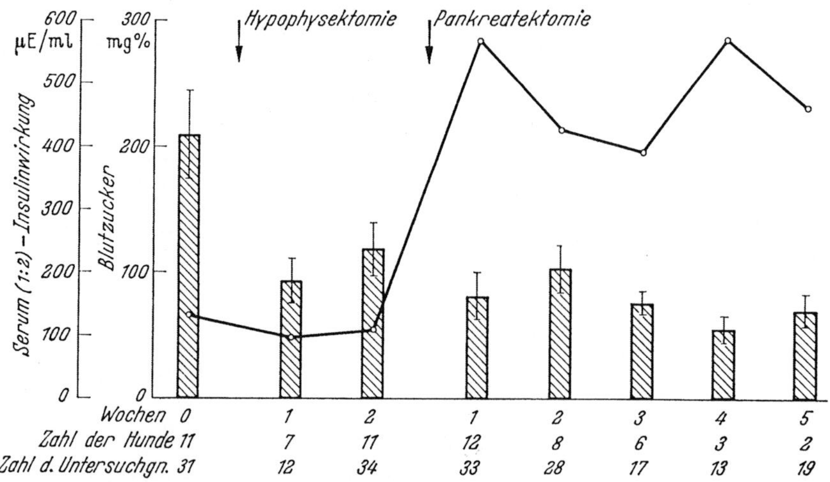

Abb. 1. Einfluß von Hypophysektomie und Pankreatektomie auf den Blutzucker und die mit dem Rattenfettgewebe gemessene Serum (1:2)-Insulinwirkung bei Hunden: Signifikanter Abfall des Insulins nach Hypophysektomie, keine wesentliche Verstärkung des Abfalls nach Pankreatektomie

klassische Versuchsanordnung der Endokrinologie — Exstirpation einer innersekretorischen Drüse, Beobachtung der Ausfallserscheinungen und Wiederherstellung der Ausgangssituation durch Substitution mit dem verantwortlichen Hormon — wurde hierbei praktiziert. Abb. 1 zeigt in der Mittelwertskurve von 11 Hunden das Verhalten der Insulinaktivitäten (bestimmt mit der Fettgewebsmethode) während der ersten 2 Wochen nach der Hypophysenentfernung (linker Teil der

Abb.). Das Seruminsulin fällt auf etwa die Hälfte des Ausgangswertes ab. Es zeigt damit einen Rückgang, der auch nach Pankreatektomie nicht stärker beobachtet werden kann, bzw. durch die nach der Hypophysektomie erfolgende Pankreasentfernung nicht wesentlich weiter verstärkt wird [SCHÖFFLING et al. (1963/64)].

Der gleiche signifikante Abfall des Insulinspiegels nach Entfernung des Vorderlappens der Hypophyse liegt somit übereinstimmend für Ratten, Hunde und

Versuchstier	Methode	HVL / Pankreas	(HVL) ⊗ / Pankreas	HVL / Pankreas ⊗	⊗ / Pankreas ⊗
Affe	Fett	a	a		
	Muskel				
	Immun				
Hund	Fett	d,e,g,h l,m,n,o,p t	t	d,e g,h n / l,m o,p t	t
	Muskel	f,i,g, t		f,i,q [A]	t
	Immun	b,e		b,e	
Katze	Fett	p		p	p
	Muskel	k,r		k,r	r
	Immun				
Ratte	Fett				
	Muskel	c,k,s	k	c [A], s [A]	c [A], s [A]
	Immun				

● Normale ILA ◑ Verringerte ILA ○ Keine ILA ⊖ Unterschiedl. große, teilw. fehlende ILA [A] Alloxanbehandlung

Untersucher: a BALL u. KNOBIL (1963); b BERSON (1962); c BORNSTEIN u. PARK (1953); d EGDAHL u. GOLDBERG (1962); e GOLDBERG u. EGDAHL (1961); f GROEN et al. (1952); g LEONARDS (1959); h LEONARDS et al. (1962); i OKUMURA (1960); k RANDLE u. YOUNG (1956); l SAMAAN et al. (1963); m SAURE (1963); n SHEPS et al. (1960); o SLATER et al. (1961); p STEINKE et al. (1962); q TAKEUCHI et al. (1957); r VALLANCE-OWEN u. LUKENS (1957); s WHITNEY u. YOUNG (1957); t eigene Untersuchungen

Abb. 2. Übersicht über die Befunde des Insulinspiegels im Blute von vier Tierarten (Affe, Hund, Katze und Ratte) vor und nach Hypophysektomie, Pankreatektomie sowie Hypophys- und Pankreatektomie durch verschiedene Untersucher(gruppen)

Affen vor [RANDLE (1955), RANDLE und YOUNG (1956), SIREK et al. (1963), SCHÖFFLING et al. (1963/64)]. In der zweiten Spalte von Abb. 2 sind die von den verschiedenen Autoren mit verschiedenartiger Methodik erhobenen Befunde zusammengestellt.

Auf die tägliche Injektion von 0,25 mg STH (STH hum. Merck-Präparation) pro kg Körpergewicht kommt es erst nach mehr als 10 Tagen kontinuierlicher Behandlung zu einem Wiederanstieg in den Bereich der Ausgangswerte hinein

(Abb. 3). Das gleiche wurde bei hypophysektomierten Ratten von Randle und Young (1956) mit Hilfe der Bestimmung der Insulinwirkung am Rattendiaphragma beschrieben. Der Abfall des Seruminsulingehaltes auf etwa die Hälfte des Ausgangswertes allein nach Entfernung der Hypophyse stimmt weiter mit den Messungen des extrahierbaren Insulingehaltes hypophysenloser Tiere überein. Ein Rückgang des Pankreasinsulins [Campbell et al. (1959)] auf etwa 50% des Ausgangswertes [Dulin und Miller (1959)] wurde festgestellt.

Daraus ergibt sich, daß die Entfernung der Hypophyse keine Hemmung der Insulinsekretion oder -aktivität, sondern tatsächlich eine verminderte Produktion von Pankreasinsulin bewirkt. Der fehlende Blutzuckerabfall nach Sulfonylharnstoffen einige Wochen nach der Hypophysektomie läßt sich befriedigend mit dem

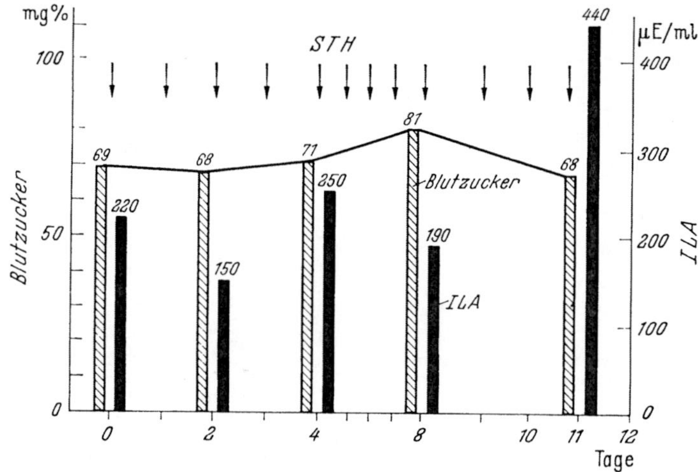

Abb. 3. Das Verhalten der Seruminsulinwirkung und des Blutzuckers beim hypophysektomierten Hund unter Behandlung mit 0,25 mg STH/kg Körpergewicht täglich: Erst nach mehr als 10 Tagen erfolgt Wiederanstieg des Insulins in den Bereich der Ausgangswerte

Mangel an mobilisierbarem Pankreasinsulin erklären [Dulin und Miller (1959)]. Die Fähigkeit des Wachstumshormons, die Ausgangssituation wieder herzustellen, weist andererseits darauf hin, daß es das STH und nicht die Gesamtheit der kontrainsulären Hypophysenhormone ist, die für die normale Insulinproduktion und -sekretion des B-Zellen-Apparates verantwortlich zeichnet.

Entfernt man einem hypophysenlosen Hund auch noch die Bauchspeicheldrüse, so fällt auf die einmalige Injektion von 5 mg STH/kg Körpergewicht der Blutzuckerspiegel nicht ab [Sirek et al. (1963)]. Bei der hohen Empfindlichkeit der Houssay-Präparation gegenüber blutzuckersenkenden Reizen zeigt diese Beobachtung, daß STH als solches nicht hypoglykämisch wirkt.

Führt man denselben Versuch unmittelbar nach Entfernung der Bauchspeicheldrüse oder bis zu 4 Tagen nach der letzten Injektion von Insulin beim pankreaslosen Hund durch, so fällt der Blutzucker ab [Kurtz et al. (1951), Sirek und Best (1956)]. Später als 96 Std nach der letzten Insulininjektion veränderte sich der Blutzuckerspiegel des pankreaslosen Hundes nach einmaliger Gabe von Wachstumshormon bei unseren Untersuchungen dagegen nicht mehr.

Allgemein erklärt man diesen Blutzuckerabfall des pankreaslosen Organismus nach STH-Injektion mit der Aktivierung von Restquantitäten von Insulin im

Gewebe durch das Hypophysenhormon. Die feste Bindung von Insulin wurde ja nicht nur von STADIE et al. (1949, 1951) am isolierten Muskel gezeigt. Die Aktivierung des fixierten Insulins durch STH wurde auch bei alloxan-diabetischen, hypophysektomierten bzw. hypophysektomierten und eviszerierten Ratten demonstriert [KRAHL (1951/52), PARK et al. (1952), OTTAWAY (1953)].

Bei allen diesen Untersuchungen liegen offenbar ganz andere Verhältnisse vor als bei der Houssay-Präparation. Wie Abb. 1 zeigt, waren im Blute dieser Tiere noch durchaus meßbare Insulinaktivitäten vorhanden. Sie waren nicht nur am Fettgewebe, sondern auch am Rattenzwerchfell nachzuweisen und darüber hinaus auch noch mit Salzsäure-Alkohol aus dem Serum zu extrahieren und mit einem Antiserum gegen Insulin zu hemmen [SCHÖFFLING et al. (1963/64)]. Trotzdem können diese Insulinquantitäten nicht biologisch aktiv sein, da nach Entfernung des Pankreas als zweitem Schritt der Blutzuckeranstieg (Abb. 1, re. Teil) bzw. Wachstumshormon keine Aktivierung etwa am Gewebe fixierter Insulinquantitäten bewirkte [SIREK et al. (1963)].

Zusammenfassend läßt sich somit sagen, daß allein die Entfernung der Hypophyse zu einer markanten Verringerung der zirkulierenden Insulinaktivitäten führt, die durch Wachstumshormon wieder in den Normbereich gehoben werden kann. Mit dieser Beobachtung wird jedoch nur der Effekt des Wachstumshormons auf die laufende Insulinproduktion der Inselzellen belegt. Der Beweis für eine direkte Stimulierung der Insulinsekretion steht noch aus. Im Gegenteil spricht der Blutzuckerabfall des pankreaslosen Tieres, das mit exogenem Insulin kürzlich behandelt worden war, in Übereinstimmung mit Untersuchungen am isolierten Gewebe dafür, daß Wachstumshormon auch das im peripheren Gewebe gebundene Insulin zu aktivieren vermag. Das beim Houssay-Hund dagegen noch nachweisbare Insulin im Blute wird von STH nicht aktiviert.

b) Wachstumshormon und die Insulinsekretion des intakten Tieres, der total oder partiell isolierten Bauchspeicheldrüse sowie von Inselschnitten in vitro

Wenden wir uns nunmehr dem Versuch des positiven Nachweises einer Stimulierung der Insulinsekretion durch STH zu, so überwiegen sowohl bei Verwendung älterer als auch neuerer Methoden der Insulinbestimmung eigentlich die enttäuschenden Resultate. Eine Ausnahme machen lediglich frühe Untersuchungen von RANDLE (1954) sowie RANDLE und YOUNG (1956). Mit der Zwerchfellmethode ließ sich zwar kein gesicherter, aber doch ein wahrscheinlicher Anstieg der Insulinaktivitäten im peripheren Blute von Ratten feststellen. Eine signifikante Erhöhung der Plasmainsulinaktivität fand sich lediglich bei Katzen. Sie fehlte bei pankreaslosen Tieren. Da diese pankreatektomierten Katzen nach Behandlung mit Insulin *und* STH keine höheren Insulinwerte aufwiesen, schlossen RANDLE und YOUNG, daß STH die Freisetzung von Insulin aus den B-Zellen selbst stimulieren müsse, und nicht etwa die Geschwindigkeit des Verbrauchs und des Abbaus des Inselhormons durch Wachstumshormon beeinflußt werde. Weitere Untersuchungen konnten nicht erfolgreich abgeschlossen werden [[GARDINER et al. (1960)].

Die Versuche am Ganztier wurden ergänzt durch Arbeiten mit dem isolierten Pankreas und Schnitten von Inselgewebe. So beobachteten ANDERSON und LONG (1947) am perfundierten Rattenpankreas zwar nach Glucose, nicht aber nach STH einen Anstieg der Insulinwirkung; es kam im Gegenteil sogar zu einer Unterdrückung der Insulinabgabe, wenn das Perfusat einen sonst stimulierend wirkenden hohen Glucosespiegel aufwies. Mit der gleichen Technik der Insulinbestimmung stellten Anderson et al. (1957) nach Tolbutamid jedoch eine sichere Steigerung der Insulinabgabe fest. Obwohl bei diesen späteren Untersuchungen das gegenüber STH so viel empfindlichere Pankreas des Hundes benutzt wurde, fehlte auch dieses Mal nach Wachstumshormon die Anregung der Insulinsekretion.

Mit einer immunchemischen Methode der Messung des Seruminsulins kamen Grodsky et al. (1963) wieder am isolierten Rattenpankreas nach STH zu dem gleichen negativen Resultat. Schließlich konnten auch Candela et al. (1963) in

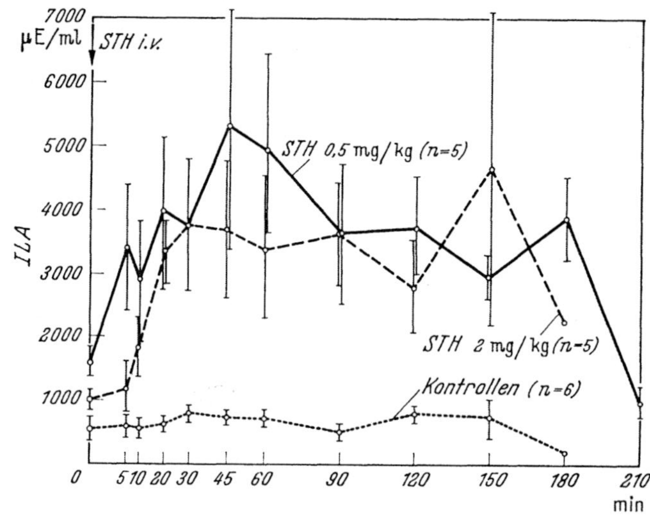

Abb. 4. Insulinaktivitäten (Mittelwerte in µE/ml) im ausfließenden Pankreasvenenblut von Hunden nach i.v.-Injektion von STH human in 2 verschiedenen Dosen: Kein sicherer Unterschied zwischen den beiden STH-Dosen, jedoch signifikante Differenz der nach STH gemessenen Werte gegenüber Ausgangswerten und den mit NaCl-Lösung injizierten Kontrollhunden

vitro keine Freisetzung von Insulin aus Schnitten der Langerhansschen Inseln von Enten nach Zusatz von STH beobachten.

Um nach solchen negativen Erfahrungen den direkten Nachweis der Förderung der Insulinsekretion durch Wachstumshormon zu erbringen, wandten wir eine Methode an, die wir bereits früher mit Erfolg zur Demonstration der von der Höhe des Blutzuckers abhängigen Insulinsekretion der Bauchspeicheldrüse des Hundes benutzt hatten [Müller-Ruchholtz et al. (1962), Pfeiffer (1962)]. Dieser Versuch war erfolgreich. Unmittelbar nach der i.v.-Injektion von menschlichem STH kam es zu einem markanten Anstieg der Aktivitäten im ausfließenden Pankreasvenenblut, der von der absoluten Höhe der den Tieren gegebenen Dosis von Wachstumshormon (0,5 mg und 2,0 mg STH hum./kg Körpergewicht) offensichtlich unabhängig war (Abb. 4). Das gleiche Resultat wurde bei Messung der Sekretion durch Umrechnung der Aktivitäten auf die pro Minute ausfließende Blutmenge erhalten (Abb. 5).

Diese beinahe sofort nach STH einsetzende Stimulierung der Insulinsekretion der partiell isolierten, in situ belassenen Bauchspeicheldrüse des Hundes verdient

Beachtung. Sie hatten wir nur nach Rastinon und hohen Dosen Glucose beobachten
können [MÜLLER-RUCHHOLTZ et al. (1962), PFEIFFER (1962)]. Beide Male hielt die

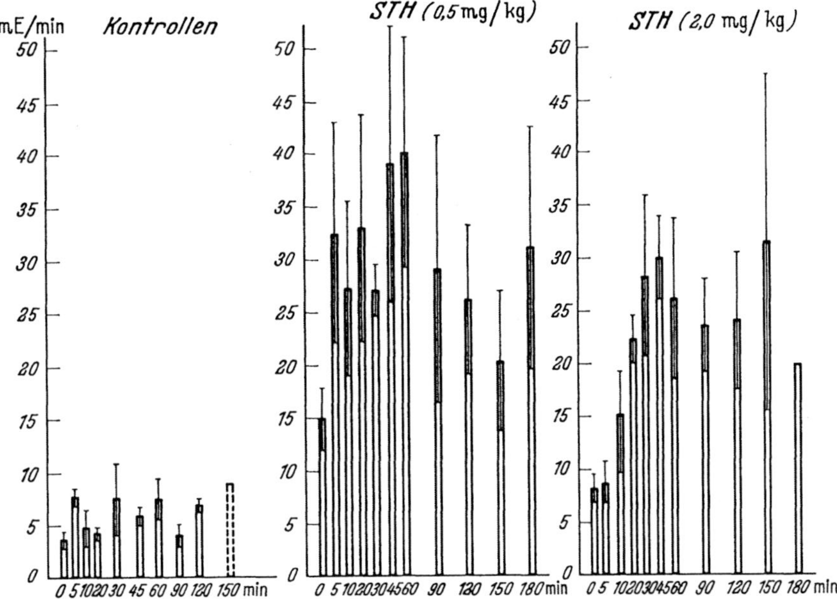

Abb. 5. Insulinsekretion (Mittelwerte in mE/min) im ausfließenden Pankreasvenenblut von Hunden nach i.v.-
Injektion von STH human in 2 verschiedenen Dosen: Kein sicherer Unterschied zwischen den beiden STH-Dosen,
jedoch signifikante Differenz der nach STH gemessenen Werte gegenüber Ausgangswerten und den mitNaCl-
Lösung injizierten Kontrollhunden

maximale Mehrsekretion jedoch nur kurzfristig an. Auf STH kam eine über die
Dauer des Versuches konstant extrem erhöhte Insulinabgabe zustande. Dies wurde
bei vergleichender Messung der ge-
samten produzierten und in der Fla-
sche aufgefangenen Insulinaktivität
offenbar. Die Additionskurve ergibt
(Abb. 6), daß STH von allen geprüften Verbindungen die absolut stärkste
Anregung der Insulinsekretion bewirkt.

Der Schluß, daß das Wachstumshormon allein über eine direkte Anregung der Insulinsekretion aus den
Inseln den Blutzucker senkt, ist jedoch
nicht haltbar. Bei Vergleich der Insulinaktivitäten, die nach Glucose,

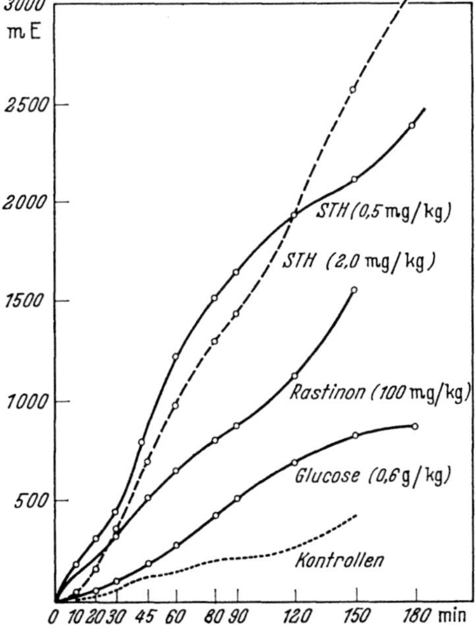

Abb. 6. Gesamte Plus-Sekretion von Pankreasinsulin
(gemessen im gesammelten Blut der Pankreasvene
über den Zeitraum von 150—180 min angegeben
in mE) von Hunden nach Injektion von STH human
(0,5 und 2,0 mg/kg) i.v., sowie Rastinon (100 mg/kg)
und Glucose (0,6 g/kg) zum Vergleich (Mittelwerte
von jeweils 5 Hunden pro Gruppe): Absolut stärkste
Anregung der Insulinsekretion durch STH

Rastinon und STH im *peripheren* Venenblut unserer Hunde bestimmt werden konnten, läßt sich nach Wachstumshormon eine Erhöhung der Insulinwirkung auch in der Peripherie feststellen (Abb. 7). Da wir das gesamte Pankreasvenenblut nach außen abgeleitet hatten, war dies nur damit zu erklären, daß das STH das im Gesamtorganismus verbliebene Insulin zu aktivieren vermocht hatte. Dieses Insulin ist im Gegensatz zu den Verhältnissen beim hypophysenlosen und pankreatektomierten Hund sicher noch biologisch aktiv. Hierfür spricht der primäre Blutzuckerabfall, dem die Erhöhung der (freigesetzten ?) Insulinaktivitäten erst folgte.

Die gleichen Resultate konnten bei immunologischer Bestimmung des Seruminsulins im ausfließenden Pankreasvenenblut erhalten werden.

Fassen wir diesen zweiten Abschnitt zusammen, so hat er uns im Gegensatz zu zahlreichen Voruntersuchungen und unter Verwendung einer bestimmten Technik der schonenden Isolierung des Pankreas gezeigt, daß menschliches Wachstumshormon unmittelbar und anhaltend die Insulinsekretion der Bauchspeicheldrüse des Hundes zu stimulieren vermag. Das Hypophysenhormon übertraf dabei hinsichtlich Quantität und Dauer der Anregung der Plus-Sekretion die mit gleicher Technik geprüfte insulinstimulierende Kapazität

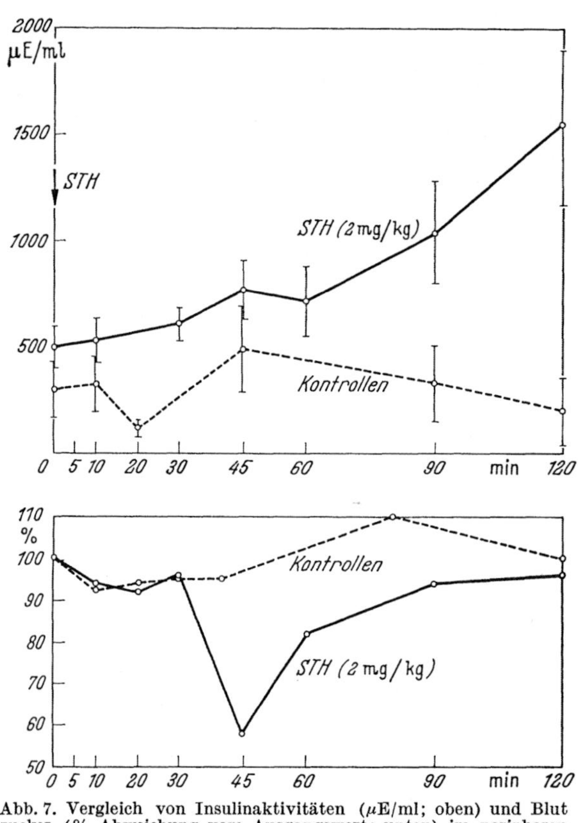

Abb. 7. Vergleich von Insulinaktivitäten (µE/ml; oben) und Blutzucker (% Abweichung vom Ausgangswert; unten) im peripheren Venenblut bei Hunden nach 2,0 mg STH human/kg Körpergewicht i.v.: Maximaler Abfall des Blutzuckers 45 min nach Injektion ohne entsprechenden Anstieg der Insulinaktivität

von Rastinon und hoher Glucosekonzentration. Trotz Ableitung des Pankreasvenenblutes nach außen führte das STH aber auch bei uns zu Blutzuckerabfall und Anstieg der Insulinwirkung im peripheren Kreislauf. Die Fähigkeit des Wachstumshormons, im Organismus vorhandene und möglicherweise am Gewebe fixierte Insulinquantitäten zu aktivieren, wurde damit auch bei unserem Vorgehen offenbar.

c) Wachstumshormon und Insulinaktivitäten im Blute von Stoffwechselgesunden und diabetischen Menschen sowie Insulomkranken

Die grundsätzlich gleiche Beobachtung läßt sich auch beim Menschen machen. Hier war unmittelbar nach i.m.- und i.v.-Injektion von 10 mg STH hum. bei stoffwechselgesunden Menschen ein Blutzuckerabfall auf etwa 75% des Ausgangswertes eine Stunde post op. festzustellen, ohne daß zu diesem Zeitpunkt und noch bis zu 4 Std danach die Insulinaktivitäten meßbar anstiegen [PFEIFFER et al. (1961 b, c)]. Gleichartige Beobachtungen machten auch ZAHND et al. (1960) bei gesunden Probanden.

Auch diese Beobachtungen stimmen mit Tierexperimenten an normalen Ratten [MILMAN und RUSSELL (1950)], normalen Katzen [COTES et al. (1949)], sowie kurzfristig hypophysektomierten Affen und Hunden [KNOBIL et al. (1961), KURTZ et al. (1951), SIREK und BEST (1956)] überein. Da der Anstieg der Glucose im Blute den adäquaten Reiz für die Freisetzung von Insulin aus den Inselzellen darstellt [GRAFE und MEYTHALER (1927), LUKENS und DOHAN (1942), METZ (1960), MÜLLER-RUCHHOLTZ et al. (1962), PFEIFFER (1962), SELTZER (1962), GRODSKY et al. (1962)], nahm man für den Tierversuch schon seit langem an, daß der primäre Blutzuckerabfall nach STH ohne Sekretion von Pankreasinsulin zustande komme.

Für diese Annahme spricht auch noch eine andere Beobachtung aus der Klinik. So ließ sich bei jugendlichen, mit exogenem Insulin behandelten Insulinmangeldiabetikern, die also kein mobilisierbares Pankreasinsulin aufweisen, ebenfalls nach STH zuerst eine Hypoglykämie und erst später ein Anstieg des Blutzuckers feststellen [DOMINGUEZ et al. (1960), PFEIFFER et al. (1961 b, c)].

Dehnt man bei Pankreasgesunden den Beobachtungszeitraum über 8 Std aus und infundiert man gleichzeitig langsam 20 mg STH, so steigt nach 4 Std der Seruminsulinspiegel an [PFEIFFER et al. (1960/61)]. Dieses Verhalten war auch bei hypophysären Zwergen festzustellen. Unter der Dauerbehandlung hypophysärer Zwerge mit menschlichem Wachstumshormon lagen dann die im nüchternen Zustand abgenommenen sowie vor allem die postprandialen Blutzuckerwerte über denen der Vorperiode, obwohl die Insulinaktivitäten im Nüchternblute ebenfalls eine (allerdings nicht signifikante) Erhöhung aufwiesen [PFEIFFER et al. (1961 b, c)].

Altersdiabetiker unterscheiden sich während der ersten 8 Std einer Infusion von menschlichem STH (20 mg) nicht prinzipiell von den Stoffwechselgesunden oder den hypophysären Zwergen [PFEIFFER et al. (1960/61 a)]. Etwa von der 4. Std nach Beginn der Infusion ab stiegen die Insulinaktivitäten gleichmäßig an, ebenso aber auch der Blutzucker. Am Tage der Infusion, dem ersten Tage des Versuchs, kam es dann auch zu einer entsprechend vermehrten Harnzuckerausscheidung. Wurden die Injektionen mit Wachstumshormon, wenn auch mit einer geringeren Dosis und in Form der i.m.-Injektion, am zweiten Tage fortgesetzt, so waren eine extreme Hyperglykämie, stark vermehrte Glykosurie sowie das Auftreten von Ketonkörpern im Harn zu beobachten. Zur gleichen Zeit fielen die Insulinaktivitäten gegenüber der Vorperiode deutlich ab. In Übereinstimmung mit DOMINGUEZ et al. (1960) und in gewissem Gegensatz zu LUFT et al. (1959) war dieses Verhalten von Blutzucker und Harnzuckerausscheidung bei allen Altersdiabetikern zu beobachten und nicht nur auf hypophysektomierte jugendliche Diabetiker beschränkt [PFEIFFER et al. (1960/61 a)].

Bemerkensweiterweise hatte Rastinon i.v. am Morgen des dritten Tages des Versuches, also am Tage nach der Dekompensation des Stoffwechsels, denselben prompten Abfall des Blutzuckers bei gleichzeitigem Anstieg des Insulinspiegels zur Folge, wie wir es auch sonst bei diesen Kranken sehen [PFEIFFER et al.(1960/61)]. Die nachfolgende Injektion von nur 2,5 mg STH hum. führte aber dann wieder zu derselben Dekompensation des Stoffwechsels bei Abfall der Insulinaktivitäten wie am Vortage. Dieser dem Wachstumshormon antagonistische Effekt des Sulfonylharnstoffs stimmt mit den Erfahrungen überein, die man bei der gleichzeitigen Behandlung von Hunden mit Tolbutamid und Wachstumshormon bzw. NNR-Hormon machte [Mirsky et al. (1959), Loubatières (1960), Loubatières et al. (1962)].

Bei Insulomkranken führte die Infusion von 20 mg menschlichem Wachstumshormon dagegen nicht zu einem Anstieg des Seruminsulinspiegels [Pfeiffer (1962)]. Es blieb bei der primären Hypoglykämie, die oftmals so stark war, daß Zucker gegeben werden mußte. Setzte man die Behandlung mit Wachstumshormon fort, so war eine deutliche Besserung der Anfälle zu verzeichnen, da der Blutzucker nunmehr im Vergleich zur Ausgangsperiode erhöht war.

Zusammenfassend ließ sich somit bei gesunden Menschen und hypophysären Zwergen nach Injektion oder Infusion von menschlichem STH eine initiale Phase des primären Blutzuckerabfalls ohne entsprechende Erhöhung des Insulinspiegels beobachten; sie muß wie im Tierversuch mit der Aktivierung von Insulin erklärt werden, das im Gewebe fixiert war. In einer zweiten Phase kommt es dann bei Vorliegen von pankreatischen Insulinreserven zu einem Anstieg des Seruminsulins, der aber von einer Erhöhung des Blutzuckers begleitet oder gefolgt wird. Dieses Verhalten ist nicht nur einige Stunden nach Gabe des STH, sondern — bei hypophysären Zwergen — auch unter der Dauerbehandlung mit dem Hypophysenhormon zu verfolgen. Das grundsätzlich gleiche Verhalten war auch bei Zuckerkranken zu demonstrieren. Hier fiel auch beim jugendlichen Insulinmangeldiabetes der Blutzucker auf STH zuerst ab, und erst später war eine Hyperglykämie festzustellen. Der Altersdiabetiker reagierte dagegen auf STH zuerst wie der Stoffwechselgesunde. Erst die Fortsetzung der Therapie führte zur Dekompensation der Zuckerkrankheit bei stark verminderten Seruminsulinwerten. Auch in dieser Phase der Dekompensation führte Rastinon schnell zu einer Erhöhung des Seruminsulins und zu einem Abfall des Blutzuckers. Insulomkranke ließen dagegen eine Stimulierung der Sekretion von Pankreasinsulin vermissen. Es blieb bei der starken primären Hypoglykämie, und erst bei Fortsetzung der Therapie führte die Erhöhung des Nüchternblutzuckers ohne entsprechenden Anstieg der Seruminsulinwirkung zu einer Besserung der Anfälle.

2. Besprechung der Ergebnisse

Die Beziehungen zwischen Wachstumshormon und Insulin wurden bisher allein anhand des Blutzuckers verfolgt. Die hieraus abgeleiteten Hypothesen reichten von der Annahme einer direkten Stimulierung der Insulinsekretion [Milman et al. (1951)], einer Aktivierung des im Blute zirkulierenden oder am Gewebe haftenden Insulins [Park et al. (1952), Krahl (1952), Ottaway (1953)] über eine Hemmung der Insulinsekretion durch das Wachstumshormon [Anderson und Long (1947)] bis zu einer Anregung eines blutzuckersteigernden und damit wieder Insulin mobilisierenden Faktors (Glucagon oder Serotonin ?) durch das Wachstumshormon [Bornstein et al. (1951), Foa et al. (1953), Sirek (1957), Young (1963)].

Auch die vorliegenden Untersuchungen haben die Wirkung des Wachstumshormons auf Produktion und Sekretion von Pankreasinsulin sowie das im Blute zirkulierende plasmatische Inselhormon nicht eindeutig mit dem einen oder anderen

Effekt erklären können. Vier verschiedene, nur z. T. miteinander verbundene Wirkungen waren nachzuweisen. Sie sollen der Übersicht halber als Sofort-Effekte und Spät-Wirkungen beschrieben werden [vgl. PFEIFFER (1957), SIREK und SIREK (1964)].

1. So ergab sich als Sofort-Effekt eine Förderung oder Aktivierung des im Gewebe haftenden Insulin durch das Wachstumshormon. Sie war bei Menschen und Tieren aus der initialen Hypoglykämie abzulesen, die beim Stoffwechselgesunden, hypophysären Zwerg und Insulomkranken ohne Anstieg des Seruminsulins zustande kam bzw. ebenfalls am pankreaslosen Tier und beim jugendlichen Insulinmangeldiabetes unter der Voraussetzung ausreichender Substitution mit exogenem Insulin auftrat. Der gleiche primäre Blutzuckerabfall nach STH war auch bei Hunden mit partiell isoliertem Pankreas zu beobachten, obwohl das Pankreasinsulin in toto über die Pankreasvene nach außen abgeleitet wurde. Diesen Blutzuckerabfall hatten wir bei Verwendung der gleichen Operationstechnik nach Sulfonylharnstoffen niemals feststellen können [PFEIFFER (1962), MÜLLER-RUCHHOLTZ et al. (1962), DITSCHUNEIT et al. (1962b)].

2. und 3. Als früher Spät-Effekt, d. h. einige Stunden nach Zufuhr des Wachstumshormons auftretend, war dagegen eine Hemmung der peripheren Glucoseverwertung aus einem Anstieg des Blutzuckers abzulesen gewesen, die bei vorhandenen pankreatischen Insulinreserven von einer Erhöhung des Seruminsulins begleitet war. Von diesem Verhalten gab es zwei Ausnahmen: Einerseits den Altersdiabetiker, dessen Erschöpfung der endogenen Insulinproduktion aus einem Abfall des anfänglich ebenfalls erhöhten Seruminsulins bei gleichzeitiger diabetischer Hyperglykämie und Ketose ersichtlich wurde; andererseits den Insulomkranken, der überhaupt nicht mit einem Anstieg des Seruminsulins auf STH reagierte, so daß die Behandlung mit Wachstumshormon bei ihm allein einen Anstieg des Blutzuckers und damit eine wesentliche Besserung der hypoglykämischen Anfälle herbeiführte.

Nach wie vor blieb es jedoch unklar, ob die Anregung der Insulinsekretion durch das Wachstumshormon unabhängig von der die zweite Phase auszeichnenden Hemmung der peripheren Zuckerverwertung zustande kam oder als Folge der normalerweise die Insulinsekretion stimulierenden Hyperglykämie. Zwar hatten wir am Pankreas des Hundes eine nahezu unmittelbar nach STH-Injektion einsetzende Anregung der Insulinabgabe nachweisen können, die auch quantitativ alles übertraf, was in gleichlaufenden Untersuchungen durch Hyperglykämie oder Sulfonylharnstoffe zu erzeugen war. Zumindest für das gegenüber der diabetogenen Wirkung von Wachstumshormon so empfindliche Inselzellsystem des Hundes ist eine direkte Stimulierung der Insulinsekretion wahrscheinlich. Auf der anderen Seite reagierte das Inselzelladenom auf STH nicht mit der Freisetzung von Pankreasinsulin. Da Insulome häufig auch auf Glucosezufuhr nicht oder nur verspätet mit Insulinabgabe antworten [vgl. WHIPPLE (1942), PFEIFFER et al. (1959b, c), DITSCHUNEIT et al. (1961), PFEIFFER (1962)], ist man versucht, hieraus auf den nur indirekten Effekt zu schließen, den STH via Blutzucker auf die Insulinsekretion der B-Zellen ausübt.

Wie stark in der zweiten Phase die Beeinträchtigung des Glucoseabbaus im peripheren Gewebe durch Wachstumshormon beeinflußt wird, läßt sich durch den Vergleich der Wirkungen von Wachstumshormon und Sulfonylharnstoffen beim

Altersdiabetes erkennen. Während das Wachstumshormon infolge der doppelten Belastung der Stimulierung der Insulinsekretion und der Beeinträchtigung der Zuckerverwertung zum Versagen des Insulinnachschubs und damit zur Dekompensation des Stoffwechsels führte, hat Rastinon mit seiner alleinigen Anregung der Insulinsekretion ohne jeden Einfluß auf die Glucoseutilisation in der Peripherie prompt einen Abfall des Blutzuckers auch nach vorhergegangener Dekompensation zur Folge. Würde das Hypophysenhormon allein die Insulinsekretion kontrollieren, hätte es den gleichen Effekt wie das Sulfonylharnstoffpräparat aufweisen müssen.

Diesen mehr oder minder schnell einsetzenden Wirkungen des Wachstumshormons auf die Insulinsekretion ist endlich noch sein offenbar übergeordneter, „chronischer" eigentlicher Spät-Effekt auf die Produktion des Inselhormons hinzuzufügen. Hierfür spricht, daß nach Entfernung der Hypophyse übereinstimmend bei verschiedenen Tierarten und auch unter Anwendung verschiedener Methoden der Seruminsulin-Bestimmung die Insulinaktivitäten auf etwa die Hälfte des Ausgangswertes abfielen. Allein Wachstumshormon war bei gleichmäßiger Zufuhr in der Lage, den Blutinsulinspiegel zu normalisieren. Da bei hypophysenlosen Tieren auch der Insulingehalt der Bauchspeicheldrüse zumindest auf die Hälfte des Ausgangswertes vermindert ist und die Insulinreserven nach Adrenalektomie weitaus weniger abfallen als nach Entfernung der Hypophyse [vgl. Dulin und Miller (1959)], muß das Hypophysenhormon als entscheidender Faktor für die normale Produktion von Pankreasinsulin angesehen werden. Auch hier muß es offen bleiben, ob ein direkter „troper" Effekt des STH auf die Vitalität der B-Zellen vorliegt, oder erst die allgemeine Höherstellung des Blutzuckerniveaus indirekt die laufende Anregung der Insulinsekretion und damit auch -produktion bewirkt.

Auch diese Frage wird erst beantwortet werden können, wenn wir die Steuerung des Stoffwechsels von Zucker, Fett und Eiweiß durch Wachstumshormon und Insulin besser übersehen. Es ist durchaus möglich, daß die Einflüsse, die die beiden Hormone aufeinander ausüben, in Form indirekter Relationen über einen oder mehrere der von ihnen gesteuerten Metaboliten verlaufen. Derartige Schaltstellen sind schon heute in Form des synergistischen Effektes beider Hormone auf den Eiweißanbau, des Antagonismus von Glucose und Fettsäurenoxydation [Randle und Hales (1963)] sowie der schnellen Mobilisation von hypophysärem STH, die der Blutzuckerabfall nach Insulin auslöst [Roth et al. (1963)], bekannt. Die Schwierigkeit der Unterscheidung von primären und sekundären Prozessen, von indirekten und direkten Regulationsvorgängen wird freilich auch bei einer derartigen Betrachtung nicht geringer.

Literatur

Anderson, E., and J. A. Long: Endocrinology 40, 92 (1947).
— F. Wherry, R. W. Bates, and J. Cornfield: Proc. Soc. exp. Biol. (N. Y.) 94, 321 (1957).
Bornstein, J., E. Reid, and F. G. Young: Nature (Lond.) 168, 903 (1951).
Campbell, J., L. Chaikof, G. A. Wrenshall, and R. Zemel: Canad. J. Biochem. 37, 1313 (1959).
— I. W. F. Davidson, and H. P. Lei: Endocrinology 46, 588 (1950).
Candela, L. R., R. R. Candela, D. Martin-Hernandez, and T. Castilla-Cortaza: In: Perspectives in biology, p. 105, dedicated to B. A. Houssay. Ed. by C. F. Cori, V. G. Foglia, L. F. Leloir, and S. Ochoa. Amsterdam, London, New York: Elsevier, Publ. Co. 1963.

Cotes, P. M., E. Reid, and F. G. Young: Nature (Lond.) **164**, 209 (1949).

Ditschuneit, H., J. D. Faulhaber u. E. F. Pfeiffer: Atompraxis 8, 172 (1962a).

— E. F. Pfeiffer, R. Cuendet, H. Kolb, Ch. Wahl u. W. H. Rott: I. Symp. Dtsch. Diab. Kom., Düsseldorf 26.—27. 10. 1962b, p. 37. In: K. Jahnke u. K. Oberdisse, Herausg.: Fortschritte der Diabetesforschung. Stuttgart: Thieme 1963.

— — u. K. Schöffling: Verh. dtsch. Ges. inn. Med. **67**, 359 (1961).

Dominguez, J. M., E. Greenberg, A. G. Pazianos, B. S. Ray, and O. H. Pearson: I. Internat. Endokrin. Kongr., Kopenhagen Juli 1960, Abstr. Nr. 121.

Dulin, W. E., and W. L. Miller: Diabetes 8, 199 (1959).

Foa, P. P., E. B. Magid, M. D. Glassman, and H. R. Weinstein: Proc. Soc. exp. Biol. (N. Y.) **83**, 758 (1953).

Gardiner, D. C., D. Martin-Hernandez, and F. G. Young: Lancet **1960**, 1269.

Grafe, E., u. F. Meythaler: Naunyn-Schmiedebergs Arch. exp. Path. Pharmak. **125**, 181 (1927).

Grodsky, J. M., A. A. Batts, L. L. Bennett, C. Vcella, N. B. McWilliams, and D. F. Smith: J. Physiol. (Lond.) **205**, 638 (1963).

Knoeil, E., G. R. Best, and H. M. Goodman: Endocrinology **68**, 723 (1961).

Krahl, M. E.: Ann. N. Y. Acad. Sci. **54**, 649 (1951).

— Science **116**, 524 (1952).

Kurtz, M., R. C. de Bodo, S. P. Kiang, and A. Ancowitz: Proc. Soc. exp. Biol. (N. Y.) **76**, 21 (1951).

Loubatières, A. L.: I. Intern. Congr. Endocrinol., Kopenhagen 1960.

— M. M. Mariani, C. Fruteau de Laclos et R. Alric: XXII Int. Congr. Physiol. Sci., Leiden, September 1962, Abstr. 431

Luft, R.: 11. Symp. Dtsch. Ges. Endokrinol., Düsseldorf, 5.-7. 3. 1964 (im Druck).

— D. Ikkos, C. A. Gemzell, and H. Olivecrona: Acta Endocr. (Kbh.) **32**, 330 (1959).

Lukens, F. D. W., and F. C. Dohan: Endocrinology **30**, 175 (1942).

Martin, D. B., A. E. Renoldt, and Y. M. Dagenais: Lancet **1958** II, 76.

Melani, F., H. Ditschuneit, H. H. Ditschuneit, A. Mucci u. E. F. Pfeiffer: X. Symp. Dtsch. Ges. Endokrinol., Wien 7.-9. 3. 1963, p. 252. Berlin-Göttingen-Heidelberg: Springer 1964.

Metz, R.: Diabetes **9**, 89 (1960).

Milman, A. E., P. de Moor, and F. D. W. Lukens: Amer. J. Physiol. **166**, 354 (1951).

—, and J. A. Russell: Endocrinology **47**, 114 (1950).

Mirsky, I. A., S. Gitelson, and G. Perisutti: Endocrinology **64**, 766 (1959).

Müller-Ruchholtz, W., H. Ditschuneit, W. Dettwyler, W. Böhm, H. Friedrich, E. Ohebschalon, R. Petzoldt, R. Sartory, and E. F. Pfeiffer: Acta Endocr. Congr., Genf 1962, Abstr. Nr. 118.

Ottaway, J. H.: Brit. med. J., 1953 II 357.

— Biochim. biophys. Acta (Amst.) **11**, 443 (1953).

Park, C. R., D. H. Brown, M. Cornblath, W. H. Daughaday, and M. E. Krahl: J. biol. Chem. **197**, 151 (1952).

Pfeiffer, E. F.: Dtsch. med. Wschr. **82**, 1789 (1957).

— 4. Congr. Internat. Diab. Fed., Genf, 10.-14. 7. 1961, p. 671

— I. Symp. Dtsch. Diab. Kom., Düsseldorf, 25.-26. 10. 1962. In: Fortschritte der Diabetesforschung, Oberdisse-Jahnke, Herausg., p. 18. Stuttgart: Thieme 1963.

— H. Ditschuneit u. R. Ziegler: 7. Symp. Dtsch. Ges. Endokrinol., Homburg/Saar 21.-23. 4. 1960. p. 206, Berlin-Göttingen-Heidelberg: Springer 1961.

— — — Klin. Wschr. **39**, 415 (1961a).

— — — E. Böhle u. R. Biegler: Med. Welt **16**, 865 (1961b).

— — — — — 8. Symp. Dtsch. Ges. Endokrinol., München 1.-3. 3. 1961 c, p. 398, Berlin-Göttingen-Heidelberg: Springer 1962.

— — — — — Verh. dtsch. Ges. inn. Med. **67**, 367 (1961c).

— M. Pfeiffer, H. Ditschuneit u. Chang-Su Ahn: Ann. N. Y. Acad. Sci. **82**, 479 (1959).

— — — — Klin. Wschr. **37**, 1239 (1959).

Randle, P. J.: Brit. med. J. **1954** I, 1237.

— Lancet **1954** I, 441.

Randle, P. J. In: Hypophyseal growth-hormone, nature and actions. p. 413. Ed. by R. W. Smith, O. H. Gaebler, and C. N. H. Long. New York: Blakiston Div., McGraw-Hill & Co. 1955.

—, and C. N. Hales: Vortr. European Symposium on Diab. mell., Genf 28.-30. 11. 1963.

—, and F. G. Young: J. Endocr. **13**, 335 (1956).

Renold, A. E., D. B. Martin, Y. M. Dagenais, J. Steinke, R. Nickerson, and M. C. Sheps: J. clin. Invest. **39**, 1487 (1960).

Roth, J., S. M. Glick, R. S. Yalow, and S. A. Berson: Science **140**, 987 (1963).

Schöffling, K., A. Sirek, H. Ditschuneit, R. Petzoldt, J. Beyer, O. V. Sirek, E. F. Pfeiffer, and C. H. Best: Diabetes 1964 (im Druck).

— — — — O. V. Sirek u. E. F. Pfeiffer: 1. Symp. Internaz. sul Diabete. Modena, 21.-22. 9. 1963 (im Druck).

Seltzer, H. S.: J. clin. Invest. **41**, 289 (1962).

Sirek, A.: Nature (Lond.) **179**, 376 (1957).

— K. Schöffling, and H. Ditschuneit: 23. Ann. Meet. Amer. Diab. Ass., Atlantic City June 1963, Abstr.

— — M. Webster, and O. V. Sirek: 1963 (im Druck).

Sirek, O. V., and C. H. Best: Amer. J. Physiol. **185**, 557 (1956).

—, and A. Sirek: Physiology of Growth Hormone. Ergebn. inn. Kinderheilk. **21**, 217 (1964).

Stadie, W. C., N. Haugaard, A. G. Hills, and J. B. Marsh: Amer. J. med. Sci. **218**, 275 (1949).

— —, and J. B. Marsh: J. biol. Chem. **188**, 167 (1951).

— — — J. biol. Chem. **189**, 53 (1951).

— — —, and A. G. Hills: Amer. J. med. Sci. **218**, 265 (1949).

Whipple, A. O.: New Engl. J. Med. **226**, 515 (1942).

Yalow, R. S., and S. A. Berson: J. clin. Invest. **39**, 1157 (1960).

Young, F. G.: Lancet **1937 II**, 372.

— In: Perspectives in biology, p. 105, dedicated to B. A. Houssay. Ed. by C. F. Cori, V. G. Foglia, L. F. Leloir, and S. Ochoa. Amsterdam, London, New York: Elsevier Publ. Co. 1963.

Zahnd, G. R., J. Steinke, and A. E. Renold: Proc. Soc. exp. Biol. (N. Y.) **105**, 455 (1960).

Diskussion

D. Knorr (München):

Sie führten aus, daß STH das Hormon sei, welches uns in die Lage versetze zu hungern, d. h. den Energiebedarf aus den Fettdepots zu decken und berichteten, daß Sie aus dieser Sicht jetzt das Verhalten des STH bei Fettsüchtigen untersuchen.

Wäre die Fettsucht eine Störung der STH-Sekretion, so müßte es sich nach diesen Ausführungen um einen STH-Mangel handeln, welcher den Fettsüchtigen hindert sein Depotfett anzugreifen. Diese Annahme steht jedoch im Widerspruch mit der klinischen Erfahrung, daß Kinder mit Präpubertätsfettsucht fast regelmäßig zu groß sind und zu unserer biochemischen Beobachtung, daß die Nüchternwerte der unveresterten Fettsäuren bei fettsüchtigen Kindern höher liegen als bei normalgewichtigen.

E. F. Pfeiffer:

Ich glaube nicht, daß man Fälle von Pubertätsfettsucht ohne weiteres mit erwachsenen Adipösen vergleichen kann. Bei den Jugendlichen ist es ja doch oft die durch fehlgeleitete Mütter erzwungene Nahrungsaufnahme, die die Fettsucht herbeiführt, und daß vermehrte Calorienaufnahme auch bei intaktem endokrinen System eine Fettsucht bewirkt, kann man kaum bezweifeln. Überhaupt scheint es gefährlich, eine „endokrine Fettsucht" auch nur am Rande jetzt schon zu erwägen. Hierzu reichen unsere Kenntnisse doch bei weitem nicht aus. Diskutiert werden muß jedoch die Möglichkeit, daß Insulin das Hormon darstellt, das bei Nahrungsaufnahme den Transport der Nahrungsstoffe in die Zelle hinein gewährleistet, STH das Hormon des Hungers, das die Einschmelzung der Fettdepots bei fehlender Nahrungszufuhr besorgt. Vielleicht ist der Blutzuckerspiegel verantwortlich dafür, ob STH gleichzeitig nach seiner Sekretion auch noch die Abgabe von Insulin — direkt oder indirekt — bewirkt, und umgekehrt [vgl. GORDON u. CHERKES: Proc. Soc. exp. Biol. (N. Y.) 97, 150 (1958)].

Z. LARON (Israel):

How do you explain the water and sodium retaining actions of growth hormone? We see water retention also after administration of anabolic steroids and androgens: does this mean that water is needed for the anabolic process itself? Next it is thought that after the injection of insulin it is the ensuing hypoglycemia which stimulates the secretion of GH. But here we have heard that the stimulus induced by insulin on GH may be ad irect one? Is any experimental data available as to give information on this?

E. F. PFEIFFER:

As regards question 1 nothing more than the mere fact is known that GH induces some kind of water retention. Clinically, that water retention is only a slight one, and cannot be compared with the water and sodium retaining activity of, e.g., DOC. As to question 2 asking for experimental evidence of direct action of insulin on STH-secretion in my oppision nothing is known supporting this idea. SH-secretion is likewise stimulated by the hypoglycemia induced by fasting (proof: no release of free fatty acids in subjects suffering from hypopituitarism). Hence, it seems likely that it is th low blood sugar value and not the insulin itself which promotes STH-release.

H. REINAUER (Düsseldorf):

1. Bei alloxan-diabetischen Ratten wird die Glucoseutilisation durch einen Faktor gehemmt, (nach den bisherigen Untersuchungen ein Lipoproteid), der in vivo bei Vorhandensein von Somatotropin und NNR-Hormonen gebildet wird. Die Hemmung der Glucoseutilisation unterbleibt, wenn die Hypophyse bzw. STH oder die Nebennieren bzw. Cortisol fehlen (KIPNIS, 1959; KIPNIS and CORI, 1960; PARK u. Mitarb. 1959—1961). Der Vortragende führt die Hemmung der Glucoseverwertung bei Diabetes allein auf die Erhöhung der freien Fettsäuren zurück. Welche Beziehungen bestehen zwischen diesen beiden Hemmfaktoren?

2. Besteht die eiweißanabole Wirkung von STH auch in Abwesenheit von Insulin oder erfolgt die proteinanabole Wirkung von STH über Insulin?

E. F. PFEIFFER:

Zu 1.: Über die Beziehungen zwischen den wohl von VALLANCE-OWEN zuerst beschriebenem Hemmfaktor beim Alloxan-Diabetes und der Hemmung der Glucoseutilisation durch erhöhte freie Fettsäuren ist noch nichts bekannt. Die Konzeption des Glucose-Fettsäuren-Cyclus geht über den Insulinhemmer des „Synalbumins" offenbar weit hinaus und umfaßt eine Fülle von endokrin gesteuerten und nicht-gesteuerten Situationen, die alle durch abgestoppte Fettsäurenoxydation ausgezeichnet sind. Wo hier Ursache, wo Wirkung zu suchen sind, bleibt zu klären.

Zu 2.: STH wirkt nur Eiweiß anbauend in Gegenwart von Insulin.

Aus der Universitäts-Kinderklinik Hamburg-Eppendorf
(Direktor: Prof. Dr. K. H. Schäfer)

Ätiopathogenese und Klinik des hypophysären Zwergwuchses

Von

J. R. Bierich

Mit 10 Abbildungen

Referat

I. Ätiopathogenese

Eine im Wachstumsalter auftretende Insuffizienz der Adenohypophyse führt zum klinischen Bilde des hypophysären Zwergwuchses. Je nach Lokalisation und Umfang des pathologischen Prozesses kommt es neben der Störung des Wachstums zum Ausfall einer oder mehrerer glandotroper Hormone. Wie Berblinger schon 1931 ausgesprochen hat, ist „das Wesentliche, daß die Funktion nicht nur der eosinophilen Epithelien, sondern zugleich auch der basophilen oder überhaupt aller Vorderlappenzellen weitgehend eingeschränkt ist". Der Funktionsausfall betrifft bei schweren Destruktionen die gesamte Adenohypophyse. Oligo- oder monosymptomatischen Formen, bei denen der Defekt auf die Bildung des Wachstumshormons und u. U. der Gonadotropine beschränkt ist, begegnet man vornehmlich bei den erblichen Formen der Erkrankung.

Ätiologisch lassen sich folgende Formen auseinanderhalten: 1. Angeborene, genetisch bedingte Defekte, 2. Dystopien des Hinterlappens, 3. Tumoren, 4. symptomatische Hypophysendefekte bei Tuberkulose, Lues oder Reticulose an Hirn- und Schädelbasis, 5. Hypophysenzerstörungen infolge akuter Schädel- und Hirntraumen, 6. die idiopathische Form. Tab. 1 gibt Anhaltspunkte für die Häufigkeit der verschiedenen Formen. Die Daten entstammen 5 in den letzten Jahren publizierten größeren Serien von hypophysären Zwergen, bei denen die Diagnose als

Tabelle 1

Autor	idiopathisch	Tumor	genet.	HHL-Dystopie	Trauma
Martin u. Wilkins (1957)	19 (18 m, 1 w)	7			
v. d. Werff ten Bosch (1962)	18 (17 m, 1 w)	3			1
Kogut (1963).	13 (8 m, 5 w)	2			
Prader (1964)	25 (17 m, 8 w)	7	4	5	3
Bierich (1964)	21 (13 m, 8 w)	9	9		
Total: 146	96 (73 m, 23 w)	28	13	5	4

gesichert betrachtet werden kann. An der Spitze steht zahlenmäßig die Gruppe der idiopathischen Fälle. Dabei möchte ich auf das starke Überwiegen des männlichen Geschlechts aufmerksam machen; die Relation männlich: weiblich beträgt rund 3 : 1. Als 2. Gruppe rangieren die Tumoren. Auch die genetischen Fälle sind nicht ganz selten.

Pathologisch-anatomisch besitzen wir über die *genetisch* bedingten Fälle bis jetzt nur ungenügende Informationen. KOCH hat bei einer Sektion eine normalgroße Hypophyse gefunden, in der der Vorderlappen weitgehend durch Cysten ersetzt und die Epithelzellen stark dezimiert waren. Ob die angeborenen Hypoplasien der Hypophyse, die MOSIER, BLIZZARD u. ALBERTS und REID beschrieben haben, extreme Varianten dieser Krankheitsform darstellen, ist noch unklar. Hier würden sich Analogien zu den Hypoplasien finden, die SMITH u. McDOWELL, KEMP, FRANCIS u. a. beim erblichen hypophysären Zwergwuchs der Maus beobachtet haben.

Bei den seltenen Fällen von *Hypophysenhinterlappendystopien,* wie sie von PRIESEL, APITZ und HEDINGER mitgeteilt worden sind, ist die Lagebeziehung von Hinterlappen und Vorderlappen gestört. Bei der hier interessierenden tuberalen Form liegt der Hinterlappen am Tuber cinereum und ist ohne Kontakt mit dem Vorderlappen, der damit seine Verbindung mit dem Zwischenhirn verloren hat.

Unter den *Tumoren,* die zum hypophysären Zwergwuchs führen, rangieren die Kraniopharyngeome an erster Stelle. Sie entstehen öfter supra- als intrasellär aus dem Epithel der Rathkeschen Tasche. In der Regel sind sie nicht solide gebaut, sondern aus mehrkammrigen Cysten zusammengesetzt, die mit cholesterinhaltiger Flüssigkeit gefüllt sind und deren Wandungen in mehr als 80% verkalken; sie sind röntgenologisch daher meist gut erfaßbar. Von Bedeutung ist die Feststellung, daß eine Reihe von Fällen beschrieben worden ist, denen oberhalb und außerhalb der Hypophyse lokalisierte Tumoren zugrunde lagen und bei denen die Hypophyse intakt gefunden wurde. Wie bei den Hinterlappendystopien war bei diesen als hypothalamisch bedingt anzusprechenden Fällen das Entscheidende die Trennung von Zwischenhirn und Vorderlappen.

Pathologisch-anatomische Befunde von *idiopathischen* Zwergwuchsfällen liegen nur in geringer Anzahl vor; insgesamt habe ich 9 Mitteilungen zusammenstellen können (SIMMONDS, STERNBERG, ALTMANN, CERANKE, HEWER, FARINA, RUSS-FIELD u. REINER, STRANSKY u. LARA, SANDISON). Gemeinsame Züge sind 1. die Kleinheit der Hypophyse, die in SIMMONDs Fall 0,2 g, in HEWERs Fall 0,3 g, in STRANSKYs Fall 0,06 g wog; 2. die relative Intaktheit des Hinterlappens, der größer als der Vorderlappen ist; 3. die starke Verminderung der chromophilen Zellen; HEWER fand in seinem Fall z. B. nur 0,2% acidophile Zellen; 4. wurden in einigen Fällen cystische Entartungen und eine beträchtliche Einlagerung von Bindegewebe registriert, was den Verdacht auf vernarbende Prozesse nahelegt.

Die *Pathogenese* der idiopathischen Form des hypophysären Zwergwuchses war bis vor kurzem unklar. Unter anderem hat man Infektionen verschiedener Art und Anlagestörungen vermutet. SIMMONDS hat für seinen Fall perinatal entstandene embolische Infarkte und Nekrosen für wahrscheinlich gehalten. BERBERICH hat 1926 im Rahmen der Untersuchungen von SCHWARTZ u. Mitarb. über geburtstraumatische Hirnläsionen eine Mitbeteiligung der Hypophyse in einem Großteil der Fälle gefunden, vor allem starke venöse Hyperämie und Stase mit begleitendem

Zellschwund. Später haben Hooft u. Mitarb., Svilokos u. Zanetti und Albeaux-Fernet u. Mitarb. über einzelne wahrscheinlich geburtstraumatisch bedingte Fälle berichtet. Wie mir Herr Prader mitgeteilt hat, war in dem Zürcher Krankengut von 25 idiopathisch-hypophysären Zwergen die Geburtsanamnese 19 mal abnorm und deutete in 10 Fällen mit großer Sicherheit auf ein Trauma hin.

Unsere eigenen Untersuchungen und die von van der Werff ten Bosch, die beide zuerst 1962 mitgeteilt wurden, sprechen in demselben Sinne. Die genaue Analyse der Geburtsakten der Patienten ergab aber in beiden Serien darüber hinaus die hohe Frequenz einer spezifischen Geburtskomplikation, nämlich die große Häufigkeit von Beckenendlagen. Tab. 2 gibt unser heutiges Material wieder. Man

Tabelle 2. *Angaben zur Geburt*

Name u. Geschlecht	Geb. Nr.	Lage	Dauer (Std)	Operationen	Komplikationen
A. S. m. . .	8	HHL	1		
J. G. m. . .	2	Steiß	—		
R. M. m.. .	5	Steiß	13		
H. P. L. m..	1	Fuß	—		
A. Sch. w. .	2	HHL	15		
H. J. T. m..	3	Steiß	9	Steißhaken, Veith-Smellie	Bradykardie
W. St. m. .	2	HHL	1(!)		
R. J. w. . .	5	Steiß	$7^1/_2$	Bracht, Veith-Smellie	Vorzeitige Armlösung, Klavikelfraktur
K. K. m.. .	3	Quer	10	Wendung a. d. Fuß	Schwere Geburt, Bradykardie
B. St. m.. .	2	Steiß	—	Zange	Lange Asphyxie
J. L. m. . .	4	Quer	—		Frühgeb. M VIII, Cyanose
R. B. w. . .	4	HHL	—	Wendung a. d. Fuß	Zwillingsgeburt
G. K. m. . .	1	Steiß	$2^1/_2$Tg.		Frühgeb. M VIII
B. S. m. . .	1	Steiß	—		Asphyxie, Lange Cyanose
Ch. N. w.. .	1	Steiß	30	Episiotomie	Nabelschnurumschlingung 2 Tg. Cyanose
D. P. m. . .	3	Steiß	—	Zange	Nabelschnurvorfall Asphyxie, Krämpfe
J. J. w. . .	3	Quer	—	Wendung a. d. Fuß, Bracht	Vorzeitige Placentalösung, Lange Asphyxie, lange Cyanose
K. Sch. m. .	2	HHL	6	—	Frühgeb. Mens VII Asphyxie $++$
Verdachtsfälle:					
M. Sz. w.. .	1	Steiß	7		
M. R. w. . .	1	Steiß	6		
G. F. w. . .	2	HHL	6		Blutverlust d. Mutter $++$ Asphyxie $++$

erkennt die Häufigkeit pathologischer Geburtslagen, vor allem von Steißlagen, dann auch Quer- und Fußlagen, ferner die Frequenz geburtshilflicher Operationen und schließlich die Häufigkeit asphyktischer und cyanotischer Zustände nach der Geburt, — dies übrigens auch bei 2 der in Kopflage geborenen Kinder.

Daß speziell die Geburt in Beckenendlage mit dem späteren Auftreten eines hypophysären Zwergwuchses korreliert ist, geht aus Tab. 3 hervor. Da auch die Querlagen schließlich nach Wendung in Beckenendlage geboren werden, kann man diese beiden Gruppen vereinigen. Wie sich ergibt, kamen in unserem Material die

Kinder, die später hypophysäre Zwerge wurden, in 78% der Fälle in diesen Positionen zur Welt, d. h. 19mal häufiger als gesunde. In der gleichen Weise kamen 11 von 16 Patienten, d. h. 63% der Fälle von VAN DER WERFF TEN BOSCH in Beckenendlage zur Welt.

Die Schädigung der Hypophyse geht den übrigen häufig vorkommenden Komplikationen der Beckenend- und Querlagen vollkommen parallel. Tödlicher Ausgang durch Tentoriumzerreißungen und Hirnblutungen ist bei diesen Geburtslagen 4—6mal häufiger als bei Kopflagen (BROWNE, HOLLAND, SUNDE, TOVERUD, GRAGERT, MARTIUS, NOACK). Hirnläsionen in Form spastischer Cerebralparesen und Debilität sind nach den Untersuchungen von BRANDER und von SHARPE gleichfalls wesentlich häufiger. Die Läsion der intrakraniellen Organe ist also insgesamt recht häufig.

Tabelle 3

Geburtslage	% Häufigkeit bei	
	gesunden Neugeborenen	hypoph. Zwergen
Kopflage	96	22
Beckenendlage .	3,2 ⎱ 4,0	62 ⎱ 78
Querlage	0,8 ⎰	16 ⎰

Als eigentliche Ursache der Schädigung der Hypophyse möchte ich die Unterbrechung ihrer arteriellen und portalen Gefäßversorgung ansehen, welche infolge von Verschiebungen des Gehirns durch die brüske Verformung des Hirnschädels unter der Geburt auftritt. Der als größter Teil unter schwierigen Umständen zuletzt geborene Schädel dieser Kinder zeigt nach der Geburt und in den ersten Tagen oft eine charakteristische Deformierung: Eine zurückfliehende Stirn und ein weit nach hinten ausgezogenes verlängertes Hinterhaupt. Wahrscheinlich kommt es zu Abscherungen der Gefäße und vielleicht des Hypophysenstiels selbst.

II. Klinik

Zur Klinik des hypophysären Zwergwuchses soll zunächst das somatische Erscheinungsbild der Patienten skizziert und im Anschluß daran der Ausfall der einzelnen Partialfunktionen der Adenohypophyse dargestellt werden.

Der hypophysäre Zwerg wird mit normalem Gewicht geboren und wächst in den ersten 2 Lebensjahren in der Regel mit normaler Geschwindigkeit. Allem Anschein nach geht das Wachstum des Säuglings und jungen Kleinkindes wie das des Fetus ohne wesentliche hormonale Einwirkungen auf Grund der großen Wachstumspotenz der Gewebe selbst vonstatten. Auch bei der Geburt hypophysektomierte Nagetiere wachsen in der ersten Lebenszeit normal. Etwa vom 3. Lebensjahr an verlangsamt sich die Größenzunahme, wie das Beispiel zeigt, das in Abb. 1

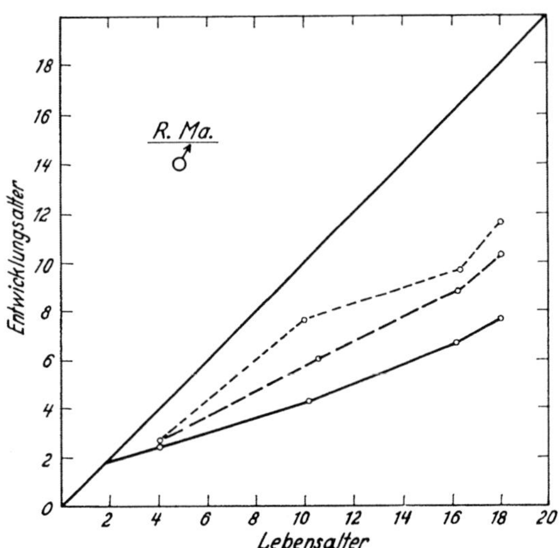

Abb. 1. Typisches Entwicklungsdiagramm bei hypophys. Zwergwuchs. Mit 2 J. einsetzende Verzögerung der Gesamtentwicklung. Längenwachstum (L) stärker beeinträchtigt als Knochenentwicklung (K) und Zahnentwicklung (Z)
------ Z ——— K ——— L

dargestellt ist. Der Pubertätswachstumsschub wird vermißt. Da der Epiphysenfugenschluß ebenfalls ausbleibt, wachsen die Patienten auch nach dem 20. Lebensjahr weiter, wenngleich meistens nur noch wenige Zentimeter. Die Skeletentwicklung ist in der Regel verzögert, obschon weniger als das Längenwachstum. Das Knochenalter überschreitet ohne Behandlung gewöhnlich nicht die Stufe eines normalen 10—13 jährigen. Röntgenologisch fallen der zierliche Bau des Skelets sowie vor allem bei älteren Patienten eine mäßige Osteoporose auf. Typisch ist ferner die mangelhafte Ausbildung der Nasennebenhöhlen. Auch die Gebißentwicklung ist retardiert, — meistens aber in geringerem Maße als die Ossifikation.

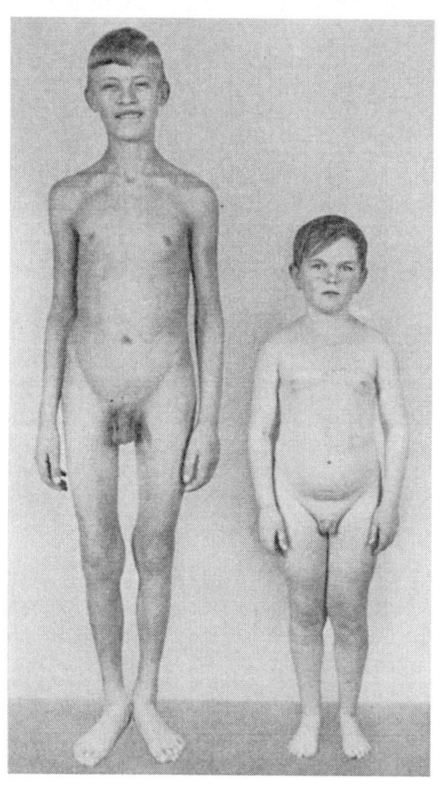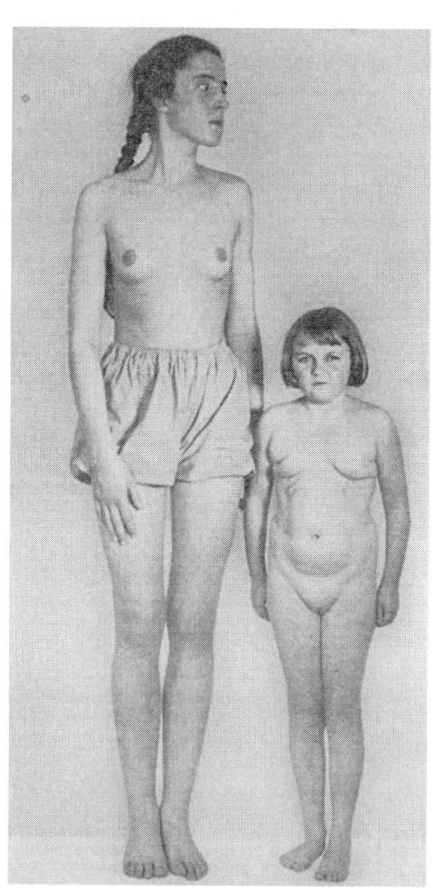

Abb. 2   Abb. 3

Abb. 2. Pat. W. St., 14¹/₂ jähr. Junge mit idiopath. hypophys. Zwergwuchs neben gleichaltrigem gesundem Jungen. Großer runder Kopf, Stammfettsucht, infantiles Genitale

Abb. 3. Pat. R. K., 14 jähr. Mädchen mit genetisch bedingtem hypophys. Zwergwuchs neben gleichaltriger Kontrolle. Puppengesicht, Stammfettsucht mit Fettmammae, fehlende sexuelle Entwicklung

Die charakteristischen Züge in der äußeren Erscheinung der Patienten gehen aus den Abb. 2 und 3 hervor. Der Zwergwuchs ist proportioniert; die Proportionen, so z. B. das Verhältnis von Ober- und Unterlänge, entsprechen annähernd dem chronologischen Alter. Allerdings ist der Kopfumfang relativ groß und die Hände und Füße sind unverhältnismäßig klein. Von unseren erwachsenen Patienten hatten mehrere Schuhgrößen von 30—34. Auffällig ist weiterhin die leichte Stammfettsucht, die so gut wie alle Patienten aufweisen. Früher wurde sie als Zeichen einer Beteiligung des Zwischenhirns betrachtet; heute interpretieren wir sie als Ausdruck des bestehenden relativen Hyperinsulinismus. Die normale Insulinsekretion erfährt keine genügende Ausbalancierung durch die Antagonisten Wachstumshormon und ACTH, was sich in einer gesteigerten Insulinempfindlichkeit manifestiert. Hieraus resultiert die vermehrte Bildung von Fett.

Kopfform und Gesichtsbildung der hypophysären Zwerge zeigen gewisse übereinstimmende Merkmale. Der Kopfumfang ist relativ groß, der Schädel kurz und der Unterkieferwinkelabstand breit. Augenbrauenwülste,Nase und Kinn sind im

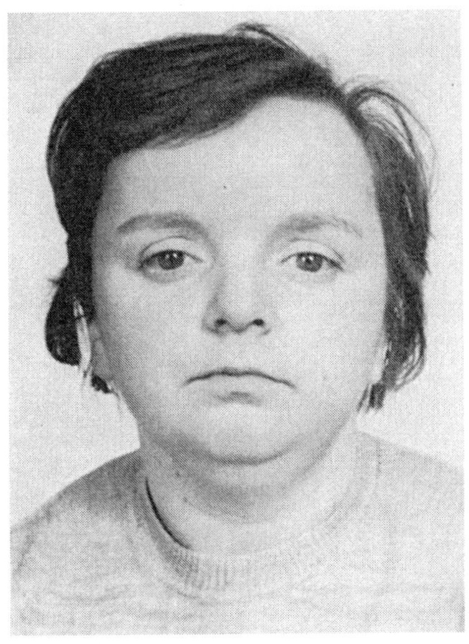

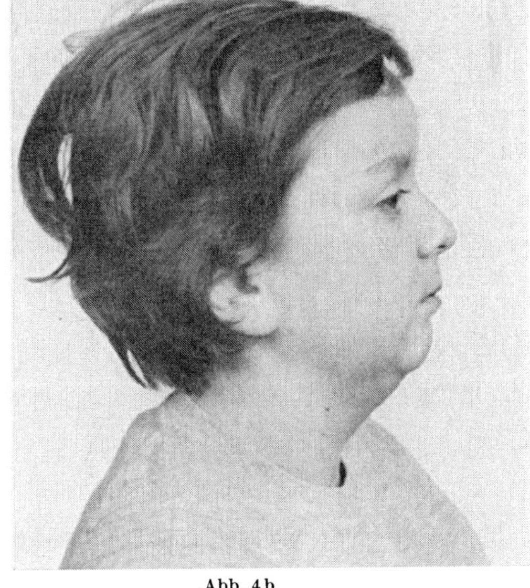

Abb. 4a   Abb. 4b

Abb. 4a und b. Pat. M. Sz., 26jähr. Mädchen mit idiopath. hypophys. Zwergwuchs. Rundes Gesicht mit kleiner Nase und wenig prominentem Kinn

Sinne der Akromikrie schwach ausgebildet. Das runde Gesicht erhält außerdem noch durch die vollen, manchmal pausbäckigen Wangen etwas Infantiles und Puppenhaftes (Abb. 4).

Die bisher geschilderten Merkmale, zu denen noch die zarte Haut und die dünnen Haare kommen, sind auf den Mangel von Wachstumshormon zurückzuführen. Der oft bestehende *thyreotrope Defekt* führt zu klinisch meistens nur wenig auffälligen Symptomen einer sekundären Hypothyreose. Hauttemperatur und Blutdruck können ebenso wie der Grundumsatz mäßig herabgesetzt sein. Manchmal besteht eine Obstipation. Die Intelligenz ist in der Regel normal. Verdächtig auf Schilddrüsenunterfunktion ist eine Retardierung des Knochenalters, die stärker ist als die des Längenalters, ferner doppelseitige perthesähnliche Veränderungen der Oberschenkelköpfe, die auf Dysgenesien der Kopfkerne beruhen, und schließlich eine Hypercholesterinämie. Erhöhte Cholesterinwerte werden bei hypophysären Zwergen allerdings auch ohne Schilddrüsenunterfunktion gelegentlich gefunden, wie HUBBLE und auch wir festgestellt haben. Für die Beurteilung der Schilddrüsenfunktion sollte man in jedem Fall jedoch die Bestimmung des eiweißgebundenen Jods im Plasma oder die Radiojodstoffwechseluntersuchung heranziehen, — schon im Hinblick auf die guten Behandlungserfolge, die man mit Thyreoidea sicca bezüglich des Wachstums erzielen kann, wenn eine Unterfunktion vorliegt.

Die klinisch markantesten Symptome der *sekundären Nebennierenunterfunktion* sind Adynamie und hypoglykämische Zustände. Die körperliche Leistungsfähigkeit ist eingeschränkt, die Patienten ermüden rasch und haben ein vermehrtes

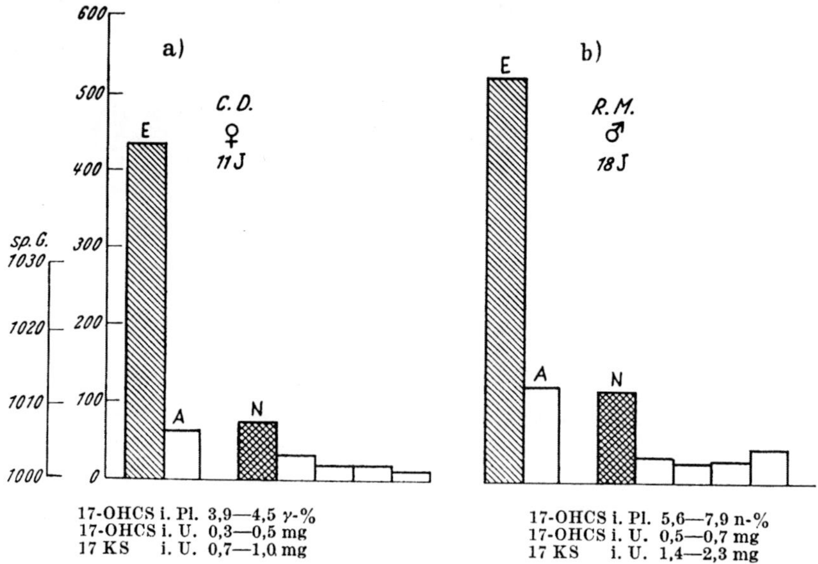

Abb. 5. Mangelhafte Wasserausscheidung im Volhardschen Wasserversuch und 1. Teil des Robinson-Power-Kepler-Tests bei 2 Pat. mit hypophys. Zwergwuchs und sekund. NNR-Insuffizienz

Schlafbedürfnis. Zum Teil läßt sich die Adynamie durch die Neigung zu Hypoglykämien erklären. Leichte, unerkannt ablaufende Hypoglykämien dürften ziemlich häufig vorkommen. Schwere Schocks, die mit Bewußtlosigkeit und Krämpfen einhergehen, werden diagnostisch oft verkannt und für epileptische Anfälle gehalten, zumal sie mit ähnlichen EEG-Veränderungen verbunden sind. Die intravenöse Verabreichung von Traubenzucker bewirkt in solchen Fällen das sofortige Sistieren der Krämpfe. Die gesteigerte Insulinempfindlichkeit der Patienten wurde bereits erwähnt. Um iatrogene Schocks zu vermeiden, muß der intravenöse Insulintest daher mit Vorsicht durchgeführt werden, am besten mit 0,05 E. Altinsulin/kg. Die Toleranzminderung beruht auf der Summation von Wachstumshormon-Defizit und Nebenniereninsuffizienz und kann durch die kombinierte Behandlung mit Wachstumshormon und Cortison normalisiert werden. Zur Vermeidung häufig auftretender hypoglykämischer Anfälle genügt jedoch im allgemeinen die Substitution mit kleinen Cortison-Dosen.

Ein weiteres Symptom der Nebennierenrindenunterfunktion ist die verzögerte Wasserausscheidung nach Wasserbelastung. Der Volhardsche Wasserversuch stellt ein einfaches diagnostisches Kriterium in dieser Hinsicht dar (Abb. 5). Sichere Aufschlüsse gewinnt man freilich nur durch die Bestimmung der Corticoide und

17-Ketosteroide, die in Blut und Harn vermindert gefunden werden. Im Plasma werden für die Porter-Silber-Chromogene oft 0-Werte registriert; nach ACTH-Gabe steigen die Steroide ungenügend an (Abb. 6). Verabreicht man 3—4 Tage lang ACTH, so normalisieren sich die Endwerte, da die atrophierte Nebennieren-rinde regeneriert. Das sicherste Kriterium für eine corticotrope Insuffizienz ist jedoch der pathologische Ausfall des Metopiron-Tests (Abb. 7).

Die letzte Funktion der Adeno-hypophyse, die zu besprechen bleibt, ist die Bildung der Gonadotropine. Bei der Mehrzahl der Patienten bleibt die sexuelle Reifung aus. Die Hoden-biopsie solcher Patienten zeigt ein völlig unreifes Parenchym mit schma-len Hodenkanälchen, fehlender

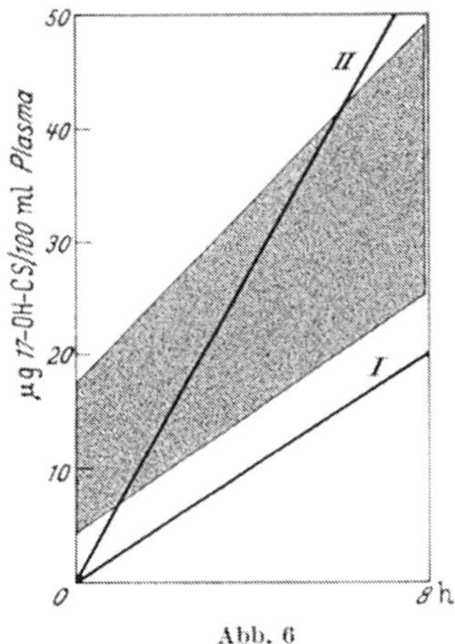

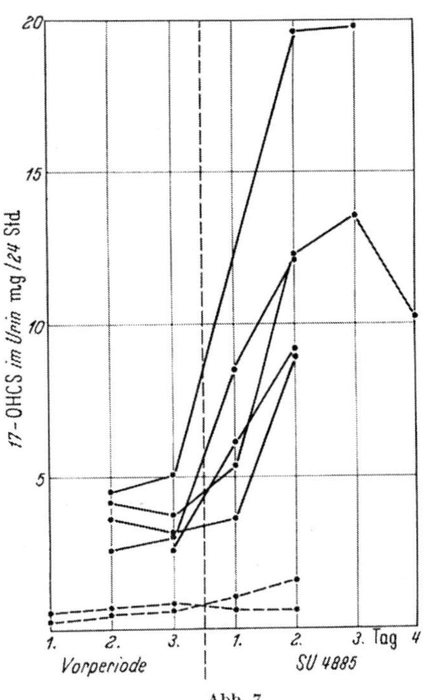

Abb. 6   Abb. 7

Abb. 6. ACTH-Test bei Pat. mit hypophys. Zwergwuchs und sekund. NNR-Insuffizienz, I vor -, II nach 3 täg. Stimulation mit Depot-ACTH. Gemessen wurden die Plasma-17-OHCS vor und 8 Std nach einer Standarddosis Depot-ACTH. Das graue Feld bezeichnet die Normalwerte (M ± 2 σ)

Abb. 7. Metopiron-Tests bei 2 hypophysären Zwergen im Vergleich zu 4 gesunden Vergleichskindern. Keine Mehr-ausscheidung von 17-OHCS während der Metopironverabreichung

Spermatogenese und fehlenden Zwischenzellen. Im Harn wird kein oder nur wenig Gonadotropin ausgeschieden. Ebenso ist die Ausscheidung der Derivate der Sexualhormone vermindert.

Diesen sexuellen Infantilismus hat man früher als besonders typisches Kri-terium des hypophysären Zwergwuchses angesehen und zusammen mit dem Offen-bleiben der Epiphysenfugen zum Kardinalsymptom des sog. infantilistischen Zwergwuchses gemacht. Heute wissen wir, daß die gonadotrope Insuffizienz ein zwar häufiges Symptom ist, das aber ebenso wenig wie die thyreotrope und cortico-trope Insuffizienz obligat ist. Hierfür möchte ich folgende Beispiele anführen. Abb. 8 zeigt das bereits in Abb. 3 demonstrierte Mädchen mit genetisch bedingtem hypophysären Zwergwuchs. Im 16. Lebensjahr traten bei ihr Mammaentwicklung und Menarche auf; das Mädchen menstruiert seitdem regelmäßig. Daß bei dieser Form der Störung der Wachstumshormondefekt häufig isoliert vorkommt, wurde bereits erwähnt. Auch bei der 26jährigen Patientin in Abb. 9 ist die sexuelle Entwicklung eingetreten, wenngleich verspätet und nicht vollständig. Immerhin erfolgen regelmäßige Menstruationen seit dem 18. Lebensjahr. Analoge Fälle finden sich in der Literatur.

Trotzdem bleibt die Häufigkeit der gonadotropen Insuffizienz bemerkenswert. Indessen muß diese Feststellung in gewissem Umfang korrigiert werden, da die

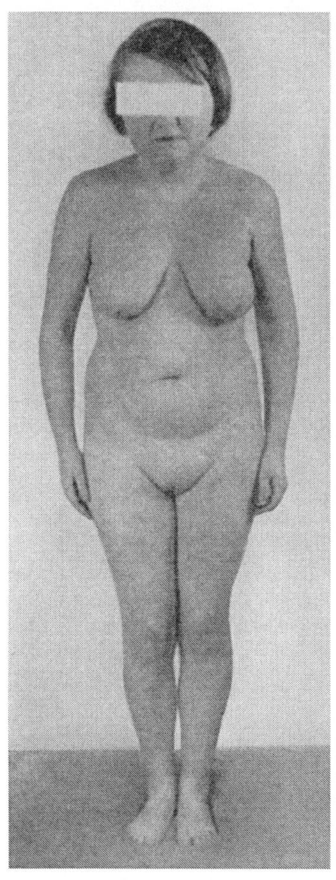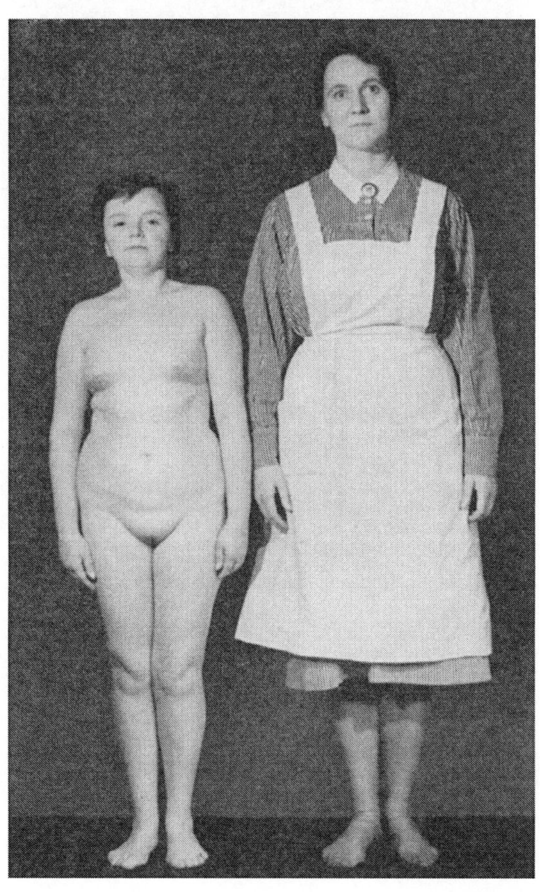

Abb. 8   Abb. 9

Abb. 8. Pat. R. K., dieselbe wie in Abb. 3, im Alter von 17 Jahren. Kräftige Mammaentwicklung, Pat. menstruiert regelmäßig. Fehlende Scham- u. Achselbehaarung

Abb. 9. Pat. M. Sz., dieselbe wie in Abb. 4. Mäßig ausgebildete Mammae, schwache Pubesentwicklung; regelmäßige Menses

Erfahrungen, die man in den letzten Jahren mit der Therapie mit anabolischen Steroiden gesammelt hat, gezeigt haben, daß eine Reihe von Patienten dabei in die Pubertät kommt. Gelingt es, das retardierte Knochenalter auf die Entwicklungsstufe von 12—14jährigen Kindern zu bringen, so setzt bei einem Teil der Patienten eine echte sexuelle Reifung ein. Der Nachweis der Gonadotropine im Harn wird positiv, und nicht nur die äußeren Genitalorgane, sondern auch die Keimdrüsen entwickeln sich. Derartige Beobachtungen sind von Weller u. Tonutti, von Prader u. Illig und von uns gemacht worden (Abb. 10). Offenbar dient dem Organismus bzw. der Hypophyse das Knochenalter als Schrittmacher und als Startsignal für die Bildung der Gonadotropine. Von einer sicheren gonadotropen Insuffizienz kann dementsprechend nur gesprochen werden, wenn das Knochenalter das 13./14. Jahr bereits überschritten hat.

Die letzte Tabelle gibt eine Übersicht über die endokrinen Verhältnisse, die wir bei 20 idiopathischen hypophysären Zwergen vorgefunden haben. Eine thyreotrope Insuffizienz bestand in 7 von 16 daraufhin geprüften Fällen, eine corticotrope Insuffizienz bei 11 von 20 Patienten, — also in jeweils rund der Hälfte der Fälle. Eine vollkommene gonadotrope Insuffizienz fanden wir bei 9, eine mäßige Unterfunktion bei 3 von den 12 Fällen, bei denen eine Beurteilung nach Maßgabe des Knochenalters möglich war. Solange wir keine klinisch brauchbare Methode zur Bestimmung des Wachstumshormons besitzen, gründet sich die Diagnose des hypophysären Zwergwuchs auf diesen Nachweis der Unterfunktion der glandotropen Hormone und auf die verminderte Insulintoleranz, die in unserem Krankengut in $^3/_4$ der Fälle festzustellen war.

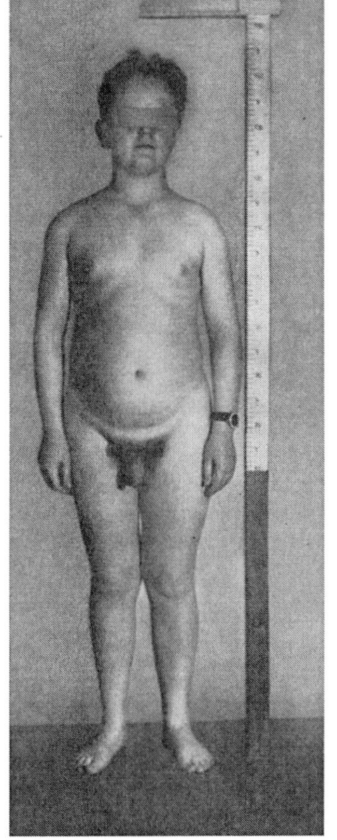

Abb. 10. Pat. W. St., derselbe wie in Abb. 2, im Alter von 18 Jahren nach Behandlung mit HCG und anabolen Steroiden. Skeletalter 14 J., FSH im Urin 6,6—12,2 ME

Tabelle 4. *HVL-Funktionen*

Name	Insulintest	Thyreotr.	Corticotr.	Gonadotr.
A. S.	—	n	↙	(↙)
J. G.	—	—	↙	↙
R. M.	n	n	↙	↙
H. P. L.	—	—	↙	↙
A. Sch.	↙	↙	n	↙
H. J. T.	↙	↙	↙	↙
W. St.	↙	n	n	(↙)
R. J.	n	n	n	↙
K. K.	↙	n	n	↙
B. St.	(↙)	n	↙	(↙)
J. L.	n	↙	n	↙
R. B.	↙	n	↙	↙
G. K.	—	—	↙	·/·
B. S.	↙	n	n	·/·
Ch. N.	↙	↙	↙	·/·
D. P.	↙	↙	↙	·/·
J. J.	↙	n	↙	·/·
K. Sch.	↙	n	n	·/·
E. G.	n	↙	n	·/·
M. Sch.	↙	↙	n	·/·

Literatur

ALBEAUX-FERNET, M., J. CHABOT, P. DESHAVES, M. GELINET, et J. D. ROMANI: Semaine Hôp. 1956, 3667.
ALTMANN, F.: Beitr. path. Anat. 85, 205 (1930).
APITZ, K.: Virchows Arch. path. Anat. 302, 554 (1939).
BERBERICH, J.: 37. Verh. d. Dtsch. Ges. Kinderheilk. 1926, 403
BERBLINGER, W.: Beitr. path. Anat. 87, 233 (1931).
BIERICH, J. R.: Acta endocr. (Kbh.) Suppl. 67, 76 (1962).
BLIZZARD, R. M., and M. ALBERTS: J. Pediat. 48, 782 (1956).
BRANDER, T.: Mschr. Geburtsh. Gynäk. 105, 205 (1937).
BROWNE, F. J.: Brit. med. J. 1922 II, 590.
CERANKE, P.: Arch. inn. Med. 29, 151 (1936).
FARINA, C.: Ateneo parmense 14, 125 (1942).
FRANCIS, T.: Investigations into the development of the pituitary at hereditary anterior pituitary dwarfism in mice, with reference to the pathogenesis of the anterior pituitary dwarfism. Kopenhagen: Ejnar Munksgaard 1944.
GRAGERT, O.: Geburtsh. u. Frauenheilk. 1, 479 (1939).
HEDINGER, CH., u. D. HÜRZELER: Acta endocr. (Kbh.) 14, 170 (1953).
HEWER, T. F.: J. Endocr. 3, 397 (1944).
HOLLAND, E.: Brit. med. J. 1922 II, 588.
HOOFT, C., L. BLANCKE et F. GARMYN: Acta paediat. belg. 15, 244 (1961).
HUBBLE, D.: Vortrag a. d. 9. Internat. Kongress f. Pädiatrie, Montreal 1959.
KEMP, T.: Acta path. microbiol. scand. Suppl. 16, 189 (1933).
— Acta path. microbiol. scand., Suppl. 26, 10 (1936).
KOCH, W.: Verh. dtsch. Path. Ges. 21, 274 (1926).
KOGUT, M. D., S. A. KAPLAN, and C. S. N. SHIMIZU: Pediatrics 31, 538 (1963).
MARTIN, M. M., and L. WILKINS: J. clin. Endocr. 18, 679 (1958).
MARTIUS, H.: In: SEITZ/AMREICH: Biol. u. Path. des Weibes 8, 435 (1955).
MOSIER, H. D.: J. Pediat. 48, 633 (1956).
PRADER, A., u. R. ILLIG: Acta endocr. (Kbh.) Suppl. 67, 23 (1962).
PRIESEL, A.: Beitr. path. Anat. 67, 220 (1920).
— Beitr. path. Anat. 70, 209 (1921).
REID, J. D.: J. Pediat. 56, 658 (1960).
RUSSFIELD, A. B., u. L. REINER: Lab. Invest. 6, 334 (1957).
SANDISON, A. T.: Arch. Dis. Childh. 33, 469 (1958).
SCHWARTZ, PH.: Mschr. Kinderheilk. 34, 511 (1926).
SIMMONDS. M.: Dtsch. med. Wschr. 1919, 487.
SMITH, P. E., u. E. C. McDOWELL: Anat. Rec. 46, 249 (1930).
STERNBERG, C.: Beitr. path. Anat. 67, 275 (1920).
STRANSKY, E., u. R. T. LARA: Philipp. J. Pediat. 10, 43 (1961).
SUNDE, A.: Acta obstet. gynec. scand. 9, 477 (1930).
SVILOKOS, N., u. P. ZANETTI: Lattante 32, 616 (1961).
TONUTTI, E., u. O. WELLER: In: TONUTTI, WELLER, SCHUCHARDT u. HEINKE: Die männliche Keimdrüse. Stuttgart: Thieme-Verlag 1960.
TOVERUD, K. U.: Acta paediat. (Uppsala) 18, 249 (1936).
VAN DER WERFF TEN BOSCH, J. J.: Ned. T. Geneesk. 106, 1282 (1962).

Diskussion

E. TONUTTI (Bonn):

Sehr aufschlußreich zur Frage der Behandlung des hypophysären Zwergwuchses mit Androgen ist das Verhalten der Gonadotropinsekretion bei Leydigzell-Tumoren. Leydigzell-Tumoren treten meist im Alter von 3—5 Jahren auf und verursachen infolge ihrer Androgenbildung eine Pseudopubertas praecox. Im Zuge des dabei auftretenden Wachstumsschubes wird das Knochenalter weit vorangetrieben. Sobald ein Knochenalter von 12—14 Jahren erreicht ist, setzt die Gonadotropinproduktion der Hypophyse ein, vorausgesetzt, daß die androgene Produktion des Tumors, gemessen an der 17-KS-Ausscheidung in einer Größenordnung bleibt, die dem normalen Erwachsenen entspricht und somit eine Blockierung der gonadotropen Partialfunktion in der Hypophyse unterbleibt. Nach der Tumorentfernung nimmt die Gonadotropinbildung, vorausgesetzt, daß das entsprechende Knochenalter erreicht ist, zu, so daß die *echte* Pubertät in Gang kommt.

Beim hypophysären Zwerg bereitet die Androgenzufuhr ebenfalls einen Wachstumsschub von etwa 20—24 cm, der in der Größenordnung somit dem der normalen Pubertät entspricht. Sobald ein Knochenalter von 13—14 Jahren erreicht ist, kann die Androgenzufuhr abgesetzt werden. Die Gonadotropinbildung kommt in Gang, wodurch die echte Pubertät ausgelöst wird und der künstlich durch Androgene eingeleitete Wachstumsschub nunmehr spontan zu Ende geführt wird. Natürlich ist dies nur bei jenen Formen von hypophysärem Zwergwuchs möglich, denen kein organischer Defekt der Hypophyse, sondern lediglich ein isolierter STH-Mangel zugrunde liegt.

G. Dhom (Würzburg):

Die acidophilen Zellen des Hypophysenvorderlappen als Quelle des Wachstumshormons erfahren — im Gegensatz zu den Mucoidzellen — erst relativ spät eine weitergehende Ausreifung. Ihr prozentualer Anteil beträgt nach eigenen Untersuchungen [Beitr. Path. Anat. **124,** 57 (1961)] bei Frühgeburten im Mittel erst 17,4%, bei Reifgeborenen 20,7%, um dann in den ersten Lebensmonaten auf über 30% anzusteigen. Obwohl STH bereits in der fetalen Hypophyse nachweisbar ist, besitzt es offenbar keinen maßgeblichen Einfluß auf das intrauterine Wachstum. Die späte Acidophilenreifung stimmt mit dieser Tatsache gut überein.

E. Werner (Berlin):

Die Unterbrechung der arteriellen und der portalen Gefäßversorgung der Hypophyse infolge eines Geburtstraumas wäre ein lokaler Prozeß, der zum Ausfall der STH-Bildung führen kann. Näherliegend als das lokale Gefäßtrauma scheint mir die bei Beckenendlage vorübergehend verminderte allgemeine Sauerstoffversorgung des Kindes während der Geburt. Dann allerdings müßten die STH-produzierenden Zellen des HVL besonders empfindlich gegen einen anoxämischen Zustand sein.

J. R. Bierich:

Es sind nicht gewöhnliche Steiß- bzw. Beckenendlagen, sondern besonders schwere, die zur Zerstörung der Hypophyse führen. Ich selbst möchte annehmen, daß eine stärkere Deformierung des Craniums mit einer Verschiebung des Schädelinhalts gegenüber der fixen Schädelbasis von Bedeutung ist. Solche Verformungen haben wir öfters bei Kindern gesehen, bei denen die Entwicklung des zuletzt geborenen Kopfes Schwierigkeiten machte und verzögert vor sich ging. Bei gewöhnlicher Steißlage haben die Kinder hingegen einen runden, wenig deformierten Kopf.

Z. Laron (Peta Tiqva):

I would like to remark that amongst familial hypopituitary dwarf males do not outmembery females. I even think that among my patients there are more girls. It is usually said that pituitary dwarfs have a normal body weight, grow at a normal rate and within normal parameters up to the age of one year. This certainly is not true for children with familial hypopituitarism. Many of them have a low birth weight under 2000 g and grow slowly from the beginning, resembling infants with primordial dwarfism.

J. R. Bierich:

Ein deutliches Überwiegen des männlichen Geschlechts besteht nur beim idiopathischen hypophysären Zwergwuchs. Der genetisch bedingte hypophysäre Zwergwuchs wird autosomal recessiv vererbt. In unserem Krankengut haben wir folgende Verhältnisse: 2 Geschwisterpaare, in denen jeweils ein Junge und ein Mädchen betroffen sind; 2 Brüder, die in einer längeren Geschwisterreihe die einzigen Betroffenen sind und schließlich 2 Brüder und eine Schwester in einer Geschwisterschaft von vier Kindern.

Hinsichtlich der Geburtsgewichte weiß ich nicht genau, ob Sie die Gewichte der idiopathischen oder der hereditären Zwerge meinen. Die Gewichte unserer idiopathischen Fälle waren normal, abgesehen von 2 Frühgeborenen, deren Untergewicht ihrer Unreife entsprach. Die Geburtsgewichte der genetisch bedingten Fälle waren, soweit wir sie exakt kennen (5 von 9 Fällen), ebenfalls völlig normal.

Aus der Universitäts-Kinderklinik Zürich

Zur Behandlung des hypophysären Zwergwuchses

Von

A. PRADER

Mit 12 Abbildungen

Referat

Beim hypophysären Zwergwuchs mangeln seit frühester Kindheit das Wachstumshormon, meist auch die Gonadotropine und häufig ACTH und TSH. Die für den Patienten besonders schwerwiegenden Symptome, deren Behebung er vom Arzt erwartet, sind der Kleinwuchs, das Fehlen der Pubertät und die geringe körperliche Leistungsfähigkeit.

Die theoretisch richtige Ersatztherapie wäre die Verabreichung aller fehlenden Hypophysenvorderlappen-Hormone. Verschiedene Schwierigkeiten stehen dem jedoch entgegen. Erstens können diese Hormone, da sie Eiweiß-Moleküle sind, nur parenteral verabreicht werden. Zweitens stehen sie nur teilweise in einer beim Menschen wirksamen und nicht antigenen Form zur Verfügung.

In der Praxis hat man sich daran gewöhnt, statt den Gonadotropinen Sexualsteroide, statt ACTH Cortison und statt TSH Thyreoidea-Hormone zu geben. Diese Therapie ist einfach, billig und innerhalb gewisser Grenzen wirksam. Sie fördert das Wachstum, ist aber nicht imstande, es ganz zu normalisieren. Ferner ist sie nicht imstande, die Fertilität zu gewährleisten. Feinere Befunde, die mit einer kombinierten Therapie von Sexualsteroiden, Cortison und Thyreoidea-Hormonen ebenfalls nicht voll korrigiert werden können, sind z. B. die Osteoporose, die mangelhafte Gegenregulation bei insulininduzierter Hypoglykämie und der mangelhafte Fettsäuren- und Ketokörper-Anstieg beim Hungern, alles Hinweise auf das Fehlen von Wachstumshormon, das sich mit der genannten Kombinationstherapie eben nicht ersetzen läßt.

Ich möchte meine Aufgabe aber weniger von der metabolischen als viel mehr von der praktischen Seite her anpacken und die folgenden therapeutischen Fragen prüfen: Was leisten Testosteron und seine anabolen Derivate, was leisten die Thyreoidea-Hormone und was leistet das menschliche Wachstumshormon für das Wachstum und die körperliche Entwicklung beim hypophysären Zwergwuchs? Die älteren Behandlungsmethoden mit Testosteron und Thyreoidea-Hormonen werde ich nur kurz und die neuere Behandlungsmethode mit menschlichem Wachstumshormon etwas genauer besprechen. Da die Mehrzahl der hypophysären Zwerge männlich ist, werde ich auf eine Besprechung der Oestrogene, die naturgemäß in der Therapie von Mädchen mit hypophysärem Zwergwuchs ihren festen

Platz haben, verzichten. Ebenso werde ich auf eine Besprechung des Cortisons verzichten und möchte nur darauf hinweisen, daß die Cortison-Dosierung besonders heikel ist, da Cortison wachstumshemmend wirkt und da bei Patienten mit Thyreoideainsuffizienz viel rascher Cushing-Symptome auftreten als bei Gesunden.

Da ich vorwiegend über eigene Erfahrungen berichten werde, möchte ich meine Mitarbeiter nennen, ohne deren Hilfe diese Untersuchungen nicht möglich gewesen wären. Es sind dies Frau Dr. J. Széky, Fräulein Dr. R. Illig, Fräulein Dr. G. Mürset, Herr Dr. H. Wagner und Herr Dr. U. Rumpf. Besonderen Dank für die Beschaffung von menschlichen Hypophysen schulde ich ferner Herrn Prof. E. Uehlinger (Pathologie-Institut Zürich), Herrn Prof. H. U. Zollinger (Pathologie-Institut St. Gallen) und Herrn Prof. Ch. Hedinger (Pathologie-Institut Winterthur).

1. Verlauf ohne Therapie

Abb. 1 zeigt die Ausgangslage, nämlich einen unbehandelten Patienten im Alter von 12 und von 41 Jahren. Außer dem Kleinwuchs fallen mit 12 Jahren der

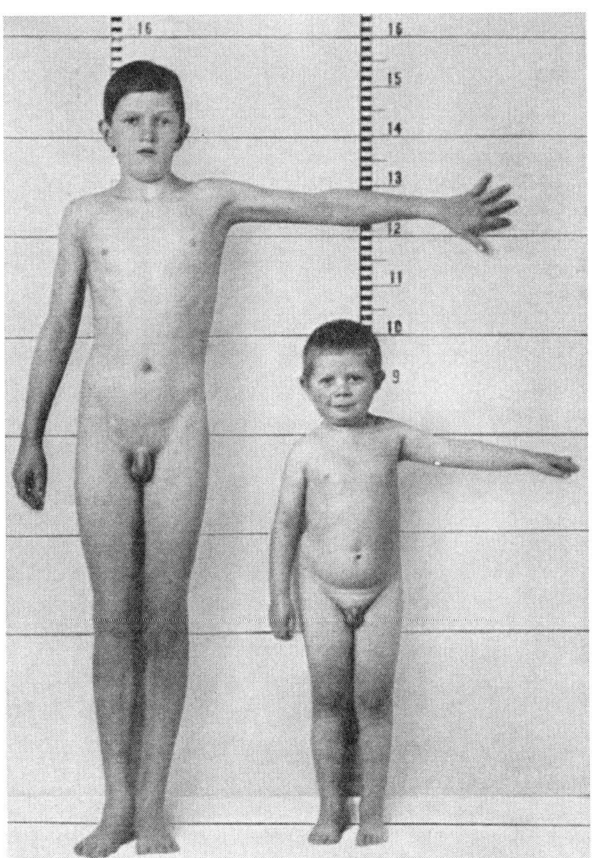
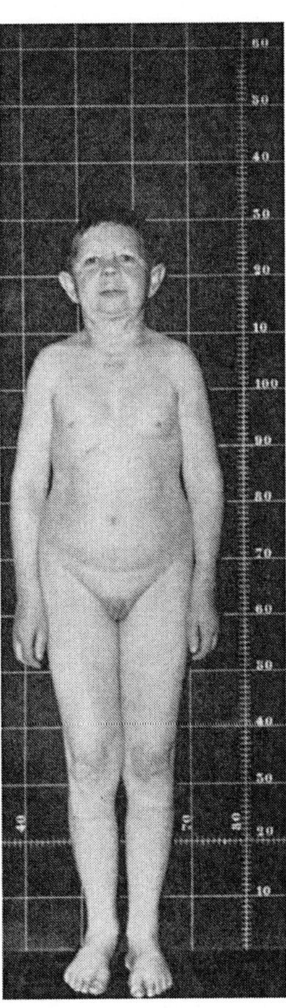

Abb. 1. Unbehandelter idiopathischer hypophysärer Zwergwuchs (B. Theophil). Links im Alter von 12 Jahren (Größe 101 cm, Knochenalter nach Tanner et al. 4 Jahre) neben gleichaltrigem gesundem Knaben. Rechts im Alter von 41 Jahren (Größe 132 cm, Knochenalter 13 Jahre)

puppenartige Aspekt von Gesicht und Körper und mit 41 Jahren das faltenreiche ältliche Gesicht und das vollkommene Fehlen der sekundären Geschlechtsmerkmale auf. Mit 12 Jahren ist der Patient nur 101 cm groß und entspricht in Größe und Knochenalter einem 4jährigen. Mit 41 Jahren ist er 132 cm groß und entspricht in der Größe einem 9jährigen und im Knochenalter einem 13jährigen, d. h. er hat noch weit offene Epiphysenfugen und könnte deshalb theoretisch noch wachsen. Das Offenbleiben der Epiphysenfugen und das Ausbleiben der Pubertät sind auf das Fehlen der Sexualsteroide, d. h. letzten Endes auf das Fehlen der Gonadotropine, und zum kleinen Teil wohl auch auf den ACTH-Mangel zurückzuführen. Ohne Therapie wird dieser 41jährige Zwerg trotz vorhandener Wachstumsmöglichkeit kaum mehr wachsen und so klein und unterentwickelt bleiben wie er ist. Unter einer Testosteronbehandlung werden dagegen nicht nur die sekundären Geschlechtsmerkmale, sondern auch ein deutlicher Wachstumsschub und der endgültige Epiphysenschluß auftreten. Damit erst wird der Patient erwachsen. Im Gegensatz zu seiner jetzigen Größe von 132 cm dürfte seine durch Testosteron erreichbare Erwachsenengröße etwa 138—140 cm betragen.

2. Therapie mit Testosteron und anabolen Steroiden

Mit diesem ersten Beispiel sind die grundsätzlichen Aspekte der Testosteron-Therapie beim hypophysären Zwergwuchs bereits aufgezeigt. Testosteron und die anabolen Steroide vermögen das Wachstum und die endgültige Größe dieser Patienten zu fördern, indem sie den sonst fehlenden Pubertätswachstumsschub ermöglichen. Eine normale Erwachsenengröße kann der Patient aber auch dann nicht erreichen, da mit dieser Therapie nur die fehlenden anabolen Sexualsteroide und nicht das fehlende Wachstumshormon ersetzt wird.

Abb. 2 illustriert diese Testosteronwirkung. Es ist die Entwicklungskurve eines jetzt 22jährigen Mannes mit hypophysärem Zwergwuchs, der seit dem 15. Jahr mit Testosteron behandelt worden ist. Der normale Wachstums- und Entwicklungsverlauf sollte bei dieser Darstellung bis zum 18. Jahr der diagonalen Linie und anschließend einer horizontalen Linie folgen. Mit 15 Jahren entsprach der Patient in seiner

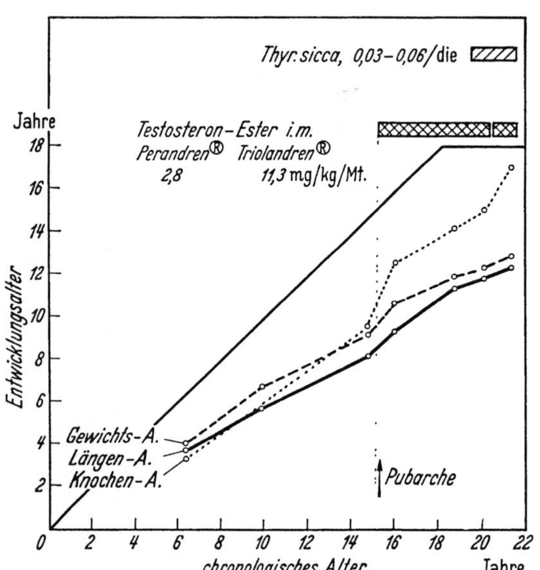

Abb. 2. Die Wirkung von Testosteron auf Wachstum und Entwicklung eines Jungen mit idiopathischem hypophysärem Zwergwuchs (H. Hanspeter). Knochenalter nach Greulich und Pyle

Größe und seiner Knochenentwicklung ungefähr einem 8jährigen. Testosteron förderte das Wachstum, beschleunigte aber das Knochenalter noch stärker. Mit 22 Jahren sind die Epiphysenfugen fast geschlossen und der jetzt 146 cm große Patient wird höchstens noch 2—3 cm wachsen. Ohne diese Behandlung hätte er wohl nur etwa 135 statt 146 cm erreicht, da seine Entwicklung wie bei dem in

Abb. 1 dargestellten Patienten kaum über das Knochenalterstadium von 14 Jahren hinausgekommen wäre. Abb. 3 zeigt die Wirkung des Testosterons beim gleichen Patienten auf die sekundären Geschlechtsmerkmale, auf den Körperbau und auf die Gesichtskonfiguration. Aus einem Kind ist ein Erwachsener, allerdings ein kleiner Erwachsener geworden.

Die neueren anabolen Steroide leisten im Endresultat nach unsern Eindrücken nicht mehr als Testosteron (PRADER; PRADER und ILLIG). In der Praxis wird man

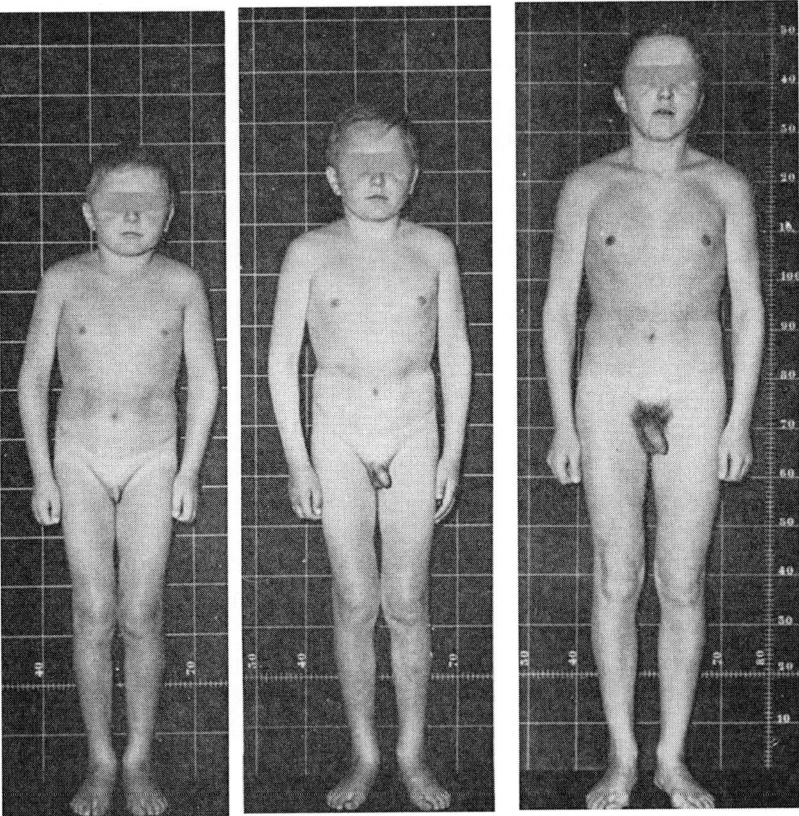

Abb. 3. Die Wirkung von Testosteron. Gleicher Patient wie in Abb. 2. Links mit $14^{11}/_{12}$ Jahren (Größe 128 cm vor Beginn der Testosteron-Behandlung. Mitte mit $15^{7}/_{12}$ Jahren (132,5 cm) und rechts mit $21^{2}/_{12}$ Jahren (150 cm) unter Testosteronbehandlung

vor dem 15. Jahr eher anabole Steroide in kleiner, nicht virilisierender Dosierung und nach dem 15. Jahr eher Testosteron in steigender virilisierender Dosierung bis zur Entwicklung aller sekundären Geschlechtsmerkmale geben. Obwohl diese Therapie die Erreichung einer normalen Erwachsenengröße nicht erlaubt und selbstverständlich auch keine Hodenreifung bewirkt, ist das Resultat in bezug auf Wachstum, Körperreifung und Körperkraft eindrücklich. Der Patient fühlt sich wohler, kräftiger und in seinem Aussehen nicht mehr so auffallend verschieden von seinen Mitmenschen. Die Testosteron- und Anabolica-Therapie ist deshalb sowohl somatisch wie auch psychologisch wertvoll und wichtig.

Eine besondere therapeutische Schwierigkeit, auf die ich noch zurückkommen werde (Abb. 6 und 8), ist die Tatsache, daß bei sehr schwerer Hypophyseninsuffizienz eine Resistenz gegenüber der Wirkung anaboler Steroide auf das Wachstum, gelegentlich auch auf die Knochenreifung und auf die Entwicklung der sekundären Geschlechtsmerkmale vorliegt (PRADER, PRADER und ILLIG; VAN DER WERFF TEN BOSCH). Es sieht aus, wie wenn ein gewisses Minimum von Wachstumshormon notwendig wäre, um die wachstumsfördernde Wirkung von anabolen Steroiden zu erlauben. Ältere tierexperimentelle Untersuchungen der Gruppe von EVANS (SIMPSON, ASLING und EVANS) und von DESAULLES sprechen ebenfalls für diese Annahme.

3. Therapie mit Thyreoidea-Hormonen

Die Wirkung von Thyreoidea-Hormonen auf das Wachstum beim hypophysären Zwergwuchs ist uneinheitlich. Die meisten erfahrenen Autoren wie MARTIN und WILKINS und DE GENNES und ROYER halten sie für unwirksam, während

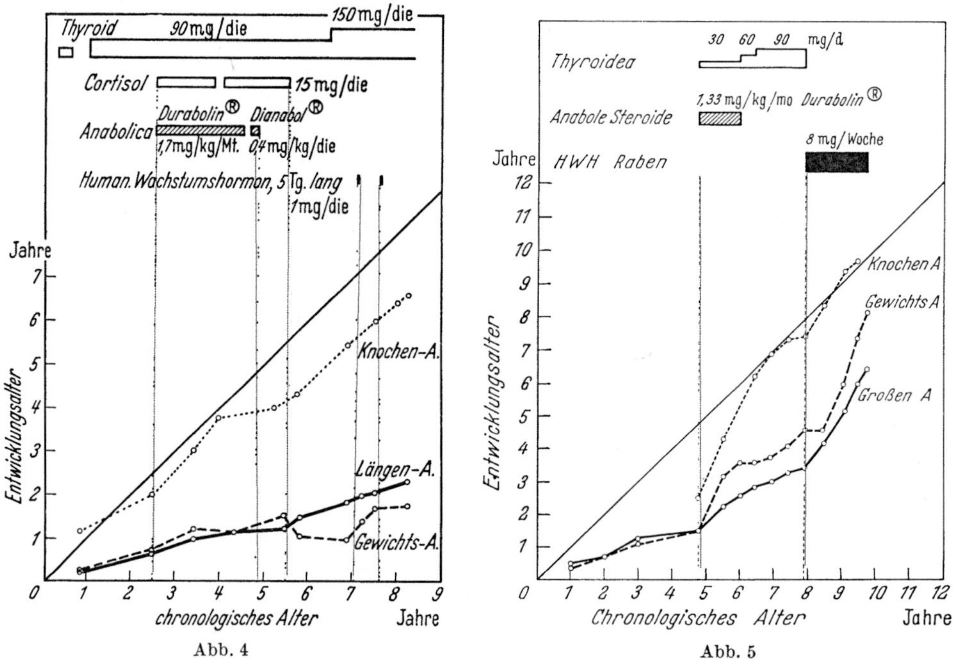

Abb. 4. Wachstum und Entwicklung eines Jungen mit idiopathischem hypophysärem Zwergwuchs und ungewöhnlich starken hypothyreoten Symptomen (H. Jürg). Knochenalter nach TANNER et al.

Abb. 5. Die Wirkung von anabolen Steroiden, Thyreoideahormonen und menschlichem Wachstumshormon vom Typus Raben auf Wachstum und Entwicklung eines Mädchens mit idiopathischem hypophysärem Zwergwuchs (S. Anna). Knochenalter nach TANNER et al.

BIERICH und wir wenigstens in einzelnen Fällen eine fördernde Wirkung auf Wachstum und Knochenreifung gesehen haben. Man würde annehmen, daß Thyreoidea-Hormone in den Fällen mit TSH-Mangel nützen würden und in den Fällen ohne TSH-Mangel unwirksam wären. Nach unserer Erfahrung besteht aber kein derart einfacher Zusammenhang. Wir sahen Patienten ohne nachweisbare Thyreoideainsuffizienz, bei denen Thyreoidea-Hormone Wachstum und Knochenreifung förderten (Abb. 5) und solche mit nachweisbarer Thyroideainsuffizienz, bei denen Thyreoidea-Hormone keinerlei Erfolg hatten oder höchstens die Knochenreifung, aber nicht das Wachstum, beschleunigten (Abb. 4).

Abb. 4 ist die Entwicklungskurve des einzigen hypophysären Zwerges in unserm Patientengut, der nicht nur testmäßig, sondern auch klinisch eine eindrückliche Hypothyreose aufwies. Mit Thyreoidea sicca gelang es, die Hypothyreose-Symptome zu beheben und die Knochenreifung fast normal zu halten, während das Wachstum extrem langsam blieb. Thyreoidea sicca hatte also keine Wirkung auf das Wachstum, aber eine deutliche Wirkung auf die Knochenreifung.

Abb. 5 ist die Entwicklungskurve eines Mädchens mit hypophysärem Zwergwuchs ohne nachweisbare Thyreoideainsuffizienz, das seit dem 5. Jahr unter wechselnder Behandlung steht. Zuerst erhielt es ein Jahr lang anabole Steroide mit deutlicher Beschleunigung des Wachstums und noch stärkerer Beschleunigung der Knochenreifung. Vom 6.—8. Jahr erhielt es ausschließlich Thyreoidea sicca. Der unter anabolen Steroiden gesehene Erfolg hielt unter dieser Therapie an, wenn auch in etwas abgeschwächter Form. Thyreoidea sicca wirkte also in diesem Fall, der allerdings einen Ausnahmefall darstellt, ähnlich wie anabole Steroide. Mit 8 Jahren wurde auch diese Therapie abgesetzt und die Behandlung mit menschlichem Wachstumshormon angefangen.

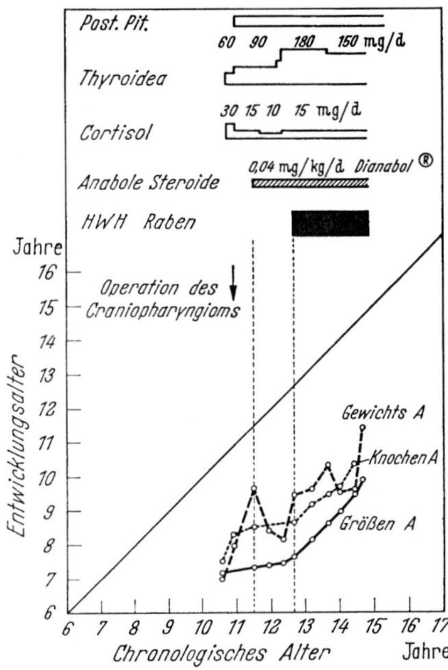

Abb. 6. Die Wirkung von anabolen Steroiden und menschlichem Wachstumshormon vom Typus Raben auf Wachstum und Entwicklung eines Mädchens mit hypophysärem Zwergwuchs nach operativer Entfernung eines Craniopharyngeoms (M. Verena). Knochenalter nach TANNER et al.

4. Therapie mit humanem Wachstumshormon (HWH)

HWH wurde 1958 von RABEN in die Therapie des hypophysären Zwergwuchses eingeführt. Leider steht es immer noch nur als Extrakt aus menschlichen Hypophysen zur Verfügung und ist deshalb kommerziell noch nicht erhältlich. Heute liegen schon zahlreiche Berichte über seine erfolgreiche Verwendung vor, doch handelt es sich immer nur um sehr kleine Patientenzahlen oder Einzelfälle ohne vergleichbare oder gar einheitliche Dosierung und Verabreichung (HUTCHINS et al. 1959, BECK et al. 1960, SHEPARD et al. 1960, LIPSETT et al. 1961, PRADER 1961, PRADER u. ILLIG 1962, RABEN 1962, VEST u. GIRARD 1962, AARSKOG 1963, DAUGHADAY u. PARKER 1963, FRANÇOIS et al. 1963 und TRAFFORD et al. 1963 u. a.).

Wir haben bisher 13 Zwerge mit HWH während einer Dauer von 4 Monaten bis 3 Jahren behandelt. Um die Ergebnisse bei den einzelnen Patienten und bei den verschiedenen Zwergwuchsformen miteinander vergleichen zu können, haben wir ausschließlich das von Raben extrahierte HWH, das ungefähr 1,5 USP-Einheiten pro mg enthält, und immer die gleiche Dosierung angewendet. Für kurzfristige Stoffwechselversuche betrug diese Dosierung 2 mg/m²/Tag und für die langfristige Wachstumsbehandlung 2 × 5 mg/m²/Woche. Wenn wir dies mit der von Parker, Utiger und Daughaday geschätzten Umsatzrate von 3 mg/m²/24 Std beim Normalen vergleichen, so liegt unsere Dosierung vermutlich im unteren physiologischen

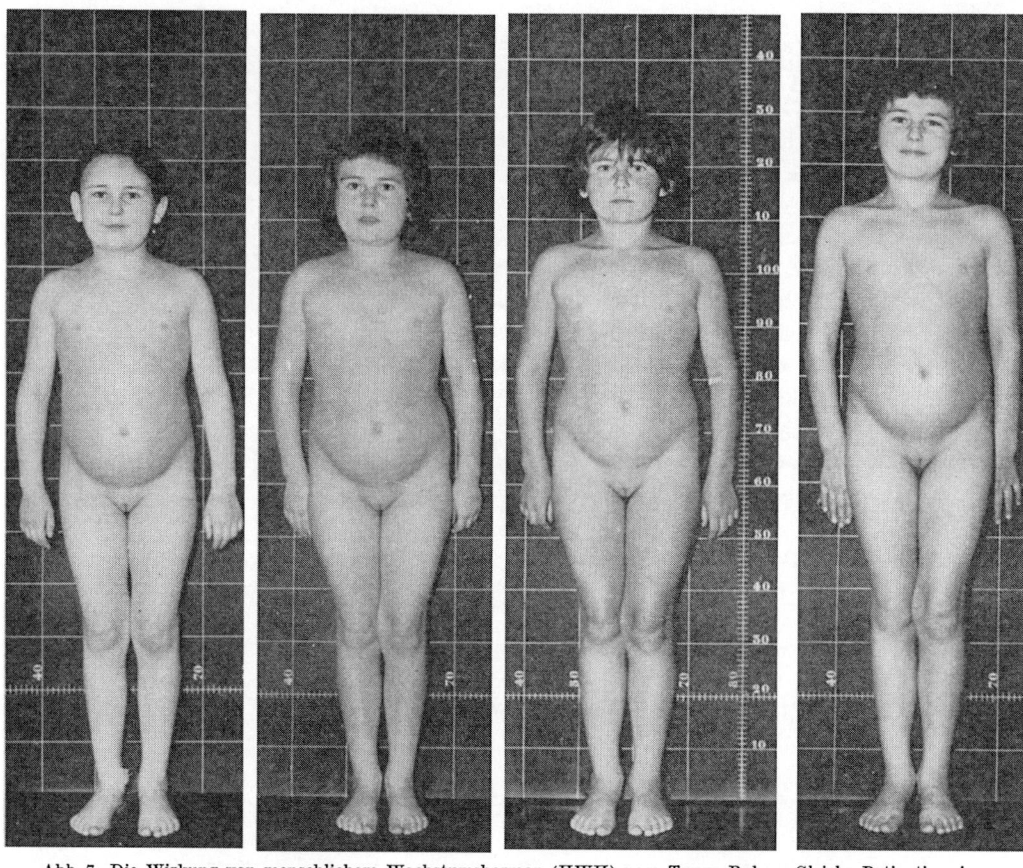

Abb. 7. Die Wirkung von menschlichem Wachstumshormon (HWH) vom Typus Raben. Gleiche Patientin wie in Abb. 6. Von links nach rechts im Alter von 10¹¹/₁₂ Jahren (Größe 122,5 cm, vor HWH), von 12⁸/₁₂ Jahren (124,5 cm, bei Beginn der HWH-Therapie), von 13⁴/₁₂ Jahren (129 cm) und von 14⁷/₁₂ Jahren (138 cm)

Bereich. In der Regel standen unsere Patienten schon vor Beginn dieser Behandlung unter einer kombinierten Ersatztherapie mit Thyreoidea sicca, Cortison und anabolen Steroiden in kleinen Dosen, zu der dann HWH noch zugefügt wurde.

Eine Ausnahme ist die oben besprochene Patientin (Abb. 5), bei der mit Beginn der HWH-Therapie alle andere Therapie abgesetzt wurde. Der Wachstumserfolg ist überaus deutlich. Die Wachstumsgeschwindigkeit stieg von knapp 4 cm pro Jahr unter Thyreoidea sicca auf über 10 cm pro Jahr und hält heute nach fast zweijähriger Therapie unvermindert an. Dies ist ein überdurchschnittlich schnelles Wachstum und stellt den besten und anhaltendsten Erfolg dar, den wir mit unserer Dosierung bisher erzielen konnten. Von Interesse ist ferner, daß das Knochenalter ebenfalls zunahm, allerdings langsamer als das Größenalter. Die Entwicklungskurve dieser Patientin (Abb. 5) zeigt deutlich, daß in der gewählten Dosierung anabole Steroide die Knochenreifung stärker fördern als das Wachstum, während umgekehrt HWH allein das Wachstum stärker fördert als die Knochenreifung.

Ein ebenfalls schöner und über 2 Jahre gleichmäßig anhaltender Erfolg ist in Abb. 6 dargestellt. Es handelt sich um eine Patientin, bei der im Alter von 11 Jahren ein Craniopharyngeom operativ entfernt wurde. Wie üblich hat dieser Eingriff die vorbestehende Hypophyseninsuffizienz noch verschlimmert. Ohne HWH ist er in 21 Monaten 2 cm und unter HWH in 23 Monaten 13,5 cm gewachsen.

Auch der in Abb. 8 dargestellte Patient hatte ein Craniopharyngeom, das mit 3 Jahren radikal operiert wurde. In den folgenden 3 Jahren ist er praktisch gar nicht gewachsen trotz sehr massiver Behandlung mit anabolen Steroiden, ein Beispiel für die erwähnte Resistenz gegenüber anabolen Steroiden bei schwerer Hypophyseninsuffizienz. Unter HWH, das wir mit einem kurzen Unterbruch während mehr als 3 Jahren verabreichten, wurde das Wachstum besser, blieb aber ungenügend. Im Gegensatz zu den beiden in Abb. 5 und 6 dargestellten Patienten flachte es allmählich ab, wurde aber nach einem Therapieunterbruch wieder schneller. Auffallend ist die starke Beschleunigung der Knochenreifung unter HWH trotz vorsichtiger Dosierung der

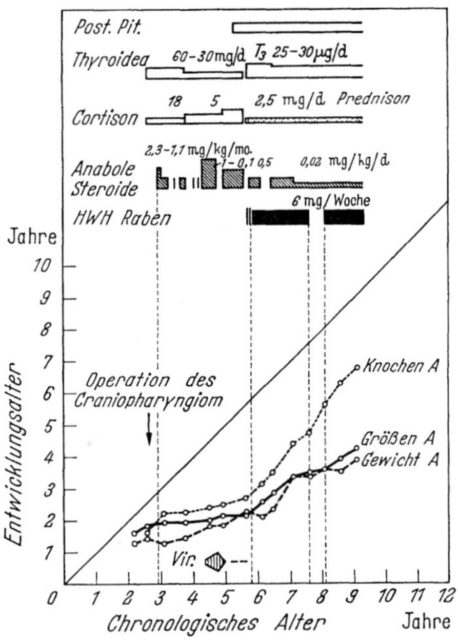

Abb. 8. Die Wirkung von anabolen Steroiden und menschlichem Wachstumshormon vom Typus Raben auf Wachstum und Entwicklung eines Jungen mit hypophysärem Zwergwuchs nach operativer Entfernung eines Craniopharyngeoms (St. René). Knochenalter nach TANNER et al.

anabolen Steroide und der Thyreoidea-Hormone. Abb. 9 zeigt den Verlauf bei diesem Patienten bildlich. Ohne HWH ist er in 37 Monaten 2,5 cm und unter HWH in 41 Monaten 16 cm gewachsen.

Einschließlich der 3 in Abb. 5, 6 und 8 dargestellten Patienten haben wir total 9 hypophysäre Zwerge während mindestens 9 Monaten behandelt. Abb. 10 zeigt die Wachstumsgeschwindigkeit dieser 9 Patienten in cm pro Jahr vor der Behandlung, im ersten Halbjahr, im ganzen ersten Jahr, im 2. Jahr und im 3. Jahr. Wenn wir die drei resistenten Fälle, die spezifische Antikörper gegen HWH entwickelten, zunächst weglassen, so betrug die Wachstumsgeschwindigkeit bei den 6 erfolgreichen Fällen vor der Behandlung im Mittel 3,4 cm, im ersten Behandlungsjahr 8,8 cm und im 2. Behandlungsjahr, wobei für die sichere Beurteilung dieses Behandlungsabschnittes allerdings noch zu wenig Fälle vorliegen, nur noch 5,5 cm.

Außer den hypophysären Zwergen haben wir vier eindeutig nichthypophysäre Zwerge behandelt und einheitlich keinen Erfolg gesehen, so daß wir die Behandlung nach 4 Monaten abbrachen.

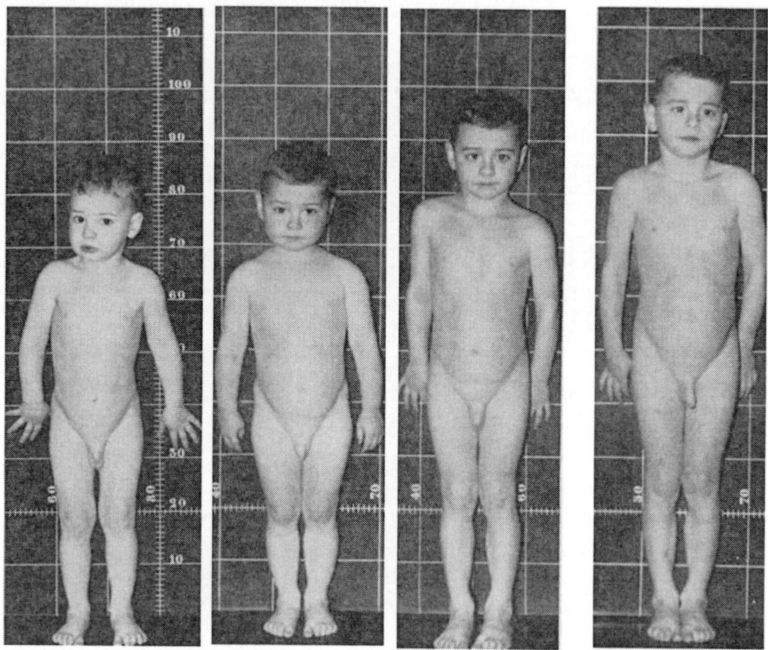

Abb. 9. Die Wirkung von menschlichem Wachstumshormon (HWH) vom Typus Raben. Gleicher Patient wie in Abb. 8. Von links nach rechts im Alter von $2\,^{7}/_{12}$ Jahren (Größe 86 cm, vor HWH), von $5\,^{8}/_{12}$ Jahren (88,5 cm, bei Beginn der HWH-Therapie), von $7\,^{3}/_{12}$ Jahren (99 cm) und von $9\,^{1}/_{12}$Jahren (104,5 cm)

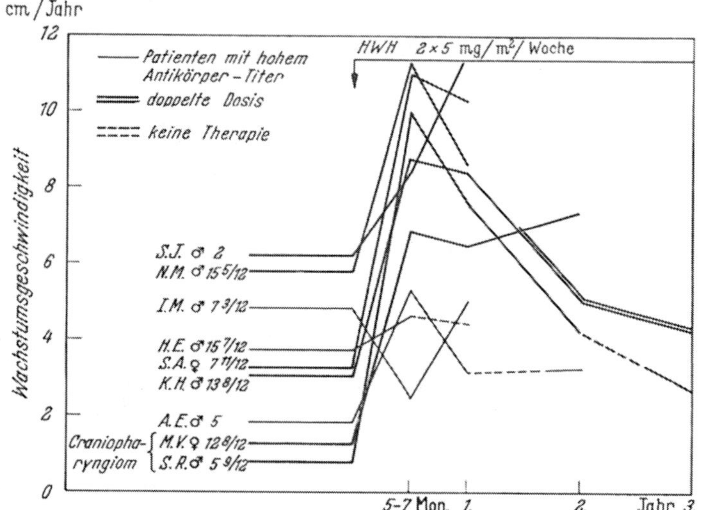

Abb. 10. Die Wachstumsgeschwindigkeit (cm/Jahr) von hypophysären Zwergen vor und während der Behandlung mit menschlichem Wachstumshormon vom Typus Raben in der Standarddosierung von 2×5 mg/m²/Woche

Obwohl unsere Erfahrung, vor allem beim nicht hypophysären Zwergwuchs, noch klein ist, möchten wir deshalb glauben, daß die HWH-Behandlung in der von uns verwendeten Dosierung nur beim hypophysären Zwergwuchs etwas nützt und beim nicht hypophysären Zwergwuchs wirkungslos ist. Dieses Resultat ist nicht überraschend, wenn unsere Dosierung, wie wir annehmen, tatsächlich im unteren physiologischen Bereich liegt und damit bestenfalls eine knapp dosierte Substitutionstherapie darstellt.

5. Standardisierter Stoffwechseltest mit humanem Wachstumshormon

Wenn HWH in dieser Dosierung nur beim hypophysären Zwergwuchs wachstumsfördernd ist, so sollte es möglich sein, in einem HWH-Kurzversuch auf Grund der Stoffwechselwirkung hypophysäre Zwerge von nicht hypophysären Zwergen

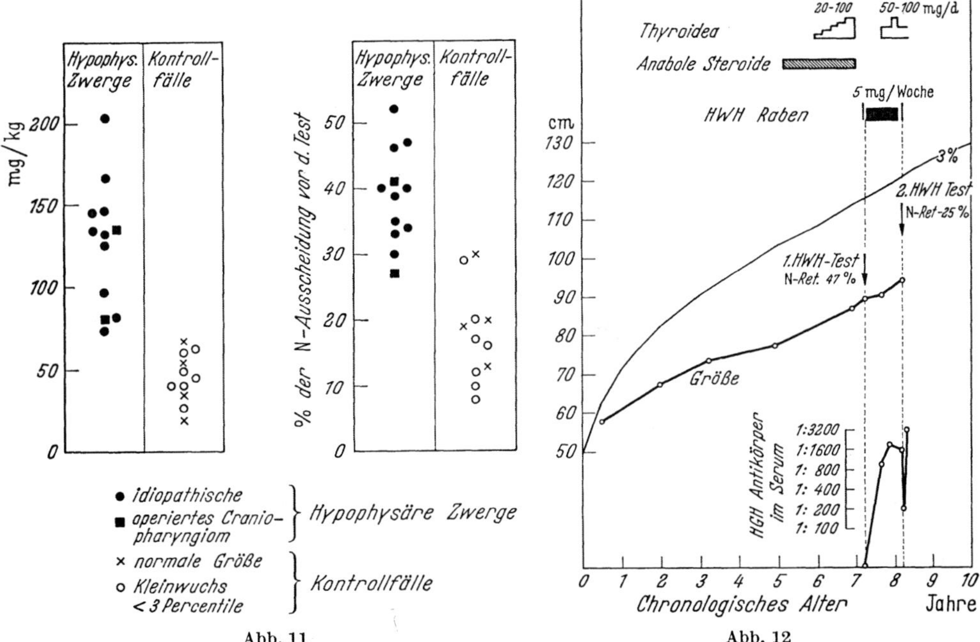

Abb. 11   Abb. 12

Abb. 11. Wirkung von menschlichem Wachstumshormon (HWH) vom Typus Raben auf die tägliche N-Retention im standardisierten HWH-Test (s. Text) bei 23 Kindern. Gemessen ist die Differenz der durchschnittlichen, täglichen N-Ausscheidung im Urin in der Vorperiode und am 2.—6. Tag der HWH-Periode

Abb. 12. Die Entwicklung von Wachstum- und Stoffwechsel-blockierenden spezifischen Antikörpern gegen menschliches Wachstumshormon als Folge der Behandlung mit menschlichem Wachstumshormon vom Typus Raben. Idiopathischer hypophysärer Zwergwuchs (I. Manuel)

zu unterscheiden und voraussagen zu können, in welchen Fällen vom HWH ein Erfolg und in welchen Fällen kein Erfolg zu erwarten ist. Wir haben deshalb einen HWH-Test entwickelt (PRADER, ILLIG und WAGNER), indem der Patient eine frei gewählte, aber streng konstant gehaltene Kost mit etwa 2 g Eiweiß pro kg erhält. Nach Einstellung auf diese Kost wird die tägliche Ausscheidung von N, Kreatin usw. bestimmt. In den 5 ersten Tagen erhält der Patient kein HWH und in den 5 folgenden unsere Standard-Dosis von 2 mg/m²/24 Std. Vor und unter HWH werden die Reaktion auf Insulin und andere Parameter geprüft, auf die ich hier nicht im einzelnen eintreten kann.

Dieser Test wurde bei 12 hypophysären Zwergen und bei 11 Kontrollen durchgeführt. Aus Zeitmangel kann ich nur die Resultate der N-Ausscheidung besprechen. Abb. 11 zeigt, daß die Zunahme der N-Retention, gemessen in mg/kg oder in Prozent der N-Ausscheidung vor dem HGH-Test bei den hypophysären Zwergen deutlich höher ist als bei den Kontrollfällen. Daß die Stoffwechselwirkung von HWH bei hypophysären Zwergen deutlicher ist als bei nicht hypophysären Zwergen, haben schon andere Autoren festgestellt (Lipsett et al. u. a.), doch wurde der Unterschied bisher nie unter streng vergleichbaren Bedingungen geprüft. Wir glauben, damit einen brauchbaren diagnostischen Test und bezüglich der HWH-Wirkung auf das Wachstum einen brauchbaren prognostischen Test geschaffen zu haben.

6. Auftreten von spezifischen Antikörpern gegen humanes Wachstumshormon

Leider haben wir gerade in dieser Beziehung einen unerwarteten Rückschlag einstecken müssen. Von den 9 während 9 Monaten und länger behandelten hypophysären Zwergen, die auf Grund dieses Tests alle einen Wachstumserfolg erwarten ließen, zeigten nämlich nur 6 einen deutlichen Erfolg und 3 gar keinen Erfolg (Abb. 10). Bei diesen resistenten 3 Fällen entwickelten sich in den ersten Behandlungsmonaten hochtitrige spezifische Antikörper gegen HWH vom 7 S-γ-Globulin-Typus, deren immunologische Eigenschaften von Frau Dr. Széky in unsern Laboratorien (Széky, Hässig und Prader) und von Dr. Touber in Amsterdam (Touber, Maingay und de Ruyter) eingehend studiert wurden (Prader, Széky et al.). Klinisch blockieren diese Antikörper die Stoffwechsel- und die Wachstumswirkung von exogenem HWH. Verschiedene Befunde weisen darauf hin, daß sie wahrscheinlich auch gegen endogenes HWH wirksam sind.

Abb. 12 ist ein Beispiel für das Auftreten solcher Antikörper. Kurvenmäßig dargestellt sind das Wachstum und der Titer der hämagglutinierenden Antikörper. HWH bewirkte keine Wachstumsbeschleunigung, löste aber die Entwicklung von spezifischen Antikörpern aus. Während die Zunahme der N-Retention im HWH-Test vor Auftreten der Antikörper 47% betrug, war sie nach Auftreten der Antikörper negativ. Im gleichen Test bewirkte HWH anfänglich eine deutliche Senkung der Kreatin-Ausscheidung und eine Normalisierung der pathologischen Insulin-Belastungskurve, während nach Auftreten der Antikörper diese Wirkung ausblieb (Prader, Széky et al.).

Im Gegensatz zu diesen 3 wachstumsresistenten Fällen entwickelten sich bei den 6 erfolgreich behandelten Fällen auch nach 2—3jähriger Therapie keine oder nur niedrigtitrige Antikörper ohne Stoffwechselauswirkung.

Die therapeutisch induzierte Entwicklung von spezifischen Antikörpern gegen ein körpereigenes Hormon ist ein faszinierendes, allgemein biologisches Problem, dessen Bearbeitung im Tierversuch erst gerade begonnen hat (Renold et al.). Es ist hier zeitlich nicht möglich, auf die damit aufgeworfenen Fragen näher einzutreten.

Zusammenfassung

Bericht über die Wirkung von Testosteron, Thyreoidea sicca und menschlichem Wachstumshormon auf das Wachstum und die Körperentwicklung beim hypophysären Zwergwuchs. Es wird ein standardisierter Kurzversuch mit menschlichem Wachstumshormon empfohlen, in dem hypophysäre Zwerge eine viel stärkere N-Retention aufweisen als nichthypophysäre Zwerge. Für die Dauertherapie wurde ausschließlich menschliches Wachstumshormon vom Typus Raben in der Standard-Dosierung von 2×5 mg/m²/Woche (1 mg = etwa 1,5 USP-E.) verwendet. 4 nichthypophysäre Zwerge wurden 4 Monate behandelt, alle ohne Wachstumserfolg. 9 hypophysäre Zwerge wurden 9—36 Monate lang behandelt. Davon zeigten 6 eine sehr deutliche Wachstumswirkung (im Mittel 3,4 cm/Jahr vor der Behandlung und 8,8 cm/Jahr im ersten Behandlungsjahr) und 3 keinerlei Wachstumswirkung. Bei den 3 resistenten Fällen traten in den ersten Behandlungsmonaten hochtitrige spezifische Antikörper gegen menschliches Wachstumshormon vom 7 S-γ-Globulin-Typus auf, die die Stoffwechsel- und Wachstums-Wirkung von exogenem menschlichem Wachstumshormon blockierten.

Literatur

AARSKOG, D.: Human growth hormone in dwarism since birth. Amer. J. Dis. Child. 105, 86 (1963).

BECK, J. C., E. E. McGARRY, I. DYRENFURTH, R. O. MORGEN, E. D. BIRD, and E. H. VENNING: Primate growth hormone studies in man. Metabolism 9, 699 (1960).

BIERICH, J. R.: Über den hypophysären Zwergwuchs im Kindesalter. Medizinische 38, 1375 (1957).

DAUGHADAY, W. H., and M. L. PARKER: Sulfation factor measurement as an aid in the recognition of pituitary dwarfism. J. clin. Endocr. 23, 638 (1963).

DESAULLES, P. A., and C. KRÄHENBÜHL: Differentiation of action of various anabolic steroids. In: Protein Metabolism, Ciba Symposium 1962.

FRANÇOIS, R., A. FREDERICH, J. BERTRAND, R. GILLY, et MORELON-BACHELOT: Action de l'hormone de croissance humaine (HGH) sur le nanisme hypophysaire. Pédiatrie 18, 49 (1963).

DE GENNES, J. L., et P. ROYER: Réévaluation, à l'occasion de la révision analytique de 26 cas, des aspects cliniques et biologiques des nanismes hypophysaires. Ann. Pédiat. 38, 240 (1962).

HUTCHINGS, J. J., R. F. ESCAMILLA, W. C. DEAMER, and CHOH HAO LI: Metabolic changes produced by human growth hormone (LI) in a pituitary dwarf. J. clin. Endocr. 19, 759 (1959).

KRAYENBÜHL, H., et A. PRADER: Traitement endocrinien des troubles de la croissance chez des malades opérés d'un craniopharyngiome. Neuro-chirurgie 8, 223 (1962).

LIPSETT, M. B., D. M. BERGENSTAL, and F. G. DHYSE: Metabolic studies with human growth hormone in dwarfism and acromegaly. J. clin. Endocr. 21, 119 (1961).

MARTIN, M. M., and L. WILKINS: Pituitary dwarfism: Diagnosis and treatment. J. clin. Endocr. 18, 679 (1958).

PARKER, M. L., R. D. UTIGER, and W. H. DAUGHADAY: Studies on human growth hormone. II. The physiological disposition and metabolic fate of human growth hormone in man. J. clin. Invest. 41, 262 (1962).

PRADER, A.: Le traitement du nanisme hypophysaire. I Coloquio Europeo de Endocrinologia. Endocrinol. Progressus Sectio II, 175 (1961).

— The influence of anabolic steroids on growth. Acta endocr. (Kbh.) (Suppl. 63), 78 (1962).

—, and R. ILLIG: Use of anabolic agents in disorders of growth. In: Protein Metabolism, Ciba-Symposium 1962, p. 383.

— — J. SZÉKY u. H. WAGNER: The effect of human growth hormone in hypopituitary dwarfism. Arch. Dis. Childh. im Druck.

— J. SZÉKY, H. WAGNER, R. ILLIG, J. L. TOUBER u. D. MAINGAY: In Vorbereitung.

RABEN, M. S.: Treatment of a pituitary dwarf with human growth hormone. J. clin. Endocr. 18, 901 (1958).

— Growth hormone. New Engl. J. Med. 266, 31 (1962).

Renold, A. E., J. Steinke, and S. Seeldrer: Immunologic studies with homologous and heterologous pancreatic insulin in the cow. Ciba Colloquium on the Aetiology of Diabetes Mellitus. London: Churchill 1964.

Simpson, M. E., C. W. Asling, and H. M. Evans: Some endocrine influences on skeletal growth and differentiation. Yale J. Biol. Med. **23**, 1 (1950).

Shepard, T. H., S. Waxman, N. Bernstein, and P. Ferrier: Human growth hormone. II. Further studies of its effect on growth in dwarfism. J. Ped. **57**, 363 (1960).

Széky, J., A. Hässig u. A. Prader: Über Antikörper gegen menschliches Wachstumshormon vom Typus Raben bei Patienten mit Zwergwuchs. Helv. paediat. Acta **17**, 411 (1962).

Tanner, J. M., R. H. Whitehouse, and M. J. R. Healy: A new system for estimating skeletal maturity from the hand and wrist, with standards derived from a study of 2600 healthy british children. Part II. The scoring system. In Vorbereitung.

Touber, J. L., D. Maingay, and H. A. de Ruyter: A radio-immuno assay for human growth hormone in serum. Acta med. Neerlandica (im Druck).

Trafford, J. A. P., D. A. Lillicrap, and M. H. Lessof: Human growth hormone in pituitary infantilism. Lancet **1963 I**, 1128.

van der Werff ten Bosch, J. J.: Diskussionsbemerkung in Protein Metabolism, Ciba-Symposium 1962, p. 401.

Vest, M., u. J. Girard: Stoffwechselwirkungen von gereinigtem menschlichem Wachstumshormon beim Zwergwuchs und bei der Anorexia nervosa im Kindesalter. Dtsch. med. Wschr. **87**, 1705 (1962).

Diskussion

C. Overzier (Mainz):

Es wird das Bild eines 20jährigen Mannes demonstriert, der ohne Mißbildungen proportioniert nur 150 cm groß und sexuellinfantil wie ein 10jähriger Knabe war. Da humane Wachstumshormone seinerzeit noch nicht zur Verfügung standen, wurde nur der hypogonadotrope Hypogonadismus behandelt (ein Jahr lang 3×1000 bis 3×3000 IE Primogonyl-Schering wöchentlich ansteigend). Der Effekt vollkommener Ausreifung bei Wachstum um 10 cm, das sich dann noch fortsetzte, wird demonstriert: Wachstum durch endogene Testosteron-Produktion.

Ufer (Berlin):

Die verblüffende chemische Ähnlichkeit von STH und Prolaktin gibt Anlaß, bei hypophysären Zwergen Prolaktin zu bestimmen. Es wäre denkbar, daß durch eine enzymatische Fehlleistung statt STH in solchen Fällen Prolaktin gebildet wird. Die vermehrte Prolaktinbildung brauchte sich klinisch nicht auszuwirken.

E. Werner (Berlin):

Wenn für die Wirkung der anabolen Steroide das STH erforderlich ist, so dürfte dies auch für das Testosteron zutreffen. Wir haben seit mehreren Jahren einen 30jährigen Patienten mit hypophysärem Minderwuchs in Behandlung. Sowohl nach einjähriger Therapie mit Anabolika als auch anschließender fast einjähriger Behandlung mit Testosteron ist der Patient kaum gewachsen und die Skeletentwicklung zeigte kein beschleunigtes Wachstum. Jetzt bekommt er STH. Die Behandlungszeit ist jedoch noch zu kurz, um etwas aussagen zu können.

Aus der Abteilung für Endokrinologie der Medizinischen Universitäts-Poliklinik Heidelberg

Zur Pathogenese des endokrinen Zwergwuchses

Von

Friedrich Bahner

Mit 1 Abbildung

Ich möchte zwei Probleme zur Sprache bringen, die Frage der Ätiologie des hypophysären Zwergwuchses und die Frage, welche Rolle der Hypothalamus beim hypophysären Zwergwuchs spielt. Ätiologisch findet man Craniopharyngeome und andere Tumoren, Entzündungen, Granulomatosen, Hypo- und Aplasien und traumatisch bedingte Infarzierungen der Hypophyse. Es ist aber auffallend, daß keineswegs alle Fälle von hypophysärem Zwergwuchs, die man nach dem Kleinwuchs und nach dem vollständigen Hypogonadismus einwandfrei zum Bild des hypophysären Zwergwuchses zu rechnen hat, deutliche Zeichen der Insuffizienz aller Hypophysenfunktionen haben. Ich möchte jetzt diejenigen Patienten näher betrachten, bei denen nach dem Radio-Jod-Test, dem PBJ, der Steroidausscheidung usw. keine Hypophyseninsuffizienz nachzuweisen ist. Sie mögen eine verminderte TSH- und ACTH-Reserve haben, aber sie haben keine aktuelle Insuffizienz. Dies stimmt auch mit dem klinischen Bild überein, denn eine ganze Anzahl dieser Zwerge sind kräftige, muntere Artisten, was sich nicht mit dem Bild der leichten Schilddrüsen- oder Nebennierenrindeninsuffizienz verträgt, wie wir es von vielen Patienten gut kennen. Worin besteht also der Unterschied zwischen den Zwergen mit nicht faßbarer und denen mit nachweisbarer Hypophyseninsuffizienz?

Hierüber kann man eine Antwort aus einem analogen Tierexperiment versuchen. Reichlin (1960) hat an Ratten Hypothalamusverletzungen mit Wachstumsausfall und anderen peripheren Drüsenausfällen erzeugt. Wenn die Hypophyse anatomisch intakt blieb, war der Wachstumsrückstand niemals so groß wie bei direktem Verlust der Hypophyse. Nun sind uns einige Fälle von hypophysärem Zwergwuchs des Menschen mit anatomisch völlig intakter Hypophyse bei zerstörtem Zwischenhirn bekannt geworden [Goldstein (1928); Berblinger (1931); Foerster, Gagel u. Mahoney (1937)]. Bei diesen Fällen lag ein Craniopharyngeom des Hypothalamus vor, die Hypophyse war intakt, das klinische Bild war das des hypophysären Zwergwuchses mit vollständigem Hypogonadismus. Man hat daher nach den Experimenten von Reichlin gute Gründe, bei den Zwergen ohne Ausfall der peripheren Drüsen vorwiegend suprasellär lokalisierte Prozesse anzunehmen. Ehe ich darauf eingehe, wie man sich die Rolle des Hypothalamus dabei denken kann, möchte ich auf eine besondere Ätiologie hinweisen.

Eine kleinere Anzahl der hierher gehörenden Zwerge hat keinen Hypogonadismus. Fertilität kommt vor. Es ist vielfach üblich, diese Zwerge für primordiale Zwerge zu halten. Nun ist aber der primordiale Zwergwuchs ein wohlumschriebenes und vom hypophysären Zwergwuchs recht klar abzutrennendes Krankheitsbild. Die primordialen Zwerge werden klein geboren, 41—44 cm lang (KIRCHHOFF, LEH-MANN u. SCHÄFER [1954)], im Gegensatz zur normalen Geburtsgröße der hypophysären Zwerge. Bei den primordialen Zwergen tritt die Pubertät zur rechten Zeit ein und sie werden dann fruchtbar. Primordiale Zwerge reagieren auf menschliches Wachstumshormon weder mit Wachstum noch mit N-Retention [LIPSETT, BERGENSTAL u. DHYSE (1961)]. Nun gibt es aber unter den fertilen Zwergen solche, die diese Bedingungen nicht erfüllen. Sie sind bei der Geburt normal groß und ihre Pubertät tritt mit ziemlicher Verspätung ein. Wir hatten Gelegenheit, einen solchen Fall zu beobachten:

Ein 40 Jahre alter Hotelportier war mit 136 cm klein, wenn auch nicht gerade sehr klein. Beide Eltern und 3 Schwestern hatten eine normale Körperlänge. Bei der Geburt war er normal groß, blieb als Kind dann im Wachstum zurück, war mit 14 Jahren 114 cm groß und trat mit etwa 18 Jahren in die Pubertät ein. Später heiratete er und hatte 1 Kind. Nach der Ausbildung der primären und sekundären Geschlechtsmerkmale, nach den normal großen Hoden konnte er der Vater des Kindes sein. Anläßlich einer Leistenbruchoperation starb er in der Narkose. Die Obduktion ergab eine Mißbildung an der Hypophyse, eine Verlagerung des Hypophysenhinterlappens an das Tuber cinereum. Diese Dystopie des Hinterlappens ist zuerst von PRIESEL (1920) beschrieben worden. Dieser Autor [PRIESEL (1927)] hat innerhalb von 8 Jahren 6 solcher Fälle mitgeteilt. Weitere Fälle stammen von APITZ (1938); BIEMOND u. HARTZ (1935); BOSISIO (1931); CORONINI zit. nach PRIESEL (1927); HAMPERL (1928); HEDINGER u. HÜRZELER (1957); LENNOX u. RUSSELL (1951) (2 Fälle); KRAUS (1926 u. 1932) (2 Fälle); LHERMITTE u. ROEDER (1922); RATZENHOFER (1938). Mit dem hier mitgeteilten Fall sind 19 Fälle bekannt geworden. Die Störung dürfte häufiger sein, als es dieser kleinen Kasuistik entspricht.

Eine Abbildung von PRIESEL (1927) zeigt die möglichen Varianten der Dystopia tuberalis (1), Dystopia infundibularis (2) und Dystopia opercularis (3):

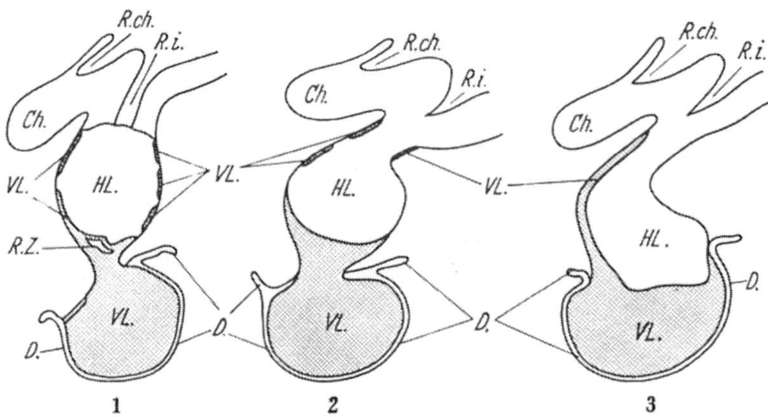

Abb. 1. Varianten der Hypophysenhinterlappen-Dystopie (nach PRIESEL, 1927)

Einige dieser Kranken waren äußerlich unauffällig, einige waren Zwerge mit Hypogonadismus. Ob ein Zwergwuchs dabei entsteht oder nicht, hängt wahrscheinlich von der Ausbildung des verbleibenden neurohypophysären Kontaktes ab, denn APITZ fand einen Fall, bei dem bei einer Dystopia infundibularis durch ein hinzukommendes Craniopharyngeom der Kontakt zwischen Vorderlappen und Hinterlappen völlig zerstört war. Dieser Mann von 46 Jahren hatte einen vollständigen Hypogonadismus, obgleich sowohl das hypothalamische Kerngebiet als auch der Hypophysenvorderlappen intakt waren und nur der neurohypophysäre Kontakt zerstört war.

Bei dem von uns beobachteten 40 Jahre alten, kleinwüchsigen, fertilen Mann, dessen pathologisch-anatomische Untersuchung ich Herrn Prof. Dr. DIEZEL vom Pathologischen Institut der Universität Heidelberg verdanke, war bei vorliegender Dystopia tuberalis der Vorderlappen in deutlichem neurohypophysärem Kontakt mit der Unterfläche des dystopen Hinterlappens. Diese Kontaktfläche entspricht dem sonstigen distalen neurohypophysären Kontakt zwischen Vorderlappen und Hinterlappen. Der proximale neurohypophysäre Kontakt zwischen einer Pars infundibularis und dem Tuber cinereum war aber nicht vorhanden. Ob der Kleinwuchs hier durch den veränderten neurohypophysären Kontakt entstanden ist oder ob im Zuge der Dystopia tuberalis Veränderungen in den hypothalamischen Kernen bestehen, muß offenbleiben. Jedenfalls beweist der Fall, daß die Prieselsche Dystopie der Abart des fertilen hypophysären Zwergwuchses zugrunde liegen kann. Ähnlich war der Fall von KRAUS (1932) eines 138 cm großen Mannes.

Wie alle diese Spielarten des hypophysären Zwergwuchses, nämlich der rein hypothalamische, der rein hypophysäre und der bei gestörtem neurohypophysärem Kontakt, auf Wachstumshormon reagieren, bleibt abzuwarten. Unterschiede des Ansprechens sind denkbar.

Überhaupt muß man sich im klaren sein, daß das Wachstumshormon mehr als nur Wachstum, also Ausdehnung in die Länge oder in die Breite durch Vermehrung der vorhandenen lebenden Substanz bewirkt. Gewiß steht etwa vom 3. Lebensjahr bis wenigstens zum Beginn der Pubertät besonders augenfällig das Wachstum unter dem Einfluß des Wachstumshormons, während danach die Sexualhormone den Wachstumsschub übernehmen. Nun ist aber ein 10jähriger Junge keineswegs nur eine größere Ausgabe eines 3jährigen. Dazwischen liegen Gestaltwandel und Funktionswandel, also ein Reifungsprozeß außerhalb der Genitalsphäre. Ich erinnere an den unverkennbaren Unterschied im Gesichtsschnitt eines etwa 20jährigen hypophysären Zwerges und einer gleichaltrigen Gonadendysgenesie. Diese hat keine infantilen Züge mehr; Mund, Nase und Augenpartie sind erwachsener, reifer als beim hypophysären Zwerg. Dies kann nicht die Wirkung der Sexualhormone sein, denn diese sind bei der Gonadendysgenesie nicht vorhanden. Der Unterschied zwischen dem Gesichtsausdruck einer Gonadendysgenesie und dem eines hypophysären Zwerges dürfte ein Ausdruck der vorpuberalen Reife sein, die unter der Wirkung des Wachstumshormons eintritt. Dafür spricht auch, daß die Ossifikation bei der Gonadendysgenesie immer einer recht fortgeschrittenen Reife entspricht.

SECKEL (1960) hat darauf hingewiesen, daß das Wachstumshormon in der Zeit nach dem 3. Lebensjahr den dann beginnenden Prozeß der endomitotischen Zellteilung fördert, wobei neue Kernklassen von mehrkernigen und von polyploiden Zellen in zahlreichen Organen, z. B. in der Leber auftreten. Hier zeigt sich das Wachstum nicht nur in der Vermehrung, sondern auch in der Differenzierung der lebenden Substanz. Es erscheint mir daher berechtigt, das Wachstumshormon als das kindliche Reifungshormon anzusehen, und wenn wir seine Wirkung auf die Körperlänge messen, dann stellen wir nur seine vergrößernde, aber nicht seine differenzierende Wirkung fest.

Von der Wirkung der Sexualhormone in der Pubertät ist uns Analoges ganz geläufig. Die Wirkung der Sexualhormone auf das Wachstum, auf die sekundären Geschlechtsmerkmale und auf Reifungsvorgänge wie den Schluß der Epiphysenfugen kann man nur als einen einheitlichen Reifungsprozeß ansehen, der zwar langsamer oder schneller, aber nur stereotyp in einer Richtung und nur in einer ganz bestimmten Aufeinanderfolge von Einzelschritten ablaufen kann. Das genaue Studium der Pubertätsentwicklung, wie es etwa von Tanner (1955) dargestellt wird, zeigt deutlich, daß die einzelnen zu beobachtenden Reifungsvorgänge ihren ganz festen Platz im Gesamtablauf des Reifungsprozesses haben. So haben Tonutti u. Weller (1960) wie auch wir beobachtet, daß die Gonadotropine auftreten, wenn ein bestimmtes Knochenalter erreicht ist. Behandelt man bei einer Reifungsverzögerung mit Testosteron, dann erscheinen Harngonadotropine in einem bestimmten Stadium der Knochenkernentwicklung. Wir beobachteten 2 Männer mit angeborenem adrenogenitalem Syndrom, bei denen nach anfänglicher Pseudopubertas praecox sich schließlich eine echte Pubertas praecox entwickelte, indem unter der Wirkung der Nebennierenrindenandrogene nach der anfänglichen Vergrößerung der sekundären Geschlechtsmerkmale und der Genitalien schließlich die Gonadotropinausschüttung und die Hodenvergrößerung eintraten. Beide Männer haben mit Sicherheit Kinder gezeugt. Auch an der Ratte konnten wir eine echte Pubertas praecox durch Testosteron erzeugen [Bahner u. Schwarz (1959)]. Diese Beispiele sollen deutlich machen, daß auch in der Pubertätsreifung viele einzelne Reifungsschritte zu einem stereotypen Gesamtgeschehen zusammengefaßt sind. Analoges müssen wir für die kindliche Entwicklung annehmen, ehe die Sexualhormone in Tätigkeit treten.

Nehmen wir die kindliche Reifungsphase und die puberale Reifungsphyse zusammen, dann erscheinen sie mir wie zwei irreversible Entwicklungsphasen, zwei Metamorphosen, so wie das Raupen- und Puppenstadium des Schmetterlings. Vieles spricht dafür, daß der Hypothalamus im Gegensatz zur Hypophyse selbst eine integrierende Tätigkeit hat, und so drängt sich nun der Gedanke auf, daß der Hypothalamus der Ort ist, von dem aus die integrierten Reifungsvorgänge ausgelöst werden. Ein intakter Hypothalamus stimuliert einen wohlgeordneten, harmonischen Ablauf der Reife. Ein Ausfall dieses „Reifezentrums" läßt die Reife auf frühkindlicher Stufe stehenbleiben, aber der resultierende Infantilismus ist harmonisch, d. h. Körpergröße und Reifungsgrad z. B. des Skelets entsprechen einander. Die peripheren Drüsen brauchen dann nicht nachweisbar insuffizient zu sein. Ist dagegen die Hypophyse selbst beschädigt, so hat man einen dysharmonischen Infantilismus, ein Durcheinander verschiedener Entwicklungsstufen [Apitz (1938)] zu erwarten, weil bei erhaltenem hypothalamischem Reifeantrieb die einzelnen Hypophysenfunktionen mit dem Fortschreiten des pathologischen Prozesses im Laufe der Zeit nacheinander aussetzen oder je nach der Struktur des Hypophysenrestes unterschiedlich stimuliert werden. Störungen der Funktion der peripheren Drüsen werden dann nachweisbar sein.

Zusammenfassung

Es wird ein Fall eines fertilen hypophysären Zwerges beschrieben, bei dem eine Prieselsche Dystopia tuberalis des Hypophysenhinterlappens vorlag.

Das Wachstum im Kindesalter beruht nicht nur auf einer Vergrößerung, sondern auch auf einer Differenzierung der Gewebe und der Gestalt. Das Wachstumshormon stimuliert nicht nur die Körperlänge, sondern auch den kindlichen Reifungsprozeß. Der Antrieb für die kindliche Reifung geht — wie für die spätere puberale Reifung — vom Hypothalamus aus. Hypophysäre Zwerge mit Zerstörung der Hypophyse haben deutliche Ausfälle der peripheren Drüsen. Beim hypophysären Zwergwuchs mit harmonischem Infantilismus ohne gleichzeitig nachweisbarer Schilddrüsen- und Nebennierenrindeninsuffizienz liegen, wie aus Tierexperimenten zu folgern ist, eher Hypothalamus- als Hypophysenschäden vor.

Literatur

APITZ, K.: Zur Pathogenese des Hypophysenkleinwuchses. Virchows Arch. path. Anat. **302**, 555 (1938).

BAHNER, F., u. G. SCHWARZ: Experimentelle Pubertas praecox bei Ratten durch Testosteron. Acta endocr. (Kbh.) **30**, 574 (1959).

BERBLINGER, W.: Zur Kenntnis des pituitären Kleinwuchses. Beitr. path. Anat. 87, 233 (1931).

BIEMOND, A., u. P. H. HARTZ: Proc. kon. med. Akad. Wet. **38**, 105 (1935).

BOSISIO, E.: Beitrag zur pathologischen Anatomie der menschlichen Hypophyse. Arch. ital. Anat. Istol. pat. **2**, 461 (1931).

CORONINI, C.: Zit. nach PRIESEL (1927).

FOERSTER, O., O. GAGEL u. W. MAHONEY: Vegetative Regulationen. Verh. dtsch. Ges. inn. Med. **49**, 165 (1937).

GOLDSTEIN, K.: Kasuistische Mitteilungen zur Klinik und pathologischen Anatomie der Nervenkrankheiten. Dtsch. Z. Nervenheilk. **103**, 225 (1928).

HAMPERL, H.: Über einen Fall von Dystopie der Neurohypophyse. Zbl. Path. **41**, 97 (1928).

HEDINGER, CHR., u. D. HÜRZELER: Hypopituitarismus bei Dystopie des Hypophysenhinterlappens. Acta endocr. (Kbh.) **14**, 170 (1953).

LENNOX, B., and D. S. RUSSELL: Dystopia of the neurohypophysis. J. Path. Bact. **63**, 485 (1951).

LHERMITTE et ROEDER: Diabète glykosurique. Presse médi. **30**, 615 (1922).

LIPSETT, M. B., D. M. BERGENSTAL, and F. G. DHYSE: Metabolic studies with human growth hormone in dwarfism and acromegaly. J. clin. Endocr. **21**, 119 (1961).

KIRCHHOFF, H. W., W. LEHMANN u. U. SCHÄFER: Klinische, erbbiologische und körperbauliche Untersuchungen bei primordialen Zwergen. Z. Kinderheilk. **75**, 243 (1954).

KRAUS, E. J.: Handbuch der speziellen pathologischen Anatomie und Histologie von HENKE-LUBARSCH 8, 828 (1926).

— Die morphologischen Veränderungen der menschlichen Hypophyse nach Zerstörung der Zwischenhirnbasis bzw. des Hypophysenstiels und deren Folgen. Virchows Arch. path. Anat. **286**, 667 (1932).

PRIESEL, A.: Ein Beitrag zur Kenntnis des hypophysären Zwergwuchses. Beitr. path. Anat. **67**, 220 (1920).

— Über die Dystopie der Neurohypophyse. Virchows Arch. path. Anat. **266**, 407 (1927).

RATZENHOFER, M.: Dystopia totalis cranialis hypophyseos. Virchows Arch. path. Anat. **301**, 17 (1938).

REICHLIN, S.: Growth and the hypothalamus. Endocrinology **67**, 760 (1960).

SECKEL, H. P. G.: Concepts relating the pituitary growth hormone to somatic growth of the normal child. Amer. J. Dis. Child. **99**, 349 (1960).

TANNER, J. M.: Growth at Adolescence. Oxford: Blackwell 1955.

TONUTTI, E., O. WELLER, E. SCHUCHARDT u. E. HEINKE: Die männliche Keimdrüse. Stuttgart: Thieme 1960.

Aus der Abteilung für Endokrinologie (Leiter: Prof. F. Bahner) der Medizinischen Universitäts-Poliklinik Heidelberg

Zur Pathogenese des Kleinwuchses bei Pseudohypoparathyreoidismus und bei Pseudo-Pseudohypoparathyreoidismus

Von

Gerhard Schwarz

Mit 2 Abbildungen

Es wird heute noch oft die Meinung vertreten, daß es bei der Nebenschilddrüsenunterfunktion eine Verminderung des Längenwachstums gibt. Diese Ansicht ist nicht richtig. Bei Untersuchung eigener und aller erreichbaren Fälle der Literatur konnten wir feststellen, daß die Nebenschilddrüsenunterfunktion im allgemeinen keine Verminderung des Längenwachstums nach sich zieht, auch wenn

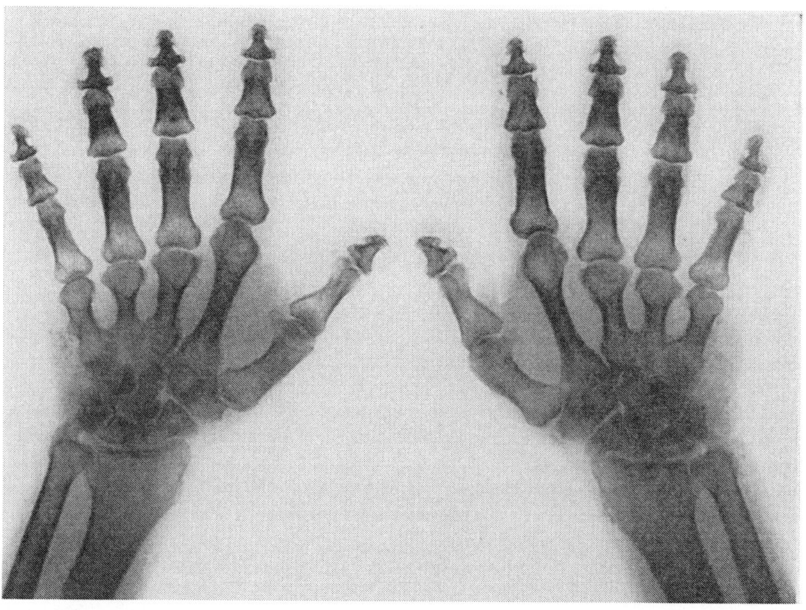

Abb. 1. Eine symmetrische Brachymetacarpie III, IV, V bei einem Fall von Pseudohypoparathyreoidismus. Die Deformierung der verkürzten Köpfchen und die unregelmäßige Begrenzung der Interphalangealgelenke sind deutlich zu erkennen

manifeste Symptome einer hypocalciämischen Tetanie vor Abschluß des Wachstums bestanden haben.

Nun gibt es aber eine Erkrankung, die die klinischen Symptome der Nebenschilddrüseninsuffizienz, also Tetanie mit Hypocalciämie und Hyperphosphatämie, hat und bei der regelmäßig ein Kleinwuchs beobachtet wird. Es handelt sich um den Pseudohypoparathyreoidismus (PH), über den ich in dieser Gesellschaft schon mehrfach berichtet habe. Die Beobachtung dieser Fälle — es sind etwa 80 in der Weltliteratur beschrieben — hat zu der obenerwähnten falschen Ansicht geführt, und ich möchte im Folgenden zeigen, daß der Kleinwuchs des PH nichts mit einer Unterfunktion der Nebenschilddrüsen zu tun hat.

Es wird das Bild eines typischen Falles demonstriert. Der Patient ist 36 Jahre alt, er hat eine Calciumkonzentration von 6,4 mg-% und eine Phosphatkonzentration im Serum von 7,4 mg-%. Seit dem 14. Lebensjahr besteht eine manifeste Tetanie und vor zwei Jahren wurde eine doppelseitige tetanische Katarakt extrahiert. Der Patient ist 152 cm groß — seine Körpergröße liegt damit weit unterhalb der Durchschnittsgröße gesunder Familienmitglieder —, er hat ein typisches Rundgesicht, und seine Extremitäten sind in Relation zur Gesamtlänge verkürzt. Sein Kleinwuchs ist also dysproportioniert. Die Dysproportion im Wachstum zeigt sich besonders deutlich an den Händen, die kurz und plump sind, weil die Mittelhandknochen verkürzt sind (Abb. 1). Die gleichen physischen Merkmale gibt es auch ohne Hypocalciämie und ohne Hyperphosphatämie, und dieses Krankheitsbild hat die umständliche Bezeichnung Pseudo-Pseudohypoparathyreoidismus (PPH) erhalten. Die folgende Tabelle zeigt die Beziehungen von Hypocalciämie und Hyperphosphatämie zum Wachstum im Schema:

Tabelle 1

	Serum-Ca	Serum-P	Kleinwuchs	Skeletanomalien
idiopathischer Hypoparathyreoidismus.	↓	↑	∅	∅
PH.	↓	↑	±	+
PPH.	normal	normal	+	+

Auf der einen Seite gibt es Kleinwuchs und Skeletanomalien mit und ohne Hypocalciämie und Hyperphosphatämie, und auf der anderen Seite gibt es bei Hypocalciämie und bei Hyperphosphatämie sowohl normales Wachstum als auch Kleinwuchs und Skeletanomalien.

Worauf beruht nun aber der Kleinwuchs von PH und PPH, nachdem eine Nebenschilddrüseninsuffizienz dafür nicht verantwortlich ist? PH und PPH sind, im Gegensatz zur echten Nebenschilddrüsenunterfunktion, genetische Erkrankungen, und bei beiden ist das gleiche Gen affiziert. Gleicher Gendefekt bedeutet gleicher basaler Stoffwechseldefekt. In diesem Sinne bilden PH und PPH also eine nosologische Einheit. Faßt man die Symptomatologie beider zusammen, so besteht obligat ein Kleinwuchs und bestimmte Skeletanomalien, und fakultativ bestehen Veränderungen der Serummineralkonzentrationen.

Das am meisten spezifische Symptom der Erkrankungen ist die Brachymetacarpie und die Brachymetatarsie. Dieses Symptom weist auf eine umschriebene Störung der epiphysären Ossifikation an den Mittelhand- und Mittelfußknochen hin. Der Kleinwuchs, d. h. die generalisierte Störung des Längenwachstums der Knochen, weist andererseits auf eine allgemeine Störung der epiphysären Ossifikation hin. An Hand von 10 ausgewählten Röntgenbildern, die von 15 eigenen Fällen stammen, wird gezeigt, daß die Störung der epiphysären Ossifikation nur scheinbar umschrieben ist, daß sie in Wirklichkeit aber alle Epiphysenfugen betrifft. Zunächst werden eine Reihe von Röntgenaufnahmen der Hände gezeigt und demonstriert, daß die Metacarpalia nicht nur verkürzt sind, sondern daß die Köpfchen der verkürzten (und auch der nicht verkürzten) Metacarpalia aufgetrieben sind und daß die Gelenkflächen abgeflacht und unregelmäßig begrenzt sind. Die gleichen Veränderungen zeigen auch die Phalangealgelenke (Abb. 1). Nach den Händen werden 4 Beckenübersichtsaufnahmen von PH und PPH-Fällen gezeigt, und hier findet man alle Schweregrade der „Hüftgelenksdysplasie". Leichte Deformierungen der Hüftköpfe, ohne Beeinträchtigung der Funktion bis

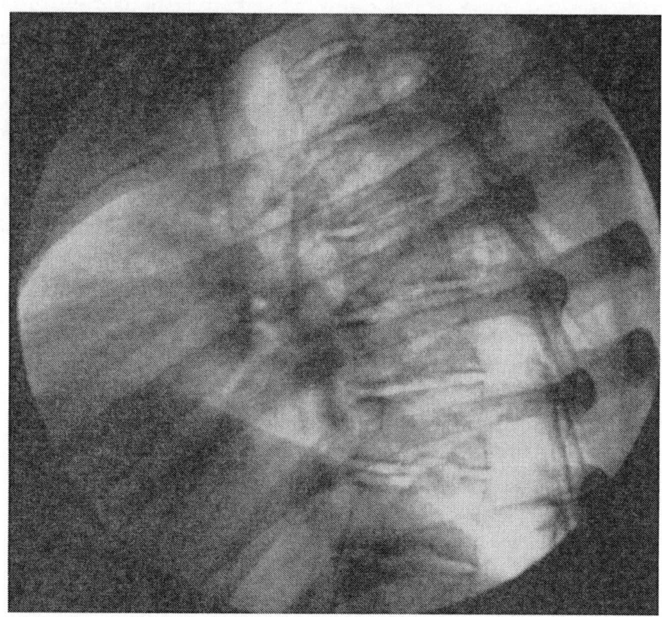

Abb. 2. Wirbelsäule mit wellig begrenzten Deckplatten eines Falles von Pseudo-Pseudohypoparathyreoidismus

zu schweren Deformierungen mit Sekundärarthrosen kann man antreffen. Schließlich kann man die Folgen der generalisierten Störung der epiphysären Ossifikation gelegentlich auch an den Wirbelkörpern erkennen. Die Abb. 2 zeigt die wellige Begrenzung der Wirbelkörper, die in der Form den Gelenkveränderungen an anderen Skeletabschnitten ähnlich ist. Dieses Bild stammt von einem Fall, der in der Kindheit offenbar Mineralveränderungen hatte, denn er war beiderseits kataraktextrahiert; als wir ihn im Erwachsenenalter sahen, waren seine Serummineralwerte normal.

Diese Röntgenbilder sollen zeigen, daß bei PH und PPH eine generalisierte Störung der epiphysären Ossifikation besteht, durch die Kleinwuchs und Skeletanomalien hervorgerufen werden. Metacarpalia und Metatarsalia werden nur bevorzugt von dieser im Prinzip generalisierten Störung befallen. Man kann den Kleinwuchs des PH und des PPH infolgedessen auch als „dysostotischen Kleinwuchs" bezeichnen.

Über die genetische Störung im Mineralstoffwechsel ist wenig bekannt, und völlig unbekannt ist es, wie diese Störung den Prozeß der epiphysären Ossifikation beeinflußt. Nur an Fällen, die Skeletaffektionen zusammen mit Hypocalciämie und Hyperphosphatämie hatten, wurde die Störung genauer studiert. Hier kann man feststellen, daß primär der Phosphatstoffwechsel gestört ist. Gelingt es nämlich, die erhöhte Phosphatkonzentration zu senken, so normalisiert sich die Calciumkonzentration von selbst. Es wird die Kurve eines Patienten mit PH gezeigt, bei dem es gelang, durch Benemid die Phosphatkonzentration zu senken, und der daraufhin auch eine normale Calciumkonzentration bekam, obgleich es vom Benemid bekannt ist, daß es nur die erhöhte Phosphatkonzentration senkt, aber keinen direkten Einfluß auf die Calciumkonzentration hat.

Hält man nun die Mineralkonzentrationen bei PH über lange Zeit normal (z. B. durch Benemid-Therapie), so kann man feststellen, daß die zuvor vorhandene Parathormonresistenz verschwindet. Die Parathormonresistenz hängt also mit den veränderten Mineralkonzentrationen zusammen. Ihre Verknüpfung geschieht über einen sekundären Hyperparathyreoidismus. Eine Parathormonresistenz im Test mit Parathormon besteht nämlich auch dann, wenn ein Organismus schon unter maximaler endogener Parathormonwirkung steht. Es wird ein Bild demonstriert, aus dem die Reaktion auf exogenes Parathormon bei verschiedener endogener Parathormonsekretion hervorgeht. Auch bei primärem und bei sekundärem Hyperparathyreoidismus gibt es eine „Parathormonresistenz". Bei PH besteht aber ein „sekundärer Hyperparathyreoidismus". Einmal findet man häufig Hyperplasien der Nebenschilddrüsen; und zum anderen sind in den letzten Jahren eine Reihe von PH-Fällen bekannt geworden, bei denen am Knochen Zeichen einer Osteitis fibrosa cystica vorhanden war. Es gibt also Befunde bei PH, die dafür sprechen, daß hier eine Überfunktion der Nebenschilddrüsen vorliegt.

Fassen wir zusammen, so besteht bei PH und bei PPH eine in ihren Einzelheiten unbekannte genetische Störung im Phosphatstoffwechsel. Diese führt durch Wirkungen am Skelet obligat zu Störungen der epiphysären Ossifikation und dadurch zum Kleinwuchs. Sie führt durch Wirkungen an der Niere fakultativ zur Hyperphosphatämie. Die Hyperphosphatämie löst über eine Senkung des Serumcalciums einen sekundären Hyperparathyreoidismus aus, der eine „Parathormonresistenz" vortäuscht. Sie ersehen aus dieser Darstellung, daß bei Pseudohypoparathyreoidismus gar keine Nebenschilddrüseninsuffizienz vorhanden ist. Diese Fälle können also nicht als Beweis dafür gelten, daß Nebenschilddrüsenunterfunktionen Veränderungen im Wachstum hervorrufen.

Ausführliche Literatur bei
SCHWARZ, G.: Pseudohypoparathyreoidismus und Pseudo-Pseudohypoparathyreoidismus (hereditärer brachymetacarpaler Kleinwuchs). Heidelberg: Springer 1964.

Forschungslaboratorium der Farmavigor S.p.A., Sesto S. Giovanni, Mailand, Italien.

Über die Wirkung des Wachstumshormons auf die Nebennierenfunktion von Ratten nach Hypophysektomie

Von

E. Mascitelli-Coriandoli und P. Lanzani

Bei Tieren kommt es nach Hypophysektomie bekanntlich zu einer Abnahme des Körpergewichts, des Eiweißgehaltes und der Synthese der Gewebseiweißstoffe, die das Wachstumshormon (GH) wieder ausgleichen kann (*1, 2*). Zusammen mit diesen Wirkungen vermag die Hypophysektomie eine Atrophie der Nebennieren hervorzurufen, der durch das Adrenocorticotropin (ACTH) vorgebeugt werden kann, welche andererseits das GH wenig zu beeinflussen vermag. Versuche mit gleichzeitiger Anwendung beider Hormone haben hingegen erwiesen, daß das GH die Wirkung des ACTH verstärkt, jedoch nicht die Konzentrationen der Corticosteroide im Plasma beeinflußt (*3, 4*).

Diese Ergebnisse haben uns Anregung zu vorliegender Arbeit gegeben, die zur Bewertung der Wirkungen des GH allein oder zusammen mit dem ACTH auf die Funktion der Nebennierenrinde von Tieren dienen soll, welche der Hypophysektomie unterzogen worden waren. Als Merkmale dieses Effekts wurden neben dem Nebennierengewicht und ihrer Fähigkeit, Corticosteroide zu synthetisieren, die Eiweißkonzentrationen, jene des Coenzyms A und die Glucose-6-Phosphatasewirkung geprüft.

1. Methodik

Rattenmännchen des Sprague-Dawley-Stammes mit einem durchschnittlichen Gewicht von 75—85 g wurden der Hypophysektomie unterzogen und in vier Gruppen geteilt. Die erste Gruppe diente zur Kontrolle und wurde mit täglichen Injektionen von physiologischer NaCl-Lösung behandelt (0,5 ml/Ratte); eine weitere Gruppe bekam tägliche intraperitoneale Injektionen von 10 mg/kg GH[1]; eine dritte Gruppe wurde mit subcutanen Injektionen von 10 E/kg ACTH[2] behandelt. Der vierten Gruppe wurden beide Hormone in Abständen von 1—2 Std einverleibt.

Nach 10 Tagen wurden die Tiere mit einer Gruppe von Ratten getötet, die nicht der Hypophysektomie unterzogen worden waren. Die Nebennieren wurden aseptisch entnommen, vom Fett befreit und gewogen. Die linke Nebenniere jeden Tieres wurde darauf homogenisiert; an einem Anteil derselben wurde der Cortico-

---

[1] N-% = 15,5; S-% = 1,0. s. Wilhelmi, A. E. (1955): in Gaebler, H., R. W. Smith, and C. H. Long: Hypophyseal growth hormone; Nature and Action. New York: McGraw-Hill, 1961.

[2] ACTH Ormonoterapia Richter S.p.A., Mailand.

steroidgehalt nach der Silberschen (*5*), von HILF u. Mitarb. abgeänderten Methode bestimmt (*6*); man schritt darauf zur Inkubation nach SCHÖNBAUM (*7*). Am Ende der Inkubationszeit wurden die Corticosteroide quantitativ bestimmt; dem Unterschied Rechnung tragend, wurde dann die in der Zeiteinheit synthetisierte Hormonmenge errechnet.

Die rechte Nebenniere wurde hingegen bei allen Tieren zur Bestimmung der Eiweißsubstanzen nach der Lowryschen Methode (*8*), zur Bestimmung des Coenzym A nach der Methode von KAPLAN und LIPMANN in der Bonomischen Abänderung (*9*) und zur Bestimmung der Glucose-6-Phosphatase nach SWANSON benützt (*10*).

Tabelle 1. *Wirkung von Wachstumshormon und ACTH auf die Nebennieren*

Parameter	Normal (12)[1]	Kontrolle (11)	GH (11)	ACTH (12)	GH + ACTH (12)
Körpergewicht g	134,6 ± 6,1[2]	78,1 ± 4,0	94,0 ± 5,1	76,5 ± 3,9	122,8 ± 5,5
Nebennieren mg/100 g . . .	37,1 ± 1,7	10,8 ± 0,5	12,4 ± 0,9	23,0 ± 1,1	27,7 ± 1,4
Protein mg/g	82,0 ± 3,8	60,7 ± 3,0	69,8 ± 3,1	75,5 ± 3,3	79,1 ± 3,9
Coenzym A Einh./mg Protein	1,37 ± 0,08	0,78 ± 0,04	0,83 ± 0,04	1,26 ± 0,06	1,18 ± 0,05
Glucose-6-Phosphatase g P/mg Protein/Std . . .	6,14 ± 0,22	4,01 ± 0,14	3,96 ± 0,12	5,43 ± 0,21	5,28 ± 0,18
Synthese der Corticoide g/mg Protein/Std . . .	1,56 ± 0,09	0,81 ± 0,04	0,77 ± 0,03	1,30 ± 0,07	1,12 ± 0,06

[1] Anzahl der Ratten.
[2] SD der Mittelwerte.

2. Ergebnisse und Besprechung

Die in der Tabelle zusammengefaßten Ergebnisse bestätigen, daß das GH eine Zunahme des Körpergewichtes der Tiere nach Hypophysektomie bewirkt, während das ACTH inaktiv ist; die gleichzeitige Zufuhr der zwei Hormone erzeugt einen Effekt, der bei weitem dem nach bloßer ACTH-Zufuhr beobachteten überlegen ist.

Das Gewicht der Nebennieren wird hingegen durch das GH wenig beeinflußt, während es durch ACTH gegenüber den Kontrolltieren mehr als verdoppelt wird. Die Verabreichung beider Hormone ruft einen weiteren Gewichtszuwachs der Drüsen hervor.

Die Eiweißkonzentration in den Nebennieren wird durch GH-Zufuhr um 14,9% gesteigert, um 23,7% durch ACTH-Behandlung; bei den mit beiden Hormonen behandelten Tieren erreicht die Steigerung 30,2%. Bezüglich des Coenzym A ist zu bemerken, daß seine Konzentrationen bei den mit GH behandelten Tieren unverändert bleiben, aber um 61,5% bei den mit ACTH behandelten Tieren zunehmen. Bei Ratten, die beide Hormone bekamen, findet man Konzentrationen, die fast jenen gleichkommen, die mit ACTH allein behandelt wurden. Analoge Ergebnisse beobachtet man bezüglich der Glucose-6-Phosphatasewirkung.

Die Bildung von Corticosteroiden ist ihrerseits gegenüber der Behandlung mit GH empfindlich; nach ACTH-Verabreichung steigt sie um 60% an. Die gleichzeitige Zufuhr der beiden Hormone bessert nicht die mit dem ACTH allein erzielten Ergebnisse.

Diese Ergebnisse sprechen für eine parallele Steigerung der Synthese der Corticosteroide und der Eiweißsynthese bei mit ACTH behandelten Tieren (*11*); es muß jedoch hervorgehoben werden, daß die Steigerung der Hormonsynthese bei weitem jene der Eiweißkonzentrationen übertrifft. Die gleichzeitige Verabreichung des GH steigert zwar die Eiweißsynthese, ruft aber keine weitere Steigerung der Synthese der Corticosteroide hervor. Es spricht dies wahrscheinlich im Sinne einer biosynthetischen Anregung verschiedener Eiweißarten seitens der beiden Hormone.

Abschließend kann man sagen, daß, während das ACTH auf die Nebenniere einwirkt, das GH nur auf die Synthese einiger Eiweißsysteme einwirken dürfte, die sich nicht an der Synthese der Corticosteroide beteiligen und folglich nicht imstande ist, sie anzuregen.

Literatur

1. Klein, E.: Dtsch. med. Wschr. 82, 484 (1957).
2. Korner, A.: In: F. Gross, Protein Metabolism: An International Symposium, 8. Berlin-Göttingen-Heidelberg: Springer-Verlag 1962.
3. Lostroh, A. J.: Proc. Soc. exp. Biol. (N. Y.) 97, 828 (1958).
4. Longcope, C., and J. W. Jailer: Proc. Soc. exp. Biol. (N. Y.) 110, 44 (1962).
5. Silber, R. H., R. D. Busch, and R. Oslapas: Clin. Chem. 4, 278 (1958).
6. Hilf, R., F. F. Burnett, and A. Borman: Cancer Res. 20, 1389 (1960).
7. Schönbaum, E.: Canad. J. Biochem. 34, 527 (1956).
8. Lowry, O. H., N. J. Rosebrough, A. L. Farr, and R. J. Randall: J. biol. Chem. 193, 265 (1951).
9. Bonomi, U.: Acta Vitaminol. 15, 151 (1961).
10. Swanson, M. A.: In: Colowick, S. P., and Kaplan, N. O., Methods in Enzymology, Vol. 2, 541. New York: Academic Press (1955).
11. Ferguson, J. J.: J. biol. Chem. 238, 2574 (1963).

Aus der Abteilung für klinische Endokrinologie (Leiter: Prof. Dr. E. F. Pfeiffer) der I. Med.
Univ.-Klinik Frankfurt/M. (Dir.: Prof. Dr. F. Hoff) und der Univ.-Frauenklinik Frankfurt/M.
(Dir.: Prof. Dr. O. Käser)

Über die Insulinwirkung im Nabelschnurvenenblut von Neugeborenen stoffwechselgesunder, diabetischer und prädiabetischer Frauen

Von

R. Morcos, G. Macht, D. Griebner, D. Berg, A. Mucci u. H. Ditschuneit

In der prädiabetischen Phase des Diabetes mellitus lassen sich mit allen uns heute zur Verfügung stehenden Methoden der quantitativen Insulinbestimmung höhere Insulinwirkungen nachweisen als bei Stoffwechselgesunden (*1, 2, 3, 4, 5*). Die oft gleichzeitig bei diesen Personen bestehende verlangsamte Geschwindigkeit der Glucoseassimilation gibt zu erkennen, daß die höhere in vitro meßbare Insulinwirkung in ihrer Wirkung auf den Glucoseumsatz in vivo geringer ist als bei Stoffwechselgesunden (*1, 2*).

Die Ursache des verminderten blutzuckersenkenden Effektes könnte durch eine verstärkte Bindung des Insulins an Plasmaproteine verursacht werden (*1, 2, 6*).

Nach den Untersuchungsresultaten von Vallance-Owen und Lilley (*5*) muß aber für die abgeschwächte Insulinwirkung auch ein im Blut zirkulierender insulinantagonistisch wirksamer Faktor diskutiert werden, der in höherer Konzentration im Blut von Prädiabetikern als von Stoffwechselgesunden auftritt.

Prädiabetische Frauen unterscheiden sich von stoffwechselgesunden Frauen außerdem durch eine höhere Rate geburtshilflicher Komplikationen, insbesondere durch Geburten von überschweren Kindern (*7, 8, 9*). Bei den Totgeburten dieser Frauen findet sich pathologisch anatomisch ein vergrößertes Inselzellsystem im Pankreas infolge Hyperplasie und Hypertrophie der β-Zellen (*10, 11*).

Histologisch weisen diese Zellen Zeichen einer funktionellen Aktivität auf, und zwischen dem Ausmaß der Hypertrophie und dem Geburtsgewicht läßt sich eine positive Korrelation herstellen (*12*). Die Annahme ist daher naheliegend, daß die Überschwere der Kinder prädiabetischer und diabetischer Mütter auf einer erhöhten Insulinsekretion des kindlichen Pankreas beruht (*13*), wobei die Frage nach der Ursache der vermehrten Stimulation des kindlichen Inselzellsystems offen bleibt. Ein dem erhöhten Insulingehalt des mütterlichen Blutes entsprechend vermehrter Übergang von Insulin von der Mutter auf das Kind via Placenta ist nach den bei Menschen und Tieren erhaltenen Untersuchungsresultaten unwahrscheinlich (*14, 15, 16*), auch spricht gegen eine derartige Annahme die kindliche Inselzellhypertrophie.

Dennoch sind in erster Linie mütterliche Faktoren als Ursache der Vergrößerung des Inselzellsystems in Betracht zu ziehen und erst in zweiter Linie eine evtl. ererbte diabetische Anlage des Kindes, da die Kinder diabetischer Väter in weit geringerem Umfange ein erhöhtes Geburtsgewicht aufweisen (*9*).

Eine erhöhte Inselzellfunktion mit vermehrter Insulinsekretion müßte bei Kindern prädiabetischer und diabetischer Mütter mit einer erhöhten Seruminsulinwirkung gekoppelt sein. Diese Annahme wird durch die von Baird u. Farquhar (*17*) bei 6 Fällen nachgewiesene erhöhte Glucoseassimilationsgeschwindigkeit unterstützt, die einen Maßstab für die im extracellulären Raum wirksame Insulinmenge darstellt. Zur Prüfung dieser Frage führten wir Seruminsulinbestimmungen im Nabelschnurvenenblut von Neugeborenen durch, deren Ergebnisse im folgenden mitgeteilt werden sollen.

1. Methodik

In die Untersuchungen wurden 115 Neugeborene einbezogen, die in 3 Gruppen unterteilt wurden. Gruppe I umfaßt 80 Kinder von gesunden Müttern, bei denen sich kein Anhalt für das Vorliegen eines Diabetes mellitus oder eines Prädiabetes fand. Das Geburtsgewicht der Kinder in dieser Gruppe lag unter 4000 g. Zu der Gruppe II zählen 19 Kinder von ebenfalls gesunden Müttern, deren Geburtsgewichte mehr als 4000 g betrugen. Die Gruppe III besteht aus 16 Kindern, deren Mütter an einem manifesten Diabetes mellitus oder einem Prädiabetes litten. In 16 Fällen wurde gleichzeitig mit der Entnahme des Blutes aus der Nabelschnur auch unmittelbar nach der Geburt und vor Lösung der Placenta Blut aus der Cubitalvene der gesunden Mütter entnommen und Zuckergehalt und Insulinwirkung der mütterlichen Blutproben mit denen der Kinder verglichen.

Die Bestimmung der Insulinwirkung erfolgte am isolierten epididymalen Rattenfettgewebe unter Verwendung von radioaktiver Glucose (*18*).

Das Serum wurde bei Kindern und Müttern mit Pufferlösung im Verhältnis 1 : 4 verdünnt.

2. Ergebnisse

80 Kinder gesunder Mütter der Gruppe I mit einem Geburtsgewicht unter 4000 g und einem mittleren Körpergewicht von 3,23 $\pm$ 0,05 kg (Mittelwert $\pm$ Standardabweichung des Mittelwertes) weisen eine mittlere Insulinwirkung von 58,8 $\pm$ 5,3 μE/ml im 1 : 4 verdünnten Serum auf.

Bei 19 mehr als 4000 g schweren Neugeborenen mit einem mittleren Geburtsgewicht von 4,4 $\pm$ 0,05 kg, deren Mütter ebenfalls keinen Anhalt für eine Stoffwechselstörung boten, fanden wir eine höhere Insulinwirkung im Nabelschnurvenenblut von 91,3 $\pm$ 17,3 μE/ml.

Der Unterschied zu der Gruppe mit kleinerem Geburtsgewicht ist deutlich, statistisch aber nicht ausreichend zu sichern $(0,1 \langle P \rangle 0,05)$.

Die 16 Kinder der diabetischen und prädiabetischen Mütter mit einem mittleren Geburtsgewicht von 3,58 $\pm$ 0,21 kg ergaben dagegen einen drei- bis vierfach höheren Mittelwert von 195 $\pm$ 50 μE/ml. Gleichartige Beobachtungen wurden vor kurzem von Stimmler u. Mitarb. mitgeteilt (*22*).

Die Regression von Blutzuckerspiegel auf die Insulinwirkung ergibt bei den insgesamt 99 Kindern gesunder Frauen eine positive, wenn auch schwache Korrelation, die statistisch zu sichern ist. Für den Regressionskoeffizienten errechnet sich ein Wert von 0,29. Dieses Resultat deckt sich mit unseren früheren Untersuchungsergebnissen (*19*) und denen von METZ (*20*) am Hund, bei denen eine positive Abhängigkeit der Insulinsekretion von der Glucosekonzentration nachgewiesen werden konnte.

Die stark erhöhten Insulinwirkungen bei den Kindern prädiabetischer und diabetischer Mütter lassen sich jedoch mit den Blutzuckerwerten nicht korrelieren, so daß als Ursache für die erhöhten Seruminsulinwirkungen dieser Kinder eine stärkere Stimulation der Insulinsekretion durch Glucose ausscheidet. Der Blutzuckermittelwert dieser Kinder liegt mit 80,8 $\pm$ 6,3 mg-% nur wenig über dem mittleren Blutzuckerwert der normalgewichtigen Kinder gesunder Mütter, der 72,4 $\pm$ 2,2 mg-% beträgt, und sogar unter dem Mittelwert von 83,3 $\pm$ 5,0 mg-% von Kindern der Gruppe II mit einem Geburtsgewicht über 4000 g.

Als Ursache der stärkeren Seruminsulinwirkung im Nabelschnurvenenblut von Kindern diabetischer und prädiabetischer Mütter sind daher andere Stimulationsfaktoren als Glucose verantwortlich zu machen.

Die Regression der Geburtsgewichte auf die Insulinwirkung läßt bei den 99 Kindern ebenfalls eine schwache positive Korrelation mit einem Koeffizienten von 0,24 erkennen. Diese Beobachtung stimmt mit der von CARDELL (*12*) nachgewiesenen positiven Korrelation zwischen Ausmaß der Inselzellhypertrophie und Geburtsgewicht gut überein und unterstützt die Annahme, daß überschweres Geburtsgewicht und Inselzellhypertrophie von Kindern diabetischer und prädiabetischer Mütter zwei voneinander direkt abhängige Größen sind.

Das Resultat der gleichzeitig bei Mutter und Kind durchgeführten Insulinbestimmung läßt erkennen, daß zumindest ein unbehinderter Übergang des Insulinmoleküls von der Mutter zum Kind nicht stattfindet. Bei 16 Müttern fanden wir im Cubitalvenenblut eine mittlere Insulinwirkung von 82 $\pm$ 16 μE/ml und bei den entsprechenden Kindern im Nabelschnurvenenblut nur einen Mittelwert von 40 $\pm$ 10 μE/ml. Zu gleichartigen Ergebnissen kamen auch BAIRD u. FARQUHAR (*12*), sowie SANTOS u. Mitarb. (*21*), die die Seruminsulinwirkung mit der Diaphragmamethode bestimmten.

Zusammenfassend ergibt sich aus unseren Untersuchungen also, daß bei Kindern von diabetischen und prädiabetischen Müttern ein Hyperinsulismus vorliegt, der auf dem Wege über einen vermehrten Glucoseumsatz als Ursache für das überschwere Geburtsgewicht dieser Kinder in Betracht zu ziehen ist. Ob die Ursache des Hyperinsulismus in einem erhöhten Gehalt des kindlichen Blutes an Wachstumshormon zu suchen ist, oder aber ob hierfür stärkere Bindungen von Insulin an Plasmaproteine oder aber Insulinantagonisten in Frage kommen, läßt sich aus diesen Untersuchungsresultaten nicht entscheiden.

Literatur

1. DITSCHUNEIT, H., E. F. PFEIFFER, R. CUENDET, H. KOLB, CH. WAHL, u. W. H. ROTT: I. Symp. Dtsch. Diab. Komitee, Düsseldorf, 1962; in: OBERDISSE, K., u. K., JAHNKE: Fortschritte in der Diabetesforschung, 37. Stuttgart: Thieme 1963.

2. — H. KOLB, CH. WAHL, R. MORCOS, W. H. ROTT u. E. F. PFEIFFER: 10. Symp. Dtsch. Ges. f. Endokrinologie, S. 260. Berlin, Göttingen, Heidelberg: Springer-Verlag 1964.

3. — Symp. sur le diabete sucre, Geneve, 28.-30. 11. 1963.

4. MELANI, F., H. DITSCHUNEIT, H. H. DITSCHUNEIT, R. PETZOLDT u. E. F. PFEIFFER: I. Intern. Symp. über Diabetes mellitus der Universität Modena, Modena, 21.-22. 9. 1963.

5. VALLANCE-OWEN, J., and M. D. LILLEY: Lancet **1961 I**, 806.

6. Antoniades, H. M., K. Gunderson, P. M. Beigelman, H. M. Pyle u. J. A. Bougasy: Diabetes 11, 261 (1962).
7. Hoet, J. P.: Diabetes 3, 1 (1951).
8. Wilkerson, H. L.: Amer. J. Physiol. 49, 1032 (1959).
9. Jackson, W. P.: Diabetes 9, 373 (1960)
10. van Beck, C.: I. Congr. Intern. Diab. Federation, Leyden, Holland, 7. 7. 1962.
11. Driscoll, S. G., K. Bernischke, and G. W. Curtis: Amer. J. Dis. Child. 100, 818 (1960).
12. Cardell, B. S.: J. Path. Bact. 66, 335 (1953).
13. Osler, M.: Acta endocr. (Kbh.) 34, 261 (1960).
14. Buse, M. G., W. Roberts, and J. Buse: J. clin. Invest. 41, 29 (1962).
15. Freinkel, N., and C. Goodner: J. clin. Invest. 39, 116 (1960).
16. Josinovich, J. B., and E. Knobel: Amer. J. Physiol. 200, 471 (1961).
17. Baird, J. P., and J. M. Farquhar: Lancet 1962 I, 71.
18. Ditschuneit, H., J. D. Faulhaber u. E. F. Pfeiffer: Atompraxis 8, 172 (1962).
19. Müller-Ruchholtz, W., H. Ditschuneit, W. Dettwyler, W. Böhm, H. Friedrich, E. Ohebshalon, R. Petzoldt, A. Sartory u. E. F. Pfeiffer: 4. Acta Endocrinologica Kongr., 3.-7. 7. 1962, Genf; Acta endocr. (Kbh.), Suppl. 67, 171 (1962).
20. Metz, R.: Diabetes 9, 89 (1960).
21. Santos, R. F., R. A. McCance, and P. J. Randle: Nature (Lond.) 176, 115 (1955).
22. Stimmler, L., J. Brazic u. D. O'Brien: Lancet 1964 I, 137.

Aus der II. Medizinischen Universitätsklinik, Hamburg-Eppendorf
(Direktor: Prof. Dr. A. JORES)

Die Pathophysiologie des Cushing-Syndroms

Von

J. TAMM

Mit 2 Abbildungen

Referat

Die vielschichtige Problematik der Pathophysiologie des Cushing-Syndroms lückenlos darzustellen, würde den Rahmen dieser Übersicht sprengen. Diese Betrachtung soll sich daher im wesentlichen auf die Pathophysiologie der Nebennierenrinden (NNR)-Regulation beim Cushing-Syndrom (CS) infolge bilateraler Hyperplasie konzentrieren. Zahlreiche experimentelle und klinische Beobachtungen der letzten Jahre erlauben es, gewisse Rückschlüsse hinsichtlich der pathogenetischen Faktoren zu ziehen, die für die Entwicklung eines CS infolge bilateraler Hyperplasie eine Rolle spielen. Diese am häufigsten vorkommende Form des CS beruht auf einer Entgleisung zentraler Regulationsmechanismen. Die Überfunktion beider Nebennieren ist deren Folge. Bevor die charakteristischen Befunde skizziert werden, die diese Ansicht stützen, sollen zum besseren Verständnis die Grundprinzipien der normalen NNR-Regulation in Kürze erörtert werden.

Glücklicherweise weist die Regulation der NNR beim Tier und beim Menschen eine Reihe gemeinsamer Merkmale auf, die es ermöglichen, aus dem Tierexperiment gewisse allgemein gültige Schlüsse zu ziehen. Die Abb. 1 gibt ein Schema wieder, das alle wesentlichen Komponenten berücksichtigt, die auf die Regulation der NNR Einfluß nehmen. Der besseren Übersichtlichkeit halber wurde das Schema stark vereinfacht. Dies betrifft vor allem die Darstellung der im ZNS lokalisierten Regulationszentren. Die Steuerung der NNR-Funktion gehorcht grundsätzlich den Prinzipien eines rückgekoppelten Regelkreises; d. h. das von den NNR abgegebene Sekretionsprodukt Cortisol bzw. Corticosteron wirkt im Sinne einer Hemmung auf bestimmte, zentral gelegene Steuerzentren ein. Das Ausmaß dieser Hemmwirkung wird nach YATES et al. (*1, 2*) bestimmt durch die Differenz, die sich aus der jeweiligen Niveaueinstellung des zentralen Kontrollelementes und aus der Cortisolkonzentration im peripheren Plasma ergibt. Ist das Kontrollelement auf ein höheres Niveau geschaltet (vereinfacht kann man sich dieses Element als eine Art Thermostaten vorstellen) und ist die Cortisolkonzentration im Plasma relativ niedriger, so wird ein stimulierender Einfluß auf die Hypophyse und die NNR zustande kommen, bis sich der Cortisolspiegel der Einstellung des Kontrollelementes angeglichen hat. Das Umgekehrte, nämlich eine Hemmung der Hypophysen-NNR-Stimulierung resultiert, wenn die Cortisolkonzentration im Plasma höher ist als die

Einstellung des Kontrollelementes. Dieses Regelprinzip nennt man nach Wagner (3) „negative Rückkopplung". Es findet sich überall im Organismus wieder, bis hinunter in den subcellulären Bereich der Enzyme, und es ermöglicht erst einen geordneten Ablauf der verschiedensten Lebensvorgänge. Yates und Urquhart (2) vertreten die Auffassung, daß die wechselnde Niveaueinstellung des zentralen Kontrollelementes nicht nur für die normale Regulierung der NNR-Tätigkeit gilt, sondern daß sie auch für die gesteigerte NNR-Funktion unter allen möglichen Stress-Situationen von entscheidender Bedeutung ist. Daß diese These von Yates stark eingeschränkt werden muß, soll weiter unten noch kurz erläutert werden. In dem Schema (Abbildung 1) weisen die direkten gestrichelten Pfeile darauf hin, daß das Kontrollelement unter manchen Stress-Bedingungen umgangen werden kann.

Welche Beweise gibt es im einzelnen dafür, daß das Schema (Abb. 1) den tatsächlichen Verhältnissen entspricht?

Durch in vivo- und in vitro-Experimente an verschiedenen Tierspecies konnte gezeigt werden, daß es auch schon auf der untersten Stufe, der NNR selbst, eine Reihe von Regelmechanismen gibt. So führt z. B. die Perfusion von Hundenebennieren mit Cortisol, sowie die Inkubation von Ratten-

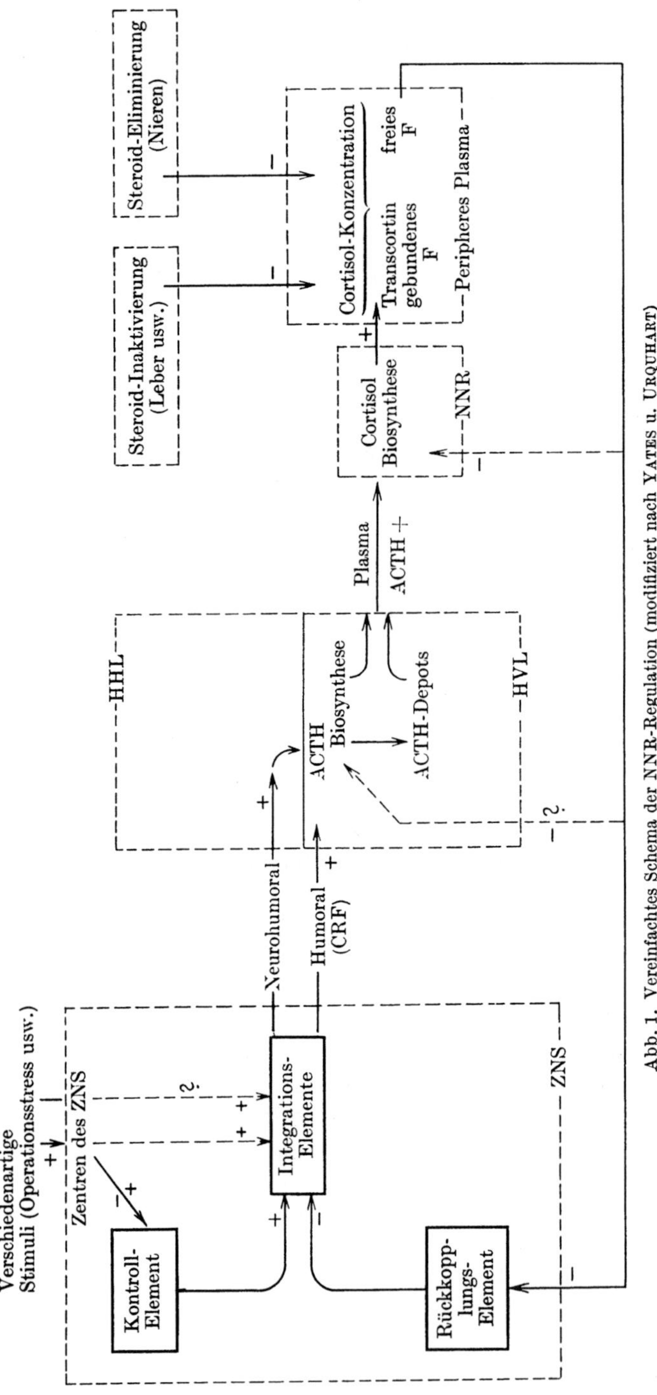

Abb. 1. Vereinfachtes Schema der NNR-Regulation (modifiziert nach Yates u. Urquhart)

nebennieren mit Corticosteron oder anderen aktiven NNR-Steroidderivaten zu einem
Absinken der endogenen Steroidbiosynthese (*4, 5, 6*). Sehr bemerkenswert erscheint
die Tatsache, daß die 11 β-Hydroxylierung in der NNR durch andere, in der NNR
entstehende Steroide wie Dehydroisoandrosteron, DHA-Sulfat, Androstendion und
Testosteron gehemmt werden kann, ein Befund, der u. U. für das AGS Bedeutung
besitzt (*7*). Dehydroisoandrosteron kann ferner unter bestimmten Bedingungen die
für die Biosynthese wichtigen TPNH-regenerierenden Reaktionen bremsen (*8*).
Eine relative Anhäufung von DPNH in der NNR führt zu einem verminderten
Umsatz der Steroidvorstufe Pregnenolon zu Corticosteron (*9*). Diese Eigenregula-
tionsmechanismen der NNR sind aber zweifellos von untergeordneter Bedeutung,
solange eine ausreichende ACTH-Stimulierung gewährleistet ist.

Die Existenz einer negativen Rückkopplung zwischen NNR einerseits und
Hypothalamus bzw. Hypophysenvorderlappen (HVL) andererseits ist durch zahlreiche Tierexperimente und durch Beobachtungen am Menschen belegt. Die Entfernung einer oder beider Nebennieren (NN) sowie der Ausfall dieser Drüsen infolge krankhafter Prozesse oder einer passageren chemischen Blockade hat einen
mehr oder minder raschen Anstieg der ACTH-Aktivität im Plasma zur Folge (*10*
bis *15*). Kompensiert man den Ausfall der endogenen Cortisolproduktion durch die
gleichzeitige Applikation physiologischer Dosen dieses Hormons, so bleibt die
Erhöhung der Plasma-ACTH-Aktivität aus (*11, 13, 16, 17, 18*). Auch bei intaktem
HVL-System kommt es unter exogener Zufuhr höherer Dosen von NNR-Steroiden
zu einem völligen Verschwinden der ACTH-Aktivität aus dem peripheren Plasma,
einer Abnahme des ACTH-Gehaltes des HVL und als Folge davon zu einer weit-
gehenden Drosselung der endogenen NNR-Steroidproduktion mit schließlicher
Atrophie beider Drüsen.

Als Angriffspunkte für die hemmende Wirkung des Plasmacortisols auf das
HVL-NNR-System kommen der HVL selbst oder eines der übergeordneten Zentren im Hypothalamus in Betracht. Die Mehrzahl der Experimente spricht dafür,
daß das Rückkopplungselement, über welches diese Hemmwirkung zustande
kommt, in der Eminentia mediana oder in deren unmittelbarer Nachbarschaft
gelegen sein muß (*19—22*). So hat beispielsweise die stereotaktische Einbringung
kleiner Dosen Cortisol in die Eminentia mediana eine deutliche Bremsung der
NNR-Tätigkeit zur Folge (*23*). Dieser Effekt fehlt dagegen, wenn man den HVL
mit Cortisol-Implantaten spickt (*24, 25*). Ein weiterer Beweis dafür, daß das
Rückkopplungselement nicht im HVL zu suchen ist, ist die Tatsache, daß decere-
brierte Hunde mit isolierter Hypophyse und einer erhöhten Basalsekretion der
NNR nicht mehr mit einer Verminderung der Cortisolbiosynthese reagieren, wenn
exogen Steroide zugeführt werden (*26*).

Die Steuerung der ACTH-Sekretion aus dem HVL durch die Eminentia
mediana, evtl. auch durch weitere Zentren in der Nachbarschaft des Infundibulums
erfolgt durch humorale, wahrscheinlich auch durch neurohumorale Vermittler.
Man bezeichnet diese Vermittlersubstanzen generell als Corticotropin-Releasing-
Factors (CRF). Substanzen mit CRF-Wirkung sind in der Eminentia mediana, im
Hypophysenhinterlappen (*27—34*), sowie im Portalvenenblut der Hypophyse verschiedener Tierarten und in der Spinalflüssigkeit des Menschen (*41, 42*) nachgewiesen worden. Adrenalektomie läßt die CRF-Aktivität in der Eminentia mediana ansteigen (*43*); exogene Steroidzufuhr bringt sie wieder zum Absinken (*44,

7*

45). Auch dies ist ein weiterer Beweis dafür, daß das Rückkopplungselement im Hypothalamus liegt. Chemisch ähneln die CRFs einmal dem α-MSH, zum anderen dem Lysin-Vasopressin (*34* bis *40*).

Die Abhängigkeit einer normalen HVL-NNR-Regulation von der Intaktheit der Eminentia mediana bzw. von einer ausreichenden Stimulierung des HVL durch CRF ist aus folgenden experimentellen Befunden zu entnehmen: Die gezielte Zerstörung der Eminentia mediana führt im Tierversuch zu einem Absinken der Steroidsekretion aus den NNR und zu einer stark verminderten Reaktion auf Stress (*20, 20a, 21, 23, 46*). Die Durchtrennung des Hypophysenstiels hat beim Tier und beim Menschen ähnliche, wenn auch nicht so ausgeprägte Folgen (*22, 47*). Die Transplantation der Rattenhypophyse unter die Nierenkapsel führt zu einer Abnahme der ACTH-Bildung. Ihre Reimplantation in die Nähe der Eminentia mediana stellt die normale ACTH-Bildung wieder her (*48, 49*). Rattenhypophysen, die unter die Nierenkapsel transplantiert wurden, behielten jedoch eine normale ACTH-Biosynthese und -Ausschüttung, wenn ein gereinigtes CRF-Präparat chronisch verabfolgt wurde. Die ACTH-Ausschüttung konnte akut durch eine erhöhte Dosis CRF gesteigert werden. Mit α-MSH bzw. Lysin-Vasopressin ließ sich dies jedoch nicht reproduzieren (*50*).

Die Doppelgleisigkeit der zentralen ACTH-Steuerung durch einen rein humoralen und einen neurohumoralen Vermittlermechanismus wurde erstmalig von Long (*51*) postuliert, da bei Ratten die Plasma-ACTH-Aktivität nach Adrenalektomie zunächst rasch anstieg, dann wieder absank, um sich später auf ein erhöhtes Niveau einzuspielen. Für diese Annahme spricht ferner auch die Beobachtung, daß bei Ratten bestimmte Stress-Reaktionen ausfallen, wenn isoliert der Hypophysenhinterlappen entfernt wurde (*52—55*).

Die Cortisolsekretion der normalen menschlichen NNR zeigt bekanntlich einen typischen Tagesrhythmus. Der Gipfel der Plasma-Cortisol-Konzentration liegt morgens gegen 6 Uhr; ihr Tiefpunkt ist nachts zwischen 0 und 2 Uhr nachweisbar (*56*). Ney et al. (*58*) und Retiene et al. (*58a*), ist es kürzlich gelungen, beim Menschen morgens ebenfalls eine höhere ACTH-Aktivität im Plasma zu finden als abends. Appliziert man beim Menschen die gleiche Menge eines CRF-Präparates zu verschiedenen Tageszeiten, so beobachtet man nachts einen relativ stärkeren Anstieg der Plasma-17-OHCS von einem tieferen Basalwert aus im Vergleich zu einem entsprechenden Versuch in den Morgenstunden (*57*). Die genannten Befunde sprechen dafür, daß die Tagesrhythmik der menschlichen NNR von einer gleichlaufenden Rhythmik des CRF-HVL-Systems verursacht wird. Die Intaktheit bestimmter zentralnervöser Strukturen, wie des Hippocampus, des N. amygdalae, des Hypothalamus, vor allem der Eminentia mediana ist die Voraussetzung für einen normalen Tagesrhythmus von HVL und NNR, wie das Tierexperiment und zahlreiche klinische Beobachtungen bei Patienten mit schweren Hirntraumata zeigen (*59—73*). Es ist anzunehmen, daß die Tagesrhythmik letzlich durch eine unterschiedliche Niveaueinstellung des Kontrollelementes im Hypothalamus zustande kommt, wie es der eingangs erwähnten Theorie von Yates und Urquhart (*2*) entsprechen würde. Es wurde oben jedoch schon angedeutet, daß diese Autoren offensichtlich nicht Recht haben, wenn sie behaupten, daß die gesteigerte NNR-Tätigkeit unter schwerem Stress auf eine rasche Höherschaltung des hypothalamischen Kontrollelementes zurückzuführen ist. Stress-Bedingungen wie z. B. Operationen oder Hypoglykämie führen sowohl beim Tier als auch beim Menschen zu einer raschen Steigerung der Cortisol- bzw. Corticosteronsekretion, auch wenn kurz vor Beginn der Stress-Einwirkung die NNR-Tätigkeit zunächst durch hohe Dosen Dexamethason völlig unterdrückt wurde (74—80). Ein erhöhter Bedarf der Peripherie an NNR-Steroiden unter Stress kann hierbei keine Rolle spielen, da dieser durch die exogene Steroidzufuhr gedeckt wäre. Die Natur hat offensichtlich für größtmögliche Sicherheit dadurch gesorgt, daß schwerer Stress direkt eine HVL-NNR-Stimulierung unter Umgehung des zentralen Kontrollelementes bewirken kann.

Welche Möglichkeiten eröffnen sich nun, um die Entstehung eines CS mit bilateraler NNR-Hyperplasie zu erklären?

Die Tatsache, daß man bei obduzierten Patienten mit einer bilateralen NNR-Hyperplasie nicht so selten kleinere Adenome des HVL findet, hat schon H. CUSHING zu der Annahme geführt, daß diese Krankheit durch einen zentralen Mechanismus ausgelöst würde. Infolge der damals noch sehr viel geringeren Kenntnisse über die Regulationsmechanismen, stellte er den Hypophysenvorderlappen in den Mittelpunkt seiner pathogenetischen Betrachtungen (Lit. s. 100). Wir stimmen mit CUSHING darin überein, daß die bilaterale NNR-Hyperplasie zentral ausgelöst ist; verlegen jedoch den Schwerpunkt unserer Aufmerksamkeit in die Region der übergeordneten hypothalamischen Zentren. Mit welchem Recht, wird zu zeigen sein. Vor einigen Jahren äußerten HORWITH und STOKES (81) in einem zusammenfassenden Artikel die Meinung, daß nicht nur das CS infolge von NNR-Tumoren eine primäre Erkrankung dieser Organe sei, sondern daß auch die Ursachen für eine bilaterale Hyperplasie zumindest teilweise in den NNR selbst zu suchen sei. Überzeugende Beweise hierfür wurden jedoch nicht beigebracht. Die Untersuchungsergebnisse, die heute vorliegen, lassen keinen Zweifel daran, daß die Voraussetzung für die Entstehung einer bilateralen NNR-Hyperplasie mit allen klinischen Folgeerscheinungen des chronischen Hypercortisolismus eine dauernd erhöhte ACTH-Konzentration im Plasma sein muß.

Die Tab. 1 gibt die Sekretionsraten der menschlichen NNR unter verschiedenen Bedingungen wieder. Die Cortisolproduktion ist bei jeder Form des CS signifikant gegenüber der Norm erhöht, bis zu excessiven Werten bei extrahypophysären, ACTH-bildenden Tumoren (82, 83). Entsprechend erhöht liegen auch die 17-OHCS-Konzentrationen im peripheren Plasma (84). Gelegentliche Ausnahmen kommen hier jedoch vor.

Die Bemühungen, bei Patienten mit CS ebenfalls erhöhte ACTH-Aktivitäten im Plasma nachzuweisen, reichen weit zurück. So weit die Literatur zugänglich war, gelang es vor fast 30 Jahren JORES als erstem, im Plasma eines solchen Cushing-Falles ACTH-Aktivitäten zu finden. Die Mehrzahl dieser Untersuchungen verlief jedoch bis in die jüngste Zeit hinein erfolglos (95, 100). LIDDLE et al.,

Tabelle 1. *Cortisol-Sekretionsraten der menschlichen Nebennierenrinde* (mg/24 Std)

Normale	NN-Hyperplasie	NN-Adenom	NN-Ca	Extrahypophysäre ACTH-bild. Tumoren	Autoren
18,8 (9—31)	37—82				BIGLIERI et al. (82)
15,7 (7—29)	64,7 (32—111)	63 (39—82)	124 (1 Fall)	206 (65—420)	BROOKS et al. (83)

Davies und Vance et al. (*11, 58, 85*) haben in letzter Zeit eindeutig zeigen können, daß beim unbehandelten CS mit bilateraler Hyperplasie eine höhere ACTH-Aktivität im peripheren Plasma im Vergleich zum Gesunden gefunden werden kann. Wie aus Tab. 2 und Abb. 2 zu ersehen ist, lassen sich beim NN-Gesunden ACTH-Aktivitäten von unter 0,5—0,7 mE/100 ml Plasma nachweisen. Diese Konzentrationen sind nur meßbar, wenn man von gereinigten Konzentraten größerer Plasmamengen ausgeht. Wie ebenfalls aus Tab. 2 zu entnehmen ist, liegen die ACTH-Aktivitäten im Plasma des unbehandelten CS je nach Untersucher in einem Bereich von

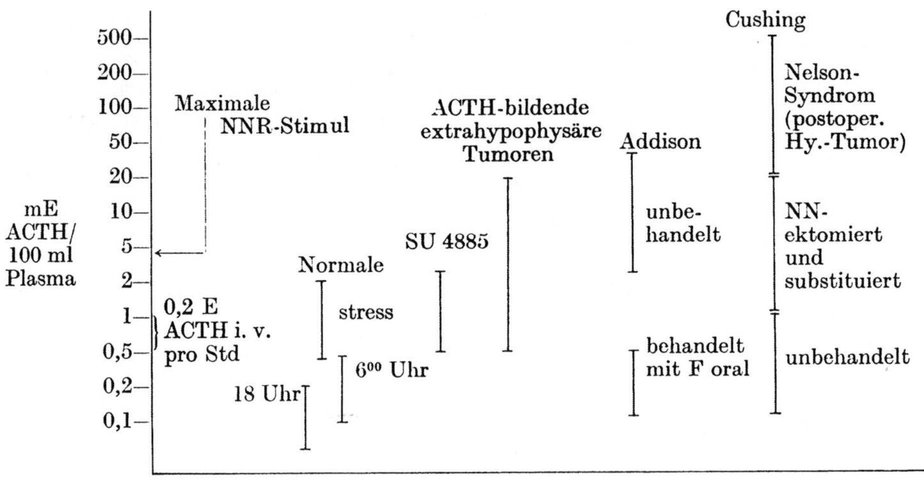

Abb. 2. ACTH-Konzentration im peripheren Plasma (modifiziert nach Liddle et al.)

0,6 bis 2,7 mE/100 ml. Auch diese Konzentrationen sind mit einfacheren Verfahren meistens nicht exakt faßbar. Diese Zahlen zeigen, daß schon eine relativ geringe Steigerung der ACTH-Aktivität im peripheren Plasma eine beträchtliche Stimulierung der Cortisolbiosynthese in den NNR hervorruft. Man muß sich vor Augen halten, daß beim NN-gesunden Menschen eine ACTH-Infusion von 0,2 E/Std die Plasma-ACTH-Aktivität auf 0,5—1,0 mE/100 ml hält und den Plasma-Cortisol-Spiegel auf 35—45 μg/100 ml steigert (*86*). Selbst Dauerinfusionen mit 0,05 E ACTH pro

Tabelle 2. *ACTH-Aktivität im peripheren Plasma* (mE/100 ml)

| Normale | Unbehandelt | Nach NN-Ektomie | | ACTH-bildende Hy.-Tumoren | Autoren |
| | | Cushing-Syndrom | | | |
		total (mit Substitution)	subtotal		
	145, <15			50, 38	Clayton (*10*)
0,7 (0,6 —1,2)	2,7 (0,9 —6,7)	1,1—3,4			Davies (*85*)
0,5 (0,4 —0,8)	0,76 (0,6 —0,8)	2 —12		12—330	Vance et al. (*11*)
6 Uhr: 0,25 (0,1—0,59)	0,62 (0,14—1,81)				Ney et al. (*58*)
18 Uhr: 0,11 (0,04—0,25)	0,33		<11—>500		Garmendia et al. (*95*)

Stunde erzeugen noch einen erhöhten Cortisolspiegel im peripheren Blut, wie er bei manchen Fällen von CS zu beobachten ist (*87*). Eine maximale Stimulierung der gesunden NNR wird bei einer ACTH-Aktivität von 2—3 mE/100 ml Plasma erreicht. Bei höherer ACTH-Aktivität besteht keine geradlinige Korrelation mehr zwischen Plasma-ACTH-Aktivität und Cortisolsekretion (*58*). Wie die Tab. 2 zeigt, findet man beim klassischen CS mit bilateraler Hyperplasie ACTH-Aktivitäten im peripheren Plasma, die unterhalb von 3 mE/100 ml liegen. Das heißt, daß die NNR dieser Patienten nicht voll stimuliert sind, und daß offenbar noch ein gewisser Hemmeffekt des endogen produzierten Cortisols auf die hypothalamischen Zentren vorhanden sein muß. Aus dem vorher Gesagten kann geschlossen werden, daß beim CS mit bilateraler Hyperplasie das hypothalamische Kontrollelement aus noch nicht erklärlichen Gründen auf ein höheres Arbeitsniveau geschaltet ist und via CRF eine leicht erhöhte ACTH-Ausschüttung mit verstärkter Stimulierung der NNR verursacht. Eine solche Dauerstimulierung durch ACTH führt nicht nur zu einer bilateralen Hyperplasie der NNR, sondern macht diese Organe auch für einen zusätzlichen exogenen ACTH-Reiz empfindlicher. Das überschießende Ansteigen der Plasma-17-OHCS unter einer ACTH-Infusion ist ja ein charakteristisches Merkmal des CS mit bilateraler NNR-Hyperplasie (Lit. s. b. *100*). Auch bei gesunden Personen kann man dieses Phänomen durch eine längere Vorbehandlung mit geringen ACTH-Dosen nachahmen (*87*).

Die Hypothese. daß das CS mit bilateraler NNR-Hyperplasie durch eine starre und höhere Einstellung des hypothalamischen Kontrollelementes verursacht sei, wird durch folgende Befunde erhärtet:

1. Um die Cortisolproduktion beim Cushing-Kranken zu bremsen, muß man wesentlich höhere Dosen exogener Steroide als beim Gesunden applizieren. LIDDLE hat dies zu einem differentialdiagnostischen Test ausgebaut (*89*).

2. Die Injektion eines Pyrogens wie Lipexal führt bei Cushing-Patienten im Gegensatz zu NN-Gesunden nicht regelmäßig zu einer Steigerung der Cortisolsekretion (*90*).

3. Die Mehrzahl der Cushing-Fälle mit bilateraler NNR-Hyperplasie läßt keine Tagesrhythmik der Plasma-17-OHCS erkennen. Zumindest wird ein stärkerer Rückgang der Plasma-17-OHCS während der Nacht vermißt (*91, 92*). In diesem Zusammenhang sind Resultate von NEY et al. und RETIENE et al. (*58, 58a*) sehr bedeutsam. Bei Fällen von CS lag auch die abendliche ACTH-Aktivität im Plasma höher als die von Gesunden (s. Tab. 2 und Abb. 2). Damit dürfte nachgewiesen sein, daß beim CS mit bilateraler NNR-Hyperplasie die zentral gesteuerte Tagesrhythmik der NNR weitgehend ausgefallen ist. Das Fehlen der Tagesrhythmik der Plasma-17-OHCS war auch zu finden bei Patienten, die Jahre vorher subtotal adrenalektomiert worden waren (*93*).

Abschließend muß in diesem Rahmen noch ein weiteres Problem kurz angeschnitten werden. Es betrifft die Tatsache, daß sich während der Schwangerschaft kein CS entwickelt, obgleich während der Gravidität die Cortisol-Konzentration im Plasma Werte erreicht, die für das CS charakteristisch sind. Der Grund hierfür ist darin zu suchen, daß bei der normalen Graviden die Konzentration des Transcortins im Serum um das Doppelte höher liegt als beim CS bzw. nicht-graviden Personen (*94*). Im Gegensatz zum CS ist bei Schwangeren der Anteil des freien Cortisols im Plasma, d. h. des biologisch aktiven Hormons, nicht pathologisch erhöht. Außerdem zeigt der Plasma-Cortisol-Spiegel in der Schwangerschaft einen normalen Tagesrhythmus (*96, 97*). Die Tatsache, daß die Erhöhung des Plasmacortisols beim CS im wesentlichen a conto des nicht-eiweißgebundenen Anteils geht, kommt auch darin zum Ausdruck, daß in der extracellulären Flüssigkeit und damit am Ort der Wirkung die Cortisolkonzentration beim Cushing-Kranken merklich höher liegt als bei Graviden und Nichtgraviden (*98*).

Zusammenfassend kann folgendes festgestellt werden:

Das CS infolge bilateraler Hyperplasie der NNR entsteht dadurch, daß das hypothalamische Kontrollelement auf ein höheres und nicht mehr variables Niveau geschaltet ist. Die Folge ist eine fortlaufende, mäßiggradige Stimulierung der ACTH-Ausschüttung aus dem HVL, die zu einer leicht erhöhten ACTH-Konzentration im Plasma führt. Diese wiederum regt beide NNR zu einer praktisch konstanten Mehrsekretion von Cortisol an, die keine Tagesrhythmik mehr aufweist. Das in das Plasma abgegebene Cortisol wird zu einem geringeren Teil an Transcortin gebunden, so daß die peripheren Gewebe dauernd einer erhöhten Cortisolkonzentration ausgesetzt sind, die die normalen Erfordernisse weit übersteigt. Die eigentliche Ursache, die für die Fehleinstellung des hypothalamischen Kontrollelementes verantwortlich ist, bleibt nach wie vor unbekannt. Die histologischen Untersuchungen am Hypothalamus von verstorbenen Cushing-Patienten haben keine eindeutigen Befunde zutage gebracht (*99*).

Literatur

1. Yates, F. E., S. E. Leeman, D. W. Glenister, and M. F. Dallman: Endocrinology **69**, 67 (1961).

2. —, and J. Urquhart: Physiol. Rev. **42**, 359 (1962).

3. Wagner, R.: Probleme und Beispiele biologischer Regelung. Stuttgart: Thieme 1954.

4. Black, W. C., R. S. Crampton, A. S. Verdesca, R. I. Nedeljkovic, and J. G. Hilton: Amer. J. Physiol. **201**, 1057 (1961).

5. Péron, F. G., F. Moncloa, and R. I. Dorfman: Endocrinology **67**, 379 (1960).

6. Fekete, G., and P. Görög: J. Endocr. **27**, 123 (1963).

7. Sharma, D. C., E. Forchielli, and R. I. Dorfman: J. biol. Chem. **238**, 572 (1963).

8. Tsutsui, E. A., P. A. Marks, and P. Reich: J. biol. Chem. **237**, 3009 (1962).

9. Koritz, S. B.: Arch. Biochem. **100**, 349 (1963).

10. Clayton, B. E.: Proc. roy. Soc. Med. **51**, 558 (1958).

11. Vance, V. K., W. J. Reddy, D. H. Nelson, and G. W. Thorn: J. clin. Invest. **41**, 20 (1962).

12. Hodges, J. R., and J. Vernikos: Acta endocr. (Kbh.) **30**, 188 (1959).

13. Kitay, J. I., D. A. Holub, and J. W. Jailer: Proc. Soc. exp. Biol. N. Y. **97**, 165 (1958).

14. Fortier, C.: Proc. Soc. exp. Biol. (N. Y.) **100**, 13 (1959).

15. Garmendia, F., W. E. Vaubel u. E. F. Pfeiffer: Klin. Wschr. **41**, 317 (1963).

16. Fortier, C.: Proc. Soc. exp. Biol. (N. Y.) **100**, 16 (1959).

17. Hodges, J. R., and J. Vernikos-Danellis: Acta endocr. (Kbh.) **39**, 79 (1962).

18. Williams, W. C., D. Island, R. A. A. Oldfield, and G. W. Liddle: J. clin. Endocr. **21**, 426 (1961).

19. de Wied, D.: Acta endocr. (Kbh.) **37**, 270 (1961).

20. Krieger, D. T., and I. H. Wagman: Acta endocr. (Kbh.) **38**, 88 (1961).

20a. Schmid, R., L. Gonzalo, R. Blobel, E. Muschke u. E. Tonutti: Endokrinologie **34**, 65 (1957).

21. McCann, S. M., and P. Haberland: Endocrinologie **66**, 217 (1960).

22. — Amer. J. Physiol. **175**, 13 (1957).

23. Smelik, P. G., and C. H. Sawyer: Acta endocr. (Kbh.) **41**, 561 (1962).

24. DAVIDSON, J. M., and S. FELDMAN: Science **137**, 125 (1962).

25. — — Endocrinology **72**, 936 (1963).

26. EGDAHL, R. H.: Recent Progr. Hormone Res. **18**, 157 (1962) (Disk.).

27. GUILLEMIN, R., W. R. HEARN, W. R. CHECK, and D. E. HOUSHOLDER: Endocrinology **60**, 488 (1957).

28. — A. V. SCHALLY, H. S. LIPSCOMB, R. N. ANDERSEN, and J. M. LONG: Endocrinology **70**, 471 (1962).

29. SCHALLY, A. V., H. S. LIPSCOMP, J. M. LONG, W. E. DEAR, and R. GUILLEMIN: Endocrinology **70**, 478 (1962).

30. — —. and R. GUILLEMIN: Endocrinology **71**, 164 (1962).

31. McCANN, S. M., and P. HABERLAND: Proc. Soc. exp. Biol. (N. Y.) **102**, 319 (1959).

32. BRIZZEE, K. R., and K. B. EIK-NES: Endocrinology **68**, 166 (1961).

33. ROYCE, P. C., and G. SAYERS: Proc. Soc. exp. Biol. (N. Y.) **102**, 447 (1960).

34. SCHALLY, A. V., and R. GUILLEMIN: Proc. Soc. exp. Biol. (N. Y.) **112**, 1014 (1963).

35. PRIVAT DE GARILHE, M., and C. GROS: Experientia (Basel) **18**, 92 (1962).

36. LI, C. H., E. SCHNABEL, D. CHUNG, and T. B. LO: Nature (Lond.) **189**, 143 (1961).

37. DE WIED, D.: Acta endocr. (Kbh.) **37**, 288 (1961).

38. McCANN, S. M.: Endocrinology **60**, 664 (1957).

39. CASENTINI, S., A. DE POLI, L. HUKOVIC, and L. MARTINI: Endocrinology **64**, 483 (1959).

40. HEARN, W. R., E. J. WEBER, P. W. RANDOLPH, and N. E. BARKS: Proc. Soc. exp. Biol. (N. Y.) **107**, 515 (1961).

41. PORTER, J. C., and H. W. RUMSFELD: Endocrinology **64**, 948 (1959).

42. EIK-NES, K. B., D. M. BROWN, K. R. BRIZZEE, and E. L. SMITH: Endocrinology **69**, 411 (1961).

43. VERNIKOS-DANELLIS, J.: 45. Meet., Endocr. Soc., Atlantic City, 1963.

44. EIK-NES, K. B., and K. R. BRIZZEE: Acta endocr. (Kbh.) **29**, 219 (1958).

45. BRODISH, A., and C. N. H. LONG: Endocrinology **71**, 298 (1962).

46. GANONG, W. F., A. M. NOLAN, A. DOWDY, and J. A. LUETSCHER: Endocrinology **68**, 169 (1961).

47. VANWYK, J. J., G. S. DUGGER, J. F. NEWSOME, and P. Z. THOMAS: J. clin. Endocr. **20**, 157 (1960).

48. NIKITOVITCH-WINER, M., and J. W. EVERETT: Endocrinology **63**, 916 (1958).

49. FOSTER, R., and I. ROTHSCHILD: Acta endocr. (Kbh.) **39**, 371 (1962).

50. CRITCHLOW, V., H. S. LIPSCOMB, and R. GUILLEMIN: J. Endocr. **25**, 465 (1963).

51. LONG, C. N. H., and M. F. M. BONNYCASTLE: Canad. J. Biochem. **35**, 929 (1957).

52. SMELIK, P. G.: Acta endocr. (Kbh.) **33**, 437 (1960).

53. NOWELL, N. W.: Endocrinology **64**, 191 (1959).

54. DE WIED, D.: Endocrinology **68**, 956 (1961).

55. ARKO, H., E. KIVALO, and U. K. RINNE: Acta endocr. (Kbh.) **42**, 293 (1963).

56. MIGEON, C. J., F. H., TYLER, J. P. MAHONEY, A. A. FLORENTIN, H. CASTLE, E. L. BLISS, and L. T. SAMUELS: J. clin. Endocr. **16**, 622 (1956).

57. CLAYTON, G. W., L. LIBRIK, R. L. GARDNER, and R. GUILLEMIN: J. clin. Endocr. **23**, 975 (1963).

58. NEY, R. L., N. SHIMIZU, W. E. NICHOLSON, D. P. ISLAND, and G. W. LIDDLE: J. clin. Invest. **42**, 1669 (1963).

58a RETIENE, K., A. ESPINOZA, Y. ABDEL RAHMAN, K. H. MARX u. E. F. PFEIFFER: 10. Symp. d. Dtsch. Ges. Endokrin. 231, Berlin-Göttingen-Heidelberg: Springer 1964.

59. MASON, J. W.: Amer. J. Physiol. **196**, 44 (1959).

60. MANDELL, A. J., L. F. CHAPMAN, W. R. RAND, and R. D. WALTER: Science **139**, 1212 (1963).

61. BRODISH, A.: 45. Meet., Endocrin. Soc., Atlantic City, 1963.

62. OPPENHEIMER, J. H., L. V. FISHER, and J. W. JAILER: J. clin. Endocr. **21**, 1023 (1961).

63. MARTINI, L., A. PECILE, S. SAITO, and F. TANI: Endokrinology **66**, 501 (1960).

64. BRODISH, A.: Endocrinology **73**, 727 (1963).

65. KNIGGE, K. M., and M. HAYS: Proc. Soc. exp. Biol. (N. Y.) **114**, 67 (1963).

66. SETEKLEIV, J., O. E. SKANG, and B. R. KAADA: J. Endocr. **22**, 119 (1961).

67. SLUSHER, M. A.: Endocrinology **67**, 347 (1960).

68. —, and J. E. HYDE: Endocrinology **68**, 773 (1961).

69. EIK-NES, K., and L. D. CLARK: J. clin. Endocr. **18**, 764 (1958).
70. PERKOFF, G. T., K. EIK-NES, C. A. NUGENT, H. L. FRED, R. A. NIMER, L. RUSH, L. T. SAMUELS, and F. H. TYLER: J. clin. Endocr. **19**, 432 (1959).
71. KRIEGER, D. T., J. WEISZ, H. KOLODNY, and H. P. KRIEGER: 45. Meet., Endocrin. Soc. Atlantic City, 1963.
72. — — — — 45. Meet., Endocrin. Soc., Atlantic City, 1963.
73. SABA, G. C., A. CARNICELLI, P. SABA, and V. MARESCOTTI: Acta endocr. (Kbh.) **44**, 413 (1963).
74. ESTEP, H. L., D. P. ISLAND, R. L. NEY, and G. W. LIDDLE: J. clin. Endocr. **23**, 419 (1963).
75. HODGES, J. R., and M. T. JONES: J. Physiol. (Lond.) **167**, 30 (1963).
76. BRINCK-JOHNSEN, T., J. H. SOLEM, K. BRINCK-JOHNSEN, and P. INGVALDSEN: Acta med. scand. **173**, 129 (1963).
77. PLAGER, J. E., G. A. BRAY, and J. E. JACKSON: Endocrinology **72**, 876 (1963).
78. SMELIK, P. G.: Acta endocr. (Kbh.) **44**, 36 (1963).
79. JAMES, V. H. T., and J. LANDON: J. Endocr. **26**, 31 (1963).
80. DICZFALUSY, E.: Recent Progr. Hormone Res. 18, 156 (1962). (Disk.)
81. HORWITH, M., and P. E. STOKES: Advanc. int. Med. **10**, 259 (1960).
82. BIGLIERI, E. G., S. HANE, P. E. SLATON, and P. H. FORSHAM: J. clin. Invest. **42**, 516 (1963).
83. BROOKS, R. V., J. DUPRÉ, A. N. GOGATE, I. H. MILLS, and F. T. G. PRUNTY: J. clin. Endocr. **23**, 725 (1963).
84. VOIGT, K. D., A. ORIOL-BOSCH u. J. TAMM: Klin. Wschr. **38**, 206 (1960).
85. DAVIES, B. M. A.: Acta endocr. (Kbh.) **45**, 55 (1964).
86. LIDDLE, G. W., D. ISLAND, and C. K. MEADOR: Recent Progr. Hormone Res. **18**, 125 (1962).
87. NUGENT, C. A., K. EIK-NES, L. T. SAMUELS, and F. H. TYLER: J. clin. Endocr. **19**, 334 (1959).
88. MEADOR, C. K., G. W. LIDDLE, D. P. ISLAND, W. E. NICHOLSON, C. P. LUCAS, J. G. NUCKTON, and J. A. LUETSCHER: J. clin. Endocr. **22**, 693 (1962).
89. LIDDLE, G. W.: J. clin. Endocr. **20**, 1539 (1960).
90. FARMER, T. A., S. R. HILL, J. A. PITTMAN, and J. W. HEROLD: J. clin. Endocr. **21**, 433 (1961).
91. DOE, R. P., J. A. VENNES, and E. B. FLINK: J. clin. Endocr. **20**, 253 (1960).
92. EKMAN, H., B. HAKANSSON, J. D. MCCARTHY, J. LEHMANN, and B. SJÖGREN: J. clin. Endocr. **21**, 684 (1961).
93. DRUCKER, W. D., and N. P. CHRISTY: 45. Meet., Endocrin. Soc., Atlantic City, 1963.
94. DOE, R. P., U. S. SEAL, and R. FERNANDEZ: 45. Meet., Endocrin. Soc., Atlantic City, 1963.
95. GARMENDIA, F., W. E. VAUBEL, H. BUBSER, and E. F. PFEIFFER: Klin. Wschr. **41**, 312 (1963).
96. COHEN, M., M. STIEFEL, W. J. REDDY, and J. C. LAIDLAW: J. clin. Endocr. **18**, 1076 (1961).
97. PEKKARINEN, A., L. RAURAMO, and B. THOMASSON: Acta endocr. (Kbh.) Suppl. 75 (1962).
98. KATZ, F. H., and I. L. SHANNON: Acta endocr. (Kbh.) **46**, 393 (1964).
99. MAREN, T. H.: J. clin. Endocr. **13**, 884 (1953).
100. TAMM, J.: Dtsch. med. Wschr. **86**, 1874 (1961).

Diskussion

E. TONUTTI (Bonn):

Ich möchte empfehlen, die Bezeichnung „Zentren" für jene Hypothalamusbereiche zu vermeiden, von denen aus eine Einflußnahme auf die glandotrope Tätigkeit der Hypophyse ausgeht. Läsionen der Eminentia mediana können auch unspezifische Schäden der Adenohypophyse zur Folge haben. Beim Meerschweinchen erlischt die ACTH-Sekretion nach Coagulation der ventro- und dorsomedialen Hypothalamuskerne, obwohl der Bereich der Eminentia mediana dabei unversehrt bleibt.

Zur Oestrogenwirkung auf die Nebennierenrinde folgende Bemerkung: Es ist seit langem bekannt, daß Oestrogene eine beträchtliche Rindenhypertrophie erzeugen. Diese bleibt jedoch beim hypophysektomierten Tier aus und stellt somit lediglich eine indirekte Wirkung dar. Ob darüber hinaus auch eine direkte Wirkung möglich ist, müßte gesondert an hypophysenlosen Tieren geprüft werden.

F. Bahner (Heidelberg):

Vielleicht ist der Tagesrhythmus des ACTH-Gehaltes und Corticoidspiegels im Blut von der Ernährung abhängig. Mirouze (Montpellier) fand bei fortlaufender Registrierung des Blutzuckers bei Diabetikern, daß nächtliche Hypoglykämien von hohen Corticoidspiegeln gefolgt sind. Man kann daher erwarten, daß die Ernährungsweise den Tagesrhythmus beeinflußt, was für die Interprätation beachtet werden muß.

Jöchle (München):

Nachdem hier übereinstimmend dargelegt wurde, daß für das Cushing-Syndrom mit beidseitiger Nebennierenhyperplasie eine zentrale Regulationsstörung, d. h. eine verminderte Ansprechbarkeit des hypothalamischen Kontrollelementes für die negative Feed-back-Wirkung der Corticoide verantwortlich zu machen ist, muß gefragt werden

1. ob die Einstellung dieses Kontrollwertes lebenslang gleich bleibt oder normalerweise — analog der Entwicklung des hypothalamischen Sexualzentrums — eine Reifung im Sinne einer Adrenarche durchläuft;

2. ob die Altersverteilung der Patienten Hinweise darauf gibt, daß sich die Fehleinstellung des hypothalamischen Kontrollelementes im Verlauf dieser Adrenarche entwickelt, d. h. einen fehlerhaften Endzustand eines an und für sich normalen Entwicklungsganges darstellt;

3. ob in dieser Sicht nicht die sog. funktionellen, passageren Cushing-Fälle Jugendlicher als extreme, aber vorübergehende Reifungsschwankungen mit letztlich doch normaler Endeinstellung im Verlauf dieser Adrenarche anzusprechen sind.

L. Zicha (Erlangen):

Es wurde bei den heutigen Vorträgen über die Pathogenese des Cushing-Syndroms insbesondere darauf hingewiesen, daß es sich bei diesem Krankheitsbild im wesentlichen um eine Störung zentraler Regulationsmechanismen handelt. Hierfür spricht auch die Beobachtungstatsache, daß der erhöhte Spiegel an freien 17 OHCS offenbar nicht in der Lage ist, den CRF-ACTH-Mechanismus zu bremsen. Es taucht nun die Frage auf, warum der Transcortinmechanismus im Gegensatz zur Schwangerschaft nicht in der Lage ist, den Anstieg an freien OHCS durch eine verstärkte Bindung zu kompensieren. Nach Sandberg und Slaunwhite besteht beim Cushing-Syndrom keine Änderung der Transcortinaktivität. Es bleibt zu diskutieren, daß hierbei Änderungen der Relation von C_{21}-freiem Alkohol und C_{21}-Acetat eine Rolle spielen, denn wir konnten zeigen, daß die Affinität von C_{21}-Acetaten von Glucocorticoiden zum Transcortin stärker ist als die der C_{21}-freien Alkohole.

D. Glaubitt (Mainz):

Über den Eiweißstoffwechsel beim Cushing-Syndrom liegen nur wenige Untersuchungen vor.
Sterling [J. clin. Invest. **39**, 1900 (1960)] stellte bei 4 Patienten eine erhöhte Umsatzgeschwindigkeit und bei 2 dieser Patienten eine Verkleinerung des Pools von ^{131}J-Humanalbumin fest. Demgegenüber konnte Hoffenberg (S. 323 in: Protein Metabolism. Ann. International Symposium. Leiden, 25. 6.—29. 6. 62. Berlin-Göttingen-Heidelberg: Springer-Verlag 1962) bei 3 Patienten mit Cushing-Syndrom nenneswerte pathologische Veränderungen der Poolgröße und der Umsatzgeschwindigkeit von ^{131}J-Humanalbumin nicht beobachten.

E. F. Pfeiffer (Frankfurt-Main):

Ich möchte die Auffassung von der möglichen Existenz zweier Regulationseinrichtungen unterstreichen, die die ACTH-Regulation kontrollieren: 1. Können nach Adrenalektomie beim M. Cushing auch ohne Hypophysentumor extrem erhöhte ACTH-Spiegel gemessen werden. 2. Fehlt der tgl. 24-Std-Rhythmus derartiger Fälle nur am Anfang völlig, während längere Zeit nach Entfernung der Nebenniere (1 Jahr in einem Fall) die ACTH-Spiegel zwar noch absolut erhöht sind, der charakteristische Rhythmus mit morgendlichem Gipfel und nächtlichen Tiefstwerten jedoch wieder nachweisbar ist. Zusammen mit dem demonstrierten Verschwinden des Rhythmus beim Normalen unter Decortintherapie läßt dies nur den Schluß zu, daß der 24 Std-Rhythmus der ACTH-Abgabe durch den hemmenden Effekt verschwindet, den ein erhöhter Corticoidspiegel von der Peripherie her auf ein spezifisches Zentrum ausübt, die absolut beim M. Cushing erhöhte ACTH-Sekretion dagegen durch eine entsprechend gesteigerte Abgabe von CRF aus dem Hypothalamus zustande kommt.

Aus der 2. Medizinischen Klinik und Poliklinik der Medizinischen Akademie Düsseldorf
(Direktor: Prof. Dr. med. K. Oberdisse)

Klinik und Differentialdiagnose des Cushing-Syndroms

Von

H. Zimmermann

Mit 8 Abbildungen

Referat

Das Cushing-Syndrom wird durch eine chronische Überproduktion von Nebennierenrinden-Steroiden, besonders jedoch von Hydrocortison, aus der Nebennierenrinde hervorgerufen. Diese Überproduktion kann durch pathogenetisch verschiedene Ursachen bedingt sein. Als häufigste Ursache muß der *Hyperfunktions-Cushing* (*1, 2, 3*) genannt werden. Er ist in etwa 70—90% (*26, 34, 40, 52, 70, 73*) der beobachteten Fälle nachweisbar. Pathologisch-anatomisch wird meist eine Hyperplasie der Nebennierenrinde gefunden, jedoch werden mitunter auch völlig unauffällige Nebennierenrinden (NNR) postoperativ nachgewiesen (*3*). Über die Pathogenese hat Herr Tamm in dem vorangegangenen Vortrag bereits ausführlich berichtet, so daß ich auf Einzelheiten nicht eingehen möchte. Es dürfte jedoch feststehen, daß der normalerweise bestehende feed-back-Mechanismus zwischen freiem Hydrocortison-Spiegel im Blut und hypothalamisch-hypophysären Zentren gestört ist. Es kommt zu einer vermehrten Stimulation Corticotropin-produzierender Zellen des HVL. Klinisch manifeste Hypophysenadenome entwickeln sich nur selten oder werden selten beobachtet. Sie treten jedoch, wie neuere Beobachtungen zeigen, häufiger postoperativ auf und führen zu Sellaerweiterungen (*3, 45, 49*). Sind primär die Sella-erweiternden Tumoren beim Cushing-Syndrom vorhanden, so handelt es sich meist um gemischtzellige chromophobe Adenome, die weitgehend autonom, jedoch oft noch unter einer gewissen Kontrolle übergeordneter Zentren (*43, 48, 49, 59*), ACTH produzieren. Sie führen durch ihre Expansion zu Gesichtsfeldeinschränkungen. Klinisch imponiert häufig eine addisonartige Pigmentierung, die beim gewöhnlichen Hyperfunktions-Cushing meist nicht vorhanden ist. Maligne Hypophysentumoren sind selten (*24, 48*). Ein Hyperfunktions-Cushing kann auch durch nicht-endokrine Tumoren verursacht werden, die (wahrscheinlich) eine ACTH-ähnliche Substanz produzieren können (*39, 41*). Seit der ersten Beschreibung eines Cushing-Syndroms bei gleichzeitig bestehendem Bronchialcarcinom durch H. Brown 1928 (*9*) wurden zahlreiche Fälle von Cushing-Syndrom bei einem Bronchialcarcinom besonders vom oat-Zelltyp beschrieben (*28, 41, 71, 76*). Verschiedene Untersucher konnten in den letzten Jahren im Blut solcher Patienten und auch in den Tumoren selbst sowie in deren Metastasen ACTH-wirksames Material isolieren und quantitativ bestimmen (*29, 39, 41, 44, 50*). Ebenso wie beim

Hyperfunktions-Cushing wird das Vorhandensein einer Hyperplasie sowie eines für die Hyperfunktion verantwortlichen Faktors diskutiert. Auch andere Tumoren als Bronchialcarcinome, insbesondere Tumoren des Thymus, des Pankreas, der Prostata sowie NNR-Resttumoren der Ovarien und der Testes können ein Cushing-Syndrom verursachen bzw. werden häufiger beim Cushing-Syndrom beobachtet, als es dem zufälligen Zusammentreffen entspricht (7, 71).

Die NNR-Überfunktionszustände durch autonome ACTH-produzierende Tumoren der Hypophyse sowie durch nicht-endokrine Tumoren bieten durch ihre Autonomie und Resistenz gegenüber den verschiedenen Belastungs- bzw. Funktionsproben mitunter differentialdiagnostische Schwierigkeiten. Auffällige Pigmentierung, Sellavergrößerung oder der Nachweis eines nicht-adrenalen Tumors gestatten jedoch meist die Abgrenzung gegenüber dem Hyperfunktions-Cushing anderer Genese.

Gegenüber den ACTH-abhängigen Überfunktionszuständen der NNR treten ACTH-unabhängige NNR-Tumoren seltener als Ursache eines Cushing-Syndroms auf. Während bei Kindern ein Cushing-Syndrom in etwa 70% der Fälle auf einen NNR-Tumor (meist Carcinome) zurückzuführen ist, liegt bei Erwachsenen die Häufigkeit von NNR-Tumoren nur bei 10—25% (34, 36, 40, 52, 71). Nach Angaben der Literatur ist das Verhältnis zwischen Adenom und Carcinom als Ursache eines

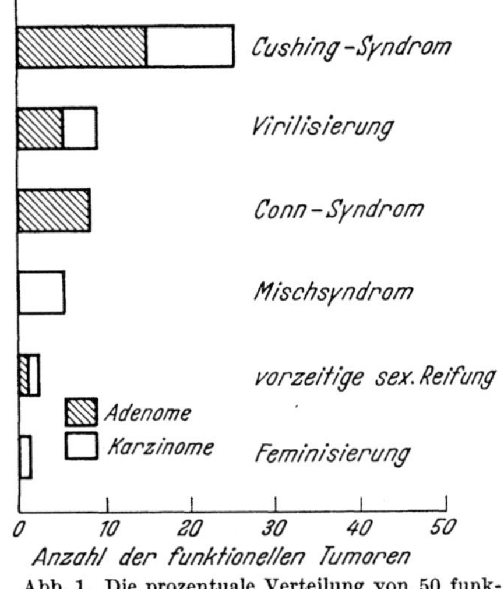

Abb. 1. Die prozentuale Verteilung von 50 funktionellen NNR-Tumoren mit Angabe ihrer endokrinen Wirkung (15)

C.-S. etwa 70/30—50/50%. Die NNR-Adenome, die meist in der Zona fasciculata liegen, kommen häufig multipel, gelegentlich jedoch auch solitär vor. Im wesentlichen funktionieren sowohl die Adenome als auch die Carcinome autonom und führen durch ihre excessive Cortisol-Überproduktion über den Reglermechanismus zu einer Atrophie des nicht befallenen NNR-Gewebes.

Adenome und Carcinome der NNR können praktisch alle Varianten der NNR-Hormone bzw. ihre Intermediärstufen produzieren und so eine Vielzahl klinischer Symptome, die von extremer Virilisierung bis zur Feminisierung reichen, hervorrufen. Auch können die Symptome des C.-S. völlig fehlen. In Abb. 1 sind die verschiedenen Wirkungen von 50 funktionellen endokrinen Tumoren nach einer Tabelle von Drx dargestellt (15).

Unabhängig von der verschiedenen Pathogenese ist jedoch immer eine vermehrte Cortisol-Produktion aus der NNR und die dadurch verursachte vermehrte Cortisol-Einwirkung auf den Organismus die Ursache des Cushing-Syndroms. Alle wesentlichen Symptome des Cushing-Syndroms lassen sich durch die Cortisol-Wirkung erklären (4) und durch eine exogen vermehrte Hydrocortison-Zufuhr imitieren. Letztlich ist diese Cortison-Überproduktion und ihre Wirkung auf den Eiweißstoffwechsel auch für den tödlichen Ausgang des unbehandelten Cushing-Syndroms verantwortlich.

1. Klinik und Symptomatologie des Cushing-Syndroms

Das Cushing-Syndrom ist eine seltene Erkrankung. In den Jahren 1956—1963 konnten wir an unserer Klinik mit einer großen endokrinen Ambulanz unter 26733 stationären Patienten nur 14mal ein Cushing-Syndrom feststellen. Das Manifestationsalter lag bei unseren Patienten zwischen dem 10.—51. Lebensjahr, im Durchschnitt bei 32 Jahren. Ein Junge mit einem Nebennieren-Carcinom war $10^{1}/_{2}$ Jahre alt. Von unseren Patienten waren 9 Frauen und 5 Männer (darunter der Junge mit dem NNR-Carcinom). Bei 11 Patienten wurde eine NNR-Hyperplasie,

Tabelle 1. *Prozentuale Verteilung der Symptome des Cushing-Syndroms zu Beginn der Erkrankung und bei der Aufnahme ins Krankenhaus* (nach SOFFER et al.)

Symptome	% Im Beginn d. Erkrankung	% Bei der Krankenhaus- aufnahme
Amenorrhoe.	38	72
Fettsucht.	34	86
Hirsutismus.	34	84
Vollmondgesicht.	22	92
Oligomenorrhoe	17	0
Cyclusstörungen	10	0
Hypertonie	10	88
Ödeme der unteren Extremitäten	10	66
Hämorrhagien.	6	68
Violette Striae.	6	50
Diabetesneigung.	4	84
Plethora	0	78
Kraftlosigkeit.	0	58
Osteoporose.	0	56
Frakturen	0	40
Psychische Veränderung	0	40
Teleangiektasien.	0	36
Büffelnacken	0	34

bei 2 Patienten ein NNR-Carcinom, bei 1 Patientin ein NNR-Adenom gefunden. Verteilung, Manifestationsalter und Pathogenese entsprechen den Angaben der Literatur (*26, 34, 40, 52, 73*).

In der *Anamnese* der Patienten finden sich vielfach bereits wichtige Hinweise. Bei einem Teil der Patienten kommt es relativ rasch zur Entwicklung der typischen Veränderungen im Habitus und in der Psyche. Meist treten frühzeitig Störungen im Sexualbereich auf, die sich bei der Frau in Oligomenorrhoe und Amenorrhoe, beim Manne in Libido- und Potenzverlust äußern. Ein wichtiges Kriterium ist die Angabe der Abnahme der körperlichen und geistigen Leistungsfähigkeit. Bei Jugendlichen und auch bei Erwachsenen spricht eine bereits seit früher Jugend vorhandene Fettsucht oder Neigung zu Adipositas im wesentlichen gegen ein Cushing-Syndrom. Dagegen deuten rasch auftretendes Vollmondgesicht und Fettansatz am Abdomen mit Aussparung der Extremitäten auf ein Cushing-Syndrom hin. Zeigt sich ein normales oder sogar gesteigertes Wachstum bei adipösen Jugendlichen, so ist ein Cushing-Syndrom fast mit Sicherheit auszuschließen.

Die Symptomatik ist im klassischen Fall so eindrucksvoll, daß die Diagnose allein auf Grund des äußeren Aspektes mit großer Sicherheit gestellt werden kann.

Es darf jedoch nicht verkannt werden, daß Ausmaß und Ausbildung der Symptome
von der Dauer der Erkrankung und damit vom Grad der Hydrocortison-Einwir-
kung auf den Organismus abhängig ist. Dies geht besonders deutlich aus einer Zusammenstellung von SOFFER (70) hervor. In Tab. 1 sind die wesentlichen Symptome des Cushing-Syndroms bei Beginn der Erkrankung und bei der Aufnahme
in das Krankenhaus aufgeführt. Die Progredienz der klinischen Erscheinungen mit
der Zeitdauer der Erkrankung ist deutlich zu erkennen. Es ist ferner ersichtlich,
daß im Beginn als auch im Verlauf der Erkrankung jedes Symptom fehlen kann.
Es kann daher kein Zweifel bestehen, daß in zahlreichen Fällen, bei denen noch kein
ausgeprägtes Cushing-Syndrom vorliegt, die Differentialdiagnose mitunter schwierig ist. Auf diese Tatsache haben verschiedene Autoren hingewiesen (65, 70, 73).

Welche Bedeutung kommt den einzelnen hauptsächlichen Symptomen für die
Differentialdiagnose des Cushing-Syndroms zu.

1. Das typische und ausgesprochene *Vollmondgesicht* ist nach übereinstimmender Ansicht
aller Autoren ein fast spezifisches Symptom. Neben der charakteristischen Fettverteilung und
der teilweise extremen Rötung ist es durch seine auffallende Ausdruckslosigkeit gekennzeichnet.
Auch beim iatrogenen Cushing-Syndrom wird es meist als erstes und oft auch einziges Symptom
beobachtet.

2. Die *Stammfettsucht* mit gleichzeitiger *Extremitäten-Atrophie* ist ein fast spezifisches
Symptom. Fettsucht mit erheblicher Übergewichtigkeit ist nicht spezifisch. Entscheidend
ist die charakteristische Fettverteilung. SOFFER (70) konnte nur bei $^1/_3$ seiner Patienten
ein Übergewicht von mehr als 10% und nur bei $^1/_5$ ein Übergewicht von mehr als 25%
feststellen. Allerdings wiesen 93% die typische Stammfettsucht auf. In unserem Patientengut waren die Verhältnisse nicht so eindeutig. 28% hatten ein Normalgewicht oder leichtes
Übergewicht bis zu 10%; 60% jedoch ein Übergewicht von mehr als 25%.

3. Die charakteristische rot-violette *Striabildung* mit etwa 10—20 cm Länge und 1—3 cm
Breite ist nach der Literatur (26, 34, 40) bei etwa 50% der Patienten vorhanden. Sie ist vorwiegend an den unteren vorderen Bauchpartien, an den Oberschenkeln, mitunter — jedoch
seltener — an den Oberarmen und Nates lokalisiert. Typische Striae wie beim Cushing-
Syndrom werden sonst nur noch in der Gravidität beobachtet. Sie unterscheiden sich sowohl
in Ausprägung, Lokalisation und Tinktion von der unspezifischen Striabildung, wie sie besonders in der Pubertäts- und Postpubertätsadipositas auftreten. Es ist schwer zu entscheiden,
welche Faktoren in der Genese die dominierende Rolle spielen. Sie können weder allein durch
Überdehnung der Haut noch allein durch die vermehrte Cortisol-Einwirkung auf die elasti-
schen Fasern der Haut sowie vermehrte Capillarfragilität erklärt werden. SIMKIN glaubt, daß
auch die rötliche Stria-Bildung bei adipösen Jugendlichen durch eine vermehrte Cortisol-
wirkung zu erklären ist (66). (Eine Abhängigkeit der Tinktion von der Höhe des Blutdruckes
konnten wir bei einigen Patienten beobachten.) Sie war bei 70% unserer Patienten mit einem
Cushing-Syndrom vorhanden, fehlte jedoch bei anderen trotz nachweisbar erheblich erhöhter
Ausscheidung der Cortisol-Metabolite. Demgegenüber konnten wir bei 71% von 117 adipösen
Patienten, bei denen kein Cushing-Syndrom verifiziert werden konnte, eine mehr oder weniger
ausgeprägte rötliche, teilweise sogar livide Striabildung feststellen. Wir billigen nur der
typischen Striabildung einen diagnostischen Wert zu. Das Fehlen spricht jedoch nicht gegen
ein Cushing-Syndrom.

4. *Hautveränderungen* verschiedener Art können zusätzlich mehr oder weniger ausgeprägt
vorhanden sein. Es finden sich Seborrhoen, Teleangiektasien, Atrophien und mitunter addisonartige Pigmentierungen. Nach Adrenalektomie tritt oft eine monatelang anhaltende Schuppung
auf. Typisch ist das Auftreten von Ekchymosen. Alle diese Veränderungen sind jedoch relativ
unspezifisch und haben nur eine die Diagnose unterstützende Bedeutung. Ihr Fehlen spricht
nicht gegen das Vorliegen eines Cushing-Syndroms.

5. *Hirsutismus* und *Virilismus*. Bei Frauen wird oft ein mäßiggradiger Hirsutismus beobachtet. Er ist besonders bei den Mischtypen des Cushing-Syndroms ausgeprägt. Stärkere
Virilisierung mit Clitoris-Hypertrophie, Tieferwerden der Stimme, Atrophie der Mammae
lenken den Verdacht auf ein NNR-Carcinom. Die Differentialdiagnose gegenüber einem AGS

sowie androgenproduzierenden Prozessen des Ovars kann gelegentlich schwierig sein. Dies ist besonders der Fall, wenn gleichzeitig eine Adipositas vorhanden ist. Rasch eintretende Virilisierung bzw. vorzeitige sexuelle Reifung bei Kindern ist immer auf ein NNR-Carcinom verdächtig.

6. Störungen im *Sexualbereich* wie Oligo- und Amenorrhoe bei Frauen sowie Libido- und Potenzverlust bei Männern sind ein fast obligates Symptom. Im allgemeinen spricht eine Gravidität gegen das Bestehen eines Cushing-Syndroms. Gelegentlich konnte jedoch die Entwicklung einer Gravidität bei bereits ausgeprägtem Cushing-Syndrom beobachtet werden (*6, 11, 18, 27*). In diesen Fällen kann das Bestehen eines Cushing-Syndroms leicht übersehen werden.

Nach Behandlung durch totale oder subtotale NNR-Resektion (*35*) konnten wir bei 2 Frauen die Entwicklung einer Gravidität, die auch normal ausgetragen wurde, beobachten. Da Störungen von Libido und Potenz sowie Oligo- und Amenorrhoe auch bei Adipösen bestehen, kommen diesen Störungen in der Differentialdiagnose nur eine zusätzliche bekräftigende Beweiskraft zu.

7. Eine ausgeprägte *Hypertonie* wurde bei 12 von unseren 14 Patienten mit einem Cushing-Syndrom gefunden. Nach der Literatur ist sie in einem Prozentsatz von 50—80% nachweisbar (*26, 34, 40, 42, 52*). Eine mäßig ausgeprägte Hypertonie konnte jedoch auch bei 57% unserer adipösen Patienten mit einzelnen Zeichen des Cushing-Syndroms festgestellt werden, so daß diesem Symptom gleichfalls nur eine zusätzliche Bedeutung zugebilligt werden kann.

8. Störungen des *Kohlenhydratstoffwechsels*, insbesondere eine pathologische Zuckerbelastungskurve, lassen sich in etwa 80% aller Fälle mit einem Cushing-Syndrom feststellen. Ein insulinbedürftiger Steroiddiabetes ist dagegen nur in einem Prozentsatz von 15—30% vorhanden. In unserem Patientengut war bei $^2/_3$ der Fälle eine pathologische Zuckerbelastung und bei 36% ein insulinbedürftiger Diabetes nachweisbar. Dem Nachweis eines Steroiddiabetes, der sich mit Hilfe von Metopiron erbringen läßt (s. u.), kommt eine hohe differentialdiagnostische Bedeutung zu, während dem pathologischen Ausfall der Zuckerbelastung nur eine untergeordnete Rolle beigemessen werden kann.

9. Dem Nachweis einer *Osteoporose* mit Fischwirbelbildung und Spontanfrakturen muß ein fast beweisender diagnostischer Wert zuerkannt werden, wenn diese mit anderen Symptomen des Cushing-Syndroms einhergeht. Ausmaß und Ausprägung der Osteoporose ist besonders von der Dauer der Erkrankung bzw. vom Ausmaß der Cortisol-Überproduktion abhängig. Bei nicht typisch ausgeprägter Osteoporose ist der Nachweis bekanntlich unsicher; durchschnittlich wird die Osteoporose bei Patienten mit einem Cushing-Syndrom in etwa 50—70% beobachtet (*26, 34, 40, 42, 52*). Sie war bei 50% unserer Patienten deutlich, bei 30% typisch. Bei den adipösen Patienten mit einzelnen Zeichen des Cushing-Syndroms konnte in keinem Fall eine Osteoporose nachgewiesen werden. Es muß betont werden, daß das Fehlen einer röntgenologisch nachweisbaren Osteoporose das Bestehen eines Cushing-Syndroms besonders im Frühstadium nicht ausschließt.

10. Neben den vorgenannten Symptomen müssen noch die im wechselnden Ausmaß vorkommenden *psychischen Veränderungen* erwähnt werden. Sie waren bei keinem unserer Patienten ausgeprägt. Ihre Häufigkeit wird nach der Literatur mit etwa 40% angegeben.

11. Seltener werden Ödeme der unteren Extremitäten, Kopfschmerzen, Polyurie und Polydypsie, hämorrhagische Diathese, Nephrolithiasis und ein mäßig ausgeprägter Exophthalmus in etwa 5% der Fälle beobachtet.

Sind die Kardinalsymptome des Cushing-Syndroms ausgeprägt und besonders die Folgen der eiweißkatabolen Stoffwechselsituation deutlich — die sich als Muskelatrophie und Adynamie, Osteoporose und Steroiddiabetes äußern —, so ist die Diagnose klinisch leicht zu stellen.

2. Adipositas mit einzelnen Zeichen des Cushing-Syndroms

Differentialdiagnostische Schwierigkeiten bestehen mitunter gegenüber adipösen Patienten, bei denen einzelne oder unsichere Symptome allein oder miteinander kombiniert auftreten. Hier erhebt sich die für die therapeutischen Konsequenzen wichtige Frage, ob bei diesen Patienten nicht ein beginnendes Cushing-Syndrom vorliegt. Wir konnten in den letzten 4 Jahren unter etwa 400 Patienten mit einer Adipositas 117 Patienten, bei denen einzelne oder mehrere unsichere Symptome ähnlich denen eines Cushing-Syndroms bestanden, stationär untersuchen.

In Tab. 2 ist die Häufigkeit der bei diesen Patienten bestehenden Symptome der bei unseren Patienten mit einem nachgewiesenen Cushing-Syndrom gegenübergestellt.

Es ist aus dieser Zusammenstellung zu ersehen, daß auch bei adipösen Patienten häufig einzelne oder mehrere unsichere Symptome des Cushing-Syndroms in mehr oder weniger ausgeprägter Form gleichzeitig vorhanden sein können. Sichere Zeichen des Eiweißkatabolismus ließen sich jedoch bei keinem der adipösen Patienten nachweisen. Sehr häufig fand sich eine allerdings meist nicht typische Striabildung, die sich durch Ausmaß, Lokalisation und Intensität der Färbung von der beim Cushing-Syndrom unterschied. Die Striae waren meist dünn, kürzer, oft nur an den Spitzen gefärbt und hinterließen weißliche Narben; gelegentlich waren jedoch Übergänge zur ausgeprägten Striabildung erkennbar. Das zweithäufigste Symptom war eine Hypertonie.

Tabelle 2. *Prozentuale Verteilung der verschiedenen Symptome bei 14 Patienten mit einem nachgewiesenen Cushing-Syndrom und bei 117 adipösen Patienten mit einzelnen Zeichen des Cushing-Syndroms, bei denen ein Cushing-Syndrom nicht verifiziert werden konnte*

	Prozent der Fälle	
	Adipöse (n = 117)	Cushing S. (n = 14)
Typisches Vollmondgesicht . .	—	93
Vollplethorisches Aussehen . .	12	93
Stammfettsucht	15	72
Adynamie u. Muskelatrophie .	—	100
Breite rote Striae	5	71
Dünne rote Striae	61	—
Störungen im Sexualbereich . .	22	93
Akne.	13	64
Hirsutismus bei Frauen. . . .	36	78
Hypertonie	57	86
Path. Glucosebelastung. . . .	14	36
Man. Diabetes.	4	36

Von den adipösen Frauen zeigten 36% einen leicht bis mäßig ausgeprägten Hirsutismus. Ferner fanden sich gelegentlich eine vorwiegend auf den Stamm lokalisierte Adipositas ohne Extremitäten-Atrophie, Störungen im Sexualbereich sowie mitunter ein pathologisches Ergebnis der Glucosebelastung und seltener ein manifester Diabetes. Ein typisches Vollmondgesicht war nie ausgeprägt, ein vollplethorisches Aussehen bei 12% der Adipösen nachweisbar.

Die *allgemeinen Laboratoriumsuntersuchungen* haben bei der Diagnose des Cushing-Syndroms nur einen untergeordneten Wert. Nur bei einem Viertel unserer Patienten mit einem Cushing-Syndrom fand sich eine ausgesprochene Leukocytose, dagegen war fast immer eine relative Lymphopenie nachweisbar. Eine typische Eosinopenie wurde bei etwa $^1/_3$ der Patienten vermißt. Auch eine Erhöhung des Hämoglobin und der Erythrocytenzahl war eindeutig nur bei 35% unserer Patienten vorhanden. Sie wurde jedoch auch bei adipösen Patienten mit einzelnen Zeichen des Cushing-Syndroms gelegentlich beobachtet (15%).

Veränderungen der Serum-Elektrolyte konnten gleichfalls nur selten festgestellt werden. Nur zwei von 14 Patienten zeigten eine eindeutige Hypokaliämie. Sie scheint jedoch häufiger latent vorzuliegen und kann postoperativ zu bedrohlichen Komplikationen führen (2, 26). Sie soll bei Patienten mit einem sog. Cortisol-

Cushing nach Hartenbach ausgeprägter vorhanden sein (26). Für die Differential-diagnose erscheint es von Bedeutung, daß beim Cushing-Syndrom durch extra-adrenale, nicht-endokrine Tumoren fast immer eine erhebliche Hypokaliämie nachweisbar ist. Diese ist offenbar allein auf die meist stärker als beim gewöhn-lichen Cushing-Syndrom gesteigerte Cortisol-Überproduktion zurückzuführen (54). Im Gegensatz zum Conn-Syndrom ist der Natriumspiegel im Serum meist normal. Wir konnten nur einmal eine Erhöhung auf 145 mval. feststellen. Eine Erhöhung des Calciumspiegels ließ sich bei keinem unserer Patienten finden, dagegen war die Ca-Ausscheidung im Harn meist erhöht.

Eine Verminderung des Gesamteiweißes im Serum konnte nur bei einer Patien-tin gesichert werden. Die Fraktionsverteilung in der Elektrophorese war uneinheit-lich und ließ keinen deutlichen Trend erkennen. Relativ häufig war der Gesamt-cholesterinspiegel erhöht.

Bei den adipösen Patienten war meist keine sichere Abweichung der Labor-befunde von der Norm festzustellen. Zu erwähnen ist noch, daß sowohl bei den Patienten mit einem Cushing-Syndrom als auch bei den adipösen Patienten mit einzelnen Symptomen des Cushing-Syndroms die Schilddrüsenfunktion bzw. die Radiojodstoffwechseldaten immer normal waren.

Sowohl die klinische Symptomatik als auch die allgemeinen Laboratoriums-untersuchungen erlauben also nicht immer, ein beginnendes oder nicht typisch aus-geprägtes Cushing-Syndrom von den benignen oligosymptomatischen Bildern ab-zutrennen. Dies ist jedoch meist mit Hilfe der Steroiddiagnostik möglich.

3. Steroiddiagnostik

Es ist nicht möglich, in der Kürze der Zeit einen umfassenden Überblick über die komplizierten Stoffwechselverhältnisse in der Sekretion sowie im Ab- und Umbau der Nebennierenrindensteroide, die das Ausscheidungsspektrum beein-flussen, zu geben.

Entsprechend der Genese des Cushing-Syndroms steht der Nachweis des gestei-gerten Cortisol-Stoffwechsels im Vordergrund der Diagnostik. Für die pathogene-tische Unterscheidung sind die schwierig durchführbaren ACTH-Bestimmungen im Blutplasma noch nicht routinemäßig geeignet. In letzter Zeit erbrachten jedoch methodische Verbesserungen die Möglichkeit, ACTH auch in kleineren Blutmengen nachzuweisen (55, 56). So konnte die bereits bekannte Rhythmik der Plasma-corticoide mit den Schwankungen der ACTH-Sekretion korreliert und gezeigt werden, daß die Aufhebung dieser Rhythmik bei Patienten mit einem Cushing-Syndrom nicht nur für die Plasmacorticoide, sondern auch für die ACTH-Sekre-tion gilt (57). Eine Kombination der ACTH-Bestimmung im Blut und der Steroid-analytik würde ohne Zweifel eine Bereicherung der Diagnostik darstellen und die Unterscheidung von ACTH-abhängigen und ACTH-unabhängigen Überfunktions-zuständen der NNR erleichtern.

Im Vordergrund steht heute noch die Steroiddiagnostik mit der Bestimmung der Metabolite des Cortisol-Stoffwechsels nach den verschiedenen Methoden: *Silber-Porter-Chromogene*, totale OHCS, 17-ketogene Steroide und Corticoide nach Staudinger und Bauer (72). Daneben haben die chromatographischen Verfahren mit Erfassung der einzelnen C_{21}-Steroidmetabolite einen hohen diagnostischen

Wert. Die Bestimmung der 17-Ketosteroide und ihre chromatographische Auftrennung liefert ebenfalls diagnostische Hinweise und ist besonders in der Differentialdiagnose des Hyperfunktions-Cushing gegenüber dem durch Nebennierenrindentumoren, insbesondere Carcinome, verursachten Cushing-Syndrom von hoher differentialdiagnostischer Bedeutung. Mit Hilfe der chromatographischen Trennung der 17-Ketosteroide unter Funktionsbelastung ist es insbesondere möglich, verschiedene mit Virilisierung einhergehende, mitunter ein Cushing-Syndrom vortäuschende Prozesse abzugrenzen (51).

Die Basalausscheidung der Corticoide und der 17-Ketosteroide hat jedoch nur bei mehrfachen Bestimmungen und bei eindeutigen Erhöhungen unter Berücksichtigung der Alters- und Geschlechtsdifferenzen einen diagnostischen Wert. In den letzten Jahren hat die Ausscheidung des freien Cortisols im Harn in der Diagnostik des Cushing-Syndroms zunehmende Bedeutung erlangt.

Die Bedeutung der Bestimmung des freien Cortisols im Harn für die Diagnose des Cushing-Syndroms beruht auf der Tatsache, daß das biologisch wirksame Cortisol frei und nicht an Transcortin oder Albumin gebunden im Plasma vorliegt. Diese ungebundene Fraktion macht etwa 10% des totalen unkonjugierten Plasmacortisols aus, das als *Silber-Porter-Chromogen* bestimmt werden kann (5, 16, 19, 69). Die Hauptmenge des Plasmacortisols ist an Transcortin mit hoher Bindungsaffinität und niedriger Bindungskapazität gebunden (14). Eine Cortisolüberproduktion aus der Nebennierenrinde wirkt sich besonders auf den Spiegel an freiem Cortisol aus. Dieses wird vermehrt durch die Niere filtriert und trotz einer Rückresorption von etwa 85% (61) ist die Ausscheidung groß genug, ein Maß für die Menge an freiem ungebundenem Cortisol im Plasma abzugeben. Mit Cortisol-Infusionen konnte

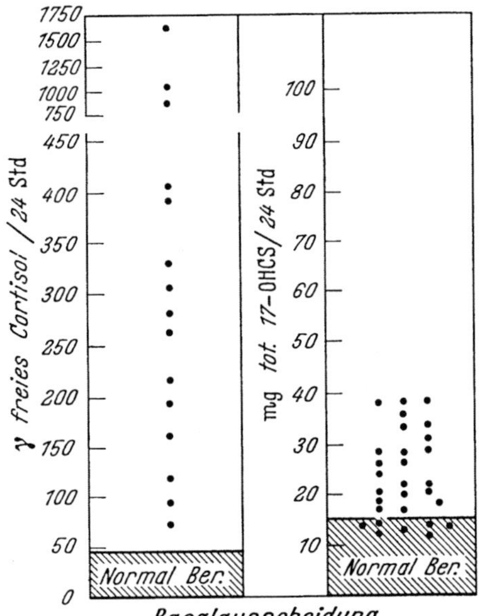

Abb. 2. Vergleichende Untersuchungen über die Ausscheidung der tot. 17 OHCS. (Methode APPLEBY u. Mitarb.) und die Ausscheidung von Cortisol (Methode NEHER u. WETTSTEIN) bei Patienten mit einem Cushing-Syndrom (20, 21)

gezeigt werden, daß es zu einem scharfen Anstieg des freien Cortisolspiegels im Harn kommt, wenn der freie Plasmacortisolspiegel gerade über der Norm liegt. Beim Cushing-Syndrom genügt offenbar allein die durch den Verlust der Tagesrhythmik verursachte dauernde leichte Erhöhung des freien Cortisolspiegels im Plasma, um zu einer vermehrten Clearance vom freien Cortisol durch die Niere zu führen.

Ähnlich wie andere Autoren konnten wir feststellen, daß die Ausscheidung an freiem Cortisol im Harn beim Cushing-Syndrom immer über die Norm, die bei der von uns verwandten Methode zwischen 14 und 30 γ lag, erhöht war (21). Wenn auch die Exkretion der 17-OHCS meist ebenfalls höher lag, so war doch die prozentuale Erhöhung der Basalausscheidung von Cortisol eindrucksvoller als die der Gruppenbestimmungen, bei denen häufiger Überlappungen mit hochnormalen Werten vorkommen können (Abb. 2).

Auch die Bestimmung des 17-OHCS-Spiegels im Blut ohne vorhergehende ACTH-Stimulierung hat offenbar nur einen begrenzten diagnostischen Wert. Der unter verschiedenen Einflüssen erhöhte Transcortingehalt führt zu Fehldeutungen, so daß die Messung der Bindungskapazität des Transcortins, die natürlich umständlicher ist, in Zweifelsfällen eine klarere Diagnostik ermöglicht. Der Verlust der Tagesrhythmik ist ein deutlicher Hinweis für das Vorliegen eines Cushing-Syndroms.

Die Bestimmung der Cortisolsekretionsrate ist ohne Zweifel eine der differenziertesten Methoden. Sie gestattet den sicheren Nachweis einer Nebennierenrindenüberfunktion. Eine Erhöhung der Cortisolsekretionsrate ist beim Cushing-Syndrom, gleich welcher Genese, immer nachzuweisen. Die Bestimmung der Cortisolsekretionsrate ist jedoch für die Routinediagnostik noch zu aufwendig und

Tabelle 3. *Cortisolsekretionsraten und Steroidbefunde in Plasma und Harn bei Patienten mit Cushing-Syndrom verschiedener Pathogenese (8)*

Normalbereich	NNR-Ca N = 1	NNR-Adenom N = 3	NNR-Hyperpl. Nr = 13	Nicht-endokrine Tumoren N = 8
17-KGS mg/24 Std: ♀ 5—18 ♂ 6—22	39	16,3 (10—21)	18,2 (8—41)	49,2 (16—83)
Cortisol-Sekretionsrate mg/24 Std: 7—29:15,7 ± 5,6	124	63 (39—82)	59,4 (32—111)	230 (65—420)
Harn-Cortisol Gamma/24 Std: 16—80	298	165 (50—302)	182 (38—588)	1525 (111—4000)
Plasma-Cortisol: 5—16 γ-%	33	20,5 (15—26)	14,5 (3—25)	30,1 (12—123)

Nach R. V. Brooks u. Mitarb. 1963.

speziell eingerichteten Kliniken vorbehalten. In Tab. 3 sind die Werte für die Cortisolsekretionsrate bei Patienten mit einem Cushing-Syndrom verschiedener Genese im Vergleich zu den mit anderen Methoden erhobenen Befunden gegenübergestellt (die Werte sind einer Arbeit von Brooks u. Mitarb. 1963 entnommen [8]). Aus dieser Tabelle ist ferner zu ersehen, daß der Cortisolstoffwechsel bei den Patienten mit einem Cushing-Syndrom durch nichtendokrine Tumoren am stärksten gesteigert ist.

Gegenüber den Basalbestimmungen der Steroide mit den üblichen Methoden haben sich die verschiedenen Belastungs-Teste in der Diagnostik und in der Differentialdiagnose des Cushing-Syndroms überlegen gezeigt.

An erster Stelle steht der standardisierte ACTH-Test mit 25 IE ACTH i.v. über 8 Std. Die früher geübte Funktionsprüfung mit Depot-ACTH ist wegen der Unsicherheit der Resorption weitgehend verlassen worden. Beim Cushing-Syndrom wird im standardisierten ACTH-Test gegenüber dem Normalen ein überschießender Anstieg besonders der nicht konjugierten 17-Hydroxycorticosteroide sowie der totalen 17-Hydroxycorticosteroide im Plasma und der — mit den verschiedenen Methoden bestimmten — Corticoide und 17-Ketosteroide im Harn gefunden (*22, 60*). Die Corticoid-Werte unter ACTH im Harn von Patienten mit Cushing-Syndrom fallen jedoch häufig in den oberen Streubereich der Corticoid-Ausscheidung unter ACTH bei adipösen Patienten. Deshalb ist der diagnostische Wert des Testes begrenzt. In der Differentialdiagnose des Hyperfunktions-Cushings gegenüber dem durch Nebennierenrinden-Tumoren verursachten Cushing-Syndrom ist von Bedeutung, daß Carcinome meist keinen Anstieg der Plasma-Steroide und 17-Ketosteroide sowie der Corticosteroide im Harn erkennen lassen. Ausnahmen wurden von VOIGT, TAMM und GALLAGHER (*23, 78*) beschrieben. Nebennierenadenome verhalten sich oft völlig uneinheitlich, so daß die differentialdiagnostische Abgrenzung mitunter schwierig ist. Eine unserer Patientinnen mit einem Nebennierenrindenadenom zeigte eine überschießende Ausscheidung unter ACTH. Unter ACTH ist auch die Ausscheidung des freien Cortisols prozentual meist excessiver gesteigert als die Ausscheidung der mit den verschiedenen Methoden bestimmten Corticoide (s. Abb. 3). Eine stärkere Steigerung der Cortisolausscheidung konnte auch von anderen Autoren festgestellt werden (*13, 25, 58*).

Mit Hilfe des Metopiron-Testes kann bei der Erkennung von Überfunktionzuständen der Nebennierenrinde gleichfalls ein wichtiger Beitrag geliefert werden. Seit Einführung des Metopiron-Testes durch LIDDLE (*36, 37*) wurde die Brauchbarkeit dieses Testes zur Unterscheidung der verschiedenen Störungen des Hypophysenvorderlappen-Nebennierenrindensystems vielfach bestätigt. Durch Metopiron wird die 11-β-Hydroxylase gehemmt und dadurch die Cortisolproduktion blockiert. Der erniedrigte Cortisolspiegel im Blut führt bei intaktem Reglermechanismus zu einer vermehrten ACTH-Ausschüttung des HVL. Der Metopirontest hat sich

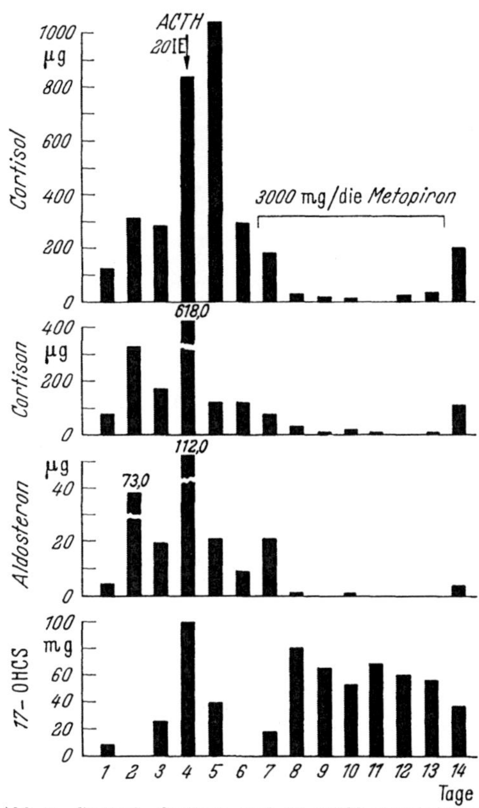

Abb. 3. Cortisol-, Cortison- und 17 OHCS-Ausscheidung im Urin unter ACTH-Belastung und SU 4885 bei der Pat. Mo. (*22*)

vornehmlich zur Erkennung von Unterfunktionszuständen der corticotropen Funktion des HVL eingeführt. Es sei an dieser Stelle auf die Untersuchungen von ECKLER und BIERICH, ferner von SCHRÖDER sowie von KLEINFELDER und HUSMAN (*17, 32, 63*) hingewiesen. Im Gegensatz zu den Unterfunktionszuständen des HVL, bei denen der Nachweis des Anstiegs von 11 Desoxy-Cortisol und 11 Desoxy-Corticosteron spezifischer und prozentual deutlicher ist, kann bei der Differenzierung von Überfunktionszuständen die Bestimmung der totalen 17-OHCS, der Silber-Porter-Chromogene und der BTC-Corticoide benutzt werden (*30, 38*). In Abb. 4 ist die Ausscheidung der Tetrazoliumblau-reduzierenden Corticoide bei verschiedenen Patienten unter Basalbedingung sowie unter 25 IE ACTH i.v. in 8 Std und 3000 mg Metopiron in 24 Std (alle 2 Std 250 mg) dargestellt. Bei den Werten unter Metopiron handelt es sich um den maximalen Anstieg unter Metopiron, der meist am Tage nach der Metopiron-Gabe beobachtet wurde.

Mit Metopiron ist es ferner möglich, den Steroid-Diabetes gegenüber den anderen Diabetesformen abzugrenzen. In Abb. 5 ist die Kohlenhydratbilanz vor,

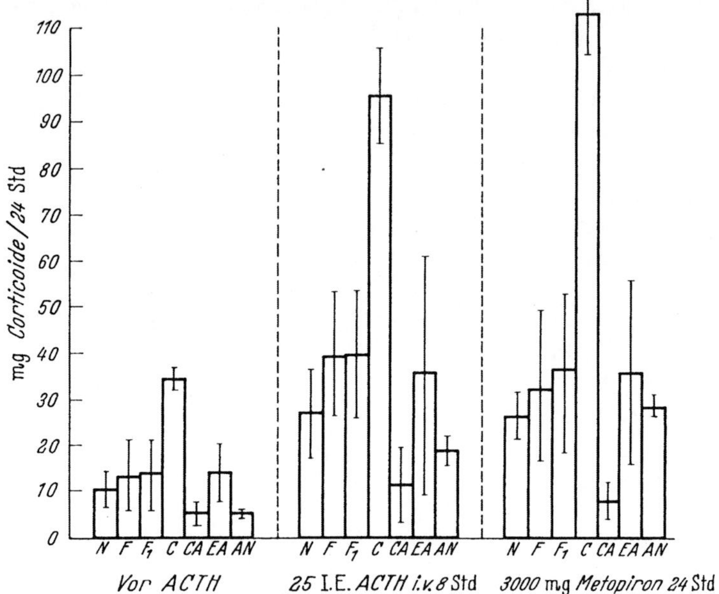

Abb. 4. Corticoidausscheidung bei Patienten mit verschiedenen Krankheitsbildern vor und unter ACTH und Metopiron

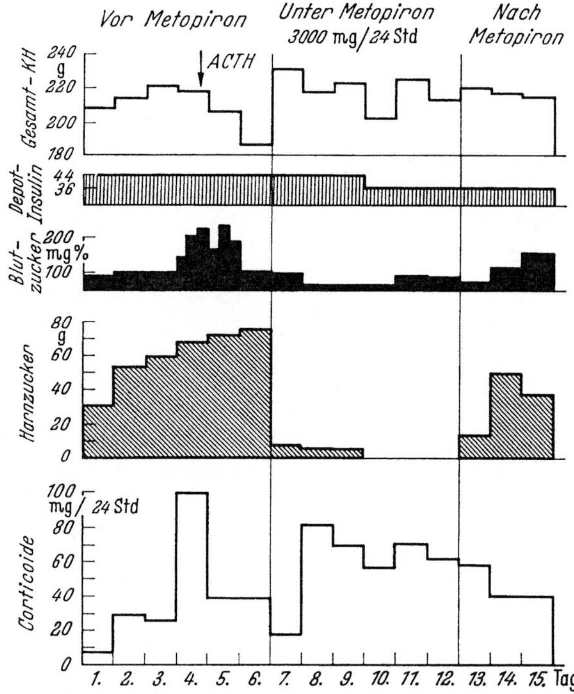

Abb. 5. Darstellung der Kohlenhydratbilanz vor, unter und nach Gabe von Metopiron bei einer Patientin mit einem Cushing-Rezidiv und Steroiddiabetes

unter und nach Gabe von Metopiron bei einer Patientin mit einem Cushing-Rezidiv und einem Steroid-Diabetes nach subtotaler Adrenalektomie dargestellt. Im Gegensatz zum Steroid-Diabetes konnten wir bei 4 Patienten mit einem unkomplizierten Diabetes mellitus unter mehrtägiger Gabe von 3000 mg Metopiron in der oben angegebenen Dosierung keinen sicheren Einfluß auf die KH-Bilanz beobachten.

Der Hemm-Test mit Dexamethason hat in der Differentialdiagnose des Cushing-Syndroms eine immer größere Bedeutung erlangt. Seit der Einführung dieses Testes durch LIDDLE (*90*) sind zahlreiche Bestätigungen über die Brauchbarkeit, insbesondere in Verbindung mit der Messung der Silber-Porter-Chromogene im Harn, erfolgt. Nach dem Schrifttum und unseren eigenen Erfahrungen ist eine Abtrennung des Cushing-Syndroms gegenüber den Pseudoformen mit Hilfe dieses Testes möglich. Er gestattet ferner eine Differenzierung zwischen Nebennierenrindenhyperplasie gegenüber den Nebennierenrindentumoren. Unter Gabe von 0,5 mg Dexamethason alle 6 Std über 2 Tage fällt der 17-OHCS-Spiegel im Harn unter 4—2,5 mg/24 Std ab. Ein ungenügender Abfall ist nach der Literatur auf ein Cushing-Syndrom verdächtig. Erfolgt unter einer weiteren Gabe von 2 mg alle 6 Std über 48 Std kein eindeutiger Abfall, so besteht der Verdacht auf einen Nebennierenrindentumor.

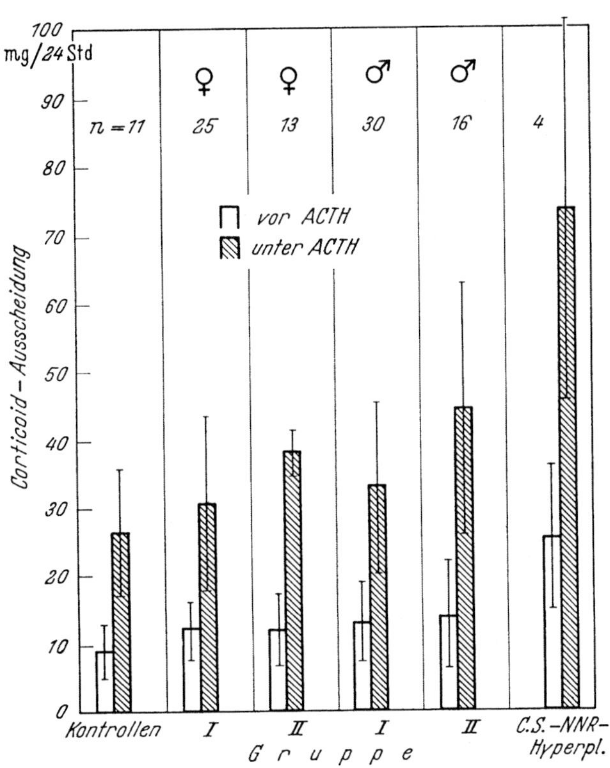

Abb. 6. Corticoidausscheidung vor und unter ACTH bei adipösen Patienten mit einzelnen oder mehreren Symptomen des Cushing-Syndroms und bei Patienten mit einem Cushing-Syndrom durch NNR-Hyperplasie

In der letzten Zeit wurde jedoch von TAMM (*75*) und SILVERMANN (*67*) über ein Versagen des Hemm-Testes berichtet, so daß gewisse Einschränkungen gemacht werden müssen. Ein vermindertes Ansprechen auf den Hemm-Test wurde auch bei autonomen ACTH-produzierenden Tumoren beobachtet, ebenso soll bei Nebennierenrindenhyperplasien durch nicht endokrine Tumoren eine verminderte Ansprechbarkeit bestehen. Eine prompte Reaktion auf den Hemm-Test in der Dosierung von 2 mg und Abfall der Silber-Porter-Chromogene im Harn ($<$ 2,5 mg/24 Std) schließt ein Cushing-Syndrom praktisch aus. Auch bei den von uns untersuchten Patienten mit einem Pseudo-Cushing-Syndrom konnte eine Normalisierung der Steroidausscheidung unter Dexamethason immer erreicht werden. Auch in der Differentialdiagnose des Nebennierenrindencarcinoms gegenüber dem adrenogenitalen Syndrom läßt sich durch die prompte Normalisierung der erhöhten 17-Ketosteroidausscheidung schon durch kleine Mengen Dexametason ein NNR-Carcinom ausschließen.

Wie bereits eingangs gesagt, ist die Differentialdiagnose zwischen Cushing-Syndrom und Adipositas, die mit einzelnen Zeichen des Cushing-Syndroms einhergeht, klinisch oft nur schwer zu stellen. Die Möglichkeit, ein beginnendes Cushing-Syndrom zu übersehen, existiert. Wir haben bei 101 Patienten — 48 Frauen und 53 Männern —, die neben einer Adipositas mehrere andere Symptome des Cushing-Syndroms, nicht aber das Vollbild aufwiesen, die Tetrazoliumblau-reduzierenden Corticoide nach Staudinger in der Modifikation von Staib und Teller im Harn vor und nach Belastung mit ACTH sowie teilweise mit Metopiron untersucht und in Zweifelsfällen den Dexamethason-Brems-Test angeschlossen. Wir teilten die Patienten in drei Gruppen. Die Kontrollgruppe bestand aus Adipösen ohne weitere Merkmale eines Cushing-Syndroms; Gruppe 1 aus Patienten mit eins bis drei Symptomen, Gruppe 2 aus Patienten mit drei bis vier Symptomen des Cushing-Syndroms. Es zeigte sich, daß bei Gruppe 1 und 2 die Basalausscheidung gegenüber der Vergleichsgruppe schwach signifikant erhöht war (p < 0,05). Eine Abhängigkeit der Basalausscheidung von der Anzahl der Symptome bestand jedoch nicht. Unter ACTH zeigte die Gruppe 1 zwar eine leichte Erhöhung der Harncorticoide gegenüber den Kontrollen, die jedoch statistisch nicht zu sichern war. Bei den Patienten mit drei bis vier Merkmalen war die Corticoidausscheidung gegenüber den Kontrollen stark signifikant erhöht (p < 0,01). Die mittlere Ausscheidung lag bei den männlichen Patienten so hoch, daß sie in den unteren Streubereich der Corticoidwerte von Patienten mit echtem Cushing-Syndrom unter ACTH fiel. Mit dem ACTH-Test war es also nicht möglich, bei diesen Patienten das Vorliegen des Cushing-Syndroms auszuschließen. In Abb. 6 ist die Basalausscheidung mit der Ausscheidung der Corticoide unter ACTH bei Patienten mit einer Adipositas mit einzelnen oder mehreren Zeichen des Cushing-Syndroms gegenüber vier Patienten mit einem Cushing-Syndrom durch Nebennierenrindenhyperplasie mit der

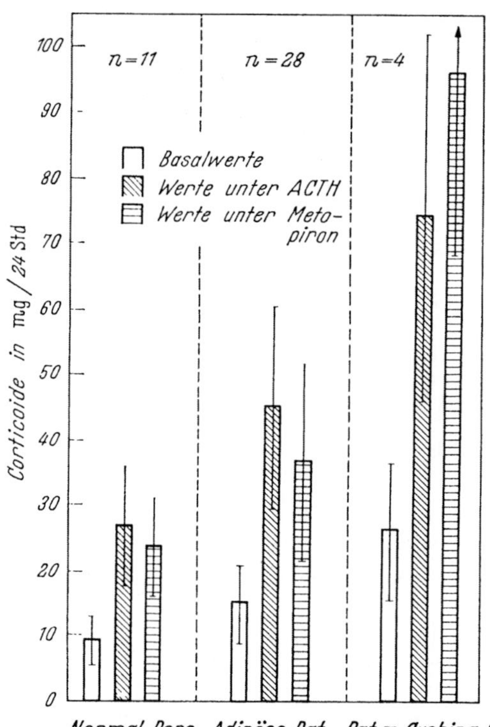

Abb. 7. Corticoidausscheidung vor und nach Funktionstesten mit ACTH und Metopiron bei Kontrollpersonen adipösen Patienten und Patienten mit einem Cushing-Syndrom

Standardabweichung dargestellt. Die Überlappung an den einzelnen Gruppen ist deutlich.

Bei 28 adipösen Patienten mit einzelnen Zeichen des Cushing-Syndroms wurden sowohl der ACTH-Test als auch eine Metopiron-Belastung durchgeführt (Abb. 7). Dabei zeigte sich das gleiche Verhalten der Corticoidausscheidung unter Basalbedingung und unter ACTH in den einzelnen Gruppen.

Auffällig war, daß die Ausscheidung der Tetrazoliumblau-reduzierenden Corticoide unter Metopiron bei den Patienten mit einem Cushing-Syndrom bei Nebennierenrindenhyperplasie im Durchschnitt höher, bei Patienten mit einer Adipositas und mehreren Zeichen des Cushing-Syndroms aber ebenso wie bei der Kontrollgruppe durchschnittlich niedriger als unter ACTH lag. Dieser Befund könnte darauf hinweisen, daß bei der Adipositas mit einzelnen Zeichen des Cushing-Syndroms keine dem Cushing-Syndrom vergleichbare Erhöhung der ACTH-Sekretion zugrunde liegt. Inwieweit diese Befunde dem durchschnittlichen Verhalten entsprechen, kann wegen der geringen Zahl der untersuchten Patienten mit einem Cushing-Syndrom durch Nebennierenrindenhyperplasie nicht entschieden werden.

Bei den einzelnen Patienten, bei denen nach dem ACTH-Test und dem Ausfall der Metopiron-Belastung der Verdacht auf einen Hypercorticismus bestand, wurde der Dexamethason-Hemmtest angeschlossen, der immer eine ausreichende Bremswirkung erkennen ließ. In Abb. 8 ist das Verhalten des Dexamethason-Hemmtestes bei adipösen Patienten und bei Patienten mit einem Cushing-Syndrom gegenübergestellt.

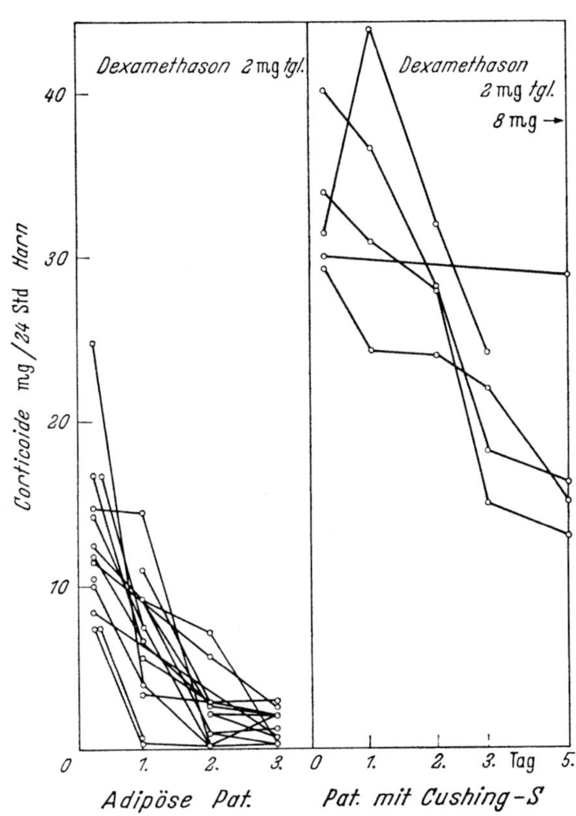

Abb. 8. Corticoidausscheidung während des Dexamethason-Hemmtestes (0,5 mg/stdl.) bei adipösen Patienten und bei Patienten mit einem Cushing-Syndrom

Auch die Bestimmung des freien Cortisols im Harn bei Patienten mit einer Adipositas mit einzelnen und mehreren Zeichen des Cushing-Syndroms ließ gegenüber Patienten mit einem Cushing-Syndrom immer eine im Normbereich liegende Ausscheidung erkennen (20).

Die Tatsache, daß bei Fettsüchtigen offenbar ein gesteigerter Cortisolstoffwechsel besteht, ist aus den Untersuchungen verschiedener Autoren (10, 46, 53,

64, 66, 74), die eine erhöhte Ausscheidung der verschieden bestimmten Corticoide fanden, bekannt. Die Plasma-Corticoide bzw. die 17—21-Hydroxy-20-Ketosteroide werden jedoch meist normal, von verschiedenen Autoren sogar erniedrigt gefunden (*46, 74*). Die Eliminationsrate für Cortisol ist erhöht. Die Untersuchungen über die Cortisolsekretionsrate bei Adipösen, die zuerst von Karl u. Mitarb. (*31, 46, 47*) sowie von anderen Autoren durchgeführt wurden, haben gezeigt, daß eine gesteigerte Cortisolproduktion besteht.

Ein echtes Cushing-Syndrom besteht offenbar bei diesen adipösen Patienten nicht; dafür spricht die von verschiedenen Autoren festgestellte normale Tages-Rhythmik sowie der meist nicht erhöhte, mitunter sogar erniedrigte Plasma-17-OHCS-Spiegel, ferner der normale Ausfall des Hemmtestes mit Dexamethason. Bei den adipösen Patienten mit einzelnen oder mehreren Zeichen des Cushing-Syndroms besteht offenbar ein milder, wahrscheinlich sekundär bedingter Hypercorticismus, der durch eine adaptive Anpassung an das gesteigerte Körpergewicht aufzufassen ist.

Faßt man die für die Diagnose und Differentialdiagnose des Cushing-Syndroms wichtigen Befunde zusammen, so muß man feststellen, daß — obwohl die Diagnose des klassischen Cushing-Syndroms leicht zu stellen ist — noch zahlreiche differentialdiagnostische Schwierigkeiten bestehen. Eine hohe Spezifität zur Erkennung des Cushing-Syndroms hat die Baselausscheidung des freien Cortisols. Die standardisierten Belastungs-Teste mit ACTH und Metopiron gestatten oft den Nachweis einer gesteigerten Nebennierenrinden-Aktivität und liefern einen Beitrag zur differentialdiagnostischen Abgrenzung der pathogenetisch verschiedenen Ursachen des Cushing-Syndroms. Die größte praktische Bedeutung in der Diagnose und Differentialdiagnose gegenüber den verschiedenen Pseudo-Formen dürfte jedoch dem Dexamethason-Hemmtest zukommen. Ein positiver Ausfall spricht eindeutig gegen das Vorliegen eines Cushing-Syndroms. Die Abgrenzung der verschiedenen Krankheitsbilder mit Hirsutismus und Virilismus bietet gewisse Schwierigkeiten.

Literatur

1. Bayer, J. M.: Med. Klin. **53**, 1715 (1958).

2. Baxyer, J. M.: Arch. klin. Chir. **291**, 531 (1959).

3. —, u. H. Rohr: Dtsch. med. Wschr. **89**, 464 (1964).

4. Beck, J. C., and E. E. McGarry: Brit. med. Bull. **18**, 134 (1962).

5. Beisel, W. R., V. L. DiRaimondo, J. Rosner, and E. G. Biglieri: Clin. Res. **10**, 114 (1962) Abstract.

6. Bergman, P. H., H. Ekman, B. Hakansson, and B. Sjogren: Acta endocr. (Kbh.) **35**, 293 (1960).

7. Berthelot, P., J. P. Benhamon, et R. Fauvert: Presse méd. **44**, 1899 (1961).

8. Brooks, R. V., J. Dupré, A. N. Cogate, J. H. Mills, and F. T. G. Prunty: J. clin. Endocr. **23**, 725 (1963).

9. Brown, W. H.: Lancet **1928 II**, 1022.

10. Cohen, H.: Brit. med. J. **1963**, 686.

11. Cope, O., and J. W. Raker: New Engl. J. Med. **253**, 119 (1955).

12. —, and E. G. Black: Brit. med. J. **1959 II**, 1117.

13. Crabbé, J., W. J. Reddy, E. J. Ross, and G. W. Thorn: J. clin. Endocr. **19**, 1185 (1959).

14. Daughaday, W. H.: J. clin. Invest. **36**, 881 (1957).

15. Dix, V. W.: Brit. J. Urol. **35**, 356 (1963).

16. Doe, R. P., H. H. Zinneman, E. B. Fink, and R. A. Ulstrom: J. clin. Endocr. **20**, 1484 (1960).
17. Eckler, E., J. R. Bierich u. D. Schönberg: Endokrinologie **42**, 321 (1962).
18. Eisenstein, A. B., R. Karsh, and J. Gall: J. clin. Endocr. **23**, 971 (1963).
19. Ekman, H., B. Hakansson, J. D. McCarthy, J. Lehmann, and B. Sjögren: J. clin. Endocr. **21**, 684 (1961).
20. Franken, H., u. H. Zimmermann: In Vorbereitung.
21. Franken, F. H., and H. Zimmermann: Acta endocr. (Kbh.) **41**, 531 (1962).
22. Froesch, E. R., E. R. Renold u. G. W. Thorn: Schweiz. med. Wschr. **89**, 623 (1959).
23. Gallagher, T. F.: IV. Int. Congress of Biochemystry. Wien 1958.
24. Haugen, H. N., and A. C. Löken: J. clin. Endocr. **20**, 173 (1960).
25. Harris, J. H., u. M. G. Crane: Metabolism **13**, 45 (1964)
26. Hartenbach, W.: Tägl. Prax. **4**, 241 (1963).
27. Hunt, A. B., and W. M. McConahey: Amer. J. Obstet. Gynec. **66**, 970 (1953).
28. Hymes, A. C., and R. P. Doe: Amer. J. Med. **33**, 398 (1962).
29. Jarett, L., P. E. Lacy, and D. M. Kipnes: Amer. J. Path. **43**, 11a (1963).
30. Kaplan, N. M.: J. clin. Endocr. **23**, 945 (1963).
31. Karl, H. J., u. L. Raith: Klin. Wschr. **29**, 702 (1961).
32. Kleinfelder, H., u. F. Husman: Dtsch. Arch. klin. Med. **208**, 227 (1962).
33. Kracht, J., and J. Tamm: Acta endocr. (Kbh.) **43**, 330 (1963).
34. Labhart, A.: Klinik d. Inneren Sekretion. Berlin-Göttingen-Heidelberg: Springer Verlag 1957.
35. — E. R. Froesch u. W. Ziegler: Schweiz. med. Wschr. **89**, 44 (1959).
36. Liddle, G. W.: J. clin. Endocr. **20**, 1539 (1960).
37. — H. L. Estep, J. W. Kendall, and W. C. Williams: J. clin. Endocr. **19**, 875 (1959).
38. — D. Island, E. M. Lance, and A. P. Harris: J. clin. Endocr. **18**, 906 (1958).
39. — — R. L. Ney, W. E. Nicholson, and N. Shimizu: Arch. int. Med. **111**, 471 (1963).
40. McCullagh, E. P.: The human adrenal cortex edited by A. R. Currie, T. Synington, and J. K. Grant. Edinburgh: E. & S. Livingstone 1962.
41. Meador, C. K., G. W. Liddle, D. P. Island, W. E. Nicholson, C. P. Lukas, J. G. Nuckton, and J. A. Luetscher: J. clin. Endocr. **22**, 693 (1962).
42. Mannix, H., and F. Glenn: J. Amer. med. Ass. **180**, 119 (1962).
43. Marks, K.: Acta Endocr. **32**, 527 (1959).
44. Marks, L. J., A. B. Russfield, and D. L. Rosenbaum: Ann. int. Med. **58**, 143 (1963); J. Amer. med. Ass. **183**, 117 (1963).
45. Mason, A. S., and D. Greenbaum: Brit. med. J. **1962 II**, 1027.
46. Migeon, C. J., O. C. Green, and J. P. Eckert: Metabolism. **12**, 718 (1963).
47. Mlynaryk, P., R. R. Gillies, B. Murphy, and C. J. Pattee: Metabolism. **22**, 587 (1962).
48. Myerson, R. M., and W. Hingston: Int. Med. **109**, 609 (1962).
49. Nelson, D. H., J. W. Meakin, and G. W. Thorn: Ann. int. Med. **52**, 560 (1960).
50. Nichols, J., and W. Gourley: J. Amer. med. Ass. **185**, 696 (1963).
51. Nocke, W., H. Zimmermann, R. Buchholz u. R. Poche: In: 10. Symp. Dtsch. Ges. f. Endokrinologie, 289. Berlin-Göttingen-Heidelberg: Springer 1964.
52. Plotz, C. M., A. J. Knowlton, and C. Ragan: Amer. J. Med. **13**, 597 (1952).
53. Poisnick, J.: J. clin. Endocr. **20**, 1360 (1956).
54. Prunty, F. T. G., R. V. Brooks, J. Dupré, T. M. G. Gimlette, J. S. M. Hutchinson, R. R. McSwiney, and I. H. Mills: Metabolism **23**, 737 (1963).
55. Retiene, K., H. Ditschuneit, M. Fischer, K. Kopp u. E. F. Pfeiffer: In: 8. Symp. Dtsch. Ges. f. Endokrinologie S. 427. Berlin-Göttingen-Heidelberg: Springer 1962.
56. — — u. E. F. Pfeifer: In: 9. Symp. Dtsch. Ges. f. Endokrinologie, Berlin-Göttingen-Heidelberg: Springer 1963.
57. — A. Espinoza, Y. A. Rahman, K. H. Marx u. E. F. Pfeifer: In: 10. Symp. Dtsch. Ges. f. Endokrinologie. Berlin-Göttingen-Heidelberg: Springer 1964. S. 231.
58. Ross, E. J.: J. clin. Endocr. **20**, 1360 (1960).
59. Salassa, R. M., T. P. Kearns, J. W. Kernohan, R. G. Sprague, and C. S. MacCarty: J. clin. Endocr. **19**, 1523 (1959).
60. Schäfer, E. L.: Dtsch. med. Wschr. **86**, 2257 (1961).

61. SCHEDL, H. P., P. S. CHEN jr., G. GREENE, and D. REDEL: J. clin. Endocr. **19,** 1223 (1959).
62. SCHNEIDER, P. B.: J. clin. Endocr. **24,** 218 (1964).
63. SCHROEDER, R.: Dtsch. med. Wschr. **87,** 237 (1962).
64. SCHTEINGART, D. E., R. J. GREGERMAN, and J. W. CONN: Metabolism **12,** 484 (1963).
65. SCOTT, H. W., G. W. LIDDLE, A. P. HARRIS, and J. H. FORSTER: Ann. Surg. **155,** 696 (1962).
66. SIMKIN, B., and R. ARCE: New Engl. J. Med. **226,** 1031 (1962).
67. SILVERMAN, ST., R. MARNELL, L. J. SHOLITON, and E. E. WERK: J. clin. Endocr. **23,** 167 (1963).
68. SLATER, I. D. H., M. HARTOGH, and R. FRASER: Brit. Med. J. **1962 I,** 1584.
69. SLAUNWHITE, W., and A. SANDBERG: J. clin. Invest. **36,** 881 (1957).
70. SOFFER, L. J., A. IANNACONE, and J. L. GABRILOVE: Amer. J. Med. **30,** 129 (1861).
71. — R. I. DORFMAN, and J. L. GABRILOVE: The Human Adrenal Gland. Philadelphia 1961.
72. STAIB, W.: Dtsch. med. Wschr. **87,** 498 (1962).
73. STIRLING, W. B.: Brit. J. Urolog. **35,** 342 (1963).
74. SZENAS, P., and C. J. PATTEE: J. clin. Endocr. **19,** 344 (1959).
75. TAMM, J., K. D. VOIGT, and J. KRACHT: Acta endocr. (Kbh.) **37,** 253 (1961).
76. THOMPSON, G. S., L. HORWICH, and J. C. DAVIS: Lancet **1962 II,** 534.
77. VEREKEI, J.: Z. ges. inn. Med. **6,** 173 (1958).
78. Zit. nach VOIGT, K. D., A. ORIOL-BOSCH u. J. TAMM: Klin. Wschr. **38,** 206 (1960).

Diskussion

E. F. PFEIFFER (Frankfurt/Main):

Wie erklären Sie sich, daß Metopiron beim Cushing-Patienten mehr tut als exogenes ACTH? Nach eigener Erfahrung ist öfter das Umgekehrte der Fall, da Metopiron in der üblichen Dosierung zwar beim Normalen zur Enthemmung der hypophysären ACTH-Sekretion via NNR ausreicht, nicht aber beim Cushing-Kranken mit hyperplastischer NNR.

H. ZIMMERMANN:

Wir haben Metopiron 250 mg alle 2 Std, in 24 Std also 3000 mg per os gegeben und die Tetrazoliumblau reduzierenden Corticoide unter und am Tage nach Metopiron gemessen. Bei 4 Patienten mit einem Cushing-Syndrom durch NNR-Hyperplasie lagen dreimal die Werte im Harn über denen nach 25 IE ACTH i.v./8 Std. Wir konnten dieses Verhalten bei früheren Untersuchungen und Bestimmung der totalen 17-OHCS nach APPLEBY nicht beobachten. Bei diesen Untersuchungen hatten wir Metopiron 750 mg alle 6 Std gegeben. Möglicherweise ist also die Blockade der NNR nach 2stündigen Gaben vollständiger. Der Schluß liegt nahe, daß beim Cushing-Syndrom doch eine stärker gesteigerte ACTH-Sekretion aus dem HVL besteht bzw. daß der Reglermechanismus höher gestellt ist. Die Frage, ob eine nur mäßig gesteigerte ACTH-Sekretion von der hyperplastischen NNR mit einer überschießenden Reaktion beantwortet wird, ist nicht zu entscheiden. Ob man aus diesen Befunden differentialdiagnostische Schlüsse ziehen darf, kann ich wegen meiner geringen Fallzahl nicht beantworten.

Aus der Neurochirurgischen Universitätsklinik und dem Max Planck-Institut für Hirnforschung, Abteilung für Tumorforschung und experimentelle Pathologie, Köln
(Direktor: Prof. Dr. W. Tönnis)

Das hypophysäre Cushing-Syndrom

Von

F. Marguth

Mit 6 Abbildungen

Referat

Gibt es ein primär hypophysäres Cushing-Syndrom oder sind die beim Hyperkortizismus zu beobachtenden Veränderungen in der Hypophyse sekundärer Natur? Bei Erörterung dieser Frage ist davon auszugehen, daß ganz verschiedenartige Prozesse mit dem klinischen Bild in ursächlichen Zusammenhang gebracht werden:

1. Die von dem amerikanischen Hirnchirurgen H. Cushing nachgewiesenen basophilen Adenome des Vorderlappens, die in der Regel mikroskopisch klein bleiben und infolgedessen keine Sellaerweiterung und kein Chiasma-Syndrom verursachen,

2. Die raumfordernden Adenome chromophober, eosinophiler und gemischtzelliger Art und

3. extraselläre cerebrale Prozesse, die ebenfalls als Ursache Cushing-ähnlicher Krankheitsbilder beschrieben worden sind.

Da basophile Adenome nicht bei allen Cushing-Fällen nachzuweisen sind — Plotz u. Mitarb. fanden sie nur bei einem Drittel, Bauer in 45% — und auch Röntgenbestrahlungen der Hypophyse nicht immer zu einer Remission des Cushing-Syndroms führten (Cope u. Raker, Labhart), geriet die ursprüngliche These von einem ACTH-produzierenden basophilen Adenom als entscheidendem ätiologischen Faktor ins Wanken. Die Verhältnisse wurden weiterhin dadurch kompliziert, daß die von Crooke entdeckten hyalinen Veränderungen in den basophilen Vorderlappenzellen nicht nur als regelmäßiger Befund bei der Cushing-Krankheit (Thompson u. Eisenhardt, Plotz), sondern auch nach Behandlung mit Cortison- und ACTH-Präparaten nachgewiesen wurden (Golden und Bondy, Kilby u. Mitarb.; Montandon). Die Zunahme der Zahl der voll granulierten normalen Basophilen beim therapeutischen Hyperkortizismus veranlaßte Montandon, die Frage aufzuwerfen, ob nicht die Basophilie und vielleicht sogar die Bildung eines basophilen Adenoms bei Nebennierenrindenhyperplasie eine *Folge* des Hyperkortizismus darstellen. Trotz diese Einwände, die gegen die primär hypophysäre Genese des Cushing-Syndroms bei Nebennierenrindenhyperplasie und Basophilie bzw. basophilem Adenom sprechen, wird man der Hypophyse jedoch nicht jede Bedeutung in der Pathogenese des Hyperkortizismus absprechen dürfen; das geht schon daraus hervor, daß die für die Rindenhyperplasie charakteristische starke Reaktion der 17-Hydroxykortikoid-Ausscheidung auf ACTH-Belastung nach Ausschaltung der Hypophyse verschwindet, sofern eine klinische Remission eingetreten ist (Jailer). Bemerkenswert bleibt allerdings die Tatsache, daß die basophilen Adenome nicht die Größe der anderen autochthonen Vorderlappengeschwülste erreichen, in der Regel auf die Sella beschränkt bleiben und daher nicht als raumfordernde Prozesse in Erscheinung treten.

Von besonderem Interesse ist nun, daß beim Cushing-Syndrom auch *raumfordernde* Hypophysenadenome beobachtet werden. Salassa u. Mitarb. konnten bei 12 von 122 Patienten mit Cushing-Syndrom und Nebennierenrindenhyperplasie Hypophysentumoren nachweisen, die zu Sellaerweiterungen und in einigen Fällen auch zu Sehstörungen geführt hatten. Bei 9 von diesen 12 Kranken war der raumfordernde Hypophysentumor bereits vor der Adrenalektomie zu erkennen, in 3 Fällen jedoch entwickelten sich Sellaerweiterung und Sehstörungen erst nach der Nebennierenentfernung. Bei keinem von 10 über längere Zeit beobachteten Patienten konnte nach der Adrenalektomie eine Rückbildung der Veränderungen im Sellabereich nachgewiesen werden; im Gegenteil, bei 7 Kranken war der tumorfördernde Einfluß der Adrenalektomie evident. Auch von Nelson wurde ein raumforderndes Hypophysenadenom beobachtet, das $3^1/_2$ Jahre nach beidseitiger Adrenalektomie in Erscheinung trat und ein Chiasmasyndrom verursachte.

Die abnorme Pigmentierung bei einem Teil dieser Patienten mit Cushing-Syndrom und Hypophysentumor sowie der Nachweis erhöhter ACTH-Spiegel im strömenden Blut, der Nelson u. Mitarb., sowie Rees und Bayliss gelang, legen die Vermutung nahe, daß diesen Cushing-Syndromen ein endokrin aktives Adenom zugrunde liegt, das übergroße Mengen von ACTH und MSH produziert. Dafür spricht auch die Beobachtung Salassas, daß trotz ausreichender Substitution mit Cortison die nach Adrenalektomie eingetretene Melanose der Haut nicht zurückging. Jailer konnte sogar im Tumorgewebe selbst große Mengen von ACTH nachweisen. Berücksichtigt man ferner, daß raumfordernde Hypophysenadenome bei Nebennierenrindentumoren einen außergewöhnlichen Befund darstellen und beim therapeutischen Hyperkortizismus nicht vorkommen, und daß eine Rückbildung des sellären Prozesses auch nach totaler Nebennierenexstirpation nicht beobachtet wurde, so wird man die primär hypophysäre Genese der Rindenhyperplasie in Fällen, die mit raumfordernden Hypophysenadenomen kombiniert sind, nicht ohne weiteres verneinen können.

Wir haben unter 570 raumfordernden Hypophysenadenomen 5 Kranke mit Cushing-Syndrom beobachten können. Besonders eindrucksvoll ist der bereits von Oberdisse mitgeteilte Fall. Es handelte sich um eine 24jährige Patientin mit dem Vollbild eines Cushing-Syndroms und einem schweren, insulinresistenten Diabetes. Die Ausscheidung der freien Kortikoide war mit $1120\,\gamma$ deutlich erhöht. Nach der Operation eines atypischen chromophoben Adenoms vom fetalen Typ verschwand der schwere, insulinresistente Diabetes, und das Cushing-Syndrom bildete sich zurück. Mit dem Tumorrezidiv entwickelte sich erneut ein schwerer Diabetes mellitus, und eine bis dahin inaktive Lungentuberkulose wurde aktiviert und führte unter dem Bild der tuberkulösen Sepsis zum Tode. Die Remission der Krankheitserscheinungen nach der Tumorentfernung und das Wiederauftreten der Symptome mit dem Rezidiv deuten darauf hin, daß hier der hypophysäre

Prozeß die entscheidende Ursache darstellte, zumal eine Nebennierenrindenfreilegung zum Ausschluß eines Tumors dem Eingriff an der Hypophyse vorausgegangen, eine Nebennierenresektion aber nicht durchgeführt worden war. Daß es mit dem Tumorrezidiv zu einer enorm gesteigerten ACTH-Ausschüttung gekommen sein mußte, kann auch aus der Aktivierung der Tuberkulose geschlossen werden. Bei einer anderen 23jährigen Patientin mit Cushing-Symptomen und raumforderndem Adenom standen die äußeren Veränderungen im Vordergrund des Krankheitsbildes. Die diabetische Stoffwechsellage und die Osteoporose fehlten

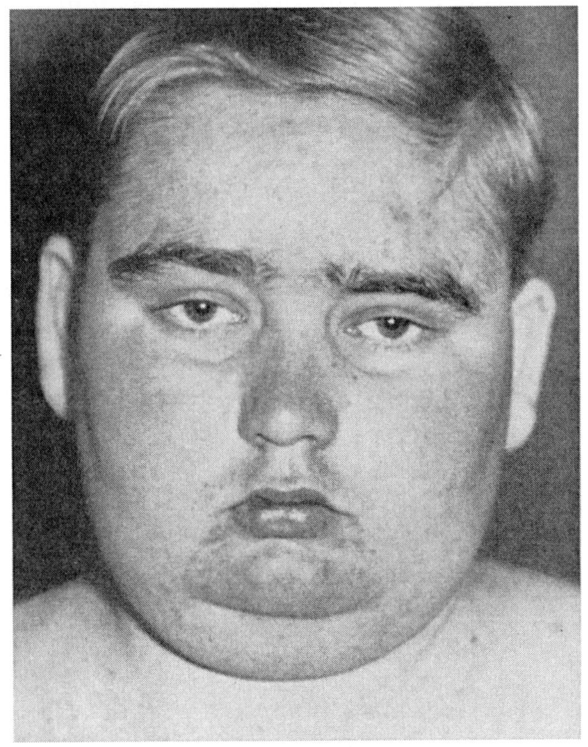

Abb. 1

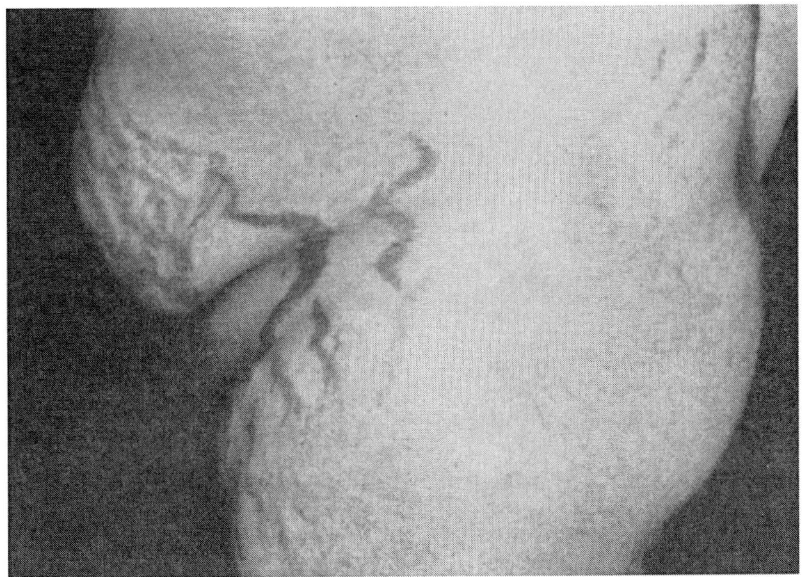

Abb. 2

Abb. 1 u. 2. 21jahr. Pat. mit voll ausgebildetem M. Cushing bei Nebennierenrindenhyperplasie und raumforderndem Hypophysentumor

allerdings. Der Blutdruck war nur bei der ersten Untersuchung auf 190/110mm Hg erhöht. Es bestand eine primäre Amenorrhoe. Die 17-Ketosteroidwerte lagen im Normalbereich, die Ausscheidung der 17-Hydroxykortikosteroide war mäßig, aber deutlich erhöht; nach mehrtägiger Behandlung mit Prednison sanken die Werte zur Norm ab, unter ACTH stiegen die 17-Hydroxykortioide von 15—25 mg auf 42 mg pro die. Ein Chiasma-Syndrom lag nicht vor, der erweiterte Türkensattel deutete jedoch auf den Hypophysentumor hin. Histologisch handelte es sich um ein Mischtypadenom mit einzelnen basophilen Zellen. Nach der Operation bildeten sich die blau-roten Striae zurück, und die Konturen der Sella, besonders im Bereich der Sattellehne, wurden wieder dichter. Die allgemeine Fettsucht ist vorerst unverändert geblieben. Eine dritte Beobachtung ist deshalb besonders erwähnenswert, weil hier, den Befunden aus der Mayo-Klinik (SALASSA u. Mitarb.) entsprechend, der selläre Prozeß nach der subtotalen Adrenalektomie allmählich zunahm und schließlich zu einer Destruktion des Türkensattels und zu einem Chiasmasyndrom führte, ohne daß das Cushing-Syndrom rezidivierte.

Es handelt sich um einen nunmehr 29 jährigen Mann, der im Alter von 12 Jahren erkrankte. Bis zu diesem Zeitpunkt hatte sich die Entwicklung normal vollzogen, dann traten Wachstumsstillstand, Gewichtszunahme und ein Vollmondgesicht auf. 2 Jahre später stellten sich diffuse Kopfschmerzen ein, und der Verdacht auf einen Hypophysentumor sei röntgenologisch bestätigt worden. Im 16. Lebensjahr kam es zu starken Rückenschmerzen entlang der Wirbelsäule. Der Junge mußte schließlich infolge Rückenschwäche am Stock gehen. Weitere 2 Jahre später erfolgte unter der Diagnose „Dystrophia adiposogenitalis" eine Behandlung mit Hypophysenextrakten, wonach eine Besserung der Schmerzen, eine deutliche Gewichtsabnahme und erneutes Längenwachstum (angeblich um 12 cm) eingetreten sein sollen. Mit 20 Jahren machte sich wieder eine krankhafte Gewichtszunahme bemerkbar, starke Rückenschmerzen kamen hinzu, und gelegentlich waren Unterschenkelödeme vorhanden. Die Gonadenfunktion kam nicht in Gang. — Der bei unserer ersten Untersuchung 21 Jahre alte, 146 cm große Mann bot eine ausgeprägte Stammfettsucht, das typische Vollmondgesicht, breite Striae im Bereich der Leisten und der Achselhöhlen, eine starke Gesichtsakne, eine Hypertrichose am Rücken, ein hypoplastisches Genitale und eine trockene, abschilfernde Haut (Abb. 1 u. 2). Augen- und nervenärztlich ließ sich nichts Krankhaftes nachweisen. Infolge der großen Schwäche in der Muskulatur der Gliedmaßen und des Rückens ging der Patient an 2 Stöcken. Röntgenologisch fanden wir eine

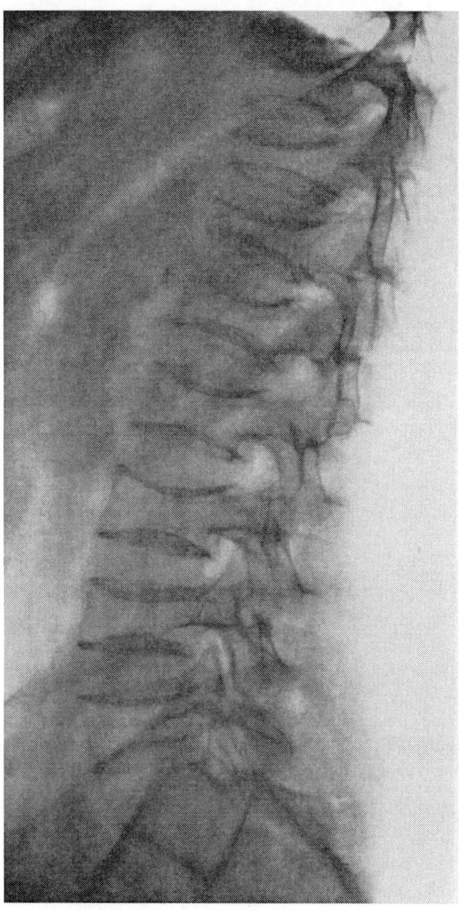

Abb. 3. Hochgradige Osteoporose der Wirbelsäule mit Fischwirbelbildung (bei Pat. der Abb. 1 u. 2)

geringe, nach den statistisch errechneten Maßen von BERGERHOFF aber eindeutige Sellaerweiterung mit Tiefstand des Sellabodens im Bereich der linken Hälfte und eine hochgradige Osteoporose der Wirbelsäule mit sog. Fischwirbelbildung (Abb. 3). Die Abweichung des Grundumsatzes betrug —1,4%, nach Eiweißbelastung +3,07%. Bei der Glucose Doppelbelastung wurden jeweils Gipfel von 180 mg-% erreicht, eine Glucosurie trat nicht ein, nach 5 Std war der Ausgangswert noch nicht erreicht. Im Differentialblutbild war eine geringe Linksverschiebung auffällig. Der Blutdruck betrug 135—155/100 mm Hg. Im Wasser- und Konzentrationsversuch wurden nach Zufuhr von 800 cm³ Flüssigkeit 1 191 cm³ Urin innerhalb von 4 Std ausgeschieden. Das maximale spezifische Gewicht betrug 1008; an anderen Tagen war eine ausgesprochene Wasserretention mit hohen spezifischen Gewichten zu verzeichnen. Die Ausscheidung der freien Kortikoide war bei mehreren Bestimmungen vorwiegend erhöht, die Ausscheidung der 17-Keto-

steroide war ebenfalls erhöht (25,4 mg); nach 2 × 25 mg Cortison i.m. täglich trat eine zunehmende Verminderung der 17-Ketosteroidausscheidung ein. Zunächst wurde die linke, stark vergrößerte Nebenniere total entfernt. In einer 2. Sitzung wurde die rechte subtotal reseziert. Innerhalb eines Jahres bildete sich das Cushing-Syndrom praktisch völlig zurück (Abb. 4). Im Februar 1964, fast 8 Jahre nach der Adrenalektomie, wurde eine deutliche Zunahme der Sellaveränderungen (Abb. 5) und eine bitemporale Hemianopsie bei beiderseits blassen Papillen festgestellt. Histologisch handelte es sich um ein basophiles Adenom.

Die Entwicklung des raumfordernden Hypophysenadenoms bis zum Chiasma-Syndrom erstreckte sich hier über 17 Jahre! Selbst unter Berücksichtigung der Tatsache, daß endokrin aktive Adenome langsamer wachsen als innersekretorisch stumme,

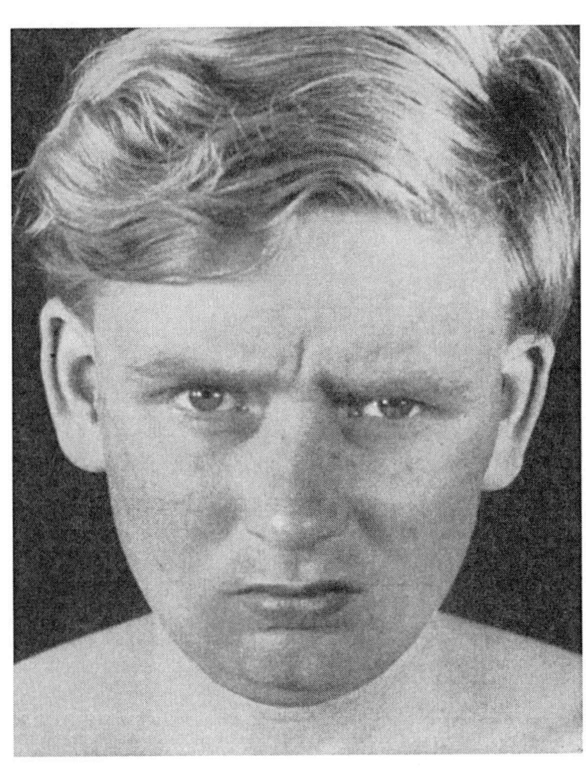

Abb. 4. Derselbe Pat. wie auf Abb. 1, 1¹/₂ Jahre nach subtotaler Adrenalektomie

worauf wir am Beispiel der hochaktiven eosinophilen Tumoren hinweisen konnten (MARGUTH), scheint uns der oben geschilderte Krankheitsverlauf außergewöhnlich zu sein. Bei den von SALASSA u. Mitarb. mitgeteilten Beobachtungen ohne Sellaerweiterung zum Zeitpunkt der Adrenalektomie beträgt die mittlere Anamnesenlänge bis zum Auftreten der röntgenologisch faßbaren Veränderungen am Türkensattel immerhin auch 5 Jahre.

Morphologisch handelt es sich bei diesen raumfordernden Tumoren mit Cushing-Syndrom vorwiegend um chromophobe Adenome, wie auch aus der Zusammenstellung von MARKS hervorgeht. Nach den Untersuchungen von PEARSE stammt

die sog. chromophobe Zelle in vielen Fällen aus dem Formenkreis der chromophilen, inkretorisch aktiven Elemente, so daß sich diese chromophoben Adenome
durchaus mit Krankheitsbildern in Einklang bringen lassen, die auf endokrine

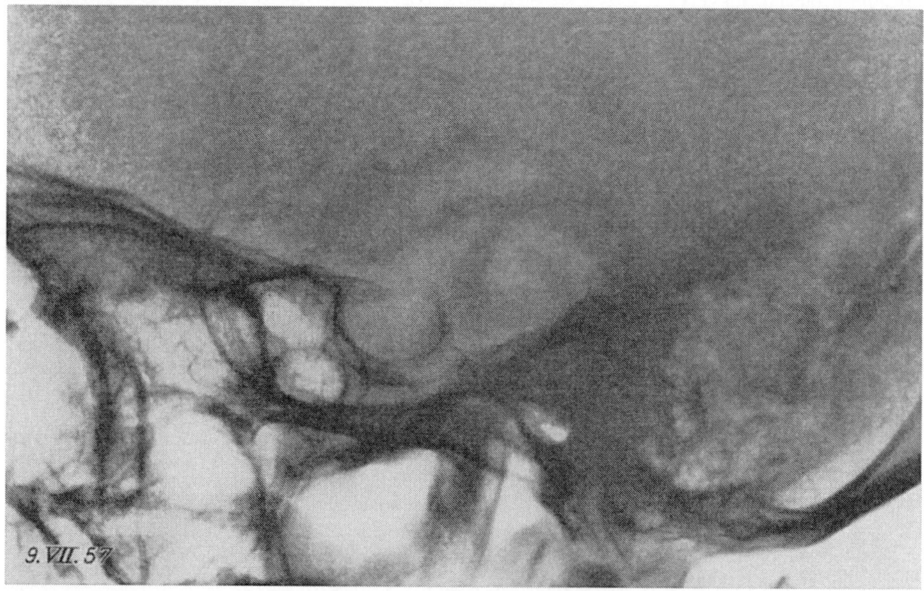

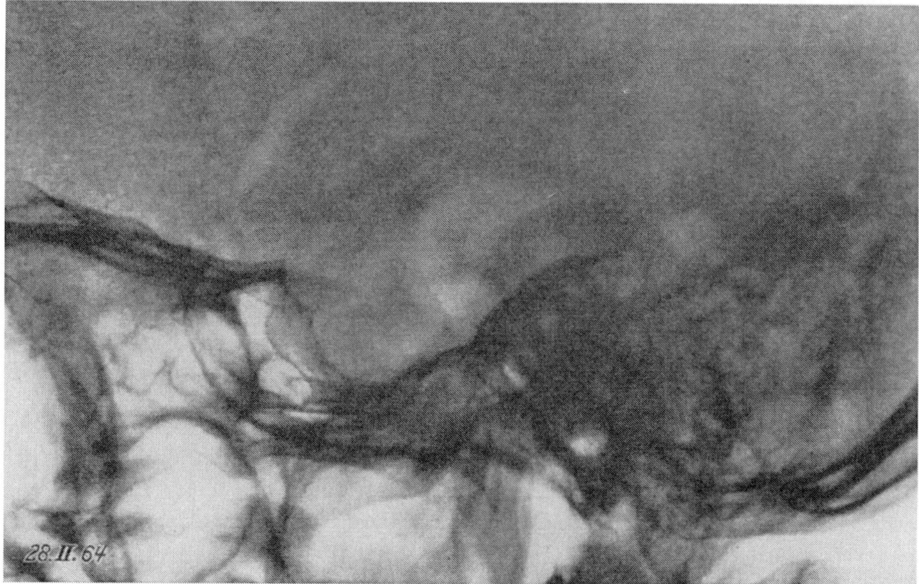

Abb. 5. Erweiterung des Türkensattels bei Pat. mit Nebennierenrindenhyperplasie. 8 Jahre nach der
Adrenalektomie (untere Sella) ist eine deutliche Zunahme der Sellaveränderungen erkennbar

Aktivität des zugrundeliegenden Prozesses hindeuten. Histologische Untersuchungen an unserem Krankengut zeigten ebenfalls, daß Zellchromophobie eines

Adenoms nicht unbedingt mit innersekretorischer Inaktivität gleichzusetzen ist (W. Müller). Aufgrund dieser Befunde und klinischer und endokrinologischer Zeichen inkretorischer Aktivität ließ sich das sog. Mischtypadenom aus der Gruppe der chromophoben, inaktiven Adenome absondern (Tönnis, Müller, Brilmayer; Brilmayer, Marguth, Müller). Orthner u. Mitarb. wiesen bereits darauf hin, daß man bei Durchsicht der Literatur den Eindruck gewinnt, daß die beim Cushing-Syndrom beobachteten raumfordernden Adenome vorwiegend dem Mischtyp zugeordnet werden können.

Im Tierexperiment kann durch Thyreoidektomie bzw. pharmakologische Inaktivierung der Schilddrüsenfunktion sowie durch Kastration eine Volumenzunahme der Hypophyse herbeigeführt werden (Lit. s. Romeis). In diesen hyperplastischen Hypophysen finden sich Zelltypen, die als „Thyreoidektomie- bzw.

Tabelle 1. *Cushingoide Fettsucht bei extrasellären cerebralen Prozessen*

Grundkrankheit	Alter	Geschl.	Pneumenceph.	Korti-koide	17-KS	RR	K.H.-Stoffwechsel	Osteoporose	Striae	Menses
Aquäduktstenose (tumorbedingt)	13	♀	Hydroceph. intern.occl.	$860\,\gamma$	2,0 mg	$^{150}/_{90}$-$^{160}/_{120}$	o.B.	keine	zahlreich, schmal	∅
Aquäduktstenose (entzündlich)	13	♀	Hydroceph. intern.occl.	23,8 mg	—	$^{115}/_{80}$	o.B.	keine	schmal, a.li.Hüfte	keine Menarche
Aquäduktstenose (Meningitis sarcomatosa)	13	♂	Hydroceph. intern.occl.	13,0 mg	4,8 mg	$^{115}/_{80}$	o.B.	BWS u. LWS (Paresen)	schmal, am re. Oberarm	
Arachnitis optochiasmatis (nach Meningitis tuberculosa)	14	♀	Ventrikel-erweiterg. (Atrophie)	$358\,\gamma$	6,8 mg	$^{120}/_{85}$	o.B.	keine	zahlreich, breit	∅

Kastrationszellen" bezeichnet werden. Die Mehrzahl der Autoren stimmt darin überein, daß diese Zellen eine besondere Aktivität besitzen. Die außerordentlich große morphologische Ähnlichkeit dieser Zellbilder mit den bei Mischtypadenomen erhobenen Befunden gibt zu der Frage Anlaß, ob nicht auch beim Menschen derartige ätiologische Faktoren bei der Entstehung der Hypophysenadenome mitwirken können. Das bedeutet für unsere Fragestellung, daß vom morphologischen Standpunkt aus und vielleicht auch aufgrund des Krankheitsverlaufes in manchen Fällen eine *reaktive* Entstehung des Hypophysentumors beim Cushing-Syndrom erörtert werden kann. Die vorhin aufgezählten klinischen und endokrinologischen Kriterien, die für die *primär* hypophysäre Genese ins Feld geführt werden, müßten dann mit einem reaktiv verursachten, aber autonom gewordenen Hypophysenadenom erklärt werden. — Wenn auch der Einfluß des Hypothalamus auf die Produktion des adrenocorticotropen Hormons im Vorderlappen heute außer Zweifel steht, so ist doch ein echtes Cushing-Syndrom bei hypothalamischen Prozessen bisher nicht erwiesen (s. Orthner, Anton, Bachmann, Graumann, Hinrichsen und Schön). Unsere Beobachtungen erstrecken sich auf 4 jugendliche Kranke. 3 Patienten hatten einen Verschlußhydrocephalus, ein Mädchen wies eine durch Hirnatrophie verursachte Ventrikelerweiterung auf (Tab. 1). Das Krankheitsbild war in allen Fällen vorwiegend gekennzeichnet durch Fettsucht, Vollmondgesicht und Striae-Bildung, wobei zu betonen ist, daß die Adipositas nicht der typischen

Cushing-Fettsucht entsprach, sondern sich über den ganzen Körper, auch die Gliedmaßen, erstreckte. Eine diabetische Stoffwechsellage war in keinem Fall vorhanden, und eine Osteoporose war nur einmal nachweisbar; allerdings bestand bei diesem Jungen gleichzeitig ein polyneuritisches Krankheitsbild mit ausgedehnten Paresen, so daß wir die Osteoporose eher mit dem neurologischen Prozeß in Zusammenhang bringen möchten. Die Ausscheidung der Nebennierenrindenmetabolyte war in 2 Fällen erhöht, bei 1 Kranken normal und bei dem 14-jährigen Mädchen (Fall 4) sogar reduziert. Bei einer Kranken hatten sich die Symptome erst nach Exstirpation der Kleinhirngeschwulst, nach Beseitigung der Liquorblockade, entwickelt. Ventrikulographisch konnte der Rückgang des

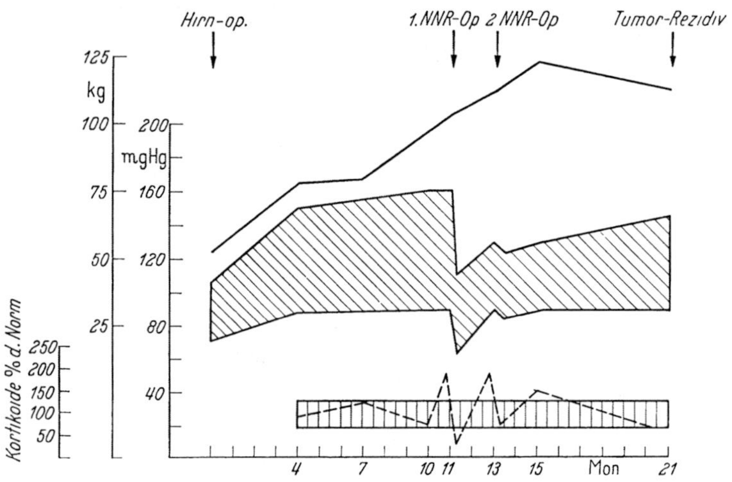

Abb. 6. Krankheitsverlauf bei Pat. mit cushingoider Fettsucht und Kleinhirngeschwulst, dargestellt am Verhalten des Blutdruckes, des Körpergewichtes und der Kortikoidausscheidung. Ansteigen von Gewicht, Blutdruck und Kortikoidwerten nach der Hirnoperation. Nach subtotaler Adrenalektomie vorübergehender Abfall der Kortikoide bei konstantem Körpergewicht. Mit dem Tumor-Rezidiv Rückgang der Kortikoidausscheidung
▨ RR ——— Körpergewicht ----- Kortikoide

Hydrocephalus nachgewiesen werden. Durch subtotale Adrenalektomie konnte das Krankheitsbild nicht beeinflußt werden (Abb. 6). Erst mit Manifestation des Geschwulst-Rezidivs waren ein geringer Gewichtsverlust sowie ein Rückgang der Kortikoidausscheidung zu verzeichnen.

Ein voll ausgeprägter Morbus Cushing lag also in keinem Fall vor. Wir glauben daher, diese Krankheitsbilder, die wir als *cushingoide Fettsucht* bezeichnen möchten, der *hypothalamischen Fettsucht* zuordnen zu müssen. Das jugendliche Erkrankungsalter dieser 4 Patienten deutet darauf hin, daß endokrine Symptome bei hydrocephaler Erweiterung des 3. Ventrikels besonders dann zu erwarten sind, wenn die Erkrankung nach Abschluß des Schädelwachstums, aber noch während der Pubertät einsetzt, was wir auch am Beispiel der Dystrophia adiposogenitalis bei Hydrocephalus internus occlusus demonstriert haben.

Literatur

Bauer, J., u. J. Jellinghaus: Das Cushing-Syndrom. Arch. inn. Med. 1, 320 (1949/51).
Brilmayer, H., F. Marguth u. W. Müller: Das Mischtypadenom und seine Abgrenzung gegen den chromophoben Hypophysentumor. Acta neuroveg. (Wien) 15, 4, 352 (1957).

COPE, O., and J. W. RAKER: Cushing's disease: Surgical experience in the care of 46 cases. New Engl. J. Med. **253**, 119 (1955).

CROOKE, A. C.: A change in the basophil cells of the pituitary gland common to conditions which exhibit the syndrom attributed to basophil adenoma. J. Path. Bact. **41**, 339 (1935).

CUSHING, H.: The basophil adenomas of the pituitary body and their clinical manifestations (pituitary basophilism). Bull. Johns Hopk. Hosp. **50**, 137 (1932).

GOLDEN, A., and P. K. BONDY: Cytologic changes in rat adenohypophysis following administration of adrenocorticotrophin or cortisone. Proc. Soc. exp. Biol. (N. Y.) **79**, 252 (1952).

JAILER, J. W., and P. H. HOLUB: The significance of pituitary tumors in Cushing's syndrome. First Internat. Congr. of Endocrinology. Copenhagen 1960, Abstracts p. 363.

KILBY, P. A., W. A. BENNET, and R. G. SPRAGUE: Anterior pituitary glands in patients treated with cortisone and corticotropin. Amer. J. Path. **33**, 155 (1957).

LABHART, A.: Klinik der Inneren Sekretion. Berlin-Göttingen-Heidelberg: Springer 1957.

MARGUTH, F.: Zur Pathogenese endokriner Funktionsstörungen bei raumfordernden intrakraniellen Prozessen. Beiheft 7, Zbl. Neurochir. (1964).

MONTANDON, A.: Quantitative und qualitative Zellveränderungen im Hypophysenvorderlappen bei therapeutischem Hypercorticismus. Virchows. Arch path. Anat. **330**, 629 (1957).

MÜLLER, W.: Zur Frage der hypophysären Tumoren von Mischtyp. Acta neuroveg. (Wien) **8**, 4, 451 (1954).

NELSON, P. H., J. W. MEAKIN, J. B. DEALY, P. D. MATSON, K. EMERSON jr., and G. W. THORN: ACTH-producing tumor of the pituitary gland. New Engl. J. Med. **259**, 161 (1958).

OBERDISSE, K.: Kohlenhydratstoffwechsel bei organischen Erkrankungen im Sellabereich. Dtsch. Arch. klin. Med. **198**, 257 (1951).

— Pathophysiologie des Hypothalamus-Hypophysen-Systems. Handb. d. Neurochirurgie IV/3, S. 80. Berlin-Göttingen-Heidelberg: Springer 1962.

ORTHNER, H., U. ANTON, R. BACHMANN, W. GRAUMANN, K. HINRICHSEN u. H. SCHOEN: Pathologische Anatomie der neuroendokrinen Erkrankungen. II. Zur Cushingschen Krankheit. Dtsch. Z. Nervenheilk. **179**, 145 (1959).

PLOTZ, CH. M., A. I. KNOWLTON, and C. RAGAN: The natural history of Cushing's syndrome. Amer. J. Med. **13**, 597 (1952).

REES, J. R., and R. I. S. BAYLISS: Cushing's syndrom with pituitary tumor and pigmentation. Proc. roy. Soc. Med. **52**, 256 (1959).

ROMEIS, B.: Hypophyse. Hdb. d. mikr. Anat. d. Menschen. Berlin: Springer 1940.

SALASSA, R. M., TH. P. KEARNS, J. W. KERNOHAN, R. G. SPRAGUE, and C. S. McCARTY: Pituitary tumors in patients with Cushing syndrom. J. clin. Endocr. **19**, 1523 (1959).

THOMPSON, K. W., and L. EISENHARDT: Further consideration of the Cushing syndrom. J. clin. Endocr. **3**, 445 (1943).

TÖNNIS, W., W. MÜLLER u. H. BRILMAYER: Zur Problematik der "mixed types" der Hypophysenadenome. Acta endocr. (Kbh.) **13**, 227 (1953).

Aus der Chirurgischen Univ.-Klinik und Poliklinik Bonn
(Direktor: Prof. Dr. A. GÜTGEMANN)

Die Therapie des Cushing-Syndroms

Von

J. M. BAYER

Mit 9 Abbildungen

Referat

Die jeweilige Therapie des Cushing-Syndroms richtet sich nach der zugrunde liegenden Ursache. Bei dem heutigen Stand der Diagnostik läßt sich letztere in der Mehrzahl der Fälle feststellen. Bei zweifelhaftem Befund ist die Freilegung der Nebennieren angezeigt.

Unbestritten ist das Vorgehen bei cortisolproduzierenden Rindentumoren. Soweit möglich kommt nur die operative Entfernung der Geschwulst in Frage. Bei all diesen Fällen ist das Rindengewebe außerhalb des Tumorbereiches, also auch die Rinde der verbleibenden kontralateralen Nebenniere atrophisch. Die perioperative, insbesondere substitutive Behandlung ist daher die gleiche wie bei der Entfernung beider Nebennieren, die noch zu erörtern sein wird.

Bei gutartigen Rindentumoren ist die operative Prognose sowohl quoad vitam als auch hinsichtlich der Remission der Cushing-Erscheinungen günstig. Die Beseitigung der Geschwulst hinterläßt jedoch vielfach langdauernde manifeste oder latente Rindeninsuffizienzen. Von 3 eigenen operierten Rindenadenomen besteht bei einer Patientin nach 6 Monaten eine schwere manifeste, bei einer weiteren Patientin noch nach 4 Jahren und 7 Monaten eine leichte manifeste und bei der 3. Patientin nach 19 Monaten eine latente Rindeninsuffizienz.

Die Rindeninsuffizienz nach Entfernung cortisolproduzierender Rindengeschwülste ist die Folge des Versagens der endogenen ACTH-Sekretion. Dies läßt sich auf der Abb. 1 erkennen. Links ist die Steroidausscheidung einer 40jährigen Frau dargestellt, bei der vor 6—7 Monaten ein Rindenadenom operiert worden war. Die Werte der 17-KS [bestimmt nach der Methode von NORYMBERSKI u. Mitarb., 1953 (35)] und der totalen 17-OHCS [bestimmt nach der Methode von APPLEBY u. Mitarb., 1955 (1)] sind erniedrigt. Der intravenöse ACTH-Test ist praktisch ohne Wirkung. Nach mehrtägiger Zufuhr von ACTH-Depot steigt die Ausscheidung der genannten Steroide zunehmend an, um nach Absetzen des Präparates wieder auf die niedrigen Ausgangswerte abzusinken. Mit anderen Worten: Die Funktion der atrophischen Nebennierenrinde läßt sich durch Zufuhr von exogenem ACTH schnell aktivieren, der Effekt bleibt jedoch bei fehlender oder ungenügender endogener ACTH-Sekretion nicht erhalten. Hieraus ergibt sich die Nutzlosigkeit einer solchen Therapie. Zweckmäßig erscheint dagegen der Versuch, durch eine unterschwellige Substitution die ACTH-Sekretion anzuregen.

Unsere Patientin benötigte im 1. Jahr nach der Operation täglich 5, später 2,5 mg Prednisolon. 4 Jahre und 7 Monate nach der Operation kam die Patientin bei fehlender Belastung während des Klinikaufenthaltes ohne Substitution aus; sie klagte jedoch zeitweise über ein leichtes Übelkeitsgefühl. Die Steroidausscheidung (Abb. 1, rechts) ist jetzt höher als bei der früheren Untersuchung, jedoch immer noch niedrig. Auf Zufuhr von Metopiron und ACTH erfolgte keine Reaktion.

Die Rindeninsuffizienz nach Entfernung cortisolproduzierender Rindenadenome betrifft offenbar nur die Cortisolbildung, während die Sekretion und die

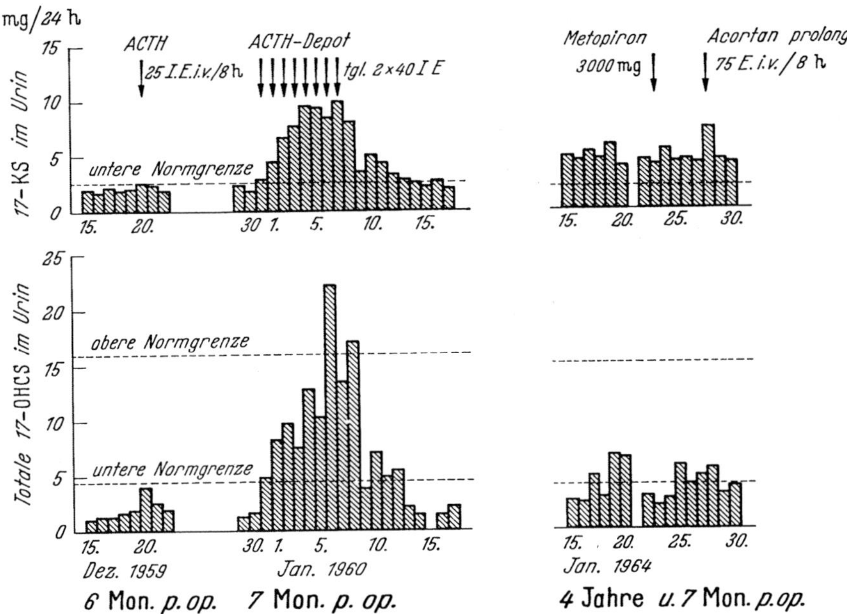

Abb. 1. Pat. M. R. (J.-Nr. 1278/59), 40 bzw. 44 Jahre. Ausscheidung der 17-KS und der totalen 17-OHCS im Urin 6—7 Monate sowie 4 Jahre und 7 Monate nach Entfernung eines Nebennierenrindenadenoms wegen eines Cushing-Syndroms. Erläuterung s. Text

Reaktion von Aldosteron erhalten bleiben. Gemeinsam mit HOLTMEIER u. BREUER (1960) (5) konnten wir bei der genannten Patientin 3—4 Monate nach der Operation nachweisen, daß sie imstande war, nach Kochsalzentzug Kochsalz in normaler Weise zu konservieren. BIGLIERI u. Mitarb. (1963) (8) beobachteten bei einem entsprechenden Fall unter den gleichen Bedingungen einen normalen Anstieg der Aldosteronausscheidung im Urin.

Die Abb. 2 zeigt das Gesicht unserer Patientin vor und ein halbes Jahr nach der Operation.

Cortisolproduzierende Rindenadenome können klein sein, so daß eine präoperative Lokalisierung nicht möglich ist. In solchen Fällen müssen gegebenenfalls beide Nebennieren freigelegt werden. Bei den seltenen diffusen kleinknotigen Adenomatosen beider Nebennieren (23, 31) ist die beiderseitige totale Adrenalektomie angezeigt.

Rindencarcinome sind in Frühstadien häufig abgekapselt und können gegebenenfalls vollständig entfernt werden. Im allgemeinen ist jedoch die Prognose schlecht. In der Mehrzahl sind die Fälle bereits inoperabel, wenn sie zur Beobachtung gelangen. Bei operabel erscheinenden Fällen gewährt der abdominothorakale Zugang die beste Übersicht und die notwendige Handlungsfreiheit. Bei nicht oder nicht radikal entfernbaren Tumoren bietet die Strahlenbehandlung nur wenig Aussicht auf Erfolg. Bei diesen Fällen sowie bei jenen mit Fernmetastasen ist die Verwendung des Adrenostaticum o, p′ — DDD zu erwägen. Eine Heilung wird damit zwar nicht erzielt, jedoch wurde in einigen Fällen ein Kleinerwerden

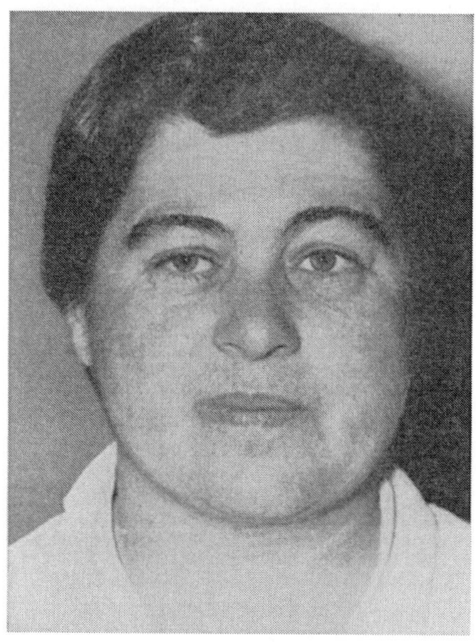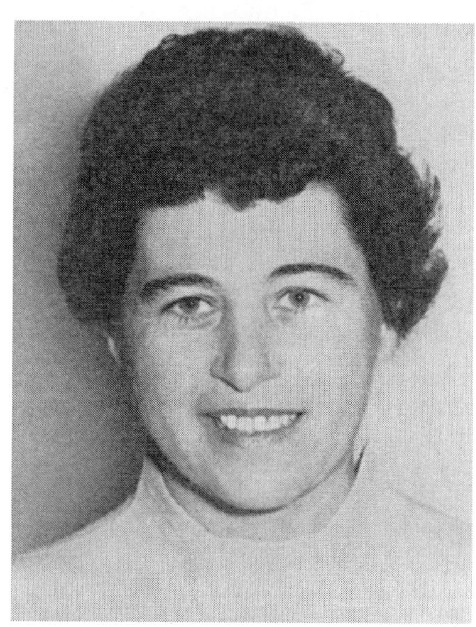

a   b

Abb. 2. Pat. M. R. (J.-Nr. 1278/59), 40 Jahre. a) Vor und b) ein halbes Jahr nach der Operation

des Tumors, eine Abnahme der Steroidausscheidung sowie ein vorübergehender Schwund der Cushing-Erscheinungen beobachtet (7).

Als Rarität sei erwähnt, daß cortisolproduzierende Rindentumoren auch mal von akzessorischem Rindengewebe ausgehen können.

Den Rindentumoren stehen die Rindenhyperplasien gegenüber. Sie bedeuten einen Sammelbegriff, keine pathogenetisch definierte Diagnose. Sie können ausgelöst sein: rein funktionell vom Hypothalmus aus, durch einen echten, autonom wachsenden Tumor des Hypophysenvorderlappens sowie durch ACTH-wirksame Stoffe sezernierende Geschwülste verschiedener nicht endokriner Organe (11, 27).

Wenn gemeinhin vom Cushing-Syndrom „infolge beiderseitiger Nebennierenrindenhyperplasie" die Rede ist, so liegt nach unseren heutigen Kenntnissen eine hypothalamische Genese vor. Etwa vorhandene Knotenbildungen im Hypophysenvorderlappen sowie die Veränderungen der Nebennierenrinde sind sekundärer Natur. In der Mehrzahl der Fälle sind die Nebennieren makroskopisch hyperplastisch. Für den Chirurgen von Bedeutung ist jedoch die Tatsache, daß auch makroskopisch und gewichtsmäßig unauffällige, u. U. sogar kleine (43) Nebennieren angetroffen werden können. Histologisch zeigt die Nebennierenrinde, wie die Untersuchungen von Kracht (1960, 1962) (22, 23), von Tonutti (1961) (46) am eigenen Operationsmaterial sowie von anderen Autoren (*17*) ergeben haben, strukturelle Veränderungen im Sinne einer mehr oder weniger ausgeprägten progressiven Transformation und damit den morphologischen Ausdruck einer ACTH-Stimulierung.

Eine kausale Behandlung des Cushing-Syndroms „infolge beiderseitiger Nebennierenrindenhyperplasie" oder des Hyperfunktions-Cushing (*3, 4, 6*), wie ich diese Form des Cushing-Syndroms bezeichne, ist zur Zeit noch nicht möglich. Dagegen bieten die Erfolgsorgane, die Hypophyse und die Nebennieren, Ansatzpunkte für eine erfolgreiche Beeinflussung der Krankheitserscheinungen.

Im Vordergrund der Behandlung stehen die Eingriffe an den Nebennieren. Die beiderseitige Teilresektion (*47*) oder die einseitige Adrenalektomie ist praktisch ohne Nutzen. Dagegen hat die radikale subtotale Adrenalektomie (*38, 42*), d. h. die totale Entfernung einer Nebenniere und die Resektion von wenigstens $^4/_5$ bis $^9/_{10}$ der 2. Nebenniere ausgezeichnete Remissionen zu verzeichnen. Ein wesentlicher Nachteil ist jedoch die hohe Zahl an Rezidiven, die mit der Länge der postoperativen Beobachtungsdauer zunimmt. SPRAGUE u. Mitarb. (1961) (*44*) sahen in ihrem Untersuchungsgut in etwa einem $^1/_3$ der Überlebenden Rückfälle. Andererseits benötigt die Hälfte der Patienten eine Substitution (*42*), ist also praktisch total adrenalektomiert. Das ideale Ziel der Erhaltung eines vitalen Nebennierenstumpfes, der die Patienten ausreichend mit Rindenhormonen versorgt ohne späteres Auftreten eines Rezidivs, ist somit kaum realisierbar. Aus diesen Gründen bevorzugt die Mehrzahl der Autoren die beiderseitige totale Adrenalektomie. Dieses Vorgehen vermeidet das Rezidiv, erfordert jedoch eine lebenslängliche Substitution.

Die Entscheidung, ob man die radikalen Nebennierenoperationen in einer oder in zwei Sitzungen durchführen soll, richtet sich nach dem Zustand der Patienten. Die einzeitige Operation erscheint logischer, sie bedeutet jedoch zweifellos eine höhere Belastung. Das zweizeitige Verfahren hat den Nachteil, daß nach Entfernung oder Resektion einer Nebenniere nicht selten Wundinfektionen auftreten und die Vornahme des 2. Eingriffes verzögern können. Wir selbst hatten bisher nur zweizeitig operiert und sind damit gut gefahren.

Die perioperative Behandlung hat folgendes zu beachten:

1. Viele Patienten haben infolge des Hochdrucks einen Herzmuskelschaden, der der vorangehenden Behandlung bedarf.

2. Zur Vermeidung postoperativer Störungen des Kreislaufes und der Wundheilung muß versucht werden, das bestehende Eiweißdefizit soweit als möglich zu verringern. Hierzu bedarf es der prä- und postoperativen Zufuhr von Plasma sowie von Blut am Operationstag (*16*). Auch die Verwendung von Anabolica ist zu empfehlen, ferner bei entsprechendem Mangel die Verabreichung von γ-Globulin. Bei 13 eigenen Patienten fanden wir vor Beginn der Behandlung 11mal eine Hypalbuminämie, 9mal eine Verminderung des Gesamteiweißes und 6mal eine Hypo-γ-Globulinämie (Prof. Dr. K.-O. VORLAENDER, Medizinische Univ.-Klinik Bonn).

3. Die Elektrolytverhältnisse müssen stabilisiert werden. Niedrige oder deutlich erniedrigte Serum-Kalium-Werte lassen sich durch Zufuhr von Kalium bei streng kochsalzarmer Kost leicht normalisieren. Stark erniedrigte Serum-Kalium-Werte (< 3 mäq/l) müssen an das Vorliegen eines Carcinoms denken lassen (*2*).

4. Bei der Infektanfälligkeit der Patienten ist die Verwendung eines Antibioticums angezeigt.

5. Die Cushing-Patienten sind durch eine erhöhte Thromboseneigung gefährdet. Dies ergibt sich aus der klinischen Beobachtung und ist bei unseren eigenen Patienten durch die Ergebnisse der gerinnungsphysiologischen Untersuchungen, die von EGLI vorgenommen wurden, belegt. Seitdem wir postoperativ mit Antikoagulantien behandeln, haben wir keine Thrombose mehr erlebt.

6. Was die Substitution mit Rindensteroiden betrifft, so hat uns die Erfahrung gelehrt, daß sie beim 1. Akt der zweizeitigen Operation im allgemeinen überflüssig ist. In jedem Fall sollte man jedoch intravenös injizierbares Hydrocortison bereit halten, um es bei Bedarf verabreichen zu können, insbesondere wenn gegen Erwarten ein Rindentumor vorliegt.

Die einzeitige beiderseitige Adrenalektomie, die 2. Sitzung der zweizeitigen Operation sowie die Entfernung eines cortisolproduzierenden Rindentumors erfordert dagegen eine intensive Substitution. Sie darf andererseits nicht zu hoch bemessen werden, da

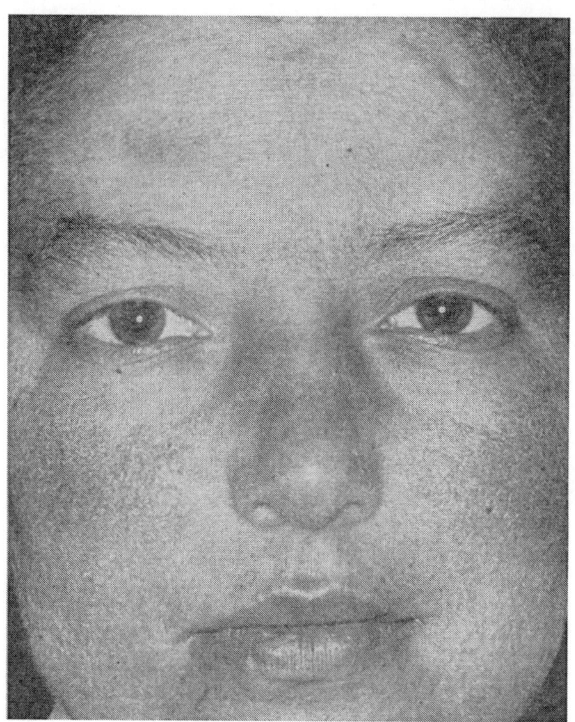

Abb. 3. Pat. A. P. (J.-Nr. 3240/57), 23 Jahre. Feinlamelläre Schuppung der Gesichtshaut während des Rückbildungsprozesses nach radikaler subtotaler Adrenalektomie (aus J. M. BAYER, Langenbecks Arch. klin. Chir. 1959, 4)

sonst u. U. der Blutdruck über Gebühr ansteigt und ein akutes Lungenödem verursachen kann. Die zahlreichen angegebenen Behandlungsschemata (*10, 15, 16, 25, 30, 37, 38, 42*) können nur als allgemeine Hinweise gelten. Als Richtschnur dient uns das Verhalten von Puls und Blutdruck sowie das Befinden der Patienten. Zweckmäßigerweise beginnt man die Substitution erst 2—3 Std vor der Operation mit einem intramuskulären Depot von 200 mg Cortisonacetat oder unmittelbar vor der Entfernung der 2. Nebenniere bzw. eines Rindentumors, wobei dann Hydrocortison im Dauertropf verabreicht werden muß. Die Gabe von DOCA erscheint zur Operation nicht unbedingt erforderlich.

Da die Patienten an den Cortisolüberschuß gewöhnt sind, muß die Reduktion von Cortison vorsichtig und schrittweise erfolgen. Reagieren die Patienten auf einen solchen Versuch mit Insuffizienzerscheinungen (Schwächegefühl, Appetitlosigkeit, Brechreiz, Erbrechen usw.), so muß auf die alte Dosis zurückgegangen werden. In akuten Fällen ist die intravenöse Zufuhr von Rindenpräparaten angezeigt. Es ist ferner daran zu denken, daß ein akuter Kochsalzmangel die gleiche Symptomatik bieten kann (*30*). Unter Umständen kann es Wochen dauern bis die erwünschte Dauersubstitution vertragen wird. Sie beträgt täglich 37,5—50 mg Cortisonacetat. Die Mehrzahl der Patienten bedarf zudem ein salzretinierendes Präparat. Als solches hat sich die tägliche Gabe von 0,1—0,2 mg Florinef® (9 α-Fluorhydrocortison) am besten bewährt. Es ist ferner darauf zu achten, daß die Patienten genügend Kochsalz zu sich nehmen.

Mit der Reduktion der Substitution beginnt die Rückbildung der Cushing-Erscheinungen. Sie geht mit einer mehr oder weniger heftigen Reaktion einher, die

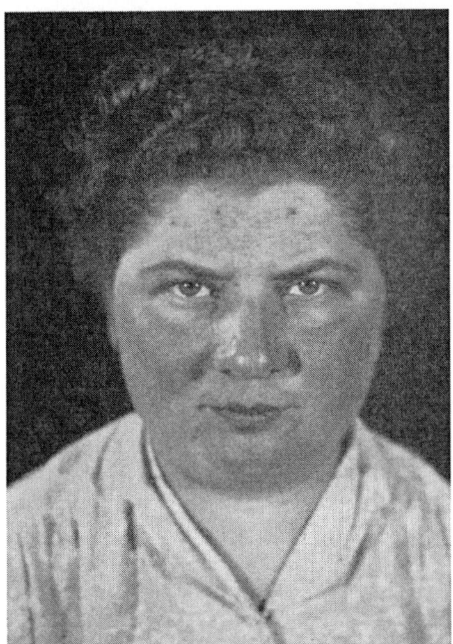 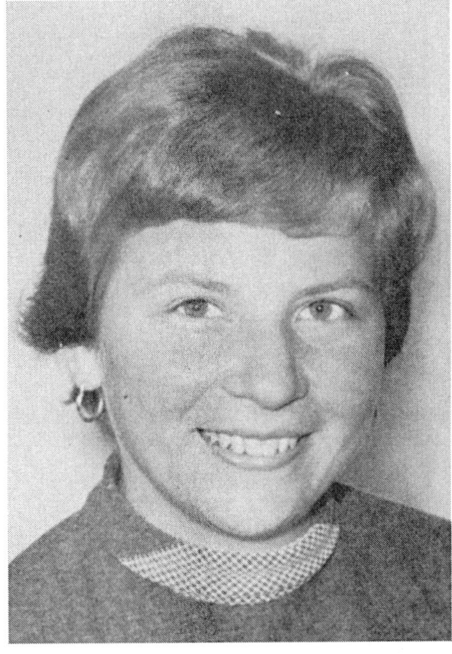

a   b

Abb. 4a u. b. Pat. K. G. (J.-Nr. 3239/57), 25 Jahre. Zustand vor a) und 1 Jahr nach radikaler subtotaler Adrenalektomie b) (aus J. M. Bayer, Langenbecks Arch. klin. Chir. 1959, 4)

offenbar mit der Neubildung protoplasmatischen Gewebes in Zusammenhang steht. Die Patienten klagen über Gelenk- und Muskelschmerzen; die Haut schält sich, juckt und zeigt u. U. ausgedehnte Erytheme. Abbildung 3 zeigt Ihnen eine fein-lamelläre Schuppung der Haut, die wir bei all unseren radikal operierten Patienten beobachtet haben.

Der Rückbildungsprozeß braucht bis zu seinem endgültigen Abschluß etwa 6 Monate bis zu 1 Jahr. Die Mehrzahl der Cushing-Erscheinungen schwinden vollständig. Die Patienten erhalten wieder ein normales Aussehen. Die Muskelkraft kehrt wieder. Nach 2—5 Monaten treten bei den Frauen erneut Cyclen auf. Schwangerschaften wurden wiederholt beschrieben. Wir selbst erlebten bei einer beiderseits total adrenalektomierten Patientin eine normale Schwangerschaft und Entbindung mit lebensfähigem Kind. Auch der Diabetes geht meist zurück. Der

Blutdruck fällt zur Norm oder wird doch in der Mehrzahl der Fälle günstig beeinflußt. Psychische Störungen und Beeinträchtigungen geistiger Funktionen werden gebessert, teilweise normalisiert. Der Hirsutismus schwindet meist, bildet sich jedoch nicht immer vollständig zurück. Die Osteoporose und etwaige Deformitäten

a

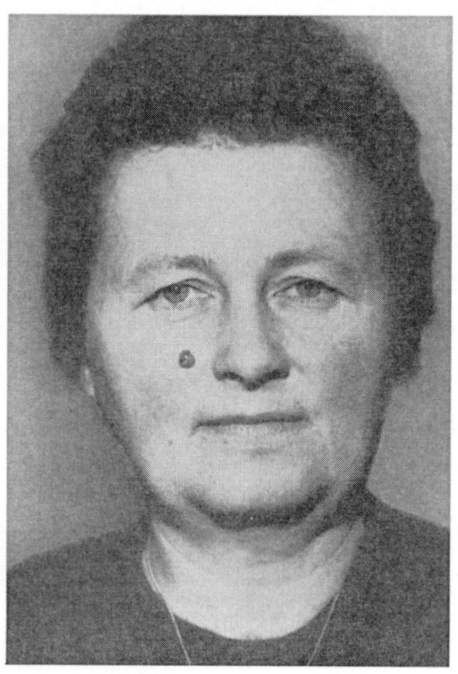

b

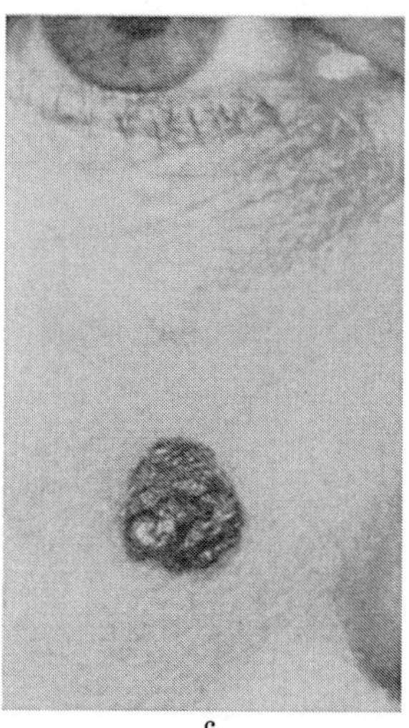

c

Abb. 5 a—c. Pat. H. H. (J.-Nr. 2275/57). a) 43 Jahre, Zustand vor der Operation. Kleine präblastomatöse Melanose an der rechten Wange. b) Pat. 5 Jahre nach radikaler subtotaler Adrenalektomie mit partieller Rindeninsuffizienz. Der Pigmentfleck an der rechten Wange ist größer und dicker geworden. c) Die präblastomatöse Melanose an der rechten Wange bei stärkerer Vergrößerung. Nach späterer Excision histologisch fragliche beginnende Malignität (Prof. Dr. G. Veltmann, Univ.-Hautklinik Bonn). Zusammenhang mit der Adrenalektomie unsicher

der Wirbelkörper kommen zum Stillstand; eine anatomische Restitution ist dagegen beim Erwachsenen nicht möglich. Bei einer Patientin, bei der das Cushing-Syndrom im Alter von 13 Jahren aufgetreten war, und die wir im Alter von 20 Jahren operierten, sahen wir dagegen einen Aufbau deformierter Wirbel und eine Rückbildungstendenz der peripheren Osteoporose. Ferner hatten sich die präoperativ noch offenen Epiphysenfugen geschlossen (Priv.-Doz. Dr. MAURER, Röntgen-Abt. der Chirurgischen Universitäts-Klinik Bonn). Im Gegensatz zur Entfernung cortisolproduzierender Tumoren sind die meisten Patienten mit Nebennierenrindenhyperplasie nach den radikalen Eingriffen an den Nebennieren trotz

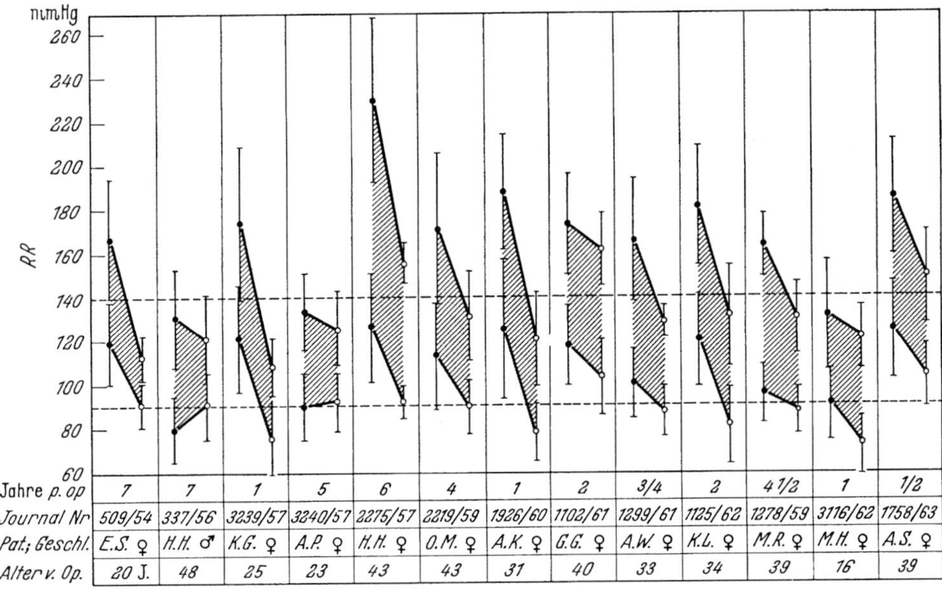

Abb. 6. Erläuterung s. Text. Blutdruckmittelwerte ± 2 s bei 13 eigenen Patienten mit Cushing-Syndrom vor und nach der Operation

scheinbar ausreichender Substitution mehr oder weniger stark pigmentiert. Wir sehen darin den klinischen Ausdruck einer Steigerung der ACTH-Sekretion, die nach Normalisierung des Cortisolspiegels im Blut auftritt (48).

Die Abb. 4 und 5 zeigen Ihnen zwei unserer Patienten vor und nach radikaler subtotaler Adrenalektomie.

Auf der Abb. 6 sind die Blutdruckmittelwerte ± 2 s bei 10 Patienten mit Hyperfunktions-Cushing und bei 3 Patienten mit Rindenadenomen vor und in einem zeitlichen Abstand von ¹/₂ bis zu 7 Jahren nach der Operation dargestellt (Dr. W. MEYER, Medizinische Univ.-Klinik Bonn, Cardiologische Abt.). Bei 10 Fällen war der Blutdruck vor der Operation erhöht. In 7 Fällen ist er postoperativ normal, in 3 Fällen noch hyperton, jedoch wesentlich gebessert. 9mal war das EKG präoperativ pathologisch verändert; 7mal war es nach der Operation normalisiert, in 2 Fällen gebessert (Dr. W. MEYER, Medizinische Univ.-Klinik Bonn, Cardiologische Abt.).

Die Ergebnisse der radikalen Nebenniereneingriffe beim Hyperfunktions-Cushing sind im Verhältnis zum präoperativen Zustand ausgezeichnet. Der Zustand vor der Erkrankung wird jedoch nur selten voll erreicht. Häufig ist die körperliche und geistige Leistungsfähigkeit doch mehr oder weniger stark vermindert, auch finden sich vielfach noch gewisse psychische Störungen. Inwieweit es sich hierbei um Defektheilungen handelt oder um die Unmöglichkeit, die Substitution den Bedürfnissen eines gewöhnlichen Tagesablaufes stets in adäquater Weise anzupassen, vermag ich vorläufig nicht zu entscheiden. Doch neige ich eher zur letzteren Meinung.

Die Operationsmortalität beträgt im Durchschnitt etwa 6% (*4*). Doch gibt es eine Reihe größerer Operationsserien ohne Todesfall (*15, 18, 36, 37*). Wir selbst

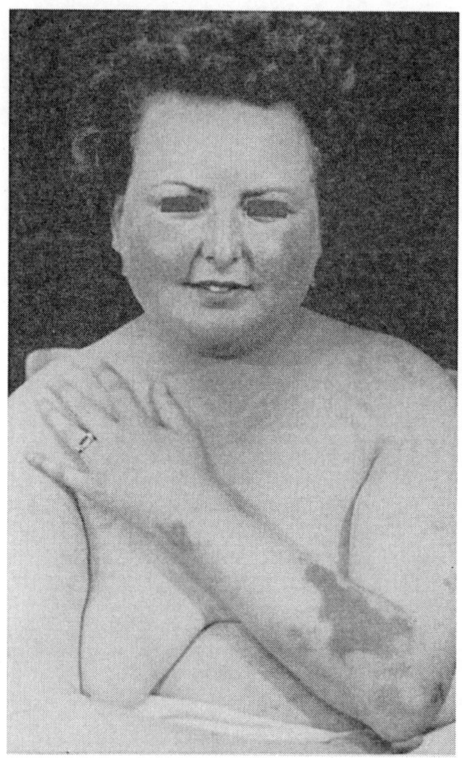 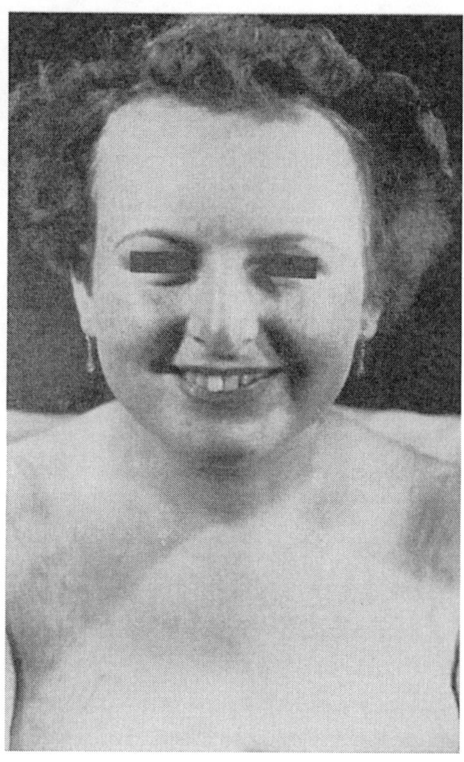

a   b

Abb. 7 a u. b. Pat. E. S. (J.-Nr. 1516/54), a) 20 Jahre, Zustand vor der Operation. b) 1 Jahr nach radikaler subtotaler Adrenalektomie mit latenter Rindeninsuffizienz (aus J. M. Bayer u. H. Rohr, Dtsch. med. Wschr. 1964, 6)

haben bei 16 operierten Cushing-Fällen (3 Rindenadenome, 13mal Hyperfunktions-Cushing) ebenfalls keinen Patienten durch die Operation verloren.

Die Notwendigkeit der Dauersubstitution nach beiderseitiger totaler Adrenalektomie läßt sich vermeiden, wenn man nach Franksson u. Mitarb. (1956, 1959) (*12, 13*) die entfernten Nebennieren in den M. sartorius einpflanzt. Das Markgewebe geht völlig, das Rindengewebe zum größeren Teil zugrunde. Es bleibt jedoch genügend Rindengewebe erhalten, das einwächst und funktioniert, so daß die Patienten nach 5—7 Monaten ohne Substitution auskommen. Sollte sich übermäßiges Rindengewebe und ein Cushing-Rezidiv entwickeln, so wäre die Möglichkeit

gegeben, an leicht zugänglicher Stelle eine Resektion vorzunehmen. Bisher war
das jedoch nicht erforderlich, obwohl FRANKSSON (*14*) inzwischen mehrere Patienten einige Jahre verfolgt hat. Die Patienten sind weiterhin latent insuffizient und
dementsprechend gefährdet. Eine unserer Patientinnen, die nach radikaler subtotaler Adrenalektomie bei üblicher Belastung keine Substitution benötigte, starb
4 Jahre nach der Operation während einer eitrigen Angina an einer akuten Rindeninsuffizienz. Die Patientin fühlte sich bis dahin wohl. Alle Ermahnungen und die

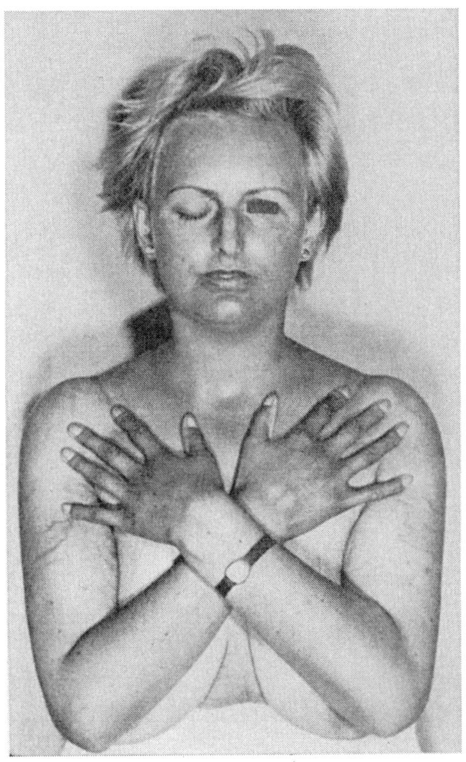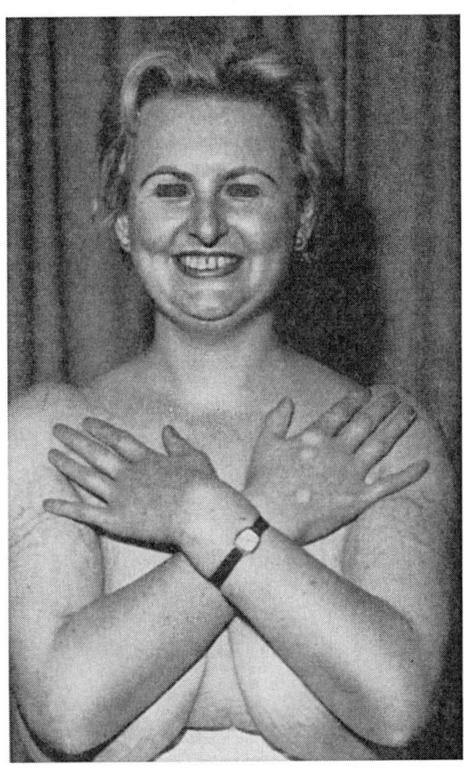

a b

Abb. 8 a u. b. Die gleiche Patientin wie in Abb. 7. a) 7 Jahre nach radikaler subtotaler Adrenalektomie. Totale
Ophthalmoplegie rechts infolge Blutung in einen supra- und rechts parasellär wachsenden Hypophysentumor.
Hyperpigmentierung der Haut, besonders deutlich an den Handrücken. b) Die Patientin 3 Monate nach Implantation von Y^{90} in die Hypophyse. Rückbildung der Augenmuskellähmungen und Schwund der Pigmentierung
(aus J. M. BAYER u. H. ROHR, Dtsch. med. Wschr. 1964, 6)

mitgegebene Ausweiskarte über die stattgehabte Operation waren in Vergessenheit
geraten.

Die beiderseitige totale Adrenalektomie galt bisher bei der Mehrzahl der
Autoren als die Behandlungsmethode der Wahl beim Hyperfunktions-Cushing.
Die zunehmende Beobachtung postoperativer Hyperphysentumoren (*6, 32, 33, 39*)
ist jedoch beunruhigend und sollte zu denken geben. Eine statistisch gesicherte
Zahl über ihre Häufigkeit ist bisher nicht bekannt. Die Angaben schwanken zwischen 5 (*28*) und 16% (*49*). In der Regel handelt es sich um chromophobe, ACTH-
sezernierende Adenome. Die Symptomatik ist charakteristisch. Sie besteht in
einer intensiven Pigmentierung der Haut und der Schleimhäute sowie in den

mechanischen Auswirkungen des Tumors. Unter 10 eigenen Fällen von Hyper-funktions-Cushing, bei denen wir eine radikale subtotale oder eine beiderseitige

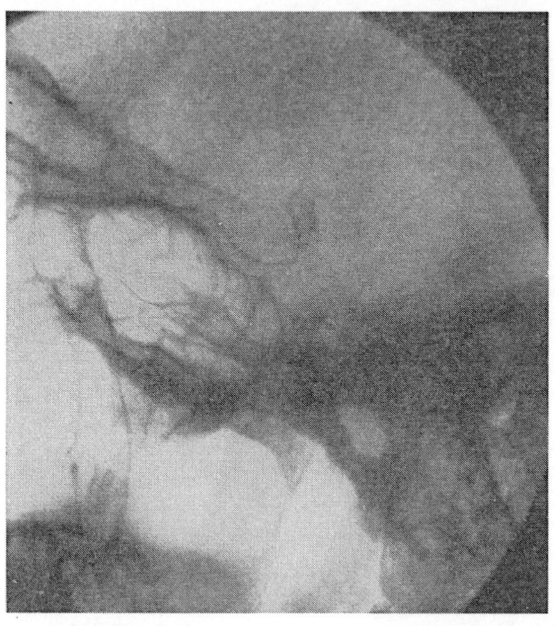

a

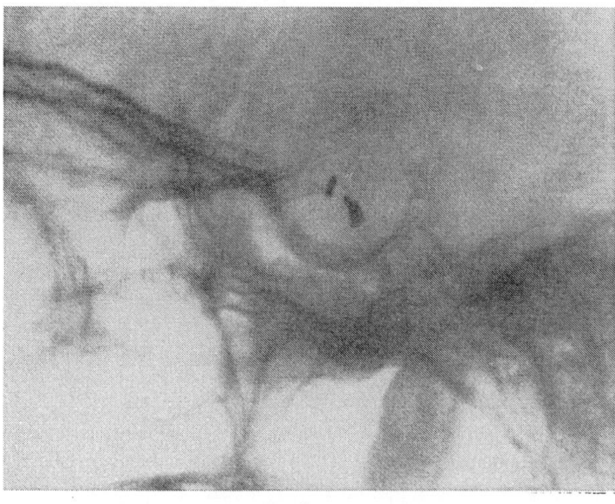

b

Abb. 9 a u. b. a) Pat. E. S., ausgeblendete Sellaaufnahme vom 5.1.1959, $3^1/_2$ Jahre nach der Adrenalektomie. Normale Form und Struktur der Sella. Keine Anzeichen für Ausweitung oder Atrophie des Dorsum. Keine Änderung gegenüber der Aufnahme vor der Operation (aus J. M. Bayer u. H. Rohr, Dtsch. med. Wschr. 1964, 6). b) Pat. E. S. Schichtaufnahme. Mittelschnitt durch die Sella (Juli 1962, 7 Jahre nach der Adrenalektomie). Ballonförmige Ausweitung der Sella mit Aufrichtung des Dorsum. Lage der Yttrium-Seeds in der Sella. (aus J. M. Bayer u. H. Rohr, Dtsch. med. Wschr. 1964, 6)

totale Adrenalektomie vorgenommen hatten, haben wir 2mal das Auftreten eines solchen Tumors beobachtet. Wir konnten sie durch Einlagerung von Au^{198} bzw. Y^{90}

in die Hypophyse erfolgreich behandeln. Die Abb. 7—9 geben den einen der beiden Fälle wieder.

Beide Patientinnen sind voll wieder hergestellt und hatten bald ihre alte Tätigkeit wieder aufgenommen. Die Ovarialfunktion ist jedoch ausgeblieben, während die der Schilddrüse erhalten blieb.

Solche Tumoren können den Cushing-Erscheinungen vorausgehen, gleichzeitig mit ihnen verbunden sein oder ihnen nachfolgen bzw. erst nach der Adrenalektomie auftreten. Mit KRACHT u. TAMM (1963) (*24*) halten wir diese Gebilde nicht für echte Tumoren, sondern für hyperplaseogene Geschwülste als Folge eines abnormen Stimulus von seiten des Hypothalamus [BAYER u. ROHR, 1964 (*6*)].

Mit besonderem Interesse verfolgen wir die Bestrebungen jener Autoren (*9, 20, 21, 45*), die den Hyperfunktions-Cushing durch primäre Einlagerung radioaktiver Substanzen in die Hypophyse erfolgreich behandeln und bei Verwendung geringer Aktivitäten keine Funktionsausfälle der Schilddrüse und der Gonaden beobachteten. Dieses Verfahren bedarf noch der weiteren Erprobung. Wir halten es jedoch bereits jetzt bei älteren Patienten und bei jenen, für die die beiderseitige totale Adrenalektomie ein zu großes Risiko bedeuten würde, für angezeigt. Herr KLEIN-FELDER wird über diesen Gegenstand noch sprechen.

Die Behandlung der Hypophyse mit konventionellen oder ultraharten Strahlungen ergibt nur zu einem geringen Teil befriedigende und länger anhaltende Remissionen. Sie wird von einigen Autoren (*19, 40, 41*) als Anfangsbehandlung bei leichten Fällen von Hyperfunktions-Cushing und bei Hypophysentumoren empfohlen. Nicht selten folgte ihr die Adrenalektomie oder Hypophysektomie nach. Von Erfolg erscheint die Bestrahlung mit schweren Kernpartikeln (Proton oder α-Teilchen) von hoher kinetischer Energie (*26*). Sie ist allerdings nur in wenigen Zentren durchführbar.

Der Verwendung von Anabolica und von Oestrogenen (*34*) kommt lediglich eine Bedeutung in der Vorbereitung zur Adrenalektomie zu. Hinsichtlich der Behandlung des Cushing-Syndroms mit o, p′-DDD möchte ich Herrn GEYER nicht vorgreifen.

Echte autonome Hypophysentumoren als Ursache eines Cushing-Syndroms sind selten. Sie bedürfen ebenso wie die hyperplaseogenen Geschwülste einer wirksamen Form der Strahlenbehandlung, gegebenenfalls eine chirurgische Therapie.

Die therapeutischen Aussichten bei Tumoren nichtendokriner Organe, die ACTH-wirksame Stoffe sezernieren, sind schlecht. Diese Fälle sind meist durch einen foudroyanten Verlauf gekennzeichnet. Auch die langsam wachsenden Tumoren dieser Art sind meist nicht radikal operabel. Jedenfalls sollte man jedes Cushing-Syndrom auch in dieser Richtung sowie nach dem möglichen Vorliegen eines endokrinaktiven Ovarialtumors untersuchen. Vielleicht gelingt es doch, den einen oder anderen dieser Fälle durch einen operativen Eingriff zu retten.

Danksagung

Die Steroidausscheidung im Urin (Abb. 1) wurde in der Chemischen Abt. der Chirurgischen Univ.-Klinik Bonn (Leiter: Prof. Dr. H. BREUER) bestimmt. Den zahlreichen Kollegen der verschiedenen Fachrichtungen danke ich für die bereitwillige Überlassung ihrer Befunde.

Literatur

1. APPLEBY, J. I., G. GIBSON, J. K. NORYMBERSKI, and R. D. STUBBS: Biochem. J. **60**, 453 (1955).
2. BAGSHAVE, K. D.: Lancet **1960 II**, 284.
3. BAYER, J. M.: Med. Klin. **53**, 1715 (1958).
4. — Langenbecks Arch. klin. Chir. **291**, 531 (1959).
5. — H. J. HOLTMEIER u. H. BREUER: Chirurg **31**, 529 (1960).
6. —, u. H. ROHR: Dtsch. med. Wschr. **89**, 464 (1964).
7. BERGENSTAL, D. M., M. B. LIPSETT, R. H. MOY, and R. HERTZ: In: Biological activities of steroids in relation to cancer, p. 463, edit.: by G. PINCUS and E. P. VOLLMER. New York and London: Academic Press 1960.
8. BIGLIERI, E. G., S. HANE, P. E. SLATON jr., and P. H. FORSHAM: J. clin. Invest. **42**, 516 (1963).
9. BROOKS, R. V., R. R. McSWINEY, D. MATTINGLY, and F. T. G. PRUNTY: J. Endocr. **19**, 366 (1960).
10. COPE, O., and J. W. RAKER: New Engl. J. Med. **253**, 165 (1955).
11. ENGEL, F. L., and L. KAHANA: Amer. J. Med. **34**, 726 (1963).
12. FRANKSSON, C., G. BIRKE, G. MOBERGER, and L.-O. PLANTIN: Acta chir. scand. **111**, 113 (1956).
13. — —, and L.-O. PLANTIN: Acta chir. scand. **117**, 409 (1959).
14. FRANKSSON, C.: persönliche Mitteilung 1964.
15. GLENN, F., R. C. KARL, and M. HORWITH: Ann. Surg. **148**, 365 (1958).
16. HARTENBACH, W.: internist. prax. **2**, 221 (1962).
17. HEDINGER, CHR.: In: A. LABHART: Klinik der inneren Sekretion. p. 339. Berlin-Göttingen-Heidelberg: Springer-Verlag 1957.
18. HORWITH, M., P. E. STOKES, and R. E. PETERSON: Acta endocr. (Kbh.) Suppl. 51, No. 187, p. 373 (1960).
19. HUME, D. M.: Diskussion zu SCOTT jr., H. W. u. Mitarb., Ann. Surg. **155**, 709 (1962).
20. KLEINFELDER, H., M. NADJMI, W. BÖRNER u. F. HUSMANN: Klin. Wschr. **40**, 1162 (1962).
21. KLOTZ, H. P., H. CHIMENES, et Mme KANOVITCH: Ann. Endocr. (Paris) **23**, 355 (1962).
22. KRACHT, J.: Virchows Arch. Path. Anat. **335**, 21 (1962).
23. —, u. J. TAMM: Virchows Arch. Path. Anat. **333**, 1 (1960).
24. — — Acta Endocr. (Kbh.) **43**, 330 (1963).
25. LABHART, A., E. R. FROESCH u. W. ZIEGLER: Schweiz. med. Wschr. **89**. 44 (1959).
26. LINFOOT, J. A., J. H. LAWRENCE, J. L. BORN, and C. A. TOBIAS: New Engl. J. Med. **269**, 597 (1963).
27. MEADOR, C. K., K. W. LIDDLE, D. P. ISLAND, W. E. NICHOLSON, C. P. LUCAS, J. G. NUCTON, and J. A. LUETSCHER: J. clin. Endocr. **22**, 693 (1962).
28. MILLS, I. H., and R. H. O. B. ROBINSON: In: Modern trends in urology (second series). p. 85—99, edit. by Sir E. RICHES, London: Butterworths 1960.
29. MOLINATTI, G. M., F. CAMANNI, and M. TEDESCHI: J. clin. Endocr. **19**, 1144 (1959).
30. MONTGOMERY, D. A. D., and R. B. WELBURN: Brit. J. Surg. **45** (190), 137 (1957).
31. MOSIER, H. D., P. J. FLYNN, D. W. WILL, and R. D. TURNER: J. clin. Endocr. **10**, 632 (1960).
32. NELSON, D. H., J. W. MEAKIN, J. B. DEALY jr., D. D. MATSON, K. EMERSON jr., and G. W. THORN: New Engl. J. Med. **259**, 161 (1958).
33. — —, and G. W. THORN: Ann. int. Med. **52**, 560 (1960).
34. NORMAN, N., and J. H. VOGT: Acta endocr. (Kbh.) **42**, 195 (1963).
35. NORYMBERSKI, J. K., R. D. STUBBS, and H. F. WEST: Lancet **1953 I**, 1276.
36. OVERTON, R. C., J. W. OVERSTREET, and R. V. FORD: Arch. Surg. **79**, 791 (1959).
37. POUTASSE, E. F., and CH. C. HIGGINS: Transact. Amer. Assoc. Genito-Urinary Surg. (1952), 116.
38. PRIESTLEY, J. I., R. G. SPRAGUE, W. WALTERS, and R. M. SALASSA: Ann. Surg. **134**, 464 (1951).
39. SALASSA, R. M., T. P. KEARNS, J. W. KERNOHAN, R. G. SPRAGUE, and C. S. MacCARTY: J. clin. Endocr. **19**, 1523 (1959).

40. Scott jr., H. W., G. W. Liddle, A. P. Harris, and J. H. Foster: Ann. Surg. **155**, 696 (1962).
41. Soffer, L. F., J. Eisenberg, H. Iannaccone, and J. L. Gabbrilove: Ciba Found. Coll. Endocr. **8**, 487 (1955).
42. Sprague, R. G., W. F. Kvale, and J. T. Priestley: J. Amer. med. Ass. **151**, 629 (1953).
43. — R. V. Randall, R. M. Salassa, D. A. Scholz, J. T. Priestley, W. Walters, and A. H. Bulbulian: A. M. A. Scientific Exhibits (1955), p. 315.
44. — R. E. Weeks, J. T. Priestley, and R. M. Salassa: In: Modern trends in endocrinology, p. 84, edit: by H. Gardiner-Hill. London: Butterworths 1961.
45. Talairach, J., G. Szikla, A. Bonis, P. Tournoux, J. Bandcaud, et E. Mempel: Presse méd. **70**, 1399, 1449, (1962).
46. Tonutti, E., u. J. M. Bayer: Endokrinologie **41**, 17 (1961).
47. Walters, W.: Lancet **1952 I**, 221.
48. Williams jr., W. C., D. Island, R. A. A. Oldfield jr., and G. W. Liddle: J. clin. Endocr. **21**, 426 (1961).
49. Zukschwerdt, L., M. G. Giebel, H. Oetjen u. J. Tamm: Schweiz. med. Wschr. **92**, 667 (1962).

Aus der I. Medizinischen Universitätsklinik Wien
(Suppl. Leiter: Prof. Dr. H. Jesserer)

Die Behandlung des Cushing-Syndroms mit o,p-DDD (2,2-bis (2-chlorophenyl-4-chlorophenyl)-1,1-dichloräthan)

Von

G. Geyer

Die Kenntnis, daß dem Insecticid DDD eine sekletive Wirkung auf die Rinde der Nebenniere zukommt, geht auf Untersuchungen von Nelson (*1*) zurück, der 1948 bei Toxicitätsprüfungen von Insecticiden nach Gabe dieses Mittels eine „cytotoxische Atrophie" der Rinde beim Hund beobachtete. Das Phänomen ist in den folgenden Jahren von einer Reihe von Autoren bei Hunden und anderen Laboratoriumstieren nachuntersucht worden, wobei die erzielten Wirkungen ziemlich ungleichmäßig waren. Auch beim Menschen ist unter der Indikation des Cushing-Syndroms (*2*) und des Diabetes (*3*) die Gabe von kommerziellem DDD schon vor Jahren versucht worden, ohne daß ein sicherer Effekt nachzuweisen war. Diese Inkonsistenz der Ergebnisse fand eine Erklärung, als 1958 Cueto und Brown (*4, 5*) das bis dahin verwendete DDD einer Fraktionierung unterwarfen, die zeigte, daß das gehandelte Präparat zu 83 % aus p,p-DDD bestand — das sich im Tierversuch als nicht toxisch für die Nebenniere erwies — und wechselnde Mengen bis zu 10 % des isomeren o,p-DDD enthielt; dieses erwies sich in Tierversuchen als reproduzierbar auf die Nebennierenrinde wirksam und ist seither mit gleichbleibendem Erfolg experimentell verwendet worden. Damit war das Bestreben, dieses Mittel auch beim Menschen therapeutisch anzuwenden, nur mehr von der Verfügbarkeit ausreichender Mengen von o,p-DDD abhängig geworden. Als das Nebennierencarcinom sich als mögliche Indikation dafür abzuzeichnen begann, konnte eine industrielle Herstellung des o,p-DDD mit den Mitteln des US. Cancer Institutes begonnen werden; es waren in der Folge R. Hertz und seine Arbeitsgruppe in Bethesda, die das o,p-DDD bei Patienten mit Nebennierencarcinomen mit Metastasen erprobten. Diese Autoren beobachteten bei einer (gemessen an der Seltenheit der Erkrankung) großen Zahl von Patienten z. T. eine objektive Regression der Metastasen, bei der überwiegenden Anzahl der Patienten eine deutliche Abnahme der vermehrten Steroidausscheidung und nur bei etwa $^{1}/_{5}$ der Probanden keinen Erfolg (*6*).

Ich habe bisher zwei weibliche Patienten mit Nebennierencarcinom mit o,p-DDD behandelt. Bei einem dieser Fälle war als Erfolg der Medikation eine Abnahme der virilen Symptomatik und ein Wiederauftreten der cyclischen Blutungen nach 1 jähriger Amenorrhoe zu verzeichnen; die Ausscheidung an 17-Ketosteroiden und 17-Hydroxycorticosteroiden nahm dabei sehr deutlich ab (*10*). Ein Rückgang der bestehenden abdominellen und Knochenmetastasen war bei diesem Falle nicht objektivierbar; ob der Tumor unbehandelt rascher gewachsen wäre, läßt sich nicht sagen, jedenfalls ist diese Patientin 2 Jahre nach Auftreten der ersten Symptome und $1^1/_2$ Jahre nach der Exstirpation des primären Nebennierentumors an der Progredienz des Tumorleidens verstorben. Eine zweite Patientin mit einem anatomisch ziemlich kleinen, jedoch schließlich in die Hohlvene eingewachsenen Nebennierencarcinom hat auf die protrahierte Gabe von o,p-DDD klinisch nicht angesprochen und auch in der Ausscheidung der Steroide keinen Therapieeinfluß erkennen lassen. Nach diesen beiden Erfahrungen bin ich mit der Empfehlung, o,p-DDD beim Nebennierencarcinom zu geben, eher zurückhaltend. Ich habe eine objektive Regression des Tumors und eine Besserung der Folgeerscheinungen der malignen Erkrankung nicht beobachten können; was vielmehr erreichbar scheint, ist eine Remission des endokrinen Symptomenbildes, während der Tumor daneben weiterwächst. Diese endokrine Remission ist erkauft mit subjektiv unangenehmen Nebenwirkungen, welche o,p-DDD in den hohen Dosen hat, die für die Beeinflussung des Nebennierencarcinoms erforderlich sind: Bei protrahierter Gabe von täglich etwa 10 g o,p-DDD sind Schwindel, Inappetenz, Nausea und Erbrechen häufige, fast obligate Nebenwirkungen. Sie beeinflussen das Befinden der Patienten unter Umständen so ungünstig, daß diese Belastung fallweise kaum mit dem in Einklang stehen wird, was mit dieser Therapie erreicht werden kann.

Es war naheliegend, o,p-DDD auch bei Hypercortizismen nicht-maligner Genese zu versuchen. Wir haben das bisher bei vier Fällen tun können und davon drei Fälle über länger als ein Jahr in Beobachtung. Zu den Nebenwirkungen des o,p-DDD ist im Hinblick auf die Behandlung dieser benignen Fälle zu sagen, daß die dabei wirksamen Dosen maximal die Hälfte der beim Nebennierencarcinom erforderlichen sind, sie liegen zwischen 3 und 5 g o,p-DDD täglich; diese Mengen des Mittels werden auch für lange Zeit praktisch immer ohne Beschwerden vertragen. Nebenwirkungen sind demnach also kein Handycap für eine Behandlung der benignen Fälle von Hypercortizismus mit o,p-DDD.

Bei unseren drei Fällen von Cushing-Syndrom auf der Basis einer Nebennieren-Hyperplasie war nun bei laufender Verfolgung der Ausscheidung der 17-Hydroxy-corticosteroide und 17-Ketosteroide im Harn festzustellen, daß die DDD-Gabe auf diese beiden Indicatoren einen sehr deutlichen Einfluß nimmt: Es kam regelmäßig zu einem markanten Rückgang der Ausscheidung dieser Steroide bis zu Werten im unteren Normalbereich (*10, 11*). Wir haben diese Medikation jeweils etwa 6 bis 8 Wochen aufrechterhalten und dann abgesetzt. Es erscheint erwähnenswert, daß wir dabei nie Zeichen einer Hypadrenie beobachtet haben.

Dieses Verhalten legt die Annahme nahe, daß die Steroidausscheidung deshalb abnimmt, weil die durch o,p-DDD geschädigte Nebenniere weniger bildet. Nun wurde neuerdings auch ein extraadrenales Eingreifen des o,p-DDD in den Steroidkatabolismus nachgewiesen: GALLAGHER u. Mitarb. (*8*) beobachteten eine Hemmung der Bildung von Pregnantriol aus seinen Vorläufern und LIDDLE u. Mitarb. (*7*) halten eine Bildung stärker polarer Steroide beim Abbau des Cortisol für möglich. Diese Annahme macht es erforderlich, zu diskutieren, ob die Drosselung der adrenalen Steroidbildung durch DDD tatsächlich so beträchtlich ist, wie die gezeigten Ausscheidungsprofile es vermuten lassen.

Die Frage, ob eine nennenswerte und therapeutisch bedeutsame Verminderung der adrenalen Cortisolsekretion mit o,p-DDD erreicht werden kann, läßt sich prinzipiell dadurch positiv beantworten, daß o,p-DDD beim Hypercortizismus zu einer Beeinflussung nicht nur der Steroidausscheidung, sondern auch der klinischen und metabolischen Symptome des Cortisolüberschusses führt: Die facies der Patienten verliert im Verlaufe einiger Wochen ihre typische Form und die teleangiektatischen Veränderungen der Haut verschwinden völlig. Auch die Hypertension geht zurück und Hyperglykämie und Glucosurie werden normalisiert. Es war bei unseren drei Cushing-Fällen somit ein eindeutiger Einfluß der Gabe des o,p-DDD in dem Sinne zu beobachten, daß eine Remission des Krankheitsbildes induziert wurde. Es mag in diesem Zusammenhange von Interesse sein, zu erwähnen, daß eine dieser drei Patientinnen eine chronische Polyarthritis hat, die ihr nicht allzuviel Beschwerden macht; wenn ihr Hypercortizismus unter dem Einfluß des o,p-DDD zurückgeht, so klagt sie regelmäßig sehr über Gelenksschmerzen und die DDD-Behandlung macht nach Eintreten ihrer Wirksamkeit regelmäßig eine Arthritistherapie erforderlich, die mit dem Vorübergehen der Cushing-Remission wieder überflüssig wird. Zusammenfassend glaube ich deshalb, daß die klinisch von uns beobachteten Besserungen des Cushing-Syndroms unter dem Einfluß von o,p-DDD die therapeutische Verwendbarkeit dieses Mittels erweisen. Sie lassen auch kaum eine andere Erklärung zu, als die, daß die adrenocorticale Cortisolsekretion durch das Mittel gedrosselt wird. Dabei soll der von Liddle beobachtete extraadrenale Effekt nicht bestritten werden — eigene Ergebnisse zur Analyse dieser Frage besitzen wir noch nicht — er kann sich aber allenfalls nur darin auswirken, daß die erhobenen Steroidprofile eine stärkere Drosselung der Nebenniere vermuten lassen, als sie tatsächlich vorliegt. Aber auch im vorliegenden Ausmaß befriedigt die mit o,p-DDD erreichte adrenocorticale Hemmung offenbar die Anforderungen der Therapie.

Welche praktische Konsequenzen haben sich für uns daraus bisher ergeben? Wir haben mit der beschriebenen Möglichkeit einer konservativen Behandlung des Cushing-Syndroms vor allem Zeit gewonnen. Wir können eine Remission erreichen für die Dauer der Medikation (die wir meist 6—8 Wochen aufrechterhalten) und für die darauffolgenden etwa 6—8 Wochen. Denn solange dauert es, bis sich wieder die ersten Anzeichen eines Wiederauftretens der klinischen Symptomatik, des Relapses also, einstellen. Nicht bei allen unseren Patienten haben diese Rezidive wieder zum gleichen Vollbild geführt, wie vor der DDD-Behandlung. Aber immerhin sind wir sicher, daß wir mit dieser Form der o,p-DDD-Medikation die Nebenniere nicht soweit schädigen können, daß eine Dauerreduktion ihrer Sekretion resultiert, die eine Dauerheilung bedeuten würde. Prinzipiell hätten wir also die Wahl zwischen einer chronisch-intermittierenden medikamentösen Behandlung und der chirurgischen Intervention. Dennoch sind wir prinzipiell für die chirurgische Intervention, aber wir glauben, daß die DDD-Behandlung es uns ermöglicht, die Operation zu einem Zeitpunkt und unter Ausgangsbedingungen durchführen zu lassen, die dem Patienten die besten Chancen geben; dies deshalb, weil man ihn in der vollen Remission als quasi-Gesunden, nur potentiell Kranken operieren lassen kann und ihm die Risiken ersparen können sollte, die sich für die postoperative Periode aus der Elektrolytstoffwechselstörung und der Anfälligkeit des peripheren Kreislaufes des im Vollbild des Cushing-Syndroms Operierten ergeben. Wir haben dieses Vorgehen bisher erst bei einer unserer drei länger behandelten Patientinnen eingehalten; sie wurde im Stadium der Vollremission operiert. Sie ist leider bereits nach der einseitigen totalen Adrenalektomie einer postoperativen Komplikation erlegen, die nicht mit dem Hypercortizismus Zusammenhang hat, nämlich einer hämorrhagischen Pankreatitis.

So wenig dieser Ausgang zur Frage aussagt, ob unser gegenwärtiges Konzept der medikamentösen Vorbereitung zur Adrenektomie richtig ist, hat er einen Einblick ermöglicht in die histologischen Veränderungen, die unsere Medikation an der Nebennierenrinde setzt. Bei unserer Patientin war die Nebenniere nicht so breit, hyperplastisch und lipoidreich, wie man dies beim unbehandelten Hypercortizismus beobachtet. Mikroskopisch waren die Zona glomerulosa intakt und auch die Zona fasciculata ziemlich ausgedehnt; es waren keine schwereren degenerativen Veränderungen an den Rindenzellen nachzuweisen. Als sehr auffälliger Befund waren jedoch teils herdförmige, teils streifige kleinzellige Infiltrate in den inneren Schichten der Rinde festzustellen, ähnlich wie sie VILAR und TULLNER (9) bei Hundeversuchen mit o,p-DDD beobachtet hatten. Färbung der Schnitte mit Sudanschwarz und Sudanrot erwies die Lipoidverteilung ausschließlich der inneren Rindenschichten verändert. Mit der von uns angewandten Dosis und Medikationsdauer des o,p-DDD wird demnach keine Atrophie der Rinde erreicht, wie das in den ursprünglichen Tierversuchen der Fall war; der Behandlungserfolg dürfte vielmehr auf eine Interferenz des Mittels mit der adrenalen Steroidbildung zurückzuführen sein.

Literatur

1. NELSON, A. A., and G. WOODARD: Arch. Path. 48, 387 (1949).
2. SHEEHAN, H. L., V. K. SUMMERS, and J. NICHOLS: Lancet 1953 I, 312.
3. TÖRNBLOM, N.: Acta med. scand. 164, 83 (1956).
4. CUETO, C., and J. H. U. BROWN: Endocrinologie 62, 326 (1958).
5. — — Endocrinologie 62, 334 (1958).
6. BERGENSTAL, D. M., R. HERTZ, M. B. LIPSETT, and R. H. MOY: Ann. int. Med. 53, 672 (1960).
7. LIDDLE, G. W.: Persönliche Mitteilung.
8. BRADLOW, H. L., S. FUKUSHIMA, B. ZUMOFF, L. HELLMAN, and T. F. GALLAGHER: J. clin. Endocr. 23, 918 (1963).
9. VILAR, O., and W. W. TULLNER: Endocrinology 65, 80 (1959).
10. GEYER, G.: Acta Endocr. 40, 332 (1962).
11. —, u. E. SCHÜLLER: Klin. Wschr. 40, 734 (1962).

Aus der II. Med. Univ.-Klinik und Poliklinik Mainz
(Direktor: Prof. Dr. P. Schölmerich)

Der Funktions-Cushing

Von

C. Overzier

Mit 2 Abbildungen

Im Rahmen der Besprechung des Cushing-Syndroms möchte ich eine Beobachtung mitteilen, die ich zuerst 1955 machte und bei insgesamt 10 Fällen verfolgen konnte.

Aus der leider großen Schar fettsüchtiger Jugendlicher werden dem Endokrinologen die Fälle mit vermeintlichen Drüsenstörungen vorgeführt. Sie erweisen sich bei dem heutigen Wohlstand in der ganz überwiegenden Zahl, man möchte sagen: fast ausschließlich, als „Jugendliche Fettsucht" durch Luxuskonsumption. Daß sie gerade im Pubertätsalter zum Arzt kommen, hat meist den äußeren Grund einer scheinbar oder auch tatsächlich verzögerten Pubertät, jedoch ist diese nicht die Ursache, eher schon eine Folge der Fettsucht. Verschwindend gering ist vergleichsweise die Zahl der Fettsüchtigen infolge eines echten Jugendlichen Cushing. Sehr selten ist auch der M. Fröhlich, der im übrigen nur unter besonderen Bedingungen der Tumoreinwirkung auf das Zwischenhirn eine mäßige Fettsucht auslösen kann und begrifflich klar von der Jugendlichen Fettsucht getrennt werden sollte.

Unter den zahlreichen Fällen, die als Jugendliche Fettsucht imponieren, habe ich nun in den letzten 8 Jahren 6 Knaben und 4 Mädchen gefunden, die eine überhöhte Corticoid-Ausscheidung im Harn aufwiesen. Wie Sie wunschgemäß in zwei Tabellen zusammengefaßt sehen, liegen diese Werte noch über der oberen Schwankungsbreite unseres Normbereichs für Corticoide, obwohl diese schon sehr weit angenommen wurde, und deutlich über unserem mittleren Normwert. Ähnliches sah ich bei den anderen Fettsüchtigen oder Normalpersonen nie. Die Werte wurden jeweils an mindestens 3 aufeinander folgenden Tagen aus der Gesamtharnausscheidung bestimmt. (Nach 40 E. Depot-ACTH erfolgte eine weitere Steigerung, die sich auch an den folgenden Tagen reproduzieren ließ.) Soweit eingesetzt, wurde ein Metopiron-Effekt erzielt. Diese Werte wurden nicht in die Kurven aufgenommen.

An dem Verlauf der Kurven aus der Patientenschar und der Verfolgung des Krankheitsbildes jeweils im Einzelfalle über Jahre ist zu erkennen, daß sowohl die Fettsucht als auch die überhöhte Corticoid-Ausscheidung mit dem Abschluß der — z. T. verlängerten — Pubertätszeit abklingt. Umgekehrt scheint der Beginn der Erscheinungen mit dem Beginn des Pubertätsalters zusammenzufallen, doch waren mir die Patienten zuvor leider nicht bekannt.

Die Ausscheidung der 17-Ketosteroide war hingegen bei den Mädchen meist normal und lag bei den Knaben eher im niedrigen Normbereich, wobei der spätere Anstieg wohl mit der auch verspäteten eigentlichen Ausreifung der Hoden erklärt

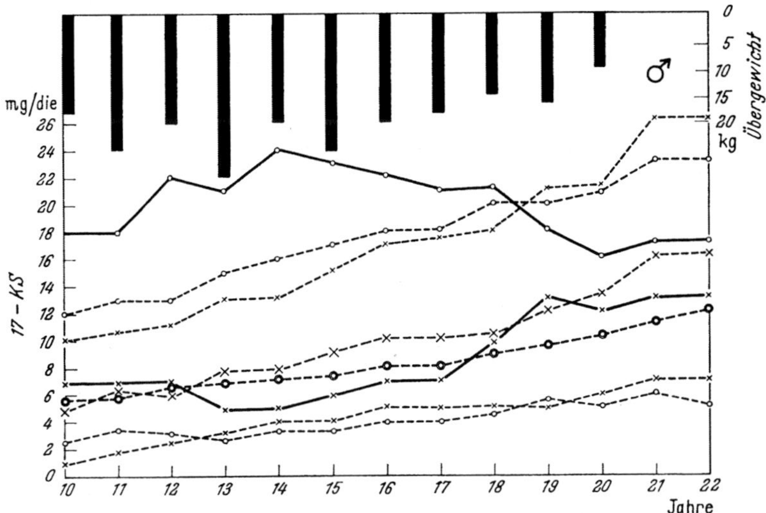

Abb. 1. Befunde bei 6 männlichen Patienten im Alter von 10—22 Jahren und einer Beobachtungszeit von 2—5 Jahren. Überhöhte Corticoid-Ausscheidung —o— während der Übergewichtigkeit und der Pubertät, normale, im ganzen etwas niedrige 17-Ketosteroid-Ausscheidung —x— mit verzögertem Anstieg. Zum Vergleich unsere Normwerte: Corticoide --- ⊚ --- und 17-Ketosteroide --- X --- und die normale Schwankungsbreite -- o --, --- x ---

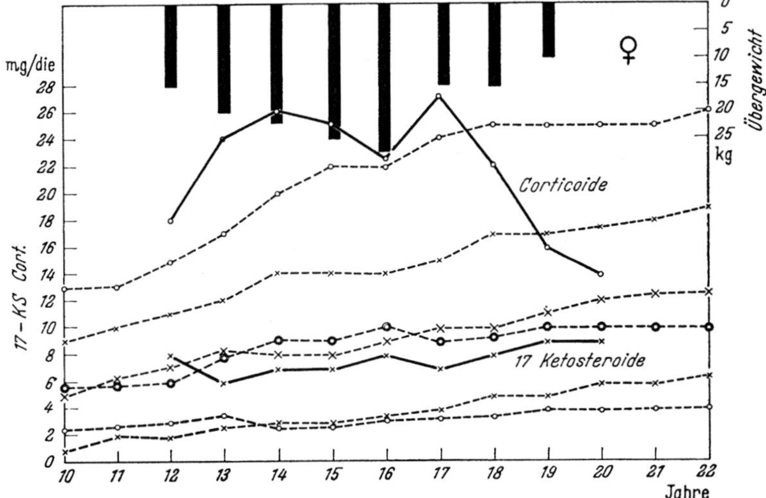

Abb. 2. Befunde bei 4 weiblichen Patienten im Alter von 12—20 Jahren und einer Beobachtungszeit von 3—5 Jahren. Überhöhte Corticoid-Ausscheidung —o— während der Übergewichtigkeit und Pubertät, normale 17-Ketosteroid-Ausscheidung —x—. Zeichen für die Normalwerte wie in Abb. 1

ist. Die Vermehrung der 17-Ketosteroid-Ausscheidung nach Depot-ACTH war nicht verwertbar gering.

Die Angabe dieser Patienten, nicht übermäßig zu essen, erschien glaubwürdiger als bei den Fällen mit Jugendlicher Fettsucht. So magerten sie auf Diät, auch klinische Diät, kaum ab und verloren ihr Übergewicht später mit Rückgang der

Corticoid-Ausscheidung auch ohne weitere Maßnahmen. Die verordnete Diät war mindercalorisch und relativ eiweißreich mit strengen Obst- bzw. Obstsafttagen und einer Begrenzung auf eine Hauptmahlzeit morgens und mittags sowie eine kleine Kost abends, ebenso wie bei allen Fettsüchtigen. Ob diese Empfehlung ambulant streng eingehalten wurde, ist selbstverständlich schwer zu sagen. Sicher scheinen diese Patienten nicht die unmäßigen Trinkgewohnheiten der übrigen Fettsüchtigen zu haben.

Klinisch fanden sich Striae, wie man diese allerdings auch bei der Jugendlichen Fettsucht nicht selten findet. Die Patienten haben ein rundes, frischrotes Gesicht, kein ausgesprochenes Vollmondgesicht, keine streng auf den Stamm begrenzte Fettsucht, keinen Hochdruck, keinen Diabetes, keine Ionen-Verschiebungen und kein eigentliches Krankheitsgefühl, häufig aber eine Akne. Und dennoch ist ihr Leiden mehr als eine einfache exogene Fettsucht. Es hängt offenbar mit den Nebennierenrinden und ihrer Pubertätsumstellung zusammen, wahrscheinlich primär, möglicherweise aber auch nur sekundär ausgelöst durch Überlastung mit Luxuskonsumption. Der Verlauf ist immer gutartig. Eine falsche Zuordnung dieser Fälle zum Jugendlichen Cushing mit operativen Konsequenzen wäre also verhängnisvoll.

Bei Jugendlicher Fettsucht fanden Ströder und Zeisel (1955) im allgemeinen die Corticoid-Werte normal, aber vereinzelt auch erhöht, während die 17-Ketosteroid-Werte immer im oberen Streubereich lagen und diesen teils sogar deutlich überschritten. Durch die Untersuchungen von Karl (Karl, 1962; Karl, Raith und Decker, 1962) ist bekannt, daß Fettsüchtige eine leicht vermehrte Tagesproduktion von Cortisol (mit Änderung des Tagesrhythmus) haben. Hier werden also bereits sicher sehr wichtige allgemeine Beziehungen zur Fettsucht aufgezeigt. Andererseits ist bemerkenswert, daß Baird (1963) nach strenger Diät Fettsüchtiger eine Verminderung der vorher erhöhten 17-ketogenen Steroide beobachtete. Befund und Verlauf der von mir beobachteten Fälle zeigen aber, daß sich diese aus dem allgemeinen Geschehen bei der Fettsucht herausheben. Wegen ihrer offenbaren Zusammenhänge mit der Pubertätsentwicklung und ihres klinisch-leichten Verlaufs kann man sie auch nicht zu den Fällen des endogenen transitorischen Cushing-Syndroms (Lit. bei Schwab und Denninger, 1959) in Beziehung setzen.

Ich möchte dieses von mir beschriebene Krankheitsbild als Funktions-Cushing bezeichnen.

Literatur

Baird, J. M.: Urinary corticoid excetion in oblese adults. Lancet **1963** II, 1022.

Jores, A.: Die Nebennieren und ihre Krankheiten, in Hdb. inn. Med. 7, 1. Teil. Berlin-Göttingen-Heidelberg: Springer 1955.

—, u. H. Nowakowski: Praktische Endokrinologie. Stuttgart: Georg Thieme 1964.

Karl, H. J.: Untersuchungen über Sekretion, Blutspiegel und Ausscheidung von Cortisol bei Fettsüchtigen. Verh. dtsch. Ges. inn. Med. **68**, 296 (1962).

— L. Raith u. W. Decker: Untersuchungen über die Cortisolsekretion beim Menschen. 9. Symposion der Dtsch. Ges. Endokrinologie 1962, p. 84. Berlin-Göttingen-Heidelberg: Springer 1963.

Labhart, A.: Klinik der inneren Sekretion. Berlin-Göttingen-Heidelberg: Springer 1957.

Overzier, C.: Wie behandelt man einen „Morbus Fröhlich"? Dtsch. med. Wschr. **83**, 2064 (1958).

Schwab, R., u. K. Denninger: Das transitorische Cushing-Syndrom. Ergebn. inn. Med. u. Kinderheilk. **12**, 563 (1959).

Ströder, J., u. H. Zeisel: Die Corticoide und neutralen C_{17}-Ketosteroide im Harn des Kindes. Z. Kinderheilk. **75**, 683 (1955).

Zondek, H.: Die Krankheiten der endokrinen Drüsen. Basel: Benno Schwabe 1953.

Diskussion

F. Bahner (Heidelberg):

Ich halte es für bedenklich, wenn man die Ausdrücke „cushingoide Fettsucht, Pseudo-
cushing oder Funktionscushing" gebraucht. Die einzigen Hinweise auf einen Hypercortizismus
sind in solchen Fällen von überwiegend jugendlicher Fettsucht die etwas erhöhte Corticoid-
ausscheidung, die kaum erhöht ist, wenn man sie auf das Körpergewicht oder die Oberfläche
bezieht, und die Striae. Aber diese Striae sind nicht nur schmaler als beim Cushing-Kranken,
sondern sie entstehen und vergehen relativ schnell. Man sieht dann frische schmale Striae an
der Grenze zur nicht befallenen Haut vordringen, während nach rückwärts von dieser Front
der Striaewanderung nur noch blasse vergangene Striae zu sehen sind. Beim Cushing-Kranken
selbst bleiben die Striae bestehen. Die entstehenden und vergehenden Striae scheinen eine für
das Jugendalter charakteristische Reaktion der Haut zu sein, wenn eine Fettsucht irgend-
welcher Ursache entsteht. Diese Erscheinung hat nichts mit einem M. Cushing zu tun und
berechtigt nicht zu Ausdrücken, in denen das Wort Cushing vorkommt. Behält man solche
Namen anstelle des einfachen Wortes Fettsucht bei, so bleibt immer wieder die Verwechselung
mit dem M. Cushing, die bis zur falsch indizierten Operation gehen kann.

J. R. Bierich (Hamburg):

Ich glaube nicht, daß die einfache Mehrausscheidung von Corticoiden bei dicken Kindern
dafür spricht, daß hier dem Cushing vergleichbare Verhältnisse vorliegen. Um bei den Harn-
corticoiden zu kommensurablen Werten zu kommen, müssen sie auf die Körperoberfläche
(nicht auf das Gewicht) bezogen werden; das haben zuerst Talbot u. Mitarb. gezeigt; wir
selbst haben es bei der Bestimmung der Normalwerte für die Porter-Silber-Chromogene im
Kindesalter bestätigt. Unter dieser Voraussetzung möchte ich annehmen, daß Sie bei Ihren
Adipösen zu normalen Werten gelangen. Zur Diagnose cushingartiger Verhältnisse müßte man
m. E. in erster Linie Hemmtests mit Dexamethason durchführen.

J. M. Bayer (Bonn):

Bei der Abgrenzung der juvenilen Fettsucht gegenüber dem Cushing-Syndrom sollte man
die Ergebnisse der Steroiduntersuchungen nicht überbewerten und nur im Zusammenhang mit
den klinischen Befunden würdigen. Die juvenile Fettsucht ist eine generalisierte Fettsucht.
Bei diesen Fällen prüfe ich vor allem die Muskelkraft. Ist diese vorhanden, so handelt es sich
wohl kaum um ein Cushing-Syndrom, bei dem doch das Schwächegefühl eines der führenden
Symptome ist. Hätte ich alle Fälle von juveniler Fettsucht, die mir als Cushing-Patienten
offeriert wurden, operieren wollen, so würde die Zahl meiner Operationen weit über 100
betragen.

C. Overzier:

Als „Funktions-Cushing" möchte ich nur jene recht seltenen Fälle verstanden wissen, die
sich aus der großen Zahl von Patienten mit jugendlicher Fettsucht durch die geschilderte deutliche Vermehrung der Corticoid-Ausscheidung eindeutig unterscheiden und sich erst in der
Verlaufsbeobachtung diagnostizieren lassen. Bei einmaliger Untersuchung kann man durchaus
im Zweifel sein, ob es sich um diese harmlose überschießende Regulation des Pubertätsalters
oder um den Beginn eines echten, klinisch in den Konsequenzen ernst zu nehmenden Cushing-
Syndroms handelt.

Aus der I. Medizinischen Klinik der Universität München
(Direktor: Prof. D. H. Schwiegk)

Die Sekretion von Corticosteron bei Fettsüchtigen und bei Patienten mit Cushing-Syndrom

Von

H. J. Karl, L. Raith und W. Fischer

Mit 2 Abbildungen

Über die tägliche Produktion von Corticosteron, einem Steroid, das den Kohlenhydratstoffwechsel und den Mineralhaushalt beeinflußt, ist beim Menschen wenig bekannt. Peterson fand bei 11 Gesunden eine Corticosteronsekretion von 1,3 bis 4,0 mg/Tag, die nach Stimulierung der Nebennierenrinde mit ACTH auf das 7- bis 10fache des Ausgangswertes anstieg. Die Corticosteronsekretion beträgt mengenmäßig beim Menschen etwa $1/_7$ der Cortisolsekretion und etwa das 15fache der Aldosteronsekretion. Ein wesentlicher Anteil der von der Nebennierenrinde des Menschen sezernierten Steroide entfällt demnach auf Corticosteron.

Beim Cushing-Syndrom mit stark gesteigerter Cortisolsekretion (Cope, Brooks, Karl) wurde in einzelnen Fällen auch eine gesteigerte Corticosteronsekretion gefunden (Biglieri). Bei Fettsüchtigen dagegen, die nach unseren Untersuchungen (Karl u. Raith) sowie Migeon, Mlynaryk, Schteingart u. Conn im Vergleich zu Normalen leicht vermehrt Cortisol sezernieren, ist bisher über die Sekretion von Corticosteron nichts bekannt.

Wir haben deshalb die Corticosteronsekretionsrate bei 7 Normalpersonen, 16 Fettsüchtigen und 3 Patienten mit Cushing-Syndrom mit einer neuen Methode untersucht. Die Bestimmung der Corticosteronsekretion erfolgte nach dem Isotopen-Verdünnungsprinzip, dessen theoretische Voraussetzungen und Fehlerquellen eingehend von Tait erörtert wurden.

Methodisch liegen die Schwierigkeiten 1. in der Reindarstellung eines geeigneten Corticosteronmetaboliten; 2. in dessen quantitativer Bestimmung und 3. in der Messung der Aktivität dieses Metaboliten, nach Verabreichung einer möglichst kleinen Dosis 4-^{14}C-Corticosteron. Wir verabreichten nur 0,2 μC (spez. Akt. 1,44 μC/mmol).

Die Sekretionsrate bestimmten wir mit Tetrahydrocorticosteron (THB), das nach Peterson neben Allo-Tetrahydrocorticosteron ein Hauptmetabolit von Corticosteron ist.

Zur quantitativen Bestimmung dieses Metaboliten war eine Nachweismethode mit so hoher Empfindlichkeit erforderlich, daß Änderungen der Sekretionsrate von Corticosteron mit der gleichen Genauigkeit wie die von Cortisol erfaßt werden können. Die Veresterung von THB mit Tritium markierter Essigsäure (Peterson) zum quantitativen Nachweis birgt den Unsicherheitsfaktor der Bildung von Di- und Triacetaten neben dem Monoacetat von THB und ist zudem sehr aufwendig (Ködding).

Die colorimetrische Bestimmung der aus Chromatographiepapier eluierten Hormone mit Blautetrazol (Izzo, Chen, Mager) liefert wegen hoher Papierleerwerte keine hinreichend genauen Ergebnisse. Erst die chromatographische Reinigung des Papiereluats an einer Silicagelsäule vor der Blautetrazol-Reaktion, die nach Seeber durchgeführt wurde, ergab bei einer meßbaren Differenz von 0,5 γ Hormon reproduzierbare Ergebnisse mit einer Extinktion der Papierleerwerte von maximal 0,015 bei einer Schichtdicke von 10 mm.

Der gesamte Arbeitsgang zur Bestimmung der Cortisolsekretionsrate war folgender:

Hydrolyse von $^1/_{10}$ der Urintagesmenge (0—24 Std nach i.v. Injektion von 0,2 μC 4-^{14}C-Corticosteron[1]) mit β-Glucuronidase (300 Fishman E/ml Urin) und Penicillinzusatz (1000 E/ml Urin) bei pH 4,6 46°C 24 Std. Anschließend 2malige Extraktion mit dem gleichen Volumen Methylenchlorid und Reinigung des Extrakts mit 2mal $^1/_{10}$ Vol. $^1/_{10}$ n NaOH und 2mal $^1/_{10}$ Vol. H_2O. Der vorgereinigte Extrakt wird anschließend zwischen Petroläther und verdünntem Äthylalkohol verteilt (Petroläther Kp. 60—80° : Äthylalkohol abs. : H_2O = 100 : 10 : 30). Aus der wäßrigen Phase erfolgt eine Rückextraktion der Hormone mit 3mal dem doppelten Volumen Methylenchlorid. Der Extrakt wird bei 40°C abgedampft und papierchromatographisch in 2 Systemen aufgetrennt. Im Vorchromatogramm (System Bush B_5) werden bei einer Laufzeit von etwa 4 Std bei 23°C die Corticosteronmetaboliten von den höher polaren Cortisol- und Cortisonmetaboliten abgetrennt. Als Leitsubstanz dienen auf Parallelstreifen Cortison und Reichstein's Subst. S. In dem Bereich zwischen Cortison und Reichstein's Subst. S werden die Hormone mit 3mal 50 ml Methanol eluiert und je die Hälfte des Eluats auf Parallelstreifen 28 Std bei 23°C rechromatographiert (System LT 21/85 BUSH). Der eine Streifen mit der Hälfte des Extrakts wird mit Blautetrazol angefärbt und die Tetrahydrocorticosteron (THB) entsprechende Zone des korrespondierenden Streifens mit der zweiten Hälfte des Extrakts mit 3mal 4 ml Methanol eluiert. Das Eluat wird an einer vorgewaschenen Silicagelsäule (2 g) adsorbiert und THB mit 20 ml Methanol/Chloroform eluiert. Nach Abdampfen des Eluats bei 40°C Lösen des Trockenrückstandes in 3 ml abs. Äthylalkohol. Davon dienen 2 ml zur colorimetrischen Bestimmung von THB mit der Tetrazolblau-Reaktion, 1 ml zur Messung der Radioaktivität im Szintillationszähler (Packard-Tri-Carb).

Berechnet wurde die Sekretionsrate von Corticosteron nach der Formel:

$$\text{Sekretionsrate [mg/Tag] von Corticosteron} = \frac{\text{Aktivität des i.v. verabreichten 4-}^{14}\text{C-Corticosteron} \times \text{Ausscheidung von THB [mg/Tag] im Urin}}{\text{Aktivität von THB in der Urintagesmenge}}$$

Nach i.v. Injektion von 0,2 μC 4-^{14}C-Corticosteron in 10 ml physiologischer NaCl-Lösung wurden in den ersten 24 Std im Urin im Mittel 74% der injizierten Radioaktivität ausgeschieden, (bestimmt wurde die Gesamtaktivität im Urin nach den methodischen Angaben von Flood). 24—48 Std nach Injektion war die Aktivität im Urin für eine genaue Messung zu gering und bei allen Untersuchten unter 5%

---

[1] 4-^{14}C-Corticosteron wurde uns von den National Institutes of Health, Bethesda, Tetrahydrocorticosteron von der Steroid Reference Collection, London, freundlicherweise zur Verfügung gestellt.

Die Aktivität des Neutralextraktes nach Hydrolyse mit β-Glucuronidase betrug im Mittel 28% der injizierten radioaktiven Dosis, ohne Unterschied zwischen Normalpersonen und Fettsüchtigen. Etwa 20% aus dem Neutralextrakt nach Fermenthydrolyse entfielen auf den Corticosteronmetaboliten THB, der bezogen auf die injizierte Aktivität im Mittel nur 5,2% ausmachte.

Ohne Hydrolyse konnten im Mittel 0,5%, nach Solvolyse im Mittel 1,3% und nach heißer Hydrolyse bei pH 1 im Mittel 15% extrahiert werden (Abb. 1). Die Werte für die Gesamtaktivität im Urin und den Neutralextrakt mit und ohne Fermenthydrolyse liegen in der gleichen Größenordnung, wie sie von Peterson angegeben wurden.

Das Ergebnis unserer Untersuchungen über die Sekretionsraten von Corticosteron bei Normalen, Fettsüchtigen und Patienten mit Cushing-Syndrom ist in der Abb. 2 zusammengefaßt.

Bei 7 Normalpersonen, Durchschnittsalter 29 Jahre, betrug die Corticosteronsekretion im Mittel 3,2 mg/Tag (2,5—3,8 mg/Tag). Peterson

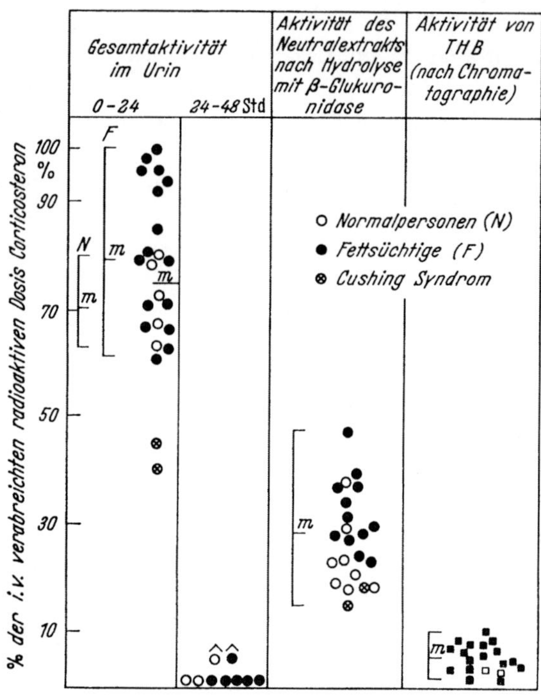

% der i.v. verabr. *radioaktiven Dosis* im Mittel:

Neutralextrakt ohne Hydrolyse n = 7   0,5% (0,3—0,9%)
Solvolyse   n = 9   1,3% (0,4—2,9%)
Hydrolyse pH 1 15' 100°C   n = 8 15,0% (7,8—25,4%)
          (nach Solvolyse)

Abb. 1. Ausscheidung der Radioaktivität in % nach i.v. Verabreichung von 0,2 μC 4-^{14}C-Corticosteron

fand bei 11 Gesunden im Mittel 2,3 mg/Tag (1,3—4,0 mg/Tag). Unsere Werte liegen im gleichen „Normalbereich".

16 fettsüchtige Patienten — Durchschnittsalter 30 Jahre — mit einem Übergewicht von 35—110%, bezogen auf das Idealgewicht, sezernierten im Mittel 4,4 mg/Tag Corticosteron mit einer Schwankungsbreite der Werte von 2,1 bis 6,7 mg/Tag. Zwischen dem Gewicht oder der Körperoberfläche der Patienten und der Corticosteronsekretion bestand keine signifikante Korrelation.

Wurde dagegen das kleine Kollektiv der Fettsüchtigen in Patienten mit normalem Blutdruck und solche mit systolischen Blutdruckwerten von 160 mm Hg und darüber unterteilt, dann zeigte sich, daß Fettsüchtige ohne Hypertonus im Mittel 3,3 mg Corticosteron, entsprechend den Normalpersonen sezernierten. Im Vergleich dazu war bei Fettsüchtigen mit erhöhtem Blutdruck die Corticosteronsekretionsrate im Mittel mit 5,5 mg/Tag signifikant erhöht.

2 Patienten mit Cushing-Syndrom und schwerer Hypertonie — einer mit Nebennierenrindenhyperplasie, der andere mit einem Nebennierenrindenadenom —

hatten mit 8,5 und 10,2 mg/Tag die höchste Corticosteronsekretionsrate aller Untersuchten. Der dritte Patient war beidseitig adrenalektomiert; bei ihm konnte weder eine Cortisol- noch Corticosteronsekretion nachgewiesen werden.

Daß beim Cushing-Syndrom neben der Cortisol- auch die Corticosteronsekretion der Nebennierenrinde gesteigert sein kann, ist verständlich. Dies ist aber offenbar nicht bei allen Patienten der Fall, denn BIGLIERI fand nur bei einem von 4 Patienten mit Cushing-Syndrom eine gesteigerte Corticosteronsekretion mit 7,3 mg/Tag.

Eine stark gesteigerte Corticosteronsekretion mit gleichzeitig leicht vermehrter Aldosteron-Produktion könnte beim Cushing-Syndrom ursächlich mit der häufig bestehenden hypokaliämischen Alkalose der Patienten in Zusammenhang stehen, da bekanntlich Corticosteron die kaliumdiuretische Wirkung von Aldosteron potenziert (THORN).

Ob bei Fettsüchtigen gleichzeitig die Cortisol- und Corticosteronsekretion leicht vermehrt ist, oder ob diese Steroide unabhängig voneinander in verschiedenen Mengen sezerniert werden, ist noch nicht untersucht und setzt die gleichzeitige Bestimmung beider Sekretionsraten voraus.

In Tierversuchen und beim Menschen wurde bei verschiedenen Formen der Hypertonie häufig eine vermehrte Sekretion einzelner Nebennierenrindensteroide, wie Aldosteron, 6 β-Hydroxycortisol und 21 β-Hydroxycortisol, nachgewiesen (LARAGH, KATZ, BESCH, WEISZ, COOPER, TOUCHSTONE). Welche Bedeutung einzelnen Nebennierenrindensteroiden in der Pathogenese des Hypertonus zukommt, wissen wir nicht. Die gesteigerte *Corticosteron*sekretion bei Fettsüchtigen mit erhöhtem Blutdruck halten wir jedoch für einen Befund, der zu weiteren Untersuchungen anregt

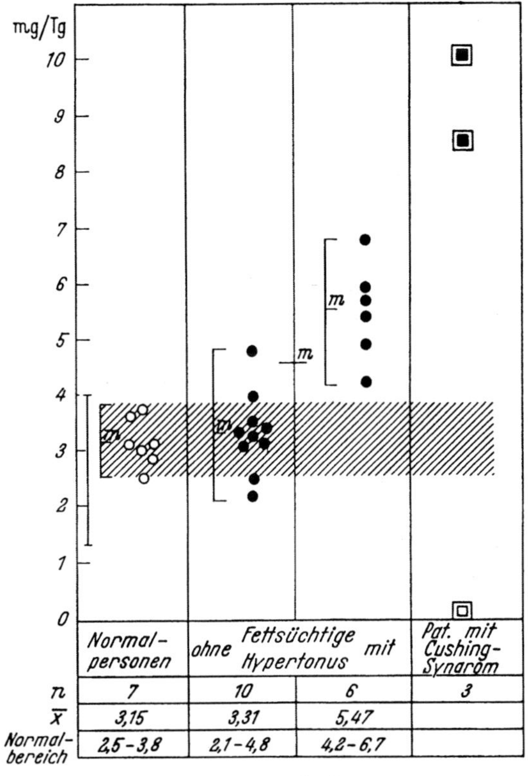

Abb. 2. Corticosteronsekretion von Normalpersonen ○ Fettsüchtigen ● und Patienten mit Cushing-Syndrom ▣ (nach beidseitiger Adrenalektomie ▢)

Literatur

BESCH, P. K., K. A. BROWNELL, F. A. HARTMANN, and D. J. WATSON: Acta endocr. (Kbh.) **39**, 355 (1962).

BIGLIERI, E. G., S. HANE, P. E. SLATON, and P. H. FORSHAM: J. clin. Invest. **42**, 516 (1963).

BROOKS, R. V., and F. T. G. PRUNTY: J. Endocr. **20**, 13 (1960).

BUSH, I. E.: The Chromatography of Steroids. New York-Oxford-London: Pergamon Press 1961.

CHEN, G., and H. E. TEWELL: Fed. Proc. **10**, 377 (1951).

COOPER, D. Y., J. C. TOUCHSTONE, J. M. ROBERTS, W. S. BLACKMORE, and O. ROSENTHAL: J. clin. Invest. **37**, 1524 (1958).

COPE, C. L., and E. G. BLACK: Clin. Sci. **17**, 147 (1958).

FLOOD, C., D. S. LAYNE, S. RAMCHARAN, E. ROSSIPAL, J. F. TAIT, and S. A. S. TAIT: Acta endocr. (Kbh.) **36**, 237 (1961).

IZZO, A. J., and E. H. KEUTMANN: Fed. Proc. **16**, 282 (1957).

KARL, H. J.: Verh. dtsch. Ges. inn. Med. **68**, 296 (1962).

—, u. L. RAITH: Klin. Wschr. **39**, 702 (1961).

— — u. W. DECKER: 9. Symp. dtsch. Ges. f. Endokrinologie. S. 84. Berlin-Göttingen-Heidelberg: Springer 1963.

KATZ, F. H., M. M. LIPMAN, A. G. FRANTZ, and J. W. JAILER: J. clin. Endocr. **22**, 71 (1962).

KÖDDING, R., W. LAMPRECHT, H. P. WOLFF, H. J. KARL u. K. R. KOCZOREK: Z. analyt. Chemie **181**, 574 (1961).

LARAGH, J. H., S. ULICK, V. JANUSZEWICZ, Q. DEMING, W. G. KELLY, and S. LIEBERMANN: J. clin. Invest. **39**, 1091 (1960).

MADER, W. J., and R. R. BUCK: Analyt. Chem. **24**, 666 (1962).

MIGEON, C. J., O. C. GREEN, and J. P. ECKERT: Metabolism **12**, 718 (1963).

MLYNARYK, P., R. R. GILLIES, B. MURPHY, and C. J. PATTEE: J. clin. Endocr. **22**, 587 (1962).

PETERSON, R. E., and C. E. PIERCE: J. clin. Invest. **39**, 741 (1963).

SCHTEINGART, D. E., R. I. GREGERMANN, and J. W. CONN: Metabolism **12**, 484 (1963).

SEEBER, E.: Naunyn-Schmiedebergs Arch. exp. Path. Pharmak. **242**, 101 (1961).

TAIT, J. F.: J. clin. Endocr. **23**, 1285 (1963).

THORN, G. W., D. JENKINS, J. C. LAIDLAW, F. C. GOETZ, F. J. DINGMAN, W. L. ARONS, D. H. P. STREETEN, and B. H. MECRACKEN: New Engl. J. Med. **248**, 232 (1953).

TOUCHSTONE, J. C., and W. S. BLAKEMORE: J. clin. Endocr. **21**, 263 (1961).

Diskussion

H. G. GOSLAR (Bonn):

Die Bestimmung von Corticosteron durch Formazanbildung aus Tetrazoliumderivaten wurde von dem Herrn Vortragenden wegen zu hoher Leerwerte — bedingt durch unspezifische Reduktion — als nicht mehr brauchbar bewertet. Bei der Verwendung von Blue Tetrazolium ist dies wohlbegründet und z. B. auch in der Enzymhistochemie wird dieses Derivat nicht mehr verwendet. Nun sind aber in den letzten Jahren eine Reihe neuer Di-Tetrazoliumderivate synthetisiert worden, welche ein niedriges Redoxpotential besitzen und bei denen in bestimmten pH-Bereichen auf dem enzymhistochemischen Sektor unspezifische Reduktionen weitgehend vermieden werden konnten (z. B. Nitro-BT, TNBT). Vielleicht lassen sich mit diesen Substanzen bei strenger pH-Wahl niedrigere Leerwerte erzielen, und ich möchte anregen, die an und für sich doch recht praktische Tetrazolium-Bestimmungsmethode mit diesen Substanzen noch einmal auf bessere Eignung durchzutesten.

Aus der II. Medizinischen Klinik der Universität München
(Direktor: Prof. Dr. Dr. G. Bodechtel)

Vergleichende Untersuchungen über das Verhalten der insulinähnlichen Aktivität im Serum beim Cushing-Syndrom und bei der Fettsucht

Von

P. Bottermann, K. Schwarz und K. Kopetz[1]

Mit 2 Abbildungen

Untersuchungen der letzten Jahre haben ergeben, daß zahlreiche Substanzen fettmobilisierend wirken [Ashmore et al., Wertheimer und Shafrir, Jeanrenaud (1—3)]. Einen lipogenetischen Effekt zeigte dagegen nur das Insulin [Jeanrenaud, Renold et al., Wertheimer (3—5)]. Es schien daher von Interesse, mögliche Zusammenhänge zwischen der gesteigerten Lipogenese durch Insulin und der Fettzunahme bei der Adipositas aufzuklären.

Tierexperimentell fand Hausberger (6—9) bei Mäusen mit Nebennierenrinden-Hyperplasie oder Nebennierenrinden-Adenomen eine gleichzeitige Hyperplasie der β-Zellen des Pankreas. Er vermutete, daß die Adipositas dieser Tiere auf einer vermehrten Freisetzung von Insulin mit gesteigerter Lipogenese beruhe. Diese vermehrte Insulin-Inkretion konnten Schwarz u. Mitarb. (10) bei Patienten mit Cushing-Syndrom nachweisen. Bei oralen Glucose-Belastungen im Staub-Traugottschen Versuch sprachen enorme Anstiege der schon im Nüchternzustand deutlich erhöhten ILA-Werte für eine gesteigerte Bildung oder Mobilisierbarkeit von Insulin.

Von diesen Befunden ausgehend untersuchten wir die "insulin-like-activity" bei 13 Patienten mit proportionierter Fettsucht und verglichen sie mit den Werten von 9 normalgewichtigen stoffwechselgesunden Personen. Wir bestimmten die ILA nach der von Froesch (11)[2] angegebenen Methode mittels Nettogasaustausch und Glucoseaufnahme des isolierten inkubierten Rattennebenhodenfettgewebes.

Dabei fanden wir, daß Adipöse schon im Nüchternserum deutlich höhere ILA-Werte aufwiesen als Normalgewichtige. Wesentlich größere Unterschiede ergaben sich nach intravenösen Belastungen mit Tolbutamid und Glucose (Abb. 1 a und 1 b). Bei beiden Versuchsgruppen stiegen die ILA-Werte im Serum an. Doch übertraf der ILA-Anstieg bei Adipösen den der Normalgewichtigen erheblich. Neben dem Unterschied beider Gruppen erkennt man auf den Abbildungen aber auch deutlich die relativ großen Differenzen in der Höhe der ermittelten einzelnen Werte. Denn

---

[1] Mit Unterstützung der Deutschen Forschungsgemeinschaft.

[2] Herrn Priv.-Doz. Dr. med. E. R. Froesch danken wir für die freundliche Einarbeitung unserer Assistentin Frau Ruth Höllerer in seinen Laboratorien.

teilweise findet sich der maximale ILA-Anstieg schon nach 10 min, teilweise erst
zu einem der späteren Untersuchungszeitpunkte. In Abb. 2 sind die Sekretions-
typen zweier adipöser Patienten gegenübergestellt. Bei dem ersten Patienten
kommt es bereits 10 min nach Tolbutamid-Gabe zum maximalen ILA-Anstieg,
beim zweiten Patienten erst nach 40 min. Man könnte fast zwischen einem „Früh-"
und einem „Spätsekretionstyp" unterscheiden. Ebenso sind auch nach Glucose-
gabe diese Früh- und Spätsekretionstypen zu beobachten.

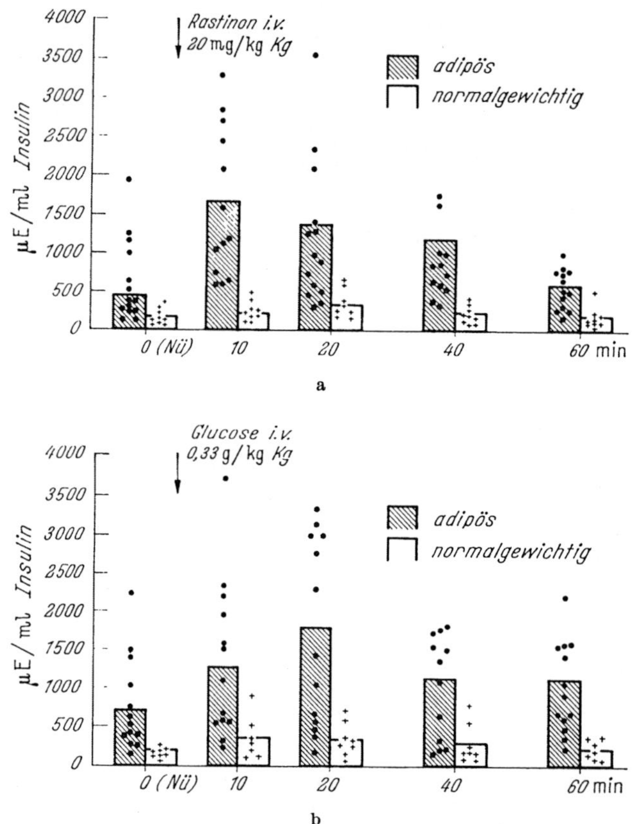

Abb. 1a u. b. a ILA nach Rastinon bei adipösen und normalgewichtigen Personen, b ILA nach Glucose bei
adipösen und normalgewichtigen Personen

Ähnliche Ergebnisse wie wir erhielten Karam u. Mitarb. (*12*), die nach oralen
und intravenösen Glucose-Belastungen bei stoffwechselgesunden normalgewichtigen und nicht diabetischen adipösen Versuchspersonen ebenfalls Insulin-Bestim-
mungen mittels einer immunologischen Methode durchführten. Auch sie fanden —
allerdings mit niedriger liegenden Absolutzahlen — bei Adipösen deutlich höhere
Insulinwerte als bei normalgewichtigen Vergleichspersonen.

Eine Deutung dieser Mehrproduktion oder -mobilisation von Insulin ist problematisch. Beim Cushing-Syndrom führt die laufende Cortisol-Überproduktion mit
entsprechend vermehrter Gluconeogenese zur Blutzuckererhöhung. Dadurch werden die β-Zellen des Inselapparates dauernd stimuliert. Das Pankreas reagiert mit

einer vermehrten Insulin-Inkretion, die nun ihrerseits durch verstärkte Lipogenese zur Adipositas führt.

Einen ähnlichen Mechanismus könnte man auch bei der sog. „alimentären" Fettsucht diskutieren. Während beim Cushing-Syndrom die überschießende Gluconeogenese die β-Zellstimulation verursachen würde, wäre bei der sog. „alimentären" Fettsucht die ständig gesteigerte Nahrungszufuhr der adäquate Reiz. Die anhaltende Stimulation der β-Zellen würde zu einer schließlich überschießenden Insulin-Inkretion bei Bewältigung der angebotenen Nahrungsmengen führen. Der Inselzellapparat würde — ungefähr einer Enzymadaptation vergleichbar — immer mehr Insulin bereitstellen, wodurch wiederum die Lipogenese und der Fettansatz gefördert würden.

Die in unseren Versuchen bei Adipösen gefundene überschießende Insulin-Inkretion nach intravenöser Tolbutamid- und Glucose-Gabe könnte in diese Richtung weisen. Sicher spielen auch noch weitere Faktoren bei einem so komplexen Vorgang wie der Entstehung einer Fettsucht, so etwa das Wachstumshormon, das Adrenalin und Noradrenalin, der Astwood-Faktor, eine Rolle. Weitere Untersuchungen, die das freie und das gebundene, das antikörperhemmbare und nicht antikörperhemmbare Insulin berücksichtigen müßten, erscheinen zur weiteren Klärung der hier angeschnittenen Fragen notwendig.

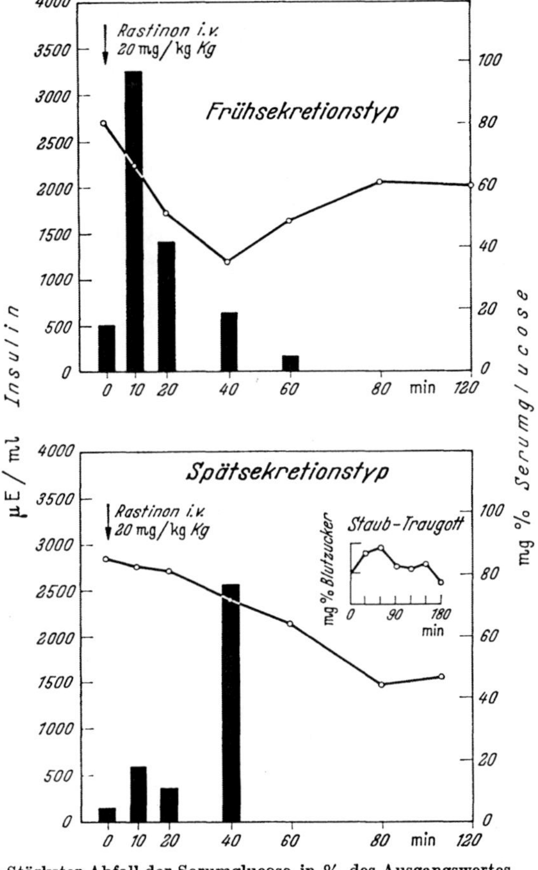

o Stärkster Abfall der Serumglucose in % des Ausgangswertes
Abb. 2. ILA nach Rastinon bei Adipositas

Zusammenfassung

Frühere Untersuchungen bei Patienten mit Cushing-Syndrom hatten nach Glucose-Belastung eine signifikante Erhöhung der "insulin-like-activity" (ILA) des Serums im Vergleich zu Normalpersonen ergeben. Von diesen Befunden ausgehend, wurde das Verhalten der ILA des Serums bei adipösen Personen nach intravenöser Glucose- und Tolbutamid-Gabe untersucht. Dabei fanden sich in der Gruppe der Adipösen deutlich höhere Anstiege der ILA-Werte als bei normalgewichtigen Vergleichspersonen. Diese Ergebnisse werden im Rahmen der hormonalen Einflüsse in der Genese der Adipositas diskutiert.

Literatur

1. Ashmore, J., G. F. Cahill jr. and A. B. Hastings: Recent Progr. Hormone Res. **16**, 547 (1960).
2. Wertheimer, E., and E. Shafrir: Recent Progr. Hormone Res. **16**, 467 (1960).
3. Jeanrenaud, J.: Metabolism **10**, 535 (1961).
4. Renold, A. E., A. J. Winegrad, J. B. Renaud, and D. B. Martin: The Mechanism of Action of Insulin, p. 153 (Herausgeber: F. G. Young) Oxford: Blackwell 1960.
5. Wertheimer, E.: Münch. med. Wschr. **100**, 1153 (1958).
6. Hausberger, F. X.: Diabetes **7**, 211 (1958).
7. — Endocrinology **65**, 165 (1959).
8. — Münch. med. Wschr. **101**, 1301 (1959).
9. — Acta endocr. (Kbh.) **37**, 336 (1961).
10. Schwarz, K., K. F. Weinges, K. P. Eymer u. K. Kopett: Verh. dtsch. Ges. inn. Med. **1962**, 289.
11. Froesch, E. R.: Persönliche Mitteilung, im Druck.
12. Karam, J. H., G. M. Grodsky, and P. H. Forsham: Diabetes **12**, 197 (1963).

Diskussion

H. Daweke (Düsseldorf):

Wir sind bei unseren Untersuchungen bei extrem Adipösen zu den gleichen Ergebnissen wie Sie gekommen. Wir bestimmten die ILA nach Martin et al. im Serum von 10 Fettsüchtigen ohne nachweisbare Störung des Kohlenhydratstoffwechsels und ohne familiäre Belastung mit Diabetes, und zwar vor und 30, 60 und 120 min nach Belastung mit 100 g Glucose per os. Im Nüchternserum war die ILA im Mittel statistisch gesichert gegenüber der Norm erhöht. Nach Belastung mit Glucose stieg der Insulinspiegel etwa parallel zum normalen Kurvenverlauf an und kehrte er dann wieder zum Ausgangswert zurück. Interessanterweise war bei einer Gruppe von fettsüchtigen Altersdiabetikern die ILA gegenüber einer Gruppe von normgewichtigen Altersdiabetikern mit klinisch gleichschwerem Diabetes ebenfalls statistisch gesichert erhöht. Eine Insulinreserve war hier allerdings nicht mehr nachweisbar.

H. Ditschuneit (Frankfurt-Main):

Hatten Sie Diabetiker oder Prädiabetiker unter Ihren Versuchspersonen?

P. Bottermann:

In den Familienanamnesen der von uns untersuchten Personen fanden sich keine Hinweise für einen Diabetes mellitus in der näheren oder entfernteren Verwandtschaft. Weiter bestimmten wir bei allen Versuchspersonen den Serumglucoseabfall nach Glucose- und Rastinon-Gabe und führten wir bei fast allen Personen einen Staub-Traugottschen Versuch durch. Auf Grund der Ergebnisse glaubten wir, einen Diabetes oder sog. „Prädiabetes" ausschließen zu können.

Aus der Endokrinologischen Abteilung
der Medizinischen Universitätsklinik St. Rafael, Löwen/Belgien
(Dir.: Prof. Dr. VANDENBROUCKE)

Extreme Hyperpigmentierung bei Cushing-Syndrom mit ACTH-produzierendem Tumor der Lunge

Von

O. STEENO[1] u. P. DE MOOR[1]

Mit 2 Abbildungen

In einigen seltenen Fällen von Cushing-Syndrom mit sog. ,,nicht endokrinen" Tumoren wurde eine extrem dunkle bis schwarze Pigmentierung der Schleimhaut und der ganzen Körperhaut beobachtet. Diese Hyperpigmentierung unterscheidet sich von der manchmal zu beobachtenden leichten Pigmentierung bei gewöhnlichen Fällen von Cushing-Syndrom wie auch von derjenigen beim Morbus Addison.

Fallbesprechung. Im März 1963 wurde dem Department für Innere Medizin der Universität Löwen ein 31 jähriger Mann überwiesen. Zum erstenmal suchte er im November 1962 wegen adynamischer Beschwerden einen Arzt auf. Seitdem war eine zunehmende Pigmentierung beobachtet worden, die von einer schweren Akne begleitet war. Im Dezember 1962 stellte sein Arzt Gewichtszunahme, ein Vollmondgesicht, Lumbalschmerzen mit Querschnittslähmung, Polydipsie, Polyurie und Diabetes fest. In den folgenden Monaten nahmen die Hauterscheinungen zu.

Die *klinische Untersuchung* (Abb. 1 und 2) ergab ein rundes, dickes Gesicht, blaue Striae, Fußödeme und fortgeschrittene Muskelatrophie der Beine. Die Haut war rauh und trocken, mit sehr dunkler und intensiv braun-schwarzer generalisierter Pigmentierung, dazu mit verschiedenen begrenzten Pigmentflecken. Die Fingernägel und die Wangenschleimhaut waren braun-schwarz. Der Blutdruck betrug 180/130. Die Leber war vergrößert und es bestand ein Ascites.

Hauthistologie. Bei weitgehend unauffälliger Oberhaut findet sich bei der Melaninfärbung eine intensive Pigmentablagerung in den Basalzellen der Epidermis.

Die pathologischen *Laborbefunde* waren wie folgt: Glykosurie mit hohen Blutzuckernüchternwerten, leicht erhöhte Calciurie, hypokaliämische Alkalose und Hypogammaglobulinämie. Eine Thoraxaufnahme erbrachte eine große rundliche Masse, die rechts parakardial und vorn lokalisiert war. Diffuse Osteoporose, ein Nierenstein rechts und eine osteoplastische Metastase im Becken waren zu sehen. Die Nebennierenschatten und die Sella turcica waren normal. Bei einer *Broncho-*

---

[1] Z. Zt. Gastarzt in der Universitäts-Hautklinik Hamburg-Eppendorf (Direktor: Prof. Dr. Dr. J. KIMMIG).

skopie war das Gebilde eines peripheren, bronchiogenen Carcinoms mit mediastinalen Lymphknotenmassen zu sehen. Es konnte keine Biopsie gemacht werden.

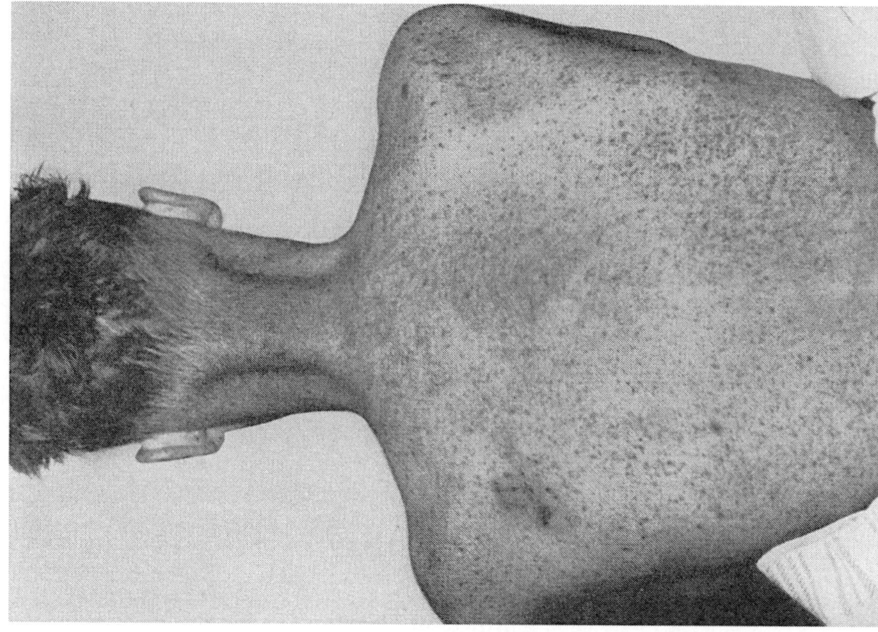

Abb. 2. Dunkle und intensiv braun-schwarze generalisierte Pigmentierung mit begrenzten Pigmentflecken

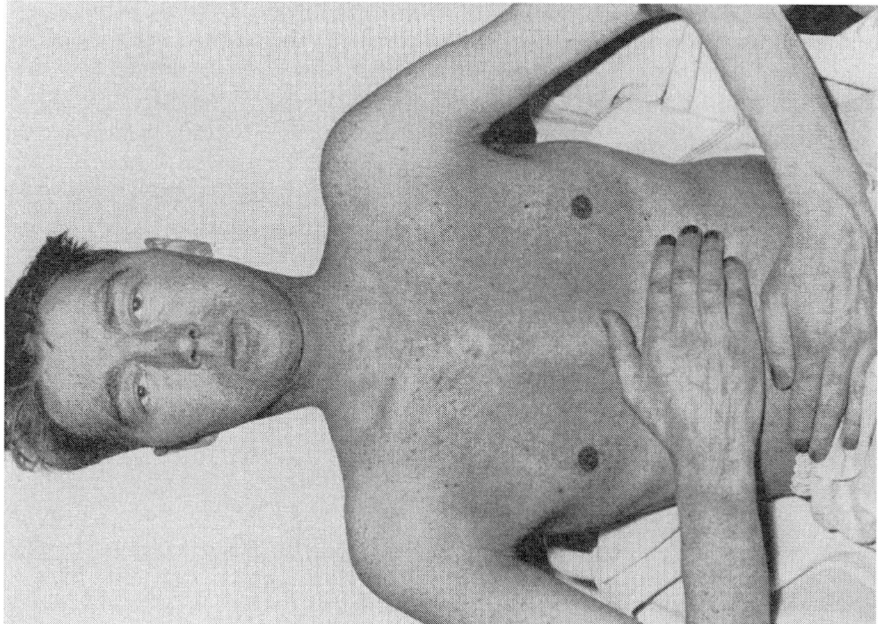

Abb. 1. Cushing-Syndrom mit ACTH-produzierendem Tumor der Lunge. Extreme Pigmentierung der Haut und der Nägel. Rauhe und trockene Haut mit schwerer Akne

Die unkonjugierten *Plasmacorticosteroide* waren extrem hoch ohne Veränderung während des Tagesablaufes. Es war kein Anstieg des Plasma-Corticosteroid-Spiegels während intravenöser ACTH-Stimulierung zu beobachten. Durch Chromatographie konnte eine Störung in der Corticosteroid-Synthese aufgezeigt werden. Die *Corticosteroide im Urin* waren deutlich erhöht, eine intravenöse ACTH-Stimulierung brachte keinen Anstieg und orale Dexamethason-Gaben (1,5 mg alle 6 Std während einer Zeitdauer von 4 Tagen) brachten kein Absinken der erhöhten Corticosteroid-Werte. Die *ACTH-Aktivität* wurde von Dr. GRANT LIDDLE von der Vanderbilt Universität an lyophilisiertem Plasma geprüft; er fand 20,1 mE pro 100 ml. Die Normalwerte in demselben Laboratorium liegen unter 0,5 mE pro 100 ml. Im gleichen Labor wurde das Plasma auf *MSH-Aktivität* hin untersucht (Froschhautmethode in vitro) und 12,5 mE pro 100 ml gefunden. Dieser Wert erklärt zweifellos die auffallende Hyperpigmentierung.

Nach einer erhöhten Zufuhr von Kalium fühlte sich der Patient besser, das Urinvolumen wurde normal und innerhalb von einigen Tagen wurde ein Gewichtsverlust von 13 Pfund beobachtet. Danach wurden Cytostatica gegeben. Vier Wochen nach der Aufnahme wurde der Patient entlassen. Er starb plötzlich zu Hause sechs Wochen später, zwei Tage vor der geplanten Neuaufnahme. Unglücklicherweise war eine postmortale Untersuchung nicht möglich.

Diskussion

Der hier beschriebene Fall weist auf einen Tumor als Ursache einer Substanz mit ACTH- und MSH-Aktivität hin. LIDDLE u. Mitarb. (*1*) haben gezeigt, daß das „Tumor-ACTH" biologisch und chemisch dem Hypophysen-ACTH in vieler Hinsicht sehr ähnlich ist. Sie fanden auch, daß beide eine Melanocyten-stimulierende Aktivität in der Froschhaut-Analyse in vitro haben. Die Frage bleibt, ob der Tumor eine Polypeptidstruktur sowohl mit ACTH- als auch MSH-Aktivität produziert, oder verschiedene Polypeptide mit entweder ACTH- oder MSH-Aktivität.

Von den Fällen mit extremer Hyperpigmentierung, die schon beschrieben wurden, waren drei Thymus-Tumoren, drei Mediastinal-Carcinome und ein Carcinom der β-Zellen des Pankreas, aber kein Fall eines sich schnell entwickelnden Prozesses (*1—6*). Dies läßt auch vermuten, daß eine über längere Zeit hohe Konzentration von zirkulierendem ACTH die Ursache der beobachteten Hyperpigmentierung sein dürfte.

Literatur

1. LIDDLE, G. W., D. P. ISLAND, R. L. NEY, W. E. NICHOLSON, and N. SHIMIZU: Arch. int. Med. **111**, 471 (1963).
2. ALBRIGHT, E. G.: J. clin. Endocr. **22**, 93 (1962).
3. ENGEL, F. L., and L. KAHANA: Amer. J. Med. **34**, 726 (1963).
4. MEADOR, C. K., G. W. LIDDLE, D. P. ISLAND, W. E. NICHOLSON, C. P. LUCAS, J. G. NUCKTON, and J. A. LUETSCHER: J. clin. Endocr. **22**, 693 (1962).
5. PRUNTY, F. T. G., R. V. BROOKS, J. DUPRÉ, T. M. D. GIMLETTE, J. S. H. HUTCHINSON, R. R. McSWINEY, and I. H. MILLS: J. clin. Endocr. **23**, 737 (1963).
6. SCHOLZ, D. A., and R. C. BAHN: Proc. Mayo Clin. **34**, 433 (1959).

Aus der Abteilung für klinische Endokrinologie (Leiter: Prof. Dr. E. F. Pfeiffer) der I. Medizinischen Universitätsklinik Frankfurt (Main) (Direktor: Prof. Dr. F. Hoff)

Qualitative und quantitative Änderungen der ACTH-Aktivitäten bei spontanen und therapeutischen Über- und Unterfunktionszuständen der Nebennierenrinde des Menschen

Von

K. Retiene, H. E. Mentzel, R. Tripp, G. Schumann und E. F. Pfeiffer

Mit 2 Abbildungen

An der hypophysären oder zentralnervösen Pathogenese des M. Cushing auf dem Boden einer bilateralen NNR-Hyperplasie wird heute nicht mehr gezweifelt. Wenn auch nach unserer Erfahrung absolut erhöhte ACTH-Spiegel sicher als Ausnahme anzusehen sind [Nelson (1960), Pfeiffer u. Mitarb. (1963)], so müssen doch die inzwischen verschiedentlich nachgewiesenen normalen bis hochnormalen ACTH-Werte im Vergleich zu den bei diesen Kranken signifikant erhöhten Cortisolkonzentrationen als zu hoch angesehen werden [Sydnor u. Mitarb. (1953), Nelson u. Meakin (1959), Davies (1964)].

Williams u. Mitarb. (1961) machten erstmals auf den verstärkten hypophysären Sekretionsdruck beim M. Cushing aufmerksam. Sie fanden nach der Adrenalektomie so hohe ACTH-Blutspiegel wie sie bei Addison-Patienten niemals nachgewiesen werden können. Den selben Befund konnten wir bestätigen, als wir bei einer Cushing-Patientin unmittelbar nach der subtotalen Adrenalektomie ACTH-Blutspiegel bis 500 mE/100 ml Plasma gemessen haben.

Eine beim M. Cushing erhöhte Produktion und Sekretion läßt sich jedoch auch noch auf andere Weise demonstrieren. Die vom Cortisol bekannte Tagesrhythmik mit einem morgentlichen Gipfel und einer tiefsten Depression um Mitternacht [Ekmann u. Mitarb. (1961)] konnten wir in gleicher Weise für das ACTH im normalen Bereich zwischen 0,2 und 1,0 mE/100 ml Plasma nachweisen. Allein durch Dauersekretion des in den Morgenstunden gemessenen Normalwertes ergibt sich eine Mehrproduktion von ACTH über 24 Std, die das quantitative Problem des M. Cushing voll erklärt. Bei den von uns untersuchten Cushing-Patienten mit steroidchemisch und operativ bestätigter NNR-Hyperplasie konnten wir nur in einem Fall erhöhte ACTH-Spiegel messen. Bei den anderen Patienten lag der ACTH-Spiegel absolut im normalen Bereich, die bei Gesunden beobachtete Tagesrhythmik war jedoch bei allen von uns untersuchten Cushing-Kranken sowohl für das Cortisol als auch für das ACTH nicht mehr nachweisbar. Es ist daher zu verstehen, daß eine starre, und nur im Einzelfall einmal erhöhte, Dauersekretion von

ACTH unlängst von uns als primäre Ursache des M. Cushing postuliert wurde [(PFEIFFER u. Mitarb. (1963), RETIENE u. Mitarb. (1963)].

Neue Untersuchungen galten der Frage, ob eine primär zentralnervöse Störung oder aber der periphere Cortisolspiegel für die Aufhebung der Tagesrhythmik beim Cushing-Kranken verantwortlich ist. Zum Vergleich der eben zitierten Befunde wurden deshalb die ACTH-Blutspiegel auch bei anderen spontanen und therapeutisch bewirkten Über- und Unterfunktionszuständen der NNR untersucht.

Wie bei früheren Untersuchungen haben wir auch diesmal alle ACTH-Blutspiegel anhand des Corticosteronanstieges im NNV-Blut hypophysektomierter Ratten gemessen. Mit einer Variation dieser erstmals von LIPSCOMB u. NELSON (1959) beschriebenen Methode lassen sich ACTH-Spiegel bis zu 0,2 mE/100 ml menschliches Plasma mit einem Genauigkeitsindex von $\lambda = 0,18$ gerade noch erfassen [RETIENE u. Mitarb. (1962)]. Die Blutentnahmen erfolgten einheitlich morgens 6 Uhr und dann folgend alle 6 Std bis um 24 Uhr. Sofort nach Entnahme wurde das Blut zentrifugiert und innerhalb von 10 min bis zur Aufarbeitung tiefgefroren. Am Versuchstag selbst mußten die Patienten Bettruhe einhalten, weil frühere Untersuchungen nur so den sicheren Nachweis der Tagesrhythmik erbracht hatten.

Abb. 1 zeigt die Ergebnisse bei einer 40jährigen Addison-Patientin, bei der die Substitutionstherapie 3 Tage vor dem Versuch unterbrochen worden war. Die ACTH-Werte sind bis auf das 30fache gegenüber der Norm erhöht. Die Tagesrhythmik mit dem hohen morgendlichen Gipfel und einer tiefsten Depression um Mitternacht ist zweifelsfrei zu erkennen.

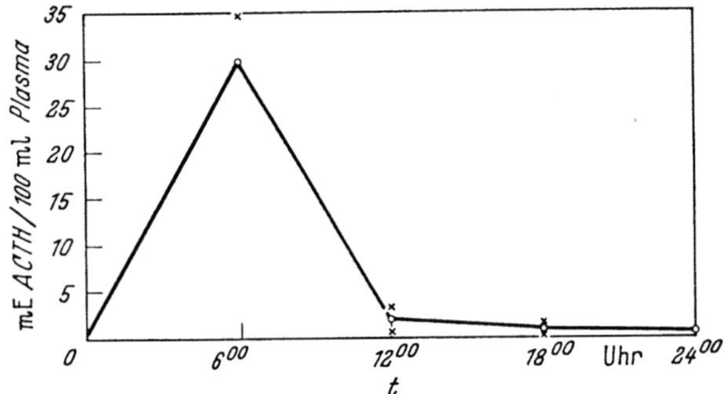
Abb. 1. Frau B. M., 42 Jahre, M. Addison. Verhalten von ACTH im Blut über 24 Std bei einer Addison-Patientin

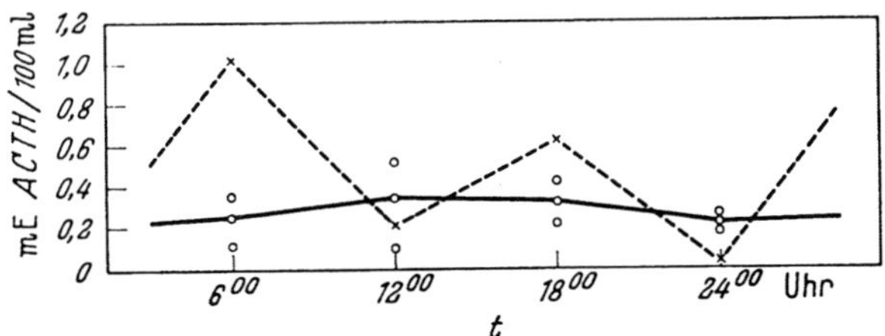
Abb. 2. Patientin Na. R., 18 Jahre, Therapie-Cushing nach 3 monatiger Gabe von Decortin bei Lebercirrhose. ACTH-Tageskurve 3 Tage nach Absetzen der Decortin-Therapie (zum Vergleich die Tagesrhythmik eines gesunden Probanden)

Bei einer 20jährigen Patientin mit einem klinisch und steroidchemisch nachgewiesenen adrenalen Hirsutismus war die Tagesrhythmik der ACTH-Sekretion ebenfalls erhalten. Bei dieser Patientin sind die Absolutwerte jedoch nur fraglich gegenüber der Norm erhöht.

Abb. 2 zeigt den ACTH-Tagesblutspiegel bei einer über Monate mit hohen Steroiddosen behandelten Patientin mit einem Therapie-Cushing bei fortgeschrittener Lebercirrhose. 3 Tage nach Absetzen der Therapie ist die Tagesrhythmik aufgehoben, die Absolutwerte jedoch gegenüber der Norm als auch gegenüber dem spontanen, mit bilateraler NNR-Hyperplasie einhergehenden M. Cushing deutlich erniedrigt und im untersten Bereich der Nachweisbarkeit.

Genau dasselbe Ergebnis hatten wir bei einem 2. Patienten mit Lebercirrhose, der im Gegensatz zur vorherigen Patientin 4 Wochen lang nur mit kleinen Mengen von 10—20 mg Decortin behandelt worden war.

Somit scheint nach diesen Untersuchungen bei jedem erhöhten Steroidspiegel im peripheren Blut die Tagesrhythmik gestört zu sein. Es muß ein spezielles, nur für die physiologische Tagesrhythmik verantwortliches Zentrum im Zwischenhirn angenommen werden, das sowohl auf exogene als auch auf endogene Steroide empfindlich reagiert. Dieses Zentrum ist beim Cushing-Kranken durch die hohen Cortisolspiegel blockiert. Als alleinige Ursache der Erkrankung nehmen wir eine aktiv gesteigerte Produktion von CRF an, die auch durch das vermehrte endogene Cortisol im Blut nur ungenügend gebremst wird. Ansonsten wären die extrem hohen ACTH-Werte nach der Adrenalektomie sowie die gelegentlich beobachteten Hypophysenadenome nach der Operation nicht denkbar.

Freilich wurden auch immer wieder qualitative Veränderungen der ACTH-Produktion beim Cushing-Syndrom diskutiert, nachdem Slaunwhite u. Sandberg (1959) eine fehlerhafte Bindung von Cortisol an Trägerproteine in Form der „Transcortin-Hypothese" beschrieben hatten. In eigenen Untersuchungen wurden deshalb je 10 cm³ Serum von Normalpersonen und Cushing-Kranken vor der Adrenalektomie entnommen und der präparativen Elektrophorese in Polyvenylchlorid unterworfen. Während der 24 stündigen Trennung wurde durch Kühlung eine ständige Temperatur von $+4°$ erhalten. Die einzelnen Serumeiweißfraktionen wurden gefriergetrocknet und nach Wiederauflösung in 3 cm³ 0,01 n HCl mit der eingangs beschriebenen Methode auf vorhandene ACTH-Aktivitäten geprüft.

10 mE exogenes ACTH konnte in Kontrollseren mit einem durchschnittlichen Verlust von 40—50% fast ausschließlich in der Albuminfraktion wiedergefunden werden.

Sowohl bei den Normalpersonen als auch bei den Cushing-Kranken war die Hauptaktivität in der α_2-Globulinfraktion nachzuweisen. Die in den anderen Fraktionen gemessene ACTH-Aktivität ist absolut sehr klein und an der Grenze der Nachweisbarkeit. Der prozentual größere Anteil der α_2-Globulinfraktion bei den Cushing-Kranken entspricht der größeren ACTH-Menge im Nativblut unserer untersuchten Kranken. Auch das nach elektrophoretischer Trennung wiedergefundene ACTH muß sich somit in der von uns vermuteten biologisch aktiven Serumfraktion wiederfinden. Die prozentuale ACTH-Aktivität der anderen Fraktionen muß dementsprechend geringer werden.

Auf keinen Fall konnten wir bei unseren Cushing-Kranken eine von der Norm abweichende Bindung von ACTH an Serumeiweißkörper feststellen. Die bisherigen rein quantitativen Überlegungen als Ursache der Pathogenese des M. Cushing sollten deshalb uneingeschränkt Gültigkeit haben.

Literatur

Davies, B. M. A.: Acta endocr. (Kbh.) **45**, 55 (1964).
Ekmann, H., B. Hakansson, I. D. McCarthy, I. Lehmann, and B. Sjögren: J. clin. Endocr. **21**, 684 (1961).
Lipscomp, H. S., and D. H. Nelson: Fed. Proc. **18**, 95 (1959).

Nelson, D. H.: Metabolism 10, 894 (1961).
— Clinical Endocrinology I, p. 626—636, E. B. Astwood Ed. New York-London: Grune & Stratton 1960.
Pfeiffer, E. F., F. Garmendia, E. Vaubel u. K. Retiene: Ergebn. inn. Med. Kinderheilk. 20, 127 (1963).
Retiene, K., H. Ditschuneit, M. Fischer, K. Kopp u. E. F. Pfeiffer: Acta endocr. (Kbh.) 41, 211 (1962).
— A. Espinoza, Y. Abdel Rahman, K. H. Marx u. E. F. Pfeiffer: 10. Symposion dtsch. Ges. Endokrinologie, Wien, 7.-9. 3. 1963, S. 231.
Slaunwhite, W. R. jr., and A. A. Sandberg: J. clin. Invest. 38, 385 (1959).
Sydnor, K. L., G. Sayers, H. Brown, and F. H. Tyler: J. clin. Endocr. 13, 891 (1953).
Williams, W. C. jr., D. Island, R. A. H. Oldfield, and G. W. Liddle: J. clin. Endocr. 21, 943 (1961).

Aus der Medizinischen Poliklinik der Universität Würzburg (Direktor: Prof. Dr. H. FRANKE)
und der Neurologischen Klinik der Universität Würzburg
(Direktor: Prof. Dr. G. SCHALTENBRAND)

Therapie des Cushing-Syndroms durch Implantation von Yttrium[90] in die Hypophyse

Von

H. KLEINFELDER, W. BÖRNER, M. NADJMI und F. HUSMANN

Mit 2 Abbildungen

In der Behandlung des Cushing-Syndroms infolge einer NNR-Hyperplasie galt die subtotale oder totale Adrenalektomie als Therapie der Wahl. Unter den Nachteilen dieses Eingriffes erscheint die Entwicklung von Hypophysentumoren im Anschluß an die Adrenalektomie, über die eine Reihe von Autoren berichten (*1, 2, 5, 10, 15, 16, 21*), besonders gravierend. In dieser Situation gewinnt ein therapeutisches Vorgehen, das zentral an der Hypophyse ansetzt, wieder mehr Interesse. Die Wirksamkeit der Röntgenbestrahlung der Hypophysengegend ist begrenzt durch die im Verhältnis zur Hypophyse höhere Strahlenempfindlichkeit der Nachbargewebe (*20*), so daß sie nur in 30—40% der Fälle eine meist nur vorübergehende Remission des Krankheitsbildes zur Folge hat (*4, 7*).

Nachdem im Jahre 1949 erstmals NORTHFIELD (*17*) und später PATTISON und SWAN (*18*) bei einem Cushing-Patienten Radonkapseln in die Hypophyse implantiert haben, berichteten MOLINATTI u. Mitarb. 1959 über ein günstiges Behandlungsergebnis, das sie bei einem entsprechenden Patienten durch die Implantation von Yttrium[90] in die Hypophyse erzielten (*13*).

Unsere bescheidenen Erfahrungen mit dieser Methode beschränken sich auf 2 Patienten mit einem Cushing-Syndrom bei Nebennierenrindenhyperplasie. In einem Fall wurde die Implantation von Yttrium[90] vor 26, im anderen Fall vor 12 Monaten vorgenommen. Die beiden Patientinnen — 27- und 33jährig — zeigten den typischen klinischen Aspekt eines Cushing-Syndromes in Verbindung mit einer Hypertonie, einer prädiabetischen Stoffwechsellage und einer Amenorrhoe. Bei einem der Fälle war eine deutliche Osteoporose mit Einbruch eines LWK nachweisbar. Nach dem Ergebnis des ACTH-, des Metopiron- und des Dexamethasontestes sowie nach dem Röntgenbefund der Nebennierenregion nach retroperitonealer O_2-Insufflation war ein Tumor der NNR auszuschließen.

Die Implantation wurde mit Hilfe des von SCHALTENBRAND (*23*) entwickelten stereotaktischen Gerätes in Lokalanaesthesie auf transorbito-sphenoidalem Wege vorgenommen. Aus der Abb. 1 ist zu erkennen, daß es am 2. und 3. Tage nach der Implantation von 2 Yttriumstäbchen mit einer Gesamtaktivität von 3,4 mC zu einem Abfall der 17-OH-CS-Ausscheidung — bestimmt nach PETERSON u. Mitarb. (*19*) — bis auf 10 bzw. 7 mg/24 Std und zu einem Abfall der 17-KS auf 5 mg/24 Std kam. 6 Monate nach der Implantation betrug die Ausscheidung der 17-OH-CS 5 mg/24 Std und 18 bzw. 24 Monate nach dem Eingriff 2,5 bzw. 2,9 mg/24 Std. Die vor der Implantation mit 52,3 mg/72 Std deutlich überhöhte Ausscheidung der nach HENKE u. Mitarb. (8) bestimmten 11-Desoxycorticosteroide (11-DO-CS) im Metopirontest betrug 2 Jahre nach dem Eingriff nur noch 1,9 mg/72 Std. Der ACTH-Test (40 E i.v. über 5 Std), der vorher einen Anstieg von 40 γ/100 ml auf 125 γ/100 ml im Plasma ergab, zeigte nach 2 Jahren einen Anstieg von 3 γ auf 6 γ/100 ml.

Die klinischen Zeichen des Cushing-Syndromes sowie die Hypertonie und die prädiabetische Stoffwechsellage bildeten sich völlig zurück. Drei Monate nach dem Eingriff traten erstmals wieder die Menses auf, die seitdem regelmäßig blieben. Die

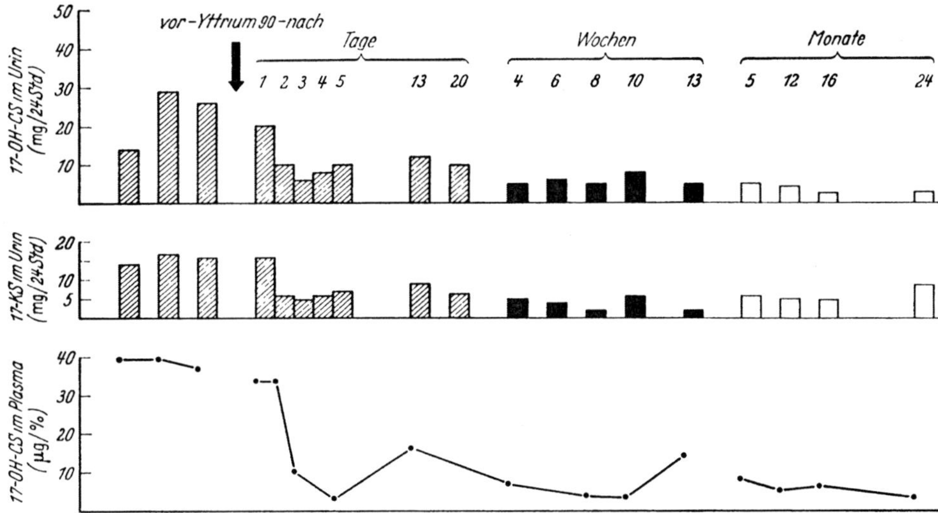

Abb. 1. Pat. S. R. Ausscheidung der 17-OH-CS und 17-KS vor und nach Implantation von 3,4 mC Y[90]

Kontrolle der Radio-Jod-Aufnahme der Schilddrüse, der J[131]-Trijodthyronin-in-vitro-Test, sowie die Bestimmung des PBJ ergaben auch 2 Jahre nach dem Eingriff keinen Anhalt für eine Schilddrüsenunterfunktion. Die Patientin fühlt sich gesund und leistungsfähig und bedarf keiner Cortisonsubstitution. Als einzige unangenehme Komplikation trat mehrere Wochen nach dem Eingriff eine Liquorfistel auf, die sich spontan wieder schloß.

Bei der zweiten Patientin P. R., 33 Jahre alt, haben wir uns auf die Implantation eines Yttriumstäbchens mit einer Aktivität von 4,0 mC beschränkt. Auch in diesem Falle kam es am 2. und 3. Tage nach der Implantation zu einem deutlichen Abfall der Ausscheidung der 17-OH-CS auf Werte um 4—6 mg/24 Std und der 17-KS auf Werte um 7—4 mg/24 Std. 1 Jahr nach der Implantation liegt die Ausscheidung der 17-OH-CS mit 4—5 mg/24 Std und die der 17-KS mit 7 mg/24 Std innerhalb der Norm.

Die Ausscheidung der 11-DO-CS im Metopirontest lag 1 Jahr nach der Implantation geringfügig unterhalb der Streubreite der Mittelwerte gesunder Versuchspersonen. Der ACTH-Test ergab zu diesem Zeitpunkt ein regelrechtes Ansprechen der NNR mit einem Anstieg der 17-OH-CS im Plasma von 18 auf 30 γ/100 ml.

Auch in diesem Fall bildeten sich die klinischen Zeichen des Cushing-Syndromes zurück, wenn auch langsamer als bei der ersten Patientin; 3 Monate nach dem Eingriff trat die Menses wieder auf, 1 Jahr nach der Implantation waren ebenfalls keine Zeichen einer Schilddrüsenunterfunktion festzustellen.

Bemerkenswert an unseren beiden Beobachtungen sind folgende Tatsachen: 1. Die verhältnismäßig geringen Dosen von Yttrium[90], die ausreichten, eine völlige Remission des Krankheitsbildes herbeizuführen; 2. der Eintritt des therapeutischen Erfolges ohne manifeste Ausfälle der Gonaden- und Schilddrüsenfunktion.

Insgesamt wurden bisher 13 Fälle von Cushing-Syndrom mitgeteilt, die mittels einer Implantation von Yttrium[90] behandelt wurden (*3, 6, 14*). Molinatti u. Mitarb., die inzwischen ihre Erfahrungen auf 9 Patienten ausdehnen konnten, implantierten Dosen zwischen 5,4 und 14,0 mC (*14*). Überraschend ist dabei, daß bei 3 ihrer Patienten kein Rückgang der Hormonausscheidung erreicht wurde. Zu einer Remission des Krankheitsbildes kam es bei diesen 3 Fällen erst, nachdem eine nochmalige Implantation von Yttrium[90] vorgenommen wurde, so daß die Patienten Gesamtdosen zwischen 17,1 und 23,8 mC erhielten. Die Autoren selbst räumen ein, daß eine exzentrische Lage der Yttriumstäbchen für das Versagen der ersten Implantation mitverantwortlich gewesen sei und zudem die errechnete Aktivitätsdosis wahrscheinlich nicht mit der wirklich implantierten übereingestimmt habe. Die von uns verabreichten Dosen lagen niedriger als die von Molinatti u. Mitarb. (*14*), Brooks (*3*) und Frey (*6*) applizierten; wir führen es darauf zurück, daß es bei unseren beiden Patienten nicht zu einer deutlichen NNR-Insuffizinez oder einer Unterfunktion der Schilddrüse und der Gonaden kam, Ausfallserscheinungen wie sie von den anderen Autoren in unterschiedlichem Anteil beobachtet wurden.

Es sei noch ein Wort dazu gestattet, warum wir bei Vornahme dieser partiellen Hypophysenausschaltung dem Yttrium[90] den Vorzug vor radioactivem Gold gaben, mit dessen Implantation Joplin u. Mitarb. (*9*) sowie der Arbeitskreis von Klotz und Talairach (*11, 12*) bei Cushing-Patienten gute Erfolge erzielten. Diesen Autoren scheint das Aurum[197] deshalb besser geeignet, weil die emittierten γ-Strahlen eine möglichst homogene Bestrahlung der Hypophyse gewährleisten und die Nekrosezone kleiner ist als beim Yttrium[90].

Es schien uns richtig, diese Nachteile des Yttrium[90] in Kauf zu nehmen, dafür aber sicher zu sein, daß bei diesen verhältnismäßig jungen Patientinnen, die Nach-

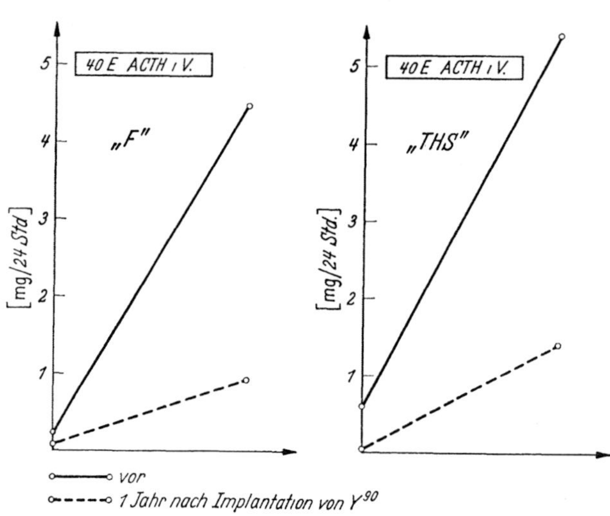

Abb. 2. Pat. R. P. Anstieg der Ausscheidung von F und THS unter ACTH vor und 1 Jahr nach Implantation von Y[90]

bargewebe, wie das Chiasma opticum, der Hypothalamus und der Hypophysenhinterlappen bei richtiger Lage des Implantates absolut geschont werden.

Literatur

1. BAHN, R. C., G. T. ROSS, and C. S. McCARTY: Proc. Mayo Clin. **35**, 623 (1960).
2. BAYER, I. M.: Langenbecks Arch. klin. Chir. **291**, 531 (1959).
3. BROOKS, R. V., R. R. McSWINEY, D. MATTINGLY, and F. T. G. PRUNTY: J. Endocr. **19**, 366 (1960).
4. DOHAEN, F. C., A. RAVENTOS, N. BOUCOT, and E. ROSE: J. clin. Endocr. **17**, 8 (1957).
5. GLENN, F., L. R. KARL, and M. HORWITH: Ann. Surg. **148**, 365 (1958).
6. FREY, E.: Oncologia (Basel) **14**, 98 (1961).
7. FREYBERG, R. H., P. E. BARKER, L. H. NEWBURH u. R. A. COLLER: Arch. intern. Med. **58**, 186 (1936).
8. HENKE, W. J., R. P. DOE, and M. E. JAKOBSON: J. clin. Endocr. **20**, 1527 (1960).
9. JOPLIN, G. F., R. FRASER, R. STEINER, J. LAWS, and E. JONES: Lancet **1961 II**, 1277.
10. KEARNS, T. P., R. M. SALASSA, J. W. KERNOHAN, S. COLLIN, C. S. McCARTY: Arch. Ophthal. **62**, 242 (1959).
11. KLOTZ, H. P., H. CHIMENES, et Mme. KANOVITCH: Sem. Hôp. Paris **38**, 3189 (1962).
12. — J. TALAIRACH, P. TOURNOUX et T. GLIMET: Journées endocr. Hóp. Biacht **5**, 224 (1961).
13. MOLINATTI, G. M., F. CAMANNI, and A. PIZZINI: J. clin. Endocr. **19**, 583 (1959).
14. — — G. ANSELMO, E. LOSANA et M. MESSINA: Journées Endocriologiyues de la Pitié 1963. Actualités Endocrinologuques, 4e série. L'Expansion édit. Paris.
15. MONTGOMERY, D. A. D., R. B. WELBOURN, W. T. CAUGHEY, and C. A. GLEADHILL: Lancet **1959 II**, 707.
16. NELSON, D. H., J. W. MEAKIN, and G. W. THORN: Ann. int. Med. **52**, 560 (1960).
17. NORTHFIELD, D. W. C.: Proc. roy. Soc. Med. **42**, 845 (1949).
18. PATTISON, R. A. D., and W. G. A. SWAN: Lancet **1958 I**, 1365.
19. PETERSON, R. E., A. KARRER, and S. L. GUERRA: Analyt. Chem. **29**, 144 (1957).
20. RASMUSSEN, T., P. V. HARPER, and T. KENNEDY: Surg. Forum **1953**, 681,
12. REES, J. R., and R. I. S. BAYLISS: Proc. roy. Soc. Med. **52**, 256 (1959).
22. SALASSA, R. M., T. P. KEARNS, J. W. KERNHOHAN, R. G. SPRAGUE, and C. S. McCARTY: J. clin. Endocr. **19**, 1523 (1959).
23. SCHALTENBRAND, G., u. P. BAILEY: Einführungen in die stereotactischen Operationen. Stuttgart: Thieme-Verlag 1959.

Aus der II. Medizinischen Universitätsklinik Hamburg-Eppendorf (Direktor: Prof. Dr. A. Jores) und der Abteilung für Endokrinologie, Hospital Santa Cruz y San Pablo, Barcelona (Direktor: Dr. J. Vilaclara)

Ein Fall von totaler Adrenalektomie bei Cushing-Syndrom ohne Notwendigkeit einer postoperativen Substitutionsbehandlung

Von

F. Morer-Fargas[1] und J. Vilaclara

Mit 1 Abbildung

Der auffällige postoperative Verlauf bei einer Patientin, die wegen eines Cushing-Syndroms mit beidseitiger NNR-Hyperplasie total adrenalektomiert wurde, läßt es berechtigt erscheinen, auf eine besondere Verhaltensweise des Organismus bei dieser Krankheit hinzuweisen.

Der Fall, über den wir berichten, ist dadurch bemerkenswert, daß sich innerhalb eines Zeitraums von mehr als 4 Jahren nach der Operation nicht nur sämtliche klinische Symptome und humorale Daten eines Morbus Cushing zurückgebildet hatten, sondern auch stärkere NNR-Insuffizienzerscheinungen entgegen der Erwartung ausgeblieben waren, obwohl eine Substitution nicht stattgefunden hatte. Vielmehr bestanden bei der Patientin nur sehr diskrete Zeichen einer Insuffizienz in Form einer verminderten Regulationsbreite der NNR unter Stress-Situationen.

Krankheitsverlauf

Die 34jährige Patientin R. A. A. wurde im August 1959 unter der Diagnose M. Cushing in der endokrinologischen Abteilung des Hospitals Santa Cruz y San Pablo in Barcelona stationär aufgenommen. Der Beginn der zur Zeit der Einweisung bereits fortgeschrittenen Erkrankung lag mehrere Jahre zurück. Anamnestisch ist besonders zu bemerken, daß im Jahre 1956 die Hypophyse bestrahlt wurde ohne Besserung des Krankheitsbildes. Danach Amenorrhoe.

Bei der Aufnahme bot die Patientin zahlreiche Symptome eines M. Cushing, nämlich: Ausgeprägtes Vollmondgesicht, Haarausfall, Hypertonie (220/130), ausgedehnte Striae, hochgradige Osteoporose mit Fischwirbelbildung, Steroiddiabetes (Blutzucker: 190 mg-%) und sehr erhöhte 17-Ketosteroid-Ausscheidung (42 mg/24 Std). Die Röntgenbilder der Sella waren unauffällig.

Auf Grund dieser Befunde konnte die Diagnose M. Cushing bestätigt werden. Als Therapie der Wahl wurde eine zweizeitige totale Adrenalektomie innerhalb von 5 Monaten durchgeführt (Oktober 1959 und Februar 1960). Beide NN wurden ohne

---

[1] Stipendiat der Stiftung "Juan March" (Madrid).

Schädigung der Kapsel entfernt, wodurch eine vollkommene Operation garantiert wird. Das Gewicht der linken NN (1. Sitzung) betrug 10,25 g das der rechten NN, 12 g.

1. Histologie

Linke NN: Es handelt sich um einen Knoten, der aus einer Proliferation von Rindenzellenelementen ohne Zellatypien besteht. Diese Elemente sind in Trabekeln von vacuolisierten Zellen angeordnet, die mit Fettfärbungen deutlich anfärbbar sind. Der Knoten ist relativ gut abgegrenzt und an manchen Stellen von einer zarten Kapsel umgeben. In den übrigen Rindenanteilen wurden diffus hyperplastische Herde beobachtet.

Rechte NN: Gesamte Rinde verbreitert, besonders zu Gunsten von Zona glomerulosa und teilweise Zona fasciculata auf Grund von kleinen, gut begrenzten hyperplastischen Knoten. Übrige Anteile unauffällig.

Der histologische Befund zeigt also deutlich das Vorhandensein einer kleinknotigen, doppelseitigen NNR-Hyperplasie mit einem Pseudoadenoma fasciculatum.

2. Postoperativer Verlauf

Nach der Operation hat die Patientin bis heute eine fast normale Leistungsfähigkeit gezeigt, ohne einer Substitution zu bedürfen. Sie kann den durch die Versorgung des Haushaltes entstehenden Anforderungen gerecht werden. Objektive Anhaltspunkte für die Entwicklung eines intrasellären Tumors liegen bisher nicht vor.

Eine Substitutionsbehandlung fand ausschließlich bei den vorübergehenden Stress-Situationen statt, die im Laufe der Zeit nach der Operation aus verschiedenen Gründen auftraten. Sie erfolgte immer durch perorale Verabreichung des Steroids 6-Methylprednisolon („Urbason" Höchst) in folgenden Dosen:
Infekte: 4 mg täglich bis zur Überwindung der Infektion.
Körperliche Anstrengungen: 4 mg 2mal wöchentlich.
Dabei ist zu betonen, daß solche Dosen nur $1/4$ der sonst in derartigen Fällen notwendigen Dosen darstellen.

Ein halbes Jahr und $3^1/_2$ Jahre nach der Operation wurden ein Thorn-Test und eine Bestimmung der 17-Ketosteroid-Ausscheidung nach ACTH-Belastung durchgeführt. Im ersten Thorn-Test wurde einen ganz geringer Abfall der Eosinophilenzahl beobachtet (von 90 auf 74 Eos/mm^3), der einer deutlichen Situation von NNR-Insuffizienz entspricht. Im zweiten Thorn-Test dagegen waren der Ausgangswert wie auch der Abfall der Eosinophilenzahl nach ACTH-Injektion als normal anzusehen (von 190 auf 100 Eos/mm^3). Bei der ACTH-Belastung wurde weder 6 Monate noch $3^1/_2$ Jahre postoperativ eine bewertbare Veränderung der 17-Ketosteroidausscheidung beobachtet.

Die 3 Monate nach der Operation durchgeführten Steroidbestimmungen zeigen Besonderheiten, die nicht leicht zu interpretieren sind (Abb. 1). Die Ausscheidung der 17-Ketosteroide betrug nur 10—20% der normalerweise bei Frauen in dem Alter unserer Patientin gefundenen Werte. Die 17-OHCS-Ausscheidung war in der konjugierten Fraktion auf ein Viertel des Normalbetrages vermindert, während die freie Fraktion nur wenig unter der unteren Normgrenze lag. Im Plasma war die freie Fraktion nur gering erniedrigt, während die konjugierte Fraktion an der unteren Grenze der Norm lag.

In den 22 Monaten postoperativ war die 17-Ketosteroidausscheidung noch eindeutig erniedrigt, etwa auf 40% der Normwerte. Auffälligerweise lag die Ausscheidung an freien 17-OHCS im unteren-, die an konjugierten im Normbereich.

		3 Monate post op. (Mai 1960)	22 Monate post op. (Dez. 1961)
Urin	17-Keto. (mg/24 Std)	1,56	3,04
	17-OHCS (mg/24 Std) Freie Konjugierte	0,12 1,32	0,27 4,90
Plasma	17-OHCS (γ/100 ml) Freie Konjugierte	5,70 10,60	

Abb. 1. Steroidwerte im Urin und Plasma nach der Operation

Diskussion

Wenn der vorliegende Fall im ganzen betrachtet wird, besteht kein Zweifel, daß das Cushing-Syndrom, der Histologie der entfernten NN nach, durch eine beidseitige NNR-Hyperplasie hervorgerufen wurde. Die Angaben dieser Untersuchung sind für die Diagnose insofern ausschlaggebend, als keinerlei lokale tumoröse Veränderungen festgestellt werden konnten. Das Vorhandensein von Pseudoadenomen als Ausdruck einer NNR-Hyperplasie kommt verhältnismäßig häufig vor.

Uns ist bis heute kein Fall von Cushing-Syndrom wegen bilateraler NNR-Hyperplasie bekannt, bei dem mehr als 4 Jahre post operationem:

1. Eine Heilung des Krankheitsbildes festzustellen ist.

2. Keine NNR-Insuffizienz aufgetreten ist, wodurch die übliche Substitutionstherapie unterblieb.

3. Kein Rezidiv aufgetreten ist. Diese bemerkenswerte Verlaufsform benötigt eine Interpretation. Dabei ergeben sich zwei Möglichkeiten:

α) Die Möglichkeit einer inkompletten Adrenalektomie

Es ist ein Prinzip der Pathophysiologie der Hormone, daß restierendes Gewebe nach Entfernung einer sekundär hyperfunktionierenden endokrinen Drüse weiterhin die Fähigkeit zur Hypertrophie besitzt und damit erneut einen Zustand von Überfunktion erzeugen kann. Dementsprechend kommt es postoperativ zu einer Hypertrophie des zurückgebliebenem NNR-Gewebes, das u. U. die volle NNR-Funktion übernehmen kann. Im vorliegenden Fall halten wir es nicht für gerechtfertigt eine inkomplette Adrenalektomie anzunehmen. Sie wird einerseits durch die Operationsprotokolle der beiden Sitzungen ausgeschlossen, andererseits durch das Gewicht der entfernten NN und ihren makroskopischen Aspekt sowie durch das Fehlen von Zeichen eines Rezidivs des Krankheitsbildes.

β) Existenz von ektopischem NNR-Gewebe und seine Bedeutung

Die Steroidbestimmungen, welche postoperativ durchgeführt wurden, beweisen das Vorhandensein schwankender Mengen an Metaboliten von NNR-Steroiden. Das weist auf eine endogene Bildung solcher Steroide (besonders Cortisol) hin. Nimmt man das Vorhandensein aberrierenden NNR-Gewebes an, dessen Existenz, Lokalisation und Struktur vielfach beschrieben wurde (Testut und Latarjet, 1951), so scheinen alle Interpretationsschwierigkeiten beseitigt zu sein.

Jedoch erscheint es auf Grund der vorgenommenen Steroidbestimmungen unmöglich, daß die nachgewiesenen endogenen Steroidmengen ausreichen könnten, ein normales Leben der Patientin, wie sie es führt, zu garantieren.

Das postoperative Vorhandensein von NNR-Metaboliten im Plasma und Urin stellt unserer Meinung nach die Kernproblematik des vorliegenden Falls dar. Tatsache ist, daß 3 Monate nach der Operation die Patientin eine gewisse Produktion von NNR-Steroiden besaß, die nach den Plasmawerten als beträchtlich, nach den Urinwerten jedoch als gering anzusehen ist. 22 Monate nach der Operation sind die im Urin nachgewiesenen Steroidmengen bis auf Normalwerte angestiegen.

Das Fehlen einer Korrelation zwischen Plasma- und Urinwerten läßt an eine stark verlängerte Halbwertszeit des Cortisols denken. Eine Leber- oder Nierenschädigung, welche diese Diskrepanz zwischen Plasma- und Urinwerten durch eine Retention der produzierten NNR-Steroiden hätte erklären können, ließ sich nicht nachweisen.

Die postoperative, endogene Produktion von Cortisol in unserem Fall und das entsprechende Überleben der Patientin ohne Notwendigkeit einer Substitution, muß auf das Vorhandensein von aberrierendem NNR-Gewebe oder von anderen, noch unbekannten mesodermalen Strukturen mit solcher Kapazität zurückgeführt werden (Soldevila und Vilaclara, 1960).

Literatur

Soldevila, J., u. J. Vilaclara: IV. Tagung der Spanischen Gesellschaft für inn. Med. Madrid, 1960.
Tamm, J., K. D. Voigt u. J. Kracht: Acta endocr. (Kbh.) **37**, 253 (1961).
Testut, L., y A. Latarjet: In: Anatomia Humana, 9. Spanische Ausgabe, Band III. Barcelona: Salvat Edit. 1951.
Tonutti, E., u. J. M. Bayer: Endokrinologie **41**, 17 (1961).
Zukschwerdt, L., M. G. Giebel, H. Oetjen u. J. Tamm: Schweiz. med. Wschr. **92**, 22 (1962).

Aus dem Pathologischen Institut der Universität Hamburg
(Direktor: Prof. Dr. Dr. h.c. C. Krauspe)

Der Hypophysenvorderlappen des Menschen bei Hypercortisolismus

Von

D. Zimmermann und J. Kracht

Mit 2 Abbildungen

Auf dem Gebiet der ACTH-Bildungsstätten sind mit neuen histochemischen Verfahren, der Immunhistologie, Elektronenmikroskopie und Autoradiographie in den letzten Jahren zwar Fortschritte erzielt worden (*1, 2, 7—14, 19, 21, 23—26*), doch kann der Komplex keineswegs als gelöst gelten. Bei der Bearbeitung des eigenen Materials von Fällen mit endogenem und exogenem Hypercortisolismus bedienten wir uns der üblichen Trichromfärbungen und besonders der von Adams und Swettenham (*1*) angegebenen Perameisensäure-Alcianblau-PAS-Orange-G-Färbung, welche die Trennung mucoider Elemente in PAS-positiv gefärbte R-Zellen und in S-Zellen erlaubt, deren mit Alcianblau gefärbte Granula einen hohen Cystingehalt anzeigen. Die α-Zellen werden mit Orange G dargestellt. Während die Existenz von zwei R-Zellen-Typen möglich, aber noch unbewiesen ist, können die S-Zellen in einen S_1- und einen S_2-Typ unterteilt werden (*21*). Die auf dieser Färbung aufgebaute Nomenklatur halten wir für die Verständigung vorläufig am besten geeignet. Nach den bisherigen Ergebnissen läßt sich beim Menschen die TSH-Produktion in die S_2-Zellen verlegen, wie dies Pearse und v. Noorden vermuteten und von uns anhand eines Falles von primärem Myxödem wahrscheinlich gemacht werden konnte. Die Zuordnung von FSH und LH zu dem S_1- bzw. einem R-Zellen-Typ ist noch nicht genügend geklärt. Als Bildungsstätten von ACTH werden R-Zellen angenommen, nachdem Pearse die S-Zellenhypothese (*2*) zugunsten der R-Zellen korrigiert hat (*21*). Dieser Auffassung entsprechen die eigenen Befunde. Aus den Untersuchungen von Dhom u. Mitarb. (*7—10*) ist zu schließen, daß die Verhältnisse in der Rattenhypophyse möglicherweise anders liegen, da dort die ACTH-Produktion in S-Zellen lokalisiert werden konnte. Siperstein (*25*) verlegt demgegenüber die ACTH-Bildung in der Rattenhypophyse anhand autoradiographischer Studien in sog. große chromophobe Zellen, doch sind diese Befunde mit den Ergebnissen der üblichen Färbemethoden nicht vergleichbar.

Wir verfügen über 5 Fälle von hypothalamisch-hypophysär vermitteltem Hypercortisolismus, 4 Beobachtungen von adrenalem Cushing-Syndrom, einem Fall von Hypercortisolismus auf dem Boden eines metastasierten Mediastinaltumors mit ACTH-ähnlicher Aktivität sowie über mehrere Hypophysen nach

ACTH- bzw. Corticoidlangzeittherapie. Zum Vergleich dienten Hypophysen von Kontrollfällen und vier Fälle mit primärer Nebennierenrindenatrophie.

Über die Befunde bei den verschiedenen pathogenetischen Varianten des Hypercortisolismus kann kaum summarisch berichtet werden, da im Einzelfall auch auf klinische Hormonalanalysen, Therapie, Krankheitsdauer usw. Bezug genommen werden müßte.

Bei hypothalamisch-hypophysärem M. Cushing wurden in 4 von 5 Fällen Adenome im Vorderlappen festgestellt, darunter dreimal in doppelter Zahl. Diese sind in der Regel klein und treiben die Hypophyse nicht oder nur wenig auf, ein Adenom war expansiv und örtlich infiltrierend gewachsen (17). Mit Ausnahme eines kleinen chromophoben Adenoms in der Nachbarschaft des infiltrierend gewachsenen bestehen alle Adenome aus R-Zellen (18) mit meist rötlich-violettem Granulafarbton und Zeichen gesteigerter Aktivität. Extraadenomatöse R-Zellen treten demgegenüber quantitativ zurück und weisen z. T. die von CROOKE (4) beschriebenen Veränderungen auf. Crooke-Zellen zeichnen sich durch eine zur Peripherie und kernwärts fortschreitende Cytoplasmahomogenesierung mit z. T. feinster Vacuolisierung und gleichzeitige Granulaverarmung bis zur völligen Entgranulierung aus. Die Anfangsstadien dieser Metamorphose werden als Para-Crooke-Zellen bezeichnet (2). Die Kerne der Crooke-Zellen sind in der Regel groß, eher hypochromatisch und weisen einen prominenten Nucleolus auf, daneben kommen auch pyknotische Kernformen vor. Crooke-Zellen sind in Adenomnähe angereichert, ohne im übrigen Vorderlappen zu fehlen. In zwei Fällen sahen wir in der Peripherie von Adenomen stufenweise Übergänge von adenomatösen R-Zellen in Crooke-Zellen (Abb. 1 u. 2). Gut granulierte S_2-Zellen waren bei M. Cushing kaum nachweisbar, vesikulierte S_2-Zellen und ähnlich auch die S_1-Zellen schienen in beiden Fällen ohne Adrenalektomie vermindert, nach Eingriffen an den Nebennieren zahlreicher zu sein. Bei endogenem Cushing-Syndrom (Nebennierenrindencarcinom, bilaterale Nebennierenrindenadenomatose (16), Fall mit vermehrter extrahypophysärer ACTH-ähnlicher Aktivität) ist das Vorkommen von Crooke-Zellen und eine mehr oder weniger ausgeprägte Verminderung von R-Zellen kennzeichnend. Als Ausnahme von der Regel fand sich in einem Fall mit metastasiertem NNR-Carcinom ein regressiv verändertes Vorderlappenadenom. Damit ergibt sich, daß Hypercortisolismus auf dem Boden vermehrter hypophysärer ACTH-Sekretion in der Regel mit einer Zunahme des R-Zellen-Potentials einhergeht, während die Varianten des endogenen Hypercortisolismus mit herabgesetzter hypophysärer ACTH-Sekretion eher ein Defizit an R-Zellen aufweisen. Diese Feststellung erscheint deshalb wesentlich, weil S_1-Zellen bei herabgesetzter corticotroper Aktivität des Hypophysenvorderlappens sich uncharakteristisch verhalten und bei bilateraler Nebennierenrindenadenomatose sogar vermehrt waren. S_2-Zellen sind bei dieser Variante in der Regel vermindert.

Das Substrat nach ACTH- bzw. Cortison-Therapie ist nicht einheitlich und unterliegt im besonderen Maße der Therapiedauer und der Dosierung. Aus diesen Gründen wird verständlich, daß es nicht zwangsläufig mit den bei adrenalem Hypercortisolismus beschriebenen Veränderungen identisch ist. Vielmehr fällt auf, daß neben Para-Crooke- und Crooke-Zellen eine scheinbar paradoxe diffuse oder auch herdförmige R-Zellen-Hyperplasie vorkommen kann. Hervorzuheben ist eine besonders nach Cortison ausgeprägte Vermehrung von S_1-Zellen und eine gering-fügige Vermehrung von S_2-Zellen, darunter auch der vesikulierten Form.

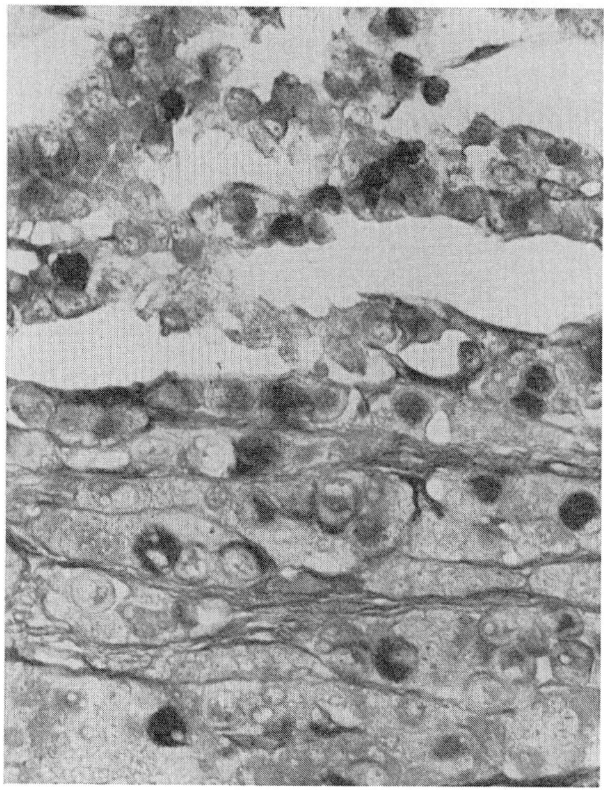

Abb. 1. ♀. 31 Jahre alt (S.-Nr. 846/58), M. Cushing. Randpartien eines R-Zellenadenoms im HVL (oben) mit Übergang in Crooke-Zellen. Perameisensäure-Alcianblau-PAS-Orange G. (560fach)

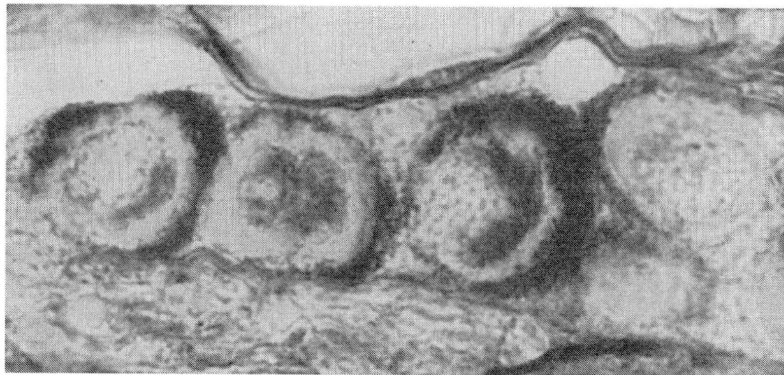

Abb. 2. Verschiedene Stadien der Cytoplasmahomogenisierung in R-Zellen (Para-Crooke- und Crooke-Zellen) in adenomnahen Anteilen des Vorderlappens (S.-Nr. 846/58). Perameisensäure-Alcianblau-PAS-Orange G (1400fach)

Die vergleichende Cytologie des Hypophysenvorderlappens bei allen Varianten des Hypercortisolismus, bei primärer Nebennierenrindenatrophie und primärem

Myxödem spricht für die Annahme der ACTH-Bildung in R-Zellen. Da ein direkter Beweis hierfür aussteht, müssen zur Begründung weitere Kriterien herangezogen werden. Wir nennen den biologischen Nachweis hoher corticotroper Aktivität in Adenomen bei M. Cushing (*3, 6, 15, 20*) und immunhistologische Befunde mit fluorescierenden ACTH-Antikörpern an Adenomzellen und extraadenomatösen R-Zellen einschließlich Crooke-Zellen (*19, 21*). Da nach PEARSE und v. NOORDEN (*21*) allerdings nur ein Teil der R-Zellen auf fluorescierende ACTH-Antikörper reagiert und sich nach eigenen Befunden färberische Unterschiede in der R-Zellen-Gruppe ergeben, muß die Möglichkeit einer künftigen Weiterdifferenzierung dieses Zelltyps in Betracht gezogen werden. Die Bedeutung der Crooke-Zellen ist trotz vieler Hypothesen noch immer ungelöst. Da ihre Beziehung zum corticotropen Sektor wohl außer Zweifel steht, scheiden Reflexionen anderer Tropine (*5, 13*) oder ihrer Erfolgsorgane bei Hypercortisolismus u. E. aus. Es ließen sich Argumente für Zeichen der Leistungssteigerung, aber auch für eine Minderung der Zellfunktion anführen, so daß die Möglichkeit einer isomorphen Reaktion in Betracht gezogen werden muß. Die größte Wahrscheinlichkeit liegt in der Annahme einer reversiblen Funktionseinschränkung dieses Zelltyps.

Die Stellung der Vorderlappenadenome bei M. Cushing ist durch die zentrale Bedeutung des Faktors CRF im Krankheitsbild und durch das Nelson-Syndrom (*27*) aktualisiert worden. Wir fassen sie heute als von hypothalamischen Reizen abhängige knotige Hyperplasien corticotropinbildender Elemente auf. Da es sich nur in Ausnahmefällen um autonome Gewächse handelt, wird ihr fakultatives Vorkommen verständlich. Auf die hiermit zusammenhängende Problematik ist von KRACHT und TAMM (*17*) eingegangen worden. Der Nachweis kleinknotiger R-Zellen-Hyperplasien nach Corticoidbehandlung steht nicht in Widerspruch zu dieser Auffassung. Hierbei könnte es sich um kompensatorische CRF-abhängige Hyperplasien auf der Basis einer jeweils nur temporären Hemmung des sonst intakten und damit zur Gegenregulation fähigen hypothalamischen Reglers handeln. Obwohl im Einzelfall nicht die Möglichkeit eines Nebenbefundes ausgeschlossen werden kann, erscheint es nicht paradox, wenn knotige R-Zellen-Hyperplasien auch nach ACTH-Therapie auftreten, wie wir dies in einem Fall beobachten konnten.

Literatur

1. ADAMS, C. W. M., and K. V. SWETTENHAM: The histochemical identification of two types of basophil cells in the normal human adenohypophysis. J. Path. Bact. **75**, 95 (1958).

2. —, and A. G. E. PEARSE: Classification of the mucoid (basophil) cells in the normal and pathological human adenohypophysis. J. Endocr. **18**, 147 (1959).

3. BAHN, R., G. T. ROSS, and C. S. McCARTY: Melanocyte stimulating hormone and ACTH-activities of pituitary tumors in patients with Cushing's syndrome. Proc. Mayo Clin. **35**, 623 (1960).

4. CROOKE, A. C.: A change in the basophil cells of the pituitary gland common to conditions which exhibit the syndrome attributed to basophil adenoma. J. Bact. Path. **41**, 339 (1935).

5. — In: Discussion on the functional activities of the basophil cells of the pituitary gland. Proc. roy. Soc. Med. **49**, 1019 (1956).

6. DAVIES, B. M. A.: Blood corticotrophin in normal adults and in patients with Cushing's syndrome. Acta endocr. (Kbh.) **45**, 55 (1964).

7. DHOM, G., u. H. U. TIETZE: Zur experimentellen Histophysiologie der β-Zellen der Rattenhypophyse nach Untersuchungen mit der Perameisensäure-Alcianblau-Reaktion. Endokrinologie **42**, 284 (1962).

8. Dhom. G., F. Burkert u. H. U. Tietze: Thyreoidektomiezellen im Hypophysenvorderlappen der Ratte nach Nebennierenblockade. Zur Frage der Bildungsstätten des ACTH. Z. Zellforsch. **57**, 679 (1962).

9. —, u. H. P. Scherer: Stress-Reaktionen am Hypophysenvorderlappen der Ratte. Virchows Arch. path. Anat. **336**, 368 (1963).

10. — Fortschritte der Histophysiologie des Hypophysenvorderlappens. Klin. Wschr. **41**, 1117 (1963).

11. Ezrin, C., H. E. Swanson, J. G. Humphrey, J. W. Dawson, and W. D. Wilson: The delta-cell of the human adenohypophysis: its response to acute and chronic illness. J. clin. Endocr. **18**, 917 (1958).

12. — — — —, and F. M. Hill: Beta and delta cells of the human adenohypophysis: Their response to adrenocortical disorders. J. clin. Endocr. **19**, 621 (1959).

13. — — — — — The cells of the human adenohypophysis in thyroid disorders. J. clin. Endocr. **19**, 958 (1959).

14. Foncin, F. F., et J. Le Beau: Etude en microscopie optique et electronique d'une tumeur hypophysaire à fonction adrenocorticotrope. C. R. Soc. Biol. (Paris) **157**, 249 (1963).

15. Jailer, J. W., and D. A. Holub: The significance of pituitary tumors in Cushing's syndrome. Advance Abstr. I. Intern. Congr. of Endocr. Copenhagen (1960), 363.

16. Kracht, J., u. J. Tamm: Bilaterale kleinknotige Adenomatose der Nebennierenrinde bei Cushing-Syndrom. Virchows Arch. path. Anat. **333**, 1 (1960).

17. — — Invasiv gewachsenes basophiles Adenom des Hypophysenvorderlappens bei Cushing-Syndrom. Acta endocr. (Kbh.) **43**, 330 (1963).

18. — Das Hypophysen-Nebennierenrinden-System bei Cushing-Syndrom. Verh. dtsch. Ges. Path. **47**, 280 (1963).

19. Leznoff, A., F. Fishman, M. Talbot, E. E. McGarry, J. C. Beck, and B. Rose: The cytological localization of ACTH in the human pituitary. J. clin. Invest. **41**, 1720 (1962).

20. Montgomery, D. A. D., R. B. Welborn, W. T. E. McCaughey, and C. A. Gleadhill: Pituitary tumors manifested after adrenalectomy for Cushing's syndrome. Lancet **1959**, 707.

21. Pearse, A. G. E., and S. van Noorden: The functional cytology of the human adenohypophysis. Canad. med. Ass. J. **88**, 462 (1963).

22. Nelson, D. H., J. W. Meakin, and G. W. Thorn: ACTH-producing pituitary tumors following adrenalectomy for Cushing's syndrome. Ann. intern. Med. **52**, 560 (1961).

23. Russell, D. S.: In: Discussion on the functional activities of the basophil cells of the pituitary gland. Proc. roy. Soc. Med. **49**, 1018 (1956).

24. Russfield, A. B.: Biological significance of aldehyd-fuchsin positive cells in the human hypophysis. Lab. Invest. **9**, 490 (1960).

25. Siperstein, E. R.: Identification of the adrenocorticotrophin producing cells in the rat hypophysis by autoradiography. J. cell. Biol. **17**, 521 (1963).

26. Swanson, H. E., and E. Ezrin: The natural history of the delta cell of the human adenohypophysis: in childhood, adulthood and pregnancy. J. clin. Endocr. **20**, 952 (1960).

Aus dem Pathologischen Institut der Universität Hamburg
(Direktor: Prof. Dr. med. Dr. med. h.c. C. KRAUSPE)

Die Nebennierenrinde bei endogenem Hypercortisolismus*

Von

J. KRACHT und D. ZIMMERMANN

Mit 2 Abbildungen

Das morphologische Substrat der Nebennierenrinde bei endogenem Hypercortisolismus erlaubt präzise Aussagen über den corticotropen Stimulierungsgrad der Rinde und damit die Differenzierung primär adrenaler von hypothalamisch-hypophysär regulierten Varianten. Die letzte Gruppe stellt das Hauptkontingent aller Fälle von M. Cushing und ist durch die diffuse oder diffus-knotige bilaterale Rindenhyperplasie gekennzeichnet. Entgegen manchen Vorstellungen muß betont werden, daß die hiermit verbundenen Strukturänderungen der Rinde unspezifisch sind und lediglich einen erhöhten corticotropen Stimulierungsgrad anzeigen. Gleichartige Veränderungen finden sich auch bei adrenogenitalem Syndrom (*17*), gelegentlich bei primärem Hyperaldosteronismus (*1*) und schließlich bei nicht-endokrinen Krankheiten. Der vermehrte corticotrope Stimulierungsgrad kommt in der progressiven Transformation der Rinde zum Ausdruck. Helle lipoidreiche, fasciculierte Strukturen stehen im Vordergrund und erreichen häufig die Kapsel, so daß eine Z. glomerulosa u. U. nicht mehr nachweisbar ist. Auch die Z. reticularis ist spongiocytär umgewandelt, doch liegen hier überwiegend kompaktzellige Elemente vor, während in der innersten Rinde und teilweise auch im Mark perivasculär häufig wieder helle Zellen angetroffen werden. Die corticotropen Wachstumsimpulse sind, gemessen an der Steroidsekretion und deren Auswirkungen, gering. Sie kommen in einer Zellkernvergrößerung in allen Zonen zum Ausdruck (*9*). Die Mitosenrate ist gering. Nach Inkubation von Operationsmaterial mit H_3-Thymidin ermittelten wir in zwei Fällen einen H_3-Index von 0,5 bzw. 1, der sich auch durch Zusatz von ACTH nicht steigern ließ. Im Autoradiogramm sind Prämitosen bevorzugt in der äußeren Z. fasciculata lokalisiert. Weitere Hinweise für corticotrope Wachstumsleistungen ergeben sich aus den nicht obligatorischen intracorticalen knotigen Hyperplasien, aus Kapseldurchbrüchen des Rindengewebes in das umliegende Fettbindegewebe (Abb. 1) und aus der Infiltration des Nebennierenmarks mit Rindenzellen. Zur Frage der funktionellen Zonierung sei kurz das enzymhistochemische Verteilungsmuster skizziert. Während die Steroid-3-β-ol-Dehydrogenase ebenso wie die Rindenlipoide in der aus hellen Zellen zusammengesetzten Fasciculata angereichert sind, dominieren die alkalische und

---

* Herrn Prof. Dr. H. W. BANSI zum 65. Geburtstag.

saure Phosphatase und die Glucose-6-Phosphatase in der kompaktzelligen Innen-
zone. Aus diesen und anderen Kriterien hat Symington (13, 14) gefolgert, daß die
Z. reticularis die aktive Zone der Biosynthese von C_{19}- und C_{21}-Steroiden außer
Aldosteron, die Z. fasciculata dagegen nur ein Reservoir für Hormonvorstufen
wäre. Diese Gruppe hat neuerdings festgestellt (7), daß die 11-β-Hydroxylase-
aktivität in hellen und kompakten Zellen gleich ist, daß beide Zelltypen auf ACTH

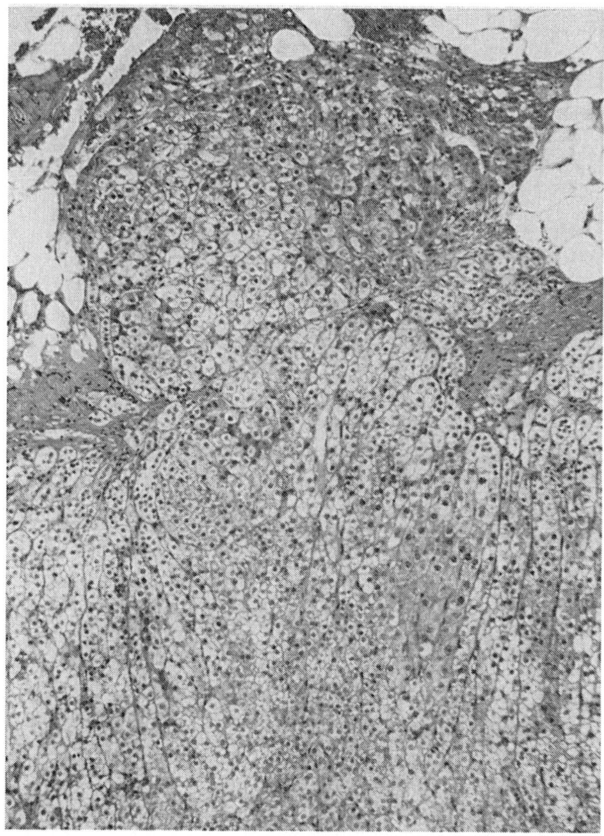

Abb. 1. Dr. I. St., ♀, 29 Jahre (Ch 5823/63): M. Cushing. Bilaterale Nebennierenrindenhyperplasie. Progressive
Transformation der äußeren Rinde. Umschriebener Kapseldurchbruch und Infiltration des angrenzenden
periadrenalen Fettbindegewebes

mit vermehrter Steroidsynthese reagieren und daß hierbei helle Zellen verstärkt
ansprechen. Da schließlich 11-β-Hydroxyandrostendion als Produkt beider Zelltypen nachgewiesen worden ist, müssen Fasciculata und Reticularis auch als
Bildungsort der Nebennierenandrogene gelten. Mit der Annahme der Bildung von
C_{19}- und C_{21}-Steroiden ausgenommen Aldosteron im Fasciculata-Reticularis-Kom-
plex hat Symington (7) seine ursprüngliche Auffassung verlassen und sich jener
von Tonutti (15, 16) genähert, der seit langem diesen Bereich als funktionelle Einheit auffaßt. Hinsichtlich der funktionstragenden Bedeutung der Z. fasciculata
stimmen wir mit Tonutti überein. Dies gilt auch für die Bewertung sog. normaler
Rindenbefunde bei M. Cushing, die wir als selten erachten, obwohl wir uns der

Diskrepanz zwischen der Geringfügigkeit der Umbauvorgänge und dem klinischen Krankheitsbild in manchen Fällen nicht verschließen können. Es bestehen somit keine überzeugenden Parallelen zwischen dem Grad morphologischer Umbauvorgänge, der klinischen Symptomatik und der Hormonalytik.

Hormonalaktive Rindenadenome oder -carcinome sind in der Regel einseitig lokalisiert, kommen in seltenen Fällen aber bilateral vor (*4, 6, 18*). Ihr Anteil an allen Nebennierenrindenadenomen oder -carcinomen bei Cushing-Syndrom wird auf weniger als 2% geschätzt (*18*). In diese Gruppe gehört die bilaterale kleinknotige Rindenadenomatose, die man prinzipiell mit der Inseladenomatose bei Hyperinsulinismus vergleichen kann. Wir haben das Substrat dieser Variante als multizentrisch entstandene, langsam wachsende, funktionell autonome Adenome in einer sekundär atrophischen Restrinde beschrieben, ihr eine Sonderstellung zuerkannt und darauf hingewiesen, daß bei dieser Konstellation sogar ein insgesamt atrophisches Organ Quelle des Hypercortisolismus ist (*10*). Dieses Substrat darf nicht mit der kompensatorischen Atrophie der kontralateralen Rinde verwechselt werden. Es bietet keine Kontraindikation, sondern vielmehr eine absolute Indikation zur Adrenalektomie. Von TONUTTI ist in diesem Zusammenhang die Frage aufgeworfen worden (*16*), inwieweit sich primär ACTH-abhängige knotige Hyperplasien funktionell autonomisieren können. Wir halten diese Möglichkeit im Falle der bilateralen Adenomatose für unwahrscheinlich, weil man schwer einsehen kann, daß sich 30—40 kleinste Adenome gleichzeitig funktionell verselbständigen, ohne wesentlich zu wachsen. Den Prinzipien der allgemeinen Pathologie entspräche eher die funktionelle oder strukturelle Entartung eines oder weniger Knoten. Für diese Annahme ließen sich Beispiele anführen. Wir verfügen über die Beobachtung eines hormonalaktiven und erfolgreich operierten Rindenadenoms bei Cushing-Syndrom, in dessen sekundär atrophischer Restrinde weitere kleinere Knoten angetroffen werden, die an der Involution nicht teilgenommen haben und sogar hyperchromatische Riesenkerne aufweisen, ohne daß Hinweise für Malignität gegeben wären. Man muß also betonen, daß hormonalaktive Knoten keineswegs immer singulär, sondern durchaus multipel, und zwar ein- und doppelseitig vorkommen können. Die Frage nach der Entartung einer hypothalamisch-hypophysär gesteuerten diffus-knotigen Rindenhyperplasie in ein hormonal aktives Rindencarcinom gehört zum gleichen Komplex. Bei einer 39 Jahre alten Frau war 7 Jahre vor dem Tode ein Cushing-Syndrom aufgetreten und 5 Jahre später ein etwa faustgroßer Nebennierenrindentumor operativ entfernt worden. Histologisch lag eine adenomatöse Hyperplasie vor, ein zu jener Zeit möglicherweise schon vorhandenes, organoid strukturiertes Nebennierenrindencarcinom wurde nicht erkannt. Zwei Jahre später Exitus an ausgedehnten Metastasen und komplizierender eitriger Meningitis (S.-Nr. 275/59). Als überraschend erwies sich das Vorliegen eines weitgehend regressiv veränderten, große Teile des Vorderlappens einnehmenden entgranulierten mucoidzelligen Adenoms. Dieser ungewöhnliche Befund gab Anlaß zur näheren Bearbeitung der Nebennieren. Während die kontralaterale Rinde zum Zeitpunkt der Sektion erwartungsgemäß das Bild der sekundären Atrophie bot, fanden sich im Operationspräparat an verschiedenen Stellen noch gut fasciculierte Anteile mit Zeichen des Kapseldurchbruchs, die strukturell nicht von ACTH-abhängigen knotigen und diffusen Hyperplasien zu trennen sind. An einzelnen Stellen konnte sekundär atrophisches Rindengewebe nachgewiesen werden. Im

Zusammenhang mit der durch eine therapeutische Hypophysenbestrahlung (6000r) oder durch Reflexion des cortisolbildenden Rindentumors erfolgten Rückbildung des Vorderlappenadenoms und einiger weniger verläßlichen klinischen

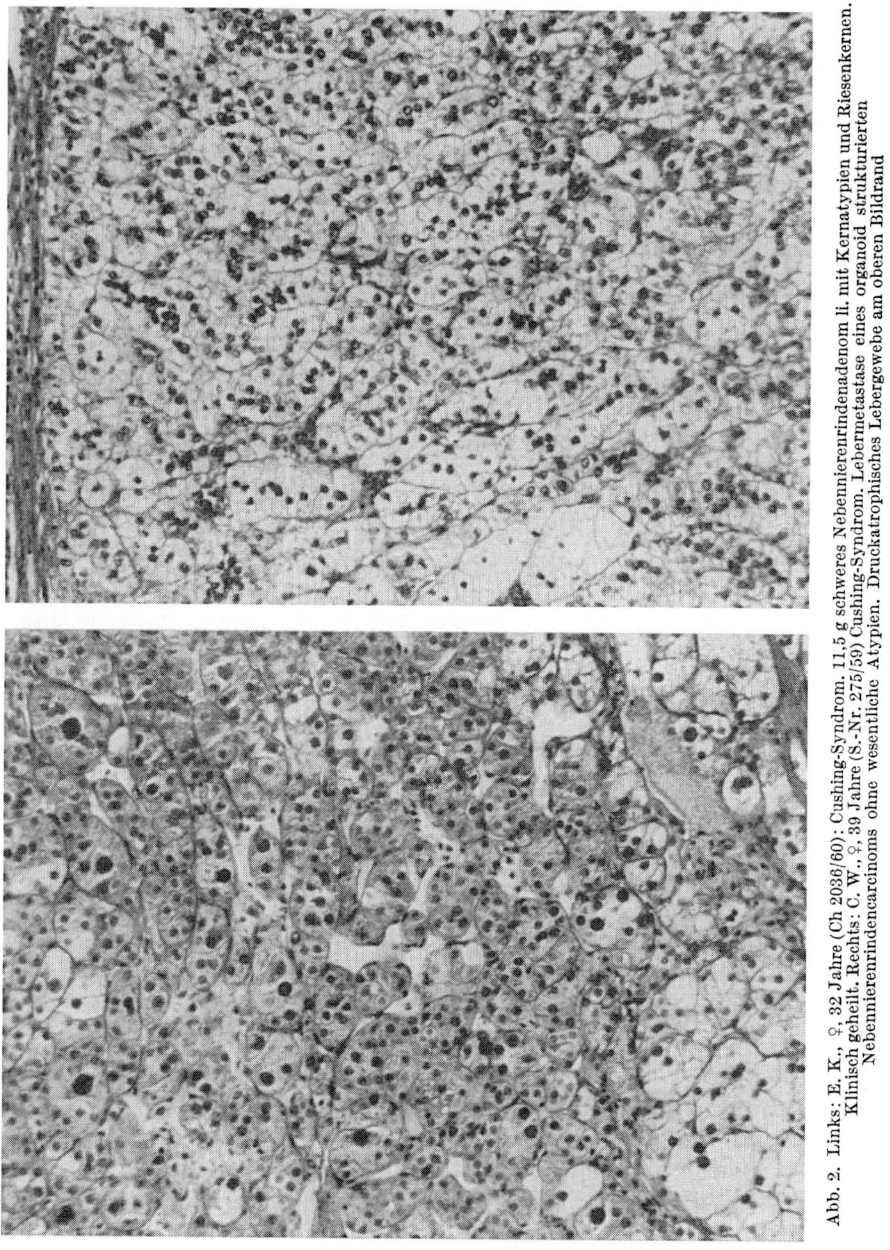

Abb. 2. Links: E. K., ♀, 32 Jahre (Ch 2036/60): Cushing-Syndrom. 11,5 g schweres Nebennierenrindenadenom li. mit Kernatypien und Riesenkernen. Klinisch geheilt. Rechts: C. W., ♀, 39 Jahre (S.-Nr. 275/59) Cushing-Syndrom. Lebermetastase eines organoid strukturierten Nebennierenrindencarcinoms ohne wesentliche Atypien. Druckatrophisches Lebergewebe am oberen Bildrand

Daten nehmen wir in diesem Fall den Übergang von M. Cushing in Cushing-Syndrom an und stellen ihm den in der Literatur mitgeteilten Fällen (*2, 3, 5, 8, 12*) zur Seite. Dieses Beispiel zeigt, daß die differentialdiagnostischen Schwierigkeiten in der Diagnose Nebennierenrindenadenom-Nebennierenrindencarcinom nicht unterschätzt werden dürfen. Sie können dann auftreten, wenn ein Carcinom organoid und wenig atypisch strukturiert ist. Das gleiche gilt für Adenome mit Kernatypien, die zwar morphologisch bereits den Verdacht auf Malignität erwecken sich aber auf Grund langfristiger Verläufe doch als gutartig erweisen (Abb. 2). Wie in der Diagnostik der Schilddrüsentumoren wird hieran die Unzulänglichkeit histologischer Malignitätskriterien deutlich.

Aus unseren Untersuchungen folgert, daß das Substrat der Nebennierenrinde beim Cushing-Symptomen-Komplex variabel ist und trotz vieler gesicherter Fakten noch eine Fülle offener Probleme in sich birgt, die der weiteren Bearbeitung bedürfen.

Literatur

1. BUCHEM, F. S. P. VAN, H. DOORENBOS, and H. S. ELINGA: Lancet **1956 II**, 335.
2. BOLINGER, R. E.: Ciba Found. Coll. Endocr. **12**, 150 (1958).
3. BRATRUD, T. E., and W. H. THOMPSON: Zit. n. HAMWI u. Mitarb.
4. CAHILL, G. F., M. M. MELICOW, and H. H. DARBY: Surg. Gynec. Obstet. **64**, 281 (1942).
5. CROOKE, H. C.: Ciba Found. Coll. Endocr. **12**, 149 (1958).
6. EMERSON, P. A.: Proc. roy. Soc. med. **56**, 165 (1963).
7. GRIFFITHS, K., J. K. GRANT, and T. SYMINGTON: J. clin. Endocr. **23**, 776 (1963).
8. HAMWI, G. J., R. M. SERBIN, and P. A. KRÜGER: New Engl. J. Med. **257**, 1153 (1957).
9. KRACHT, J.: Virchows Arch. path. Anat. **335**, 21 (1962).
10. —, u. J. TAMM: Virchows Arch. path. Anat. **333**, 1 (1960).
11. LANGERON, L., and P. LOBEAC: Ann. Med. **24**, 249 (1962).
12. LUKENS, F. D. W., and H. D. PALMER: Endocrinology **26**, 941 (1940).
13. SYMINGTON, T.: Brit. med. Bull. **18**, 117 (1962).
14. — A. R. CURRIE, V. J. O'DONNELL, J. K. GRANT, E. G. OASTLER, and W. G. WHYTE: Ciba Found Coll. Endocr. **12**, 102 (1958).
15. TONUTTI, E.: Verh. dtsch. Ges. Path. **36**, 123 (1953).
16. —, u. J. M. BAYER: Endokrinologie **41**, 17 (1961).
17. — —, u. W. SPIEGELHOFF: Endokrinologie **40**, 310 (1960).
18. WILLIAMS, R. H.: Textbook of Endocrinology, 3rd ed. Philadelphia and London: Saunders Co. 1962.

Diskussion

G. DHOM (Würzburg):

Die bilaterale kleinknotige Rindenadenomatose scheint ein besonders charakteristischer Befund beim kindlichen Cushing-Syndrom zu sein. Mir sind 5 entsprechende Beobachtungen aus der Literatur bekannt, die zweifellos hier einzuordnen sind, auch wenn die Autoren teilweise die Eigenart des Befundes nicht herausgestellt haben. Alle Fälle sind durch einen negativen ACTH-Test ausgezeichnet.

Aus dem Anatomischen Institut (Direktor: Prof. Dr. E. Tonutti) und der Chirurgischen
Klinik der Universität Bonn (Direktor: Prof. Dr. A. Gütgemann)

Kompensatorische Atrophie der Nebennierenrinde beim Conn-Syndrom

Von

E. Tonutti und J. M. Bayer

Bei 4 Fällen von primärem Aldosteronismus, denen jeweils ein Adenom der
Nebennierenrinde zugrunde lag, das vorwiegend aus Spongiocyten, ähnlich den
Zellen der Zona fasciculata aufgebaut war, zeigte die Nebennierenrinde außerhalb
des Tumorbereiches eine ausgesprochene regressive Transformation mit schmaler
Fasciculata, aber *breiter Zona glomerulosa* und reticularis. Dieser Befund deutet
auf eine verminderte ACTH-Stimulierung des Rindengewebes und findet seine
Erklärung in der Tatsache, daß die meisten Conn-Adenome keine ausschließlichen
Aldosteronbildner sind, sondern in wechselndem Ausmaße auch Corticosteron
und/oder Cortisol sezernieren.

Der Abfall der Aldosteronsekretion nach der Adenom-Extirpation, der bis auf
subnormale Werte erfolgen kann, läßt zwei eindeutige Schlußfolgerungen zu:

1. Die Aldosteronbildung kann auch von spongiocytär, nach Art der Zona
fasciculata differenzierten Zellen, aus denen die Adenome überwiegend bestehen,
erfolgen und kann daher nicht das ausschließliche Reservat von Zellen nach Art
der Zona glomerulosa sein.

2. Die Verbreiterung der Zona glomerulosa, die von manchen Autoren als
„Hyperplasie" gedeutet wurde, hat nichts mit der vermehrten Aldosteronbildung
beim Conn-Syndrom zu tun, da die Aldosteronbildung trotz des Verbleibs der ver-
breiterten Glomerulosa im Organismus nach der Adenomentfernung absinkt.

Die Verbreiterung der Zona glomerulosa ist lediglich Ausdruck verminderter
ACTH-Stimulierung des Rindengewebes, bedingt durch die gleichzeitige Gluco-
corticoidbildung der Connschen Adenome. Sie findet sich daher in gleicher Weise
bei Vorhandensein von Rindenadenomen, die ein Cushing-Syndrom auslösen und
kann als Zeichen der *„kompensatorischen Atrophie"* des Rindenorgans aufgefaßt
werden.

Ausführliche Darstellung und Schrifttumsangaben bei: Tonutti, E., u. J. M. Bayer:
Endokrinologie **45**, 276—295 (1963).

Diskussion

K. Zicha (Erlangen):

Wir hatten Gelegenheit, einen Fall zu beobachten, bei dem ein NNR-Carcinom links vorlag
mit dem klinischen Bild eines Cushing-Syndroms. Nach operativer Entfernung des Tumors
und Revision der „gesunden" Seite bildeten sich die Cushing-Symptome zurück. Ein Jahr

später kam die 18jährige Patientin mit einem Conn-Syndrom zur Wiederaufnahme. Es fand sich ein rechtsseitiger NN-Tumor, die Aldosteronausscheidung schwankte um 400 γ/24 Std. Die Patientin verstarb schließlich im Lebercoma an den Folgen von Lebermetastasen. Da auch beim 2. Klinikaufenthalt neben der vermehrten Aldosteronausscheidung die 17-OHCS in Plasma und Harn vermehrt waren, weist auch dieser Fall auf die von Ihnen erwähnten Zusammenhänge zwischen Conn-Syndrom und vermehrter Glucocorticoidproduktion hin.

Aus dem Anatomischen Institut der Universität Bonn
(Direktor: Prof. Dr. med. E. Tonutti)

Kann ACTH die cortisoninduzierte Insuffizienz des HVL-NNR-Systems beheben?

Von

M. Herrmann, J. Thomsen und F. Rentsch

Mit 2 Abbildungen

Die Wirkung von Cortison auf die NNR ist bekannt. Die erste Beschreibung der Atrophie von Ingle u. Mitarb. (*19, 20*) wurde mehrfach in der späteren Literatur (*3, 34*) bestätigt. Seit Einführung des Cortisons in die Therapie ist jedoch diese unerwünschte Begleiterscheinung ein Problem, das große Aufmerksamkeit verdient. Es ist auch heute noch keine befriedigende Möglichkeit bekannt, die NNR-Atrophie und die damit verbundene Insuffizienz der NNR und des HVL-NNR-Systems zu verhindern. Es lag nahe, ACTH, das Hormon des Hypophysenvorderlappens mit direkter Wirkung auf die NNR, zur Behandlung dieser Insuffizienzerscheinungen einzusetzen. So finden wir in der klinischen Literatur positive (*4, 9, 16, 28*) aber wesentlich mehr negative (*1, 2, 5, 6, 7, 8, 10, 24, 29, 35*) Ergebnisse dieser zusätzlichen Hormongabe. Ebenso widersprüchlich sind die Ergebnisse in Tierversuchen, wo einige Untersuchungen für (*16, 17, 26, 27*), zahlreiche aber gegen (*13, 14, 22, 23, 33, 38*) eine zusätzliche ACTH-Gabe sprechen.

Von jeder Substanz, die zu einer anhaltenden Verbesserung der durch die langfristige Cortisontherapie geschaffenen Verhältnisse verwendet werden soll, müssen zwei Eigenschaften verlangt werden:

1. Sie muß eine positive morphokinetische Wirkung auf die NNR haben.

2. Sie muß das Reaktionsvermögen des HVL-NNR-Systems und evtl. übergeordneter Zentren verbessern, was einschließt, daß eine Blockierung der Steuerungselemente bei ihrer Anwendung ausbleibt.

Unter diesen Aspekten haben wir uns erneut mit der Anwendung des ACTH zur Verhinderung der cortisoninduzierten Insuffizienz des HVL-NNR-Systems beschäftigt.

Die Untersuchungen wurden an 138 männlichen Meerschweinchen im Gewicht um 250 g ausgeführt, die in folgende Versuchsgruppen aufgeteilt wurden:
47 Tiere erhielten 28 Tage 5 mg Cortison/die i.m. (*12*).
56 Tiere erhielten 28 Tage 5 mg Cortison/die i.m. und 4. i.E. ACTH s.c./die (*37, 14*).
35 Tiere erhielten 28 Tage 4 i.E. ACTH s.c./die (*31*).
Bei allen Gruppen wurde nach Abschluß der Behandlung die 17-OHCS-Ausscheidung und das Verhalten der Kernvolumina der Zona fasciculata der NNR über einen längeren Zeitraum verfolgt. Ein Teil der Tiere der beiden erstgenannten Versuchsgruppen erhielt 24 Std vor dem Abtöten 5 d.l.m. Diphtherietoxin s.c.

Wie aus Abb. 1 hervorgeht, ist nach alleiniger, langfristiger Cortisonvorbehandlung zunächst die 17-OHCS-Ausscheidung infolge Abgabe des zugeführten Cortisons erhöht, fällt aber innerhalb 48 Std auf subnormale Werte ab. Erst am 13. Tag wird mit einer überschießenden Ausscheidung wieder die Norm erreicht. Nach alleiniger, langfristiger Vorbehandlung mit ACTH erfolgt innerhalb 24 Std nach Behandlungsende ein schneller Abfall der unter der Therapie stark erhöhten Ausscheidung auf Werte hypophysenloser Tiere. Innerhalb des Beobachtungszeitraumes von 21 Tagen wurde kein Ansteigen auf Normalwerte beobachtet. Bei

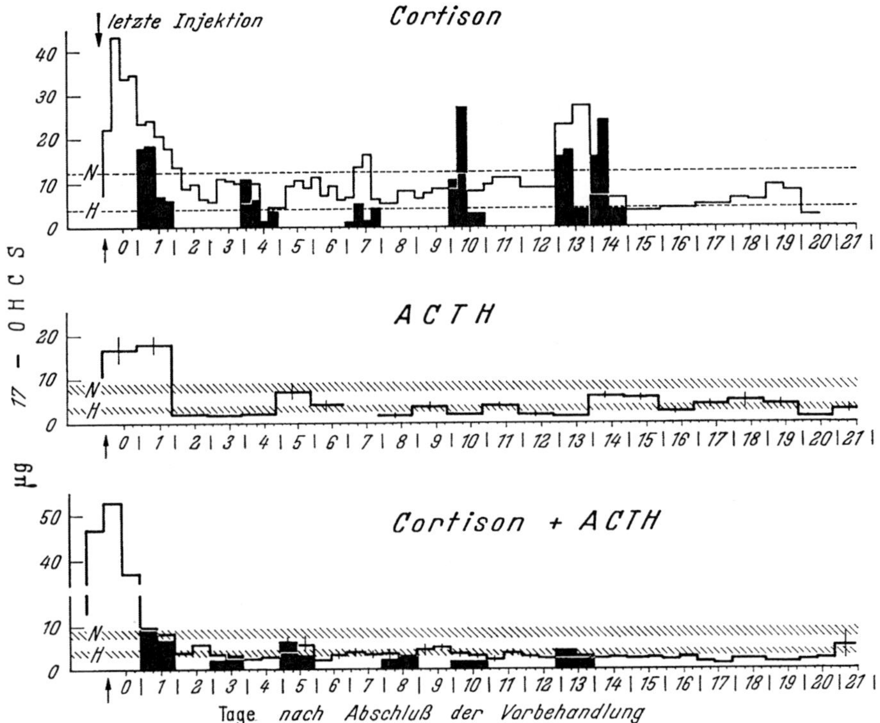

Abb. 1. Darstellung der 17-OHCS-Ausscheidung nach Abschluß einer 28 tägigen Behandlung mit Cortison, ACTH, Cortison und ACTH. Helle Säulen: Tiere ohne, dunkle Säulen: Tiere mit 5 d.l.m. Diphtherietoxinvergiftung. Durchschnittsbereiche (m ± ε): N = Normaltiere, H = hypophysenlose Tiere

kombinierter Behandlung mit Cortison und ACTH folgt nach Absetzen der Hormonzufuhr ebenfalls ein schneller Abfall auf subnormale Werte, der aber stärker ausgeprägt ist und über den Beobachtungszeitraum bis zum 21. Tag anhält.

In die Abb. 1 sind außerdem die Werte der 17-OHCS-Ausscheidung nach der Vergiftung mit 5 d. l. m. Diphtherietoxin eingetragen. Daraus ist ersichtlich, daß nach alleiniger Cortisonzufuhr erstmals am 9.—10. Tag nach Behandlungsende eine normal erhöhte Ausscheidung als Antwort auf den Stress der Diphtherietoxin-Vergiftung auftritt. Bei den einer kombinierten Behandlung mit Cortison und ACTH unterworfenen Tieren ist bis zum 13. Tag nach Abschluß der Hormonzufuhr keine Reaktion der 17-OHCS-Ausscheidung nach Diphtherietoxin festzustellen.

In Abb. 2 ist das Verhalten der Kernvolumina der Zona fasciculata aller Versuchsgruppen nach Abschluß der Vorbehandlung dargestellt. Nach langfristiger

Cortisonzufuhr entsprechen die Kernvolumina zunächst den Werten hypophysenloser Tiere. Erst am 10. Tag nach Behandlungsende wird der Normalwert erreicht,
am 13. Tag werden erheblich über die Norm hinausgehende Kernmeßwerte ermittelt. Die Behandlung mit ACTH hat bei Abschluß der Behandlung eine deutliche
Zunahme der Kerngröße bewirkt, jedoch gehen die Kernvolumina innerhalb weniger Tage auf den Normalwert und danach sogar auf subnormale Werte
zurück. Erst am 15. Tag werden wieder normale Kernwerte festgestellt. Nach

kombinierter Behandlung
mit Cortison und ACTH
findet man zunächst eben-
falls gegenüber der Norm
erhöhte Kernvolumina, je-
doch werden schon am
4. Tag Werte im Bereich
hypophysenloser Tiere ge-
messen. Bis zum Ende der
Beobachtungsperiode am
22. Tag nach Abschluß der
Vorbehandlung ist der Normalwert noch nicht erreicht.

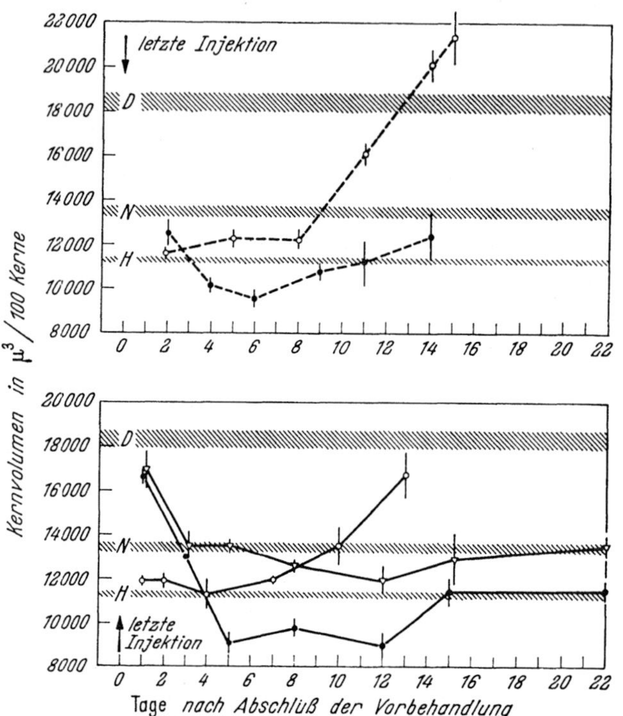

Abb. 2. Darstellung der Kernvolumina der Zona fasciculata der Nebennierenrinde nach langfristiger Behandlung mit Cortison (O—O), ACTH
(▽—▽) und Cortison + ACTH (●—●), sowie nach zusätzlicher Diphtherietoxinvergiftung nach langfristiger Behandlung mit Cortison
(O---O) und Cortison + ACTH (●----●). Durchschnittsbereiche
(m ± ε) : N = Normaltiere, H = hypophysenlose Tiere, D = Normaltiere mit 5 d.l.m. Diphtherietoxin s.c. vergiftet. Die senkrechten Linien
durch die einzelnen Punkte geben den mittleren Fehler (ε) des
Mittelwertes an

Das Verhalten des Kern-
volumens unter der Diph-
therietoxin-Vergiftung,
nach Abschluß der Vorbe-
handlung, ist ebenfalls in
Abb. 2 wiedergegeben. Während nach alleiniger Cortisonbehandlung das Kern-
volumen am 11. Tag mit
einer adäquaten Zunahme
reagiert, ist eine solche
Reaktion nach kombinier-
ter ACTH-Cortison-Behand-
lung bis zum 14. Tag nicht
festzustellen.

Die hämorrhagische Ne-
krose der NNR, die nach

Tonutti (36) streng ACTHabhängig ist, tritt nach alleiniger Cortisonvorbehandlung erstmalig am 8. Tag auf, bei kombinierter ACTH-Cortison-Behandlung war
sie bis zum 14. Tag nach Absetzen der beiden Hormone nicht zu finden.

Aus den geschilderten Ergebnissen geht hervor, daß alleinige ACTH-Zufuhr
zwar den bekannten morphokinetischen und sekretionsfördernden Effekt auf die
NNR während der Zufuhr ausübt. Kurz nach Behandlungsende sind die Auswirkungen des ACTH abgeklungen, und eine morphologisch wie sekretorisch nachweisbare Insuffizienz des Rindenorgans tritt ein, die über einen längeren Zeitraum
verfolgt werden kann. Erwartungsgemäß findet man bei zusätzlicher ACTH-
Therapie zur Cortison-Medikation zum Zeitpunkt des Behandlungsabbruches eine

entfaltete Nebennierenrinde mit erhöhtem Kernvolumen der Fasciculatazellen und eine erhöhte 17-OHCS-Ausscheidung im Harn. Die Atrophie der Nebennierenrinde und der Abfall der 17-OHCS-Ausscheidung auf subnormale Werte tritt erst im Verlauf weniger Tage *nach* Behandlungsende ein, um dann aber wesentlich länger als nach alleiniger Cortisonbehandlung bestehen zu bleiben. Innerhalb dieses Zeitraumes der NNR-Insuffizienz bleibt auch der Eintritt der hämorrhagischenNekrose der NNR auf Diphtherietoxin aus, was auf eine mangelhafte ACTH-Ausschüttung aus dem HVL schließen läßt. Dieses Verhalten möchten wir in Übereinstimmung mit den in der Literatur gemachten Angaben (*9, 10, 11, 15, 21, 23, 29, 30, 32, 33, 37*) mit einer noch stärkeren Blockierung der Steuerung des HVL-NNR-Systems im HVL und auch der übergeordneten Regulationszentren, die vorzugsweise im Hypothalamus vermutet werden, erklären.

Die tierexperimentellen Ergebnisse und auch die bekannten klinischen Ergebnisse dieser Behandlung zeigen eine so weitgehende Übereinstimmung in den negativen Auswirkungen, daß es nahe liegt, diese Therapie zur Vermeidung der unter langfristiger Cortisontherapie auftretenden NNR-Insuffizienz aufzugeben.

Literatur

1. AMATRUDA, T. T. jr., D. R. HOLLINGSWORTH, N. D. D'ESOPO, G. V. UPTON and P. K. BONDY: Clin. Res. **6**, 253 (1958).
2. — — — — — J. clin. Endocr. **20**, 339—354 (1960).
3. BENNETT, H. S.: Anat. Rec. **76**, Suppl. 2, 5 (1940).
4. BIRKE, G., G. B. DONEJ, B. OLHAGEN u. L. PLANTIN: Acta med. scand. **155**, 245—252 (1956).
5. BONDY, P. K., D. HOLLINGSWORTH, N. D. D'ESOPO, and G. V. UPTON: Clin. Res. **6**, 26—27 (1958).
6. CARREON, G. G., J. J. CANARY, R. J. MEYER, and L. H. KYLE: J. Lab. clin. Med. **56**, 235—244 (1960).
7. CHRISTY, N. P., E. Z. WALLACE, and J. W. JAILER: J. clin. Endocr. **16**, 1059—1074 (1956).
8. CRABBÉ, J., et J. W. MEAKIN: Ann. Endocr. (Paris) **22**, 576—581 (1961).
9. FERRIMAN, D., u. B. PAGE: Lancet **1960 II**, 410—411.
10. GEYER, G.: Wien. klin. Wschr. **72**, 293—301 (1960).
11. HERRMANN, M.: Endokrinologie **43**, 155—166 (1962).
12. —, u. G. WINKLER: Acta neuroveg. (Wien) **20**, 38—49 (1959).
13. — — Arzneimittel-Forsch. **12**, 720 (1962).
14. — J. THOMSEN u. G. WINKLER: Arzneimittel-Forsch. **14**, 203—204 (1964).
15. HODGES, J. R., and J. VERNIKOS: Nature (Lond.) **182**, 725 (1958).
16. HOHLWEG, W., u. U. LASCHET: Acta endocr. (Kbh.) **32**, 437—447 (1959).
17. — — u. CH. CZECZATKA: Naturwissenschaften **45**, 368 (1958).
18. HOLUB, D. A., J. W. JAILER, J. I. KITAY, and A. G. FRANTZ: J. clin. Endocr. **19**, 1540 (1959).
19. INGLE, D. J., G. M. HIGGINS, and E. C. KENDALL: Anat. Rec. **71**, 363—372 (1938).
20. —, and E. C. KENDALL: Science **86**, 245 (1937).
21. KITAY, J. I., D. A. HOLUB, and J. W. JAILER: Fed. Proc. **17**, 87 (1958).
22. KRACHT, J.: Naturwissenschaften **45**, 61 (1958).
23. — Medizinische **3**, 106—109 (1959).
24. KYLE, L. H., R. J. MEYER, and J. J. CANARY: New Engl. J. Med. **57**, 257—261 (1957).
25. LANDING, B. H., and D. FERIOZI: J. clin. Endocr. **14**, 910—921 (1954).
26. LASCHET, U., W. HOHLWEG u. D. BILZ: Symp. Dtsch. Ges. Endokr. **7**, 285—289 (1961).
27. — — u. CH. CZECZATKA: Endokrinologie **37**, 293—301 (1959).
28. MEAKIN, W. J., M. S. TANTONGCO, J. CRABBÉ, T. B. BAYLES, and D. H. NELSON: Amer. J. Med. **29**, 459—464 (1960).

29. Paris, J.: Proc. Mayo Clin. **36**, 305—317 (1961).
30. Plager, J. E., and P. Cushman: J. clin. Endocr. **22**, 147 (1962).
31. Rentsch, F.: (in Vorbereitung).
32. Salassa, R. M., W. A. Bennett, F. R. Keating jr., and R. G. Sprague: J. Amer. med. Ass. **152**, 1509—1515 (1953).
33. Schreiner, H.: Derm. Wschr. **133**, 614—615 (1956).
34. Selye, H., u. Ch. Dosne: Proc. Soc. exp. Biol. (N. Y.) **44**, 165 (1940).
35. Smyth, C. J., J. J. Bunim, W. S. Clark, D. C. Crain, F. E. Demartini, I. F. Duff, E. P. Engleman, D. C. Graham, M. M. Montgomery, B. M. Norcross, H. F. Polley, M. W. Popes, and E. F. Rosenberg: Ann. int. Med. **50**, 366—494 (1959).
36. Tonutti, E.: Verh. dtsch. Ges. Path. **36**, 123—158 (1953).
37. Thomsen, J.: Endokrinologie **45**, 296—306 (1963).
38. Vasama, Ra., L. Kalliomäki, Ri. Vasama, and E. Näätänen: Ann. med. exp. Fenn. **35**, Suppl. 1—12 (1957), 1—17.

Aus der Medizinischen Poliklinik der Universität des Saarlandes
(Direktor: Prof. Dr. H. P. Wolff)

Zum Wirkungsmechanismus von Cortisol und einiger seiner synthetischen Abkömmlinge am Fettgewebe in vitro

Von

G. Löffler, H. Schneider und K. F. Weinges

Mit 2 Abbildungen

Eigene Untersuchungen über die insulinähnliche Aktivität im Plasma bei Patienten mit Morbus Cushing hatten uns zu der Vorstellung geführt, daß Cortisol nicht nur, wie bisher vermutet wurde, über eine gesteigerte Gluconeogenese in der Leber mit erhöhtem Glucoseangebot diabetogen wirkt, sondern dieses Hormon auch eine spezifisch den Insulineffekt hemmende Wirkung auf den Glucosestoffwechsel peripherer Gewebe haben muß (*1*). Wir untersuchten daher u. a. den Glucoseumsatz im Fettgewebe unter dem Einfluß von Cortisol.

Es genügen bereits Mengen von 0,005 μg Cortisol/ml, um einen deutlich hemmenden Effekt auf die Glucoseaufnahme des isolierten Fettgewebes nachweisen zu können, die Maximalwirkung ist schon mit 0,5 μg/ml erreicht (Abb. 1a). Während die Oxydation des radioaktiv markierten 6. C-Atoms der Glucose zu CO_2 ähnlich beeinträchtigt wird, erfährt die Oxydation des ersten C-Atoms demgegenüber eine zusätzliche, statistisch signifikante Hemmung. Auffallend war dabei,

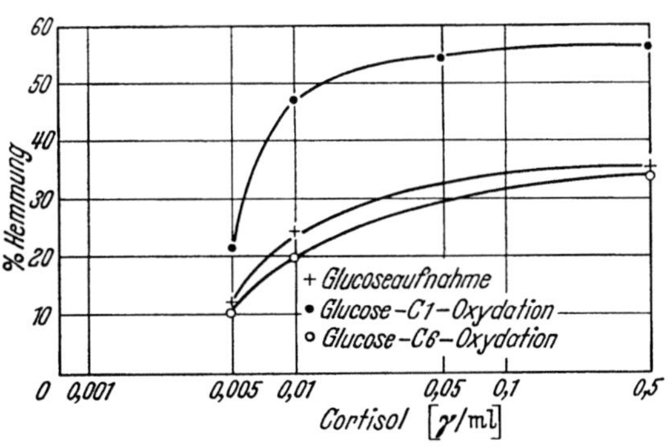

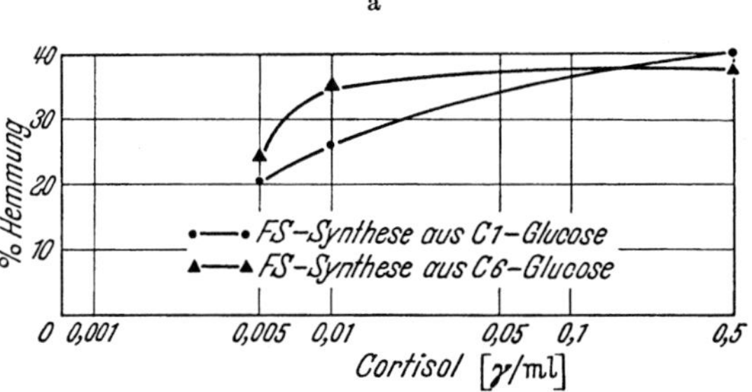

Abb. 1a u. b. Hemmung des Glucoseumsatzes des Fettgewebes und der FS-Synthese des Fettgewebes durch Cortisol in vitro. Jeder Punkt stellt den Mittelwert aus 8 Einzelbestimmungen dar. (Die Werte für C1-Oxydation liegen statistisch signifikant höher als die der C6-Oxydation (P = < 0,01). Es besteht keine statistische Signifikanz zwischen Fettsäuresynthese aus C1- und C6-markierter Glucose)

daß bis zum Eintritt der Cortisolwirkung eine Inkubationszeit von mindestens 4 Std notwendig ist, was in gutem Einklang mit der verzögerten Wirkung von

Cortisol auf die Glykogensynthese der Leber steht. Insulinkonzentrationen zwischen 100 und 1000 μE/ml waren nicht imstande, die Wirkung von Cortisol aufzuheben.

Bei der engen Verknüpfung zwischen Kohlenhydrat- und Fettstoffwechsel im Fettgewebe lag es natürlich nahe, Änderungen des Glucoseumsatzes mit solchen des Lipidstoffwechsels in Beziehung zu bringen (Abb. 1b). Untersuchungen der Fettsäuresynthese aus C 1- bzw. C 6-markierter Glucose unter Einwirkung gleicher Konzentrationen von Cortisol zeigten ebenfalls eine deutliche Hemmung der Fettsäuresynthese, die sich aber hier gleichmäßig auf den Einbau des 1. und 6. C-Atoms der Glucose erstreckte. Die gleichzeitig bestimmte Freisetzung von nicht veresterten Fettsäuren durch das Fettgewebe unterschied sich nicht signifikant von dem Leerwert. Dieser Befund wirft die grundsätzliche Frage auf, ob Cortisol in die Wechselbeziehungen zwischen Kohlenhydrat- und Fettstoffwechsel eingreift.

Glucose kann vom Organismus einmal über den klassischen Glykolysemechanismus nach Emden und Mayerhof abgebaut werden, zum anderen aber auch über den Pentosephosphatcyclus laufen, in dessen ersten Schritten es zu einer Decarboxylierung am ersten C-Atom der Glucose und zur Bildung von 2 TPNH kommt. Verknüpfungen zum Lipidstoffwechsel bestehen einmal auf der Stufe der Triosephosphate, aus denen das zur Synthese der Glycerinester notwendige Glycerin-1-Phosphat hervorgeht, zum anderen in der oxydativen Decarboxylierung von Pyruvat zu Acetyl-CoA, das für die Synthese der Fettsäuren eine zentrale Stellung einnimmt.

Die von uns erhobenen Befunde zeigen einmal eine deutliche Hemmung der Glucoseaufnahme des isolierten Fettgewebs durch Cortisol. Diese beruht nach vorläufigen Untersuchungen hauptsächlich auf einer Beeinträchtigung der Phosphorylierung von Glucose zu Glucose-6-Phosphat. Über entsprechende Ergebnisse am Zwerchfell haben Kipnis u. Mitarb. (2) berichtet. Sie geben aber außerdem Anlaß zu der Frage, ob nicht Verschiebungen im Verhältnis von Glucoseabbau über den Emden-Mayerhof-Weg oder Pentosephosphatcyclus zu Änderungen im Lipidstoffwechsel führen können, da ja möglicherweise das während des Pentosephosphatcyclus gebildete TPNH zur Fettsäuresynthese benötigt wird. Erklärlicherweise liefert der Vergleich von C-6-Glucose- und C-1-Glucose-Oxydation zur Bewertung des relativen Anteils der Glykolyse am Gesamtglucoseumsatz keine zuverlässigen Werte, da die Carboxylierungsvorgänge im Organismus mit Einbau von CO_2 in Zellbestandteile nicht berücksichtigt werden. Etwas zuverlässiger ist zu diesem

Tabelle 1. *Prozentualer Anteil der Glykolyse am Gesamt-Glucoseumsatz, berechnet aus der Fettsäuresynthese sowie Verhältnis von C6-Oxydation zur C1-Oxydation. Jeder Wert stellt den Mittelwert aus 8 Einzeluntersuchungen dar*

	leer	Cortisol (γ/ml)		
		0,005	0,01	0,5
Glykolyse in % des Glucoseumsatzes . . . . (ber. aus FS-Synthese)	53,6	53,2	49,2	55,2
$\dfrac{CO_{2C6}}{CO_{2C1}}$	1,04	1,14	1,42	1,42

Zweck die Bestimmung der Fettsäuresynthese aus C 1- bzw. C 6-markierter Glucose. Hier wird von der Vorstellung ausgegangen, daß der in die Fettsäurefraktion eingebaute Anteil von C 1-Glucose nicht über den Pentosephosphatcyclus gegangen sein kann, da es hier ja zu einer Decarboxylierung an C 1 kommt.

Unter unseren Versuchsbedingungen liegt der Anteil der Glykolyse bei 50%, er ändert sich auch unter Cortisol nicht. Der Abbau der Glucose über den Pentosephosphatcyclus erfährt offenbar durch Cortisol keine Änderung (Tab. 1). In der unteren Spalte der Tab. 1 sind die Verhältnisse von C6-Oxydation zu C1-Oxydation dargestellt. Dieses Verhältnis steigt unter Cortisol von etwa 1 auf 1,4 . Die Interpretation dieses Ergebnisses bereitet zunächst einige Schwierigkeiten.

Bei alleiniger Beschreitung des Glucoseabbaues über die Glykolyse mit anschließender, teilweiser Oxydation im Citronensäurecyclus ist ein Verhältnis von C6-Oxydation zu C1-Oxydation von 1 zu erwarten. Mit zunehmendem Anteil des Pentosephosphatcyclus am Glucoseabbau müßte relativ mehr CO_2 vom ersten C-Atom der Glucose anfallen und das Verhältnis infolgedessen unter 1 sinken. Das von uns unter Cortisol gefundene Verhältnis von 1,4 läßt den Schluß zu, daß das erste C-Atom der Glucose relativ vermehrt in gewisse Zellbestandteile eingebaut wird. Im Falle des Fettgewebes liegt es natürlich nahe, einen gesteigerten Einbau in die Glycerinfraktion der Triglyceride zu vermuten.

Für einen derartigen Mechanismus müssen 2 Voraussetzungen gemacht werden: Einmal muß die Geschwindigkeit der Äquilibrierung zwischen den 2 Triosephosphaten, Phospo-Glycerinaldehyd und Dioxyaceton-Phosphat, relativ gering im Vergleich zur Durchsatzrate der Glykolyseintermediate sein. Daß dies unter gewissen Voraussetzungen vorkommen kann, ist zwar nicht für das Fettgewebe, wohl aber für die Leber bewiesen (3). Zum zweiten müssen Mechanismen auftreten, die für die erhöhte Umsetzung der Triosen in Richtung Glycerin-1-Phosphat verantwortlich gemacht werden können. Im Falle der gesteigerten Gluconeogenese unter Cortisol in der Leber ist von vielen Autoren die Induktion von Enzymen, die für die Gluconeogenese verantwortlich sind, beschrieben worden. Wir untersuchten infolgedessen die Aktivität von Triosephosphatisomerase und Glycerophosphat-Dehydrogenase im Fettgewebe unter Cortisoleinfluß nach 5stündiger Inkubation. Es ergab sich, daß die Aktivität beider Enzyme unter Cortisol signifikant anstieg. Dieser Befund legt nahe, daß es unter Einwirkung des Hormones zu einem relativ vermehrten Angebot von Glycerin-1-Phosphat kommt, das als Acceptor für unveresterte Fettsäure in der Zelle dient und damit, trotz herabgesetztem Glucoseumsatz, eine verstärkte Lipolyse verhindert. Die durchgeführten Untersuchungen weisen darauf hin, daß Cortisol nicht nur den Glucoseumsatz am Fettgewebe hemmt, sondern wahrscheinlich auch in die Wechselbeziehungen zwischen Glucose- und Fettstoffwechsel eingreift.

Von klinischem Interesse war, welchen Einfluß synthetische Abkömmlinge des Cortisols bei diesen in vitro Untersuchungen zeigen, die sich klinisch hinsichtlich der diabetogenen Wirkung wesentlich vom Cortisol unterscheiden. Wir haben deshalb gleiche Untersuchungen mit 16-Methylenprednisolon, 6-Methylprednisolon, Dexamethason und Paramethason durchgeführt.

Die Ergebnisse sind in Abb. 2 dargestellt. Es fällt sofort auf, daß nur Cortisol den vorhin beschriebenen, besonders starken Hemmeffekt auf die C1-Oxydation von Glucose hat, während alle anderen Steroide erheblich niedriger liegen. Dagegen zeigt Cortisol bei den angewandten Konzentrationen keinen signifikanten Effekt auf die Freisetzung von unveresterten Fettsäuren und steht damit wieder im Gegensatz zu den übrigen untersuchten Hormonen, die hier zu erheblicher Lipolyse führen. Ähnliche Befunde wurden auch von Fain, Scow u. Chernick erhoben (4).

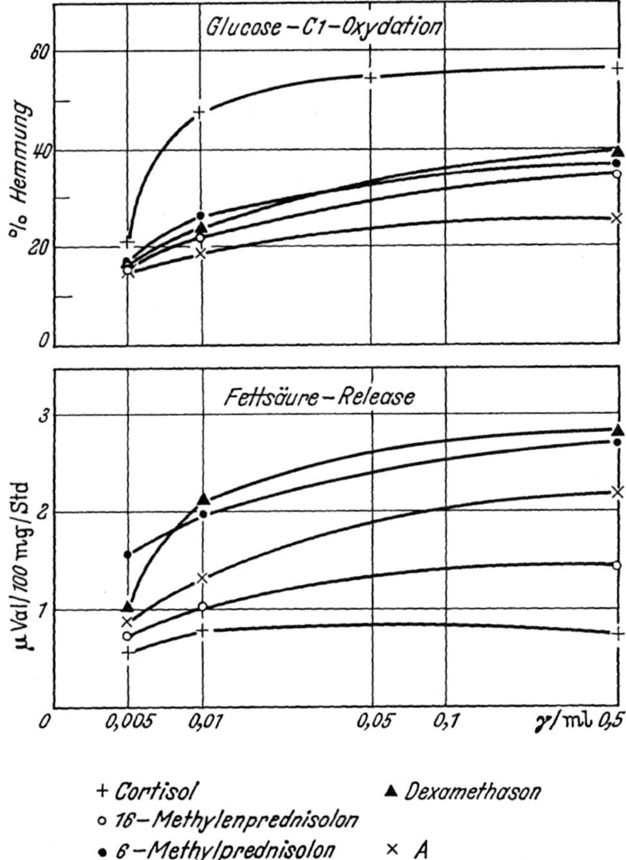

Abb. 2. Wirkung einiger Steroide auf Glucose-C1-Oxydation und Fettsäurerelease (A = Paramethason-Gruenenthal-). Jeder Punkt stellt den Mittelwert aus 8 Einzelbestimmungen dar. Die Hemmung der C1-Oxydation durch Cortisol unterscheidet sich signifikant von der Hemmung durch die anderen untersuchten Steroide (P = < 0,01). Die Steigerung des Fettsäurerelease durch Cortisol liegt statistisch signifikant tiefer wie sie durch die anderen Steroide (P = > 0,02—0,01)

Der besonders starke Fettsäurerelease könnte dadurch erklärt werden, daß die synthetischen Steroide nicht wie das Cortisol zu einem vermehrten Angebot an Glycerin-1-Phosphat führen. Diese Vermutung wird durch die Messung der Enzymaktivitäten von Triosephosphatisomerase und Glycerophosphatdehydrogenase bestärkt. Dabei war eine Aktivitätssteigerung der Enzyme, wie sie unter Cortisol gefunden wurde, unter den Einfluß der synthetischen Steroide (hier nur Dexamethason angeführt) nicht nachzuweisen.

Die klinischen Beobachtungen eines deutlich geringeren diabetogenen Effektes der synthetischen und vom Cortisol chemisch differenten Steroide kommen auch bei den experimentellen Untersuchungen zum Ausdruck. Es zeigte sich, daß die weitaus schwächere, hemmende Wirkung letzterer auf die ^{14}C-1-Glucoseoxydation mit verhältnismäßig kleinen Insulinkonzentrationen weiter zu verringern ist (3).

Literatur

1. SCHWARZ, R., K. F. WEINGES, K. P. EYMER u. K. KOPETZ: Verh. dtsch. Ges. inn. Med. 68, 289 (1962).
2. KIPNIS, D. M.: pers. Mitteilung.
3. ROSE, J. A., R. KELLERMEYER, R. STJERNHOLM, and H. G. WOOD: J. biol. Chem. 237, 3325, (1962).
4. FAIN, J. N., R. O. SCOW, and S. S. CHERNICK: J. biol. Chem. 238, 54 (1963).

Die Arbeit wurde mit Mitteln der Deutschen Forschungsgemeinschaft durchgeführt.

Wir danken für die Überlassung ausreichender Versuchsmengen von 16-Methylenprednisolon der Firma Merck, von 6-Methylprednisolon der Firma Hoechst, von Dexamethason der Firma Schering und von Paramethason der Firma Gruenenthal.

Aus der Medizinischen Poliklinik der Universität des Saarlandes
(Direktor: Prof. Dr. H. P. WOLFF)

Versuche zur Biogenese der Corticosteroide

Von

D. LOMMER, H. G. STEINACKER, M. KÖHLER, G. LÖFFLER, H. LIEBAU und
H. P. WOLFF[1]

Mit 1 Abbildung

Seit einiger Zeit befassen wir uns mit Studien über die Wirkungsweise von physiologischen und synthetischen Aktivator- und Hemmsubstanzen der Corticosteroidbiosynthese.[2] Wir verwenden dabei nebeneinander zwei verschiedene Synthesemodelle:

Die in vitro Inkubation von Nebennierenrindengewebe und die in vivo Perfusion von Nebennieren mit ^{14}C-Progesteron. Die einzelnen, aus Progesteron gebildeten, ^{14}C-markierten Corticosteroide werden durch Messung der Radioaktivität quantitativ bestimmt. Durch Zusatz der zu analysierenden Hormone in ^{3}H-markierter Form und bekannter Radioaktivität können Aufarbeitungsverluste exakt ermittelt werden.

Bei den in vitro Versuchen dienen Nebennierenrinden frisch geschlachteter Rinder als Synthesematerial, als Inkubationsmedium dient Plasma vom gleichen Tier. Die Proben werden während 2stündiger Inkubation bei 37°C mit Sauerstoff begast, dem zur Aufrechterhaltung des Plasma-pH 5% Kohlendioxyd beigemischt sind. Nach Beendigung des Versuches werden die Steroide aus dem Plasma extrahiert.

Für die in vivo Perfusion verwenden wir Hunde als Versuchstiere. Die Tiere werden etwa 3 Wochen vor dem Versuch einseitig nephrektomiert und adrenalektomiert. Bei entsprechender Fragestellung kann der Hund 2—3 Std vor dem Versuch hypophysektomiert werden. Vor dem Versuch wird die zweite Niere entfernt, der Abfluß der Nebenniere in die Vena cava unterbunden und ein PVC-Katheter in die Nebennierenvene gelegt. Das Radioprogesteron wird über die gesamte Versuchsdauer in eine der vorderen Pfotenvenen infundiert. Nach 10 min Vorinfusion beginnt die Sammlung des Nebennierenvenenblutes in 15 min-Portionen. Die Steroide werden aus dem Plasma extrahiert.

---

[1] Die in vitro Versuche wurden zusammen mit L. GÖLDEL und W. ZIMMERMANN an der 1. Medizinischen Klinik der Universität München durchgeführt. Für die Ausführung der Hypophysektomien an Hunden danken wir Herrn Dr. H.-D. HERRMANN von der Neurochirurgischen Klinik der Universität des Saarlandes.

[2] Verwendete Abkürzungen: PROG = Progesteron, DOC = Cortexon, B = Corticosteron, A = 11-Dehydrocorticosteron, ALD = Aldosteron, OH-P = 17 α-OH-Progesteron, S = 17 α-OH-Cortexon, F = Cortison, E = Corticosteron.

Aufarbeitung und Analyse der Proben folgen bei in vitro- und in vivo-Versuchen dem gleichen Schema. Nach Zugabe der zu analysierenden Steroide in tritiummarkierter Form werden die Plasmaproben mit Aceton, Äthylacetat und Chloroform extrahiert und die vereinigten Extrakte nach Trocknung über Natriumsulfat im Vakuum eingedampft. Zur Abtrennung fettiger Plasmabestandteile folgt eine Flüssigkeitsverteilung zwischen Heptan und 90%igem Methanol und anschließend Säulenchromatographie auf Kieselgel.

Die einzelnen Steroide werden in einer Kombination papierchromatographischer Systeme vom Bush-, Zaffaroni- und Mattox-Typ isoliert. Nach Vorabtrennung von PROG im System Bush A kann dieses durch Rechromatographie mit Wasser im 12Std-Durchlauf senkrecht zur ersten Laufrichtung isoliert werden. DOC und OH—P trennen sich von den hochpolaren Steroiden durch 15Std-Durchlaufchromatographie senkrecht zur ersten Laufrichtung im System Bush A. Die hochpolaren Steroide werden in zweidimensionaler Chromatographie in den Systemen Chloroform-Hexan und Formamid-Butylacetat-Wasser auf Formamid-imprägniertem Papier isoliert. Die beiden Steroide mittlerer Polarität, DOC und OH-P erhält man durch einfachen Lauf im System Formamid-Butylacetat-Wasser. Damit sind also isoliert: PROG, ALD, F, E, A, B, S, DOC und OH—P.

Aus unserem Präkursor Progesteron werden in der Nebennierenrinde durch eine Reihe von Hydroxylierungs- und Dehydrierungsreaktionen die Corticosteroide gebildet. Neun wesentliche Substanzen der Synthesekette, nämlich Progesteron, Cortexon, Corticosteron, 11-Dehydrocorticosteron, Aldosteron, 17 α-OH-Progesteron, 17 α-OH-Cortexon, Cortisol und Cortison lassen sich in unserem Analysensystem erfassen. Wir haben damit die Möglichkeit, durch Vergleich der jeweils vor und nach einer Reaktion stehenden Steroide die einzelnen Syntheseschritte zu kontrollieren. So können beispielsweise die Angriffspunkte von Aktivator- und Hemmsubstanzen in der Synthesekette ziemlich exakt ermittelt werden.

Ergebnisse einiger in vitro Experimente, die als Beispiele für die praktische Verwendbarkeit der Methode dienen mögen:

Da die einzelnen Hydroxylierungsreaktionen der Biosynthese mit molekularen Sauerstoff ablaufen, sollte Sauerstoffmangel eine beträchtliche Synthesehemmung zur Folge haben. Ein entsprechendes Experiment ergab, daß tatsächlich alle Hormone gegenüber den Kontrollproben in verminderten Mengen gebildet werden (Abb. 1a). Das Bemerkenswerte an diesem Versuch ist, daß sich die Hemmung nicht auf alle Steroide gleich stark auswirkt. Die stärkste Depression erhielten wir bei den direkten 11 β-Hydroxylierungsprodukten B und F. Die Enzyme der 11 β-Hydroxylierung sind in den Mitochondrien lokalisiert, während man die übrigen Hydroxylierungssysteme in den Mikrosomen vermutet. Wahrscheinlich findet unser Befund seine Erklärung in dieser Sonderstellung der 11 β-Hydroxylierung.

Ein anderes sehr eindrucksvolles Beispiel für die spezifische Hemmung der 11 β-Hydroxylierung gibt ein in vitro-Versuch mit Metopiron (Abb. 1b). Dem Anstau von DOC und S um etwa 100% steht eine minimale Bildung von B und F gegenüber.

Nach FARRELL bildet die Pinealisdrüse ein Carbolin mit spezifisch aldosteronotroper Wirkung, das Adrenoglomerulotropin, und ein sog. Anticorticotropin mit hemmendem Einfluß auf die Corticosteroidsynthese. Bei Versuchen mit decerebrierten Hunden konnte FARRELL nach Verabreichung von Pinealisextrakten eine

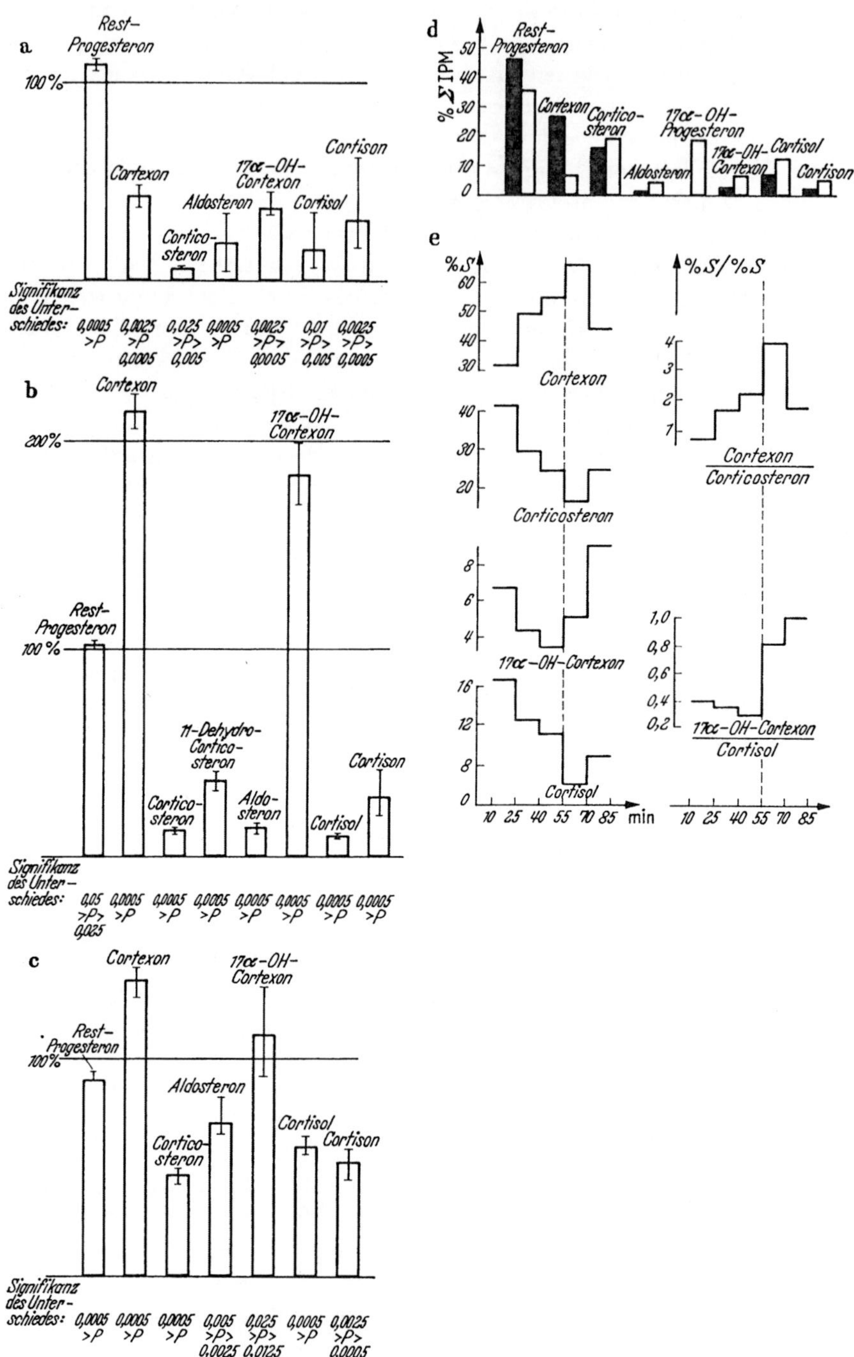

Abb. 1. Versuchsergebnisse. a, b, c: Ergebnisse aus in vitro Experimenten; d, e: Ergebnisse aus in vivo Experimenten; e: Untersuchung der Wirkungsweise eines Nebennierenrindenblockers. Die Hemmsubstanz wurde zum Zeitpunkt 55 min i.v. verabreicht. Zu d: % Σ Ipm = relative Konzentration = prozentualer Anteil an der Summe der Radioaktivität aller analysierten CS pro ml Plasma. Zu e: % S = relative Konzentration = prozentualer Anteil an der Summe der Radioaktivität aller analysierten CS ohne Rest-Progesteron pro ml Plasma

Steigerung der Aldosteronsekretion beobachten. Wir untersuchten in vitro die Wirkung eines entsprechenden Extraktes aus Pinealisdrüsen von Rindern. Eine Stimulierung der Aldosteronproduktion konnte von uns nicht beobachtet werden (Abb. 1c). Dies steht sinngemäß in Übereinstimmung mit den Befunden von Davis, die auch bei pinealektomierten und dekapitierten Hunden nach Aderlaß die übliche Zunahme der Aldosteronproduktion fand. Der von uns gleichzeitig beobachtete Anstau von DOC und S, sowie die verminderte Bildung von B und F, liefern das typische Bild einer selektiven Blockierung der 11 β-Hydroxylierung.

Befunde aus in vivo Experimenten:

In Abb. 1d ist die Verteilung der Corticosteroide im Nebennierenvenenblut von Hunden dargestellt. Die verhältnismäßig uniforme Verteilung der Aktivität auf die einzelnen Hormone deutet darauf hin, daß bei Progesteron als Präkursor die hypophysäre Steuerung der Nebennierenrinde nicht zur Wirkung kommen kann. Der Vergleich der Ergebnisse beim normalen und beim hypophysektomierten Tier bestärkt diese Annahme, die mit den bisherigen Kenntnissen von der ACTH-Wirkung übereinstimmt.

In einem anderen Perfusionsexperiment versuchten wir, den Angriffspunkt eines neuen Nebennierenrindenblockers zu ermitteln, von dem man sich spezifische Beeinflussung der 18-Hydroxylierung bzw. der 18-OH-Dehydrierung erwartete. Wir fanden jedoch wiederum eine Hemmung der 11 β-Hydroxylierung. Nach i.v. Gabe der Hemmsubstanz zum Zeitpunkt 55 min (Abb. 1e) war deutlich ein Anstieg der relativen Konzentrationen von DOC und S und ein Abfall der Konzentrationen von B und F zu beobachten. Die Ergebnisse gewinnen an Anschaulichkeit, wenn man die Quotienten aus den relativen Konzentrationen der Substanzen vor und nach der 11 β-Hydroxylierung bildet (rechts in Abb. 1e). Erstaunlicherweise ergab sich auf der Seite der 17 α-hydroxylierten Steroide eine stärkere Hemmung als bei den 17 α-Desoxysteroiden. Dies deutet darauf hin, daß die Enzyme der 11 β-Hydroxylierung für die verschiedenen Steroide unterschiedliche Substratspezifität aufweisen.

Wir hoffen, gezeigt zu haben, daß wir mit unseren Anordnungen der in vitro und in vivo Biosynthese und mit unserem Analysensystem, das die Beobachtung von neun wesentlichen Punkten der Synthesekette gestattet, geeignete Modelle für Regulationsstudien an der Corticosteroid-Biosynthese in Händen haben.

Wir danken der Firma ORGANON für die freundliche Überlassung verschiedener Pinealisextrakte.

II. Medizinische Klinik der Universität München und Departments of Biological Chemistry
and Medicine, Harvard Medical School, Boston

Zum biochemischen Mechanismus der Stimulation der Proteinsynthese der Nebenniere durch adrenocorticotropes Hormon

Von

P. C. SCRIBA und W. J. REDDY

Mit 1 Abbildung

In früheren Untersuchungen konnte von REDDY u. Mitarb. (*1, 2*) gezeigt werden, daß ACTH in vitro und in vivo die Proteinsynthese der Nebenniere beschleunigt. Versuche zur Aufklärung des biochemischen Mechanismus der Stimulation der Proteinsynthese (*2, 3*) zeigten, daß die lösliche Zellfraktion (105000 $\times$g Überstand des Homogenats) von Nebennieren mit ACTH behandelter Ratten den Einbau von C^{14}-markierten Aminosäuren in Eiweiß durch eine mikrosomenhaltige Präparation (15000 $\times$g Überstand des Homogenates) von Nebennieren ACTH-behandelter oder Kontrollratten beschleunigt. Diese Stimulation war durch einen nicht dialysierbaren Faktor in der löslichen Zellfraktion bedingt, der weder ein Aminosäuren aktivierendes Enzym noch Transfer-Ribonucleinsäure war. Die Autoren (*2*) konnten zeigen, daß die Beschleunigung der Proteinsynthese an einem Schritt nach der Bildung von Aminoacyl-Transfer-Ribonucleinsäure erfolgte.

Hinweise auf die Natur des für die in vitro Proteinsynthese geschwindigkeitsbestimmenden Faktors in der löslichen Zellfraktion ergaben folgende Versuche (*3*): Der aktive Faktor zeigte keine Artspezifität bezüglich seiner Eigenschaft, die Proteinsynthese der mikrosomenhaltigen Präparationen aus Ratten-, Schweine- oder Hundenebennieren zu beschleunigen. — Er war in der löslichen Zellfraktion der Nebennieren ACTH-behandelter Ratten in etwa doppelter Aktivität im Vergleich zu der löslichen Zellfraktion, gewonnen aus einer gleichen Gewichtsmenge Nebennieren von Kontrolltieren, zu finden. — Kontrollversuche, bei denen je 2 Einheiten eines synthetischen ACTH-Peptids [N-terminales Triacosapeptid, HOFMANN (*4*)] hypophysektomierten Ratten in 5 Std infundiert wurden, zeigten, daß die beobachtete signifikante Beschleunigung des Einbaus von C^{14}-markiertem Glykokoll in Eiweiß bei Zusatz der löslichen Zellfraktion des Nebennierenhomogenats (ACTH-Tiere: 527 $\pm$ 19,5 gegenüber Kontrolltiere: 356 $\pm$ 10,7 Impulse pro Minute und Ansatz) für ACTH spezifisch und nicht durch eine Verunreinigung (z. B. Wachstumshormon) in den benützten kommerziellen ACTH-Präparaten bedingt war. — Zentrifugation der löslichen Zellfraktion von Nebennieren ACTH-behandelter Ratten auf Saccharosedichtegradienten nach der Technik von MARTIN und AMES (*5*) ergab einen S-Wert von 6—7 und ein ungefähres Molekulargewicht von

$10-14 \times 10^4$ für den aktiven Faktor. — Folgende Versuche sprachen für die Eiweißnatur des Faktors: Es konnte gezeigt werden, daß der Faktor durch 15 min Präinkubation bei 55°C, nicht aber bei 45°C inaktiviert wurde. Auch Präinkubation mit Trypsin, nicht aber mit Ribonuclease, führte in Versuchen, bei denen die Technik der Dichtegradientenzentrifugation zur anschließenden Trennung von Trypsin und aktivem Faktor benutzt wurde, zu Inaktivation des Faktors. — Schließlich konnte der Faktor mittels eines von TAKANAMI (6) für die Reinigung

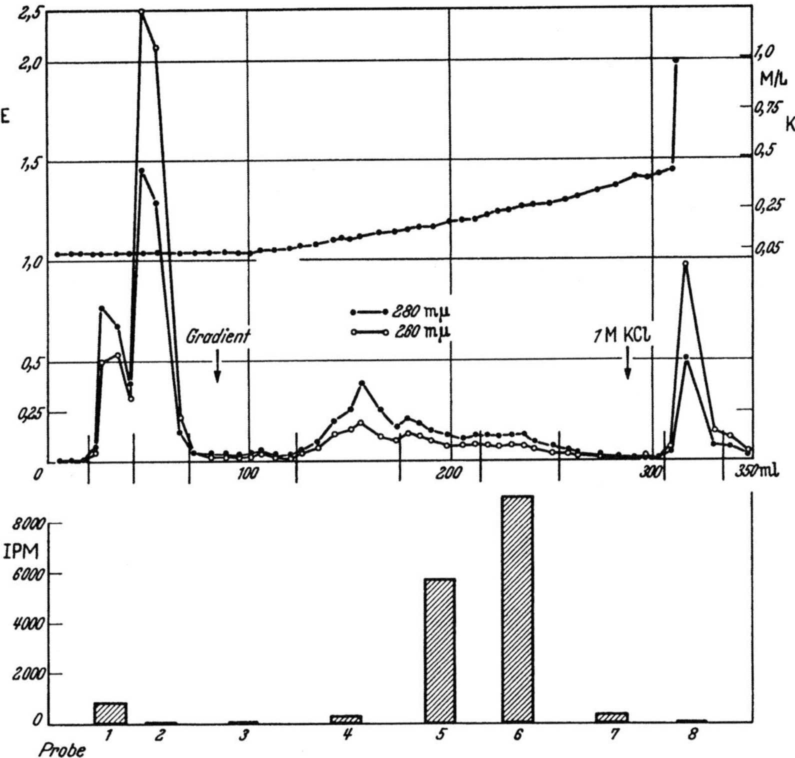

Abb. 1. Anionenaustauschchromatographie der löslichen Zellfraktion von Nebennierenhomogenaten ACTH-behandelter Ratten (oben), Beschleunigung des Einbaus von C¹⁴-markiertem Glykokoll in Eiweiß (Gesamtaktivität, unten), nach (3)

der Aminoacyltransferase aus Rattenleber angegebenen Diäthylaminoäthyl-Cellulose-Chromatographiesystems von anderen Proteinen, Nucleotiden und Nucleinsäuren getrennt werden (Abb. 1). Das Verhalten bei der Anionenaustauschchromatographie (KCl-Gradientenelution) entsprach dabei dem der Leber-Aminoacyltransferase (Elution zwischen 0,17 und 0,3 M Chlorid). Die Anreicherung lag zwischen 3- und 7fach, die Ausbeute zwischen 53 und 70%.

Die chemische Analyse der Fraktionen mit der höchsten spezifischen Aktivität (Stimulation des Einbaus von markiertem Glykokoll bzw. Leucin pro mg Eiweiß) zeigte die Proteinnatur des Materials (2,5 mg, Trockengewicht 2,6 mg), ein meßbarer Nucleinsäureanteil wurde nicht gefunden (weniger als 1% nach der Empfindlichkeit der Methode). Es wird daher vermutet, daß es sich bei dem aktiven Faktor um eine Aminoacyltransferase handelt.

Unter den gewählten Bedingungen war dagegen kein Anhalt für eine Stimulation der Proteinsynthese durch eine hochmolekulare Ribonucleinsäure (Messenger-Ribonucleinsäure) zu finden. — Die beobachtete Beschleunigung der Proteinsynthese in der Nebenniere durch ACTH und ihr biochemischer Mechanismus sind möglicherweise von Bedeutung für den Wirkungsmechanismus des Hormons, jedenfalls konnte Ferguson (7) zeigen, daß nach Blockierung der Proteinsynthese der Nebenniere durch Inkubation mit Puromycin der die Corticoidsynthese stimulierende Effekt des ACTH entfällt. Möglicherweise verursacht ACTH in der Nebenniere die Synthese eines an der Corticoidsynthese beteiligten Enzymproteins.

Literatur

1. Bransome, E. D., and W. J. Reddy: Arch. Biochem. **101**, 21 (1963).
2. Farese, R. V., and W. J. Reddy: Endocrinology **73**, 294 (1963).
3. Scriba, P. C., u. W. J. Reddy: Für Endocrinology eingereicht (1964),
Scriba, P. C., u. W. J. Reddy: Fed. Proc. **22**, Abstr. 30 (1963).
4. Hofmann, K., and H. Yajima: Recent Progr. Hormone Res. **18**, 41 (1962).
5. Martin, R. G., and B. N. Ames: J. biol. Chem. **236**, 1372 (1961).
6. Takanami, M.: Biochim. biophys. Acta (Amst.) **51**, 85 (1961).
7. Ferguson, J. J. Jr.: J. biol. Chem. **238**, 2745 (1963).

Diskussion

G. Bettendorf (Hamburg):

Ich möchte fragen, ob nur das eine Ferment unter der ACTH-Wirkung zunimmt oder ob auch andere Enzyme untersucht wurden. Bei eigenen Untersuchungen über die primäre Stoffwechselwirkung der Gonadotropine konnten wir sehr früh Änderungen bei den Nucleosidphosphaten feststellen. Bereits 2 Std nach Gonadotropininjektion traten charakteristische Enzymveränderungen auf (Schulz, Maass, Bettendorf: Klin Wschr. im Druck).

P. C. Scriba:

Kürzere Zeiten als die 5 Std-Periode der Infusion von synthetischem ACTH wurden, u. a. wegen der technischen Schwierigkeiten, nicht untersucht. Bransome und Reddy beobachteten bei in vitro-Versuchen eine Beschleunigung der Proteinsynthese 15 min nach Zugabe von ACTH. — Andere Enzymaktivitäten wurden nicht untersucht.

Aus der Chemischen Abteilung der Chirurgischen Universitäts-Klinik Bonn
(Direktor: Prof. Dr. A. Gütgemann)

Stoffwechsel von Oestron in der Nebenniere*

Von

R. Knuppen und H. Breuer

Mit 3 Abbildungen

Im allgemeinen gilt die Leber als das Hauptorgan des Zwischenstoffwechsels der Oestrogene. Die Steroide unterliegen in diesem Organ einer größeren Anzahl von Reaktionen, zu denen Oxydoreduktionen, Hydroxylierungen, Methylierung, Demethylierung, Epoxydierung sowie Sulfatierungen und Glucuronidierungen gehören (1). Hingegen ist eine physiologisch bedeutsame Reaktion, nämlich die 11β-Hydroxylierung, bisher nur in der Nebenniere nachgewiesen worden, wenn man von bestimmten Hodentumoren absieht, die ebenfalls zur 11β-Hydroxylierung befähigt sind (2, 3). Aus diesem Grunde haben wir unsere Untersuchungen zur Biogenese von 11β-substituierten Oestrogenen mit Nebennieren durchgeführt. Dabei konnten wir nach Inkubation von Oestron nicht nur 11β-Hydroxyoestron, sondern auch andere, bisher unbekannte Metaboliten isolieren, über deren Identifizierung im folgenden kurz berichtet werden soll.

Nach Inkubation von Nebennierenbrei des Rindes mit Oestron [Einzelheiten s. (4)] wurden die Extraktrückstände der präparativen Chromatographie auf formamidimprägniertem Papier im System Chloroform unterworfen (Abb. 1). Dabei wurden außer nicht-umgesetztem Oestron 6 phenolische Fraktionen erhalten, von denen im vorliegenden Fall nur Oe-4 und Oe-6 interessieren. Bei der Rechromatographie im System Monochlorbenzol-Äthylacetat (3 : 1) trennte sich Oe-4 in drei Substanzen — Oe-4 A, B und C — auf, während Oe-6 im gleichen System zwei Substanzen — nämlich Oe-6 A und B — ergab. Nachdem die Verbindungen über Celite-Säulen gereinigt worden waren, wurden ihre Schmelzpunkte ermittelt, Derivate hergestellt und I.R.-Spektren aufgenommen. Es zeigte sich, daß Oe-4 C mit 6α-Hydroxyoestron (4), Oe-6 A mit 11β-Hydroxyoestron (6) und Oe-6 B mit 16α-Hydroxyoestron (6) identisch ist.

Etwas schwieriger gestaltete sich die Identifizierung der Metaboliten Oe-4 A und Oe-4 B. Bei der Verbindung Oe-4 B nahmen wir zunächst an, es handele sich um 6β-Hydroxyoestron. Oe-4 B verhielt sich nämlich bei der Papierchromatographie ähnlich wie 6β-Hydroxyoestron; außerdem entstand bei der Wasserabspaltung als Reaktionsprodukt 6-Dehydrooestron (Abb. 2). Bei der Oxydation von Oe-4 B mit Chromtrioxyd zeigte sich jedoch, daß das entstehende Keton nur *ein* U.V.-Maximum bei 280 mμ aufwies, während das erwartete Reaktionsprodukt, nämlich

---

* Ausgeführt mit Unterstützung der Deutschen Forschungsgemeinschaft

6-Oxo-oestron, zwei Maxima bei 256 und 325 mμ besitzt. Auf Grund dieses Ergebnisses mußte es sich bei Oe-4B um eine 7-hydroxylierte Verbindung handeln, deren Identität mit 7α-Hydroxyoestron durch Infrarot-Spektroskopie eindeutig bewiesen wurde (7).

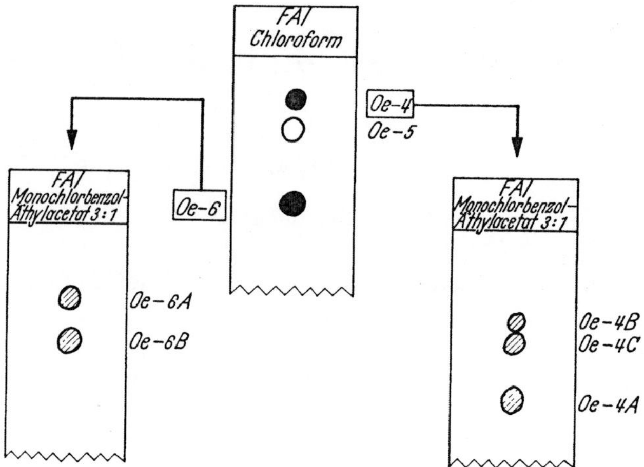

Abb. 1. Papierchromatographische Systeme zur Trennung der Oestronmetaboliten Oe-4, Oe-5 und Oe-6. Einzelheiten über die verwendeten Systeme siehe (5). FA = formamidimprägniertes Papier

Abb. 2. Verhalten von 6β-Hydroxyœstron und 7α-Hydroxyœstron bei der Wasserabspaltung und bei der Oxydation mit Chromtrioxyd. Die Aufnahme der U.V.-Spektren erfolgte in Methanol

Der Metabolit Oe-4A entstand bei allen Inkubationen in der größten Ausbeute. In papierchromatographischen Vorversuchen konnte eine Identität von Oe-4A mit den bisher bekannten Metaboliten des Oestrogenstoffwechsels und den bisher synthetisch hergestellten Oestrogenen weitgehend ausgeschlossen werden. Klassische Analysenmethoden ergaben, daß Oe-4A ein Ketol war; nach diesen Versuchen mußten folgende Hydroxylverbindungen von Oestron näher in Betracht gezogen werden: 8β-, 9α-, 12α-, 12β-, 14α-, 15α- oder 15β-Hydroxyoestron. Da Oe-4B mit HCl leicht Wasser abspaltet, wurde zunächst angenommen, die Verbindung enthielte eine tertiäre Hydroxylgruppe und wäre demnach mit 8β-, 9α- oder 14α-Hydroxyoestron identisch. Gegen diese Annahme sprach allerdings die Tatsache, daß Oe-4A ein Diacetat bildete; infolgedessen mußte Oe-4B eine sekundäre

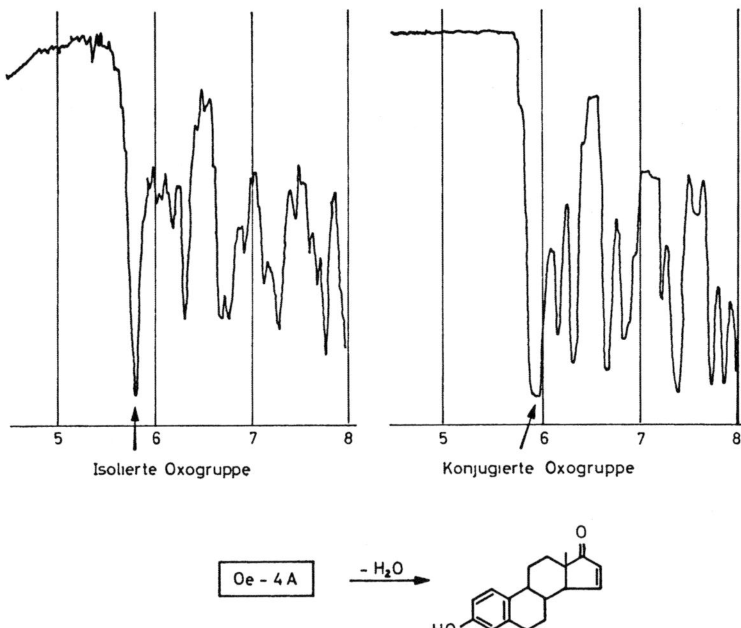

Abb. 3. Verschiebung der Carbonylbande im Infrarot nach Wasserabspaltung von Oe-4A; als Reaktionsprodukt entsteht 15-Dehydrooestron

Hydroxylgruppe enthalten, die dann auch im NMR-Spektrum nachgewiesen wurde. Nunmehr konnte es sich nur noch um ein 12- oder 15-hydroxyliertes Oestron handeln. Bei einer erneuten Untersuchung der Wasserabspaltung von Oe-4A wurden zwei Reaktionsprodukte isoliert, von denen das eine im I.R.-Spektrum eine konjugierte Ketogruppe mit intakter C-18-Methylgruppe zeigte (Abb. 3). Eine solche Verbindung kann bei der Wasserabspaltung nur aus 15-Hydroxyoestron entstehen. Die endgültige Identifizierung von Oe-4A als 15α-Hydroxyoestron erfolgte durch Vergleich mit der authentischen Substanz (*8*).

Nach den vorliegenden Ergebnissen konnten bisher folgende Hydroxylierungsprodukte von Oestron in der Nebenniere nachgewiesen werden: 6α-, 7α-, 11β-, 15α- und 16α-Hydroxyoestron; die Frage, ob auch 18-Hydroxyoestron als Metabolit entsteht (*9*), ist zur Zeit noch ungeklärt. In weiteren Versuchen ist nunmehr zu prüfen, ob neben der 11β-Hydroxylierung auch die 15α-Hydroxylierung nebennierenspezifisch ist oder ob die letztgenannte Reaktion ebenfalls in der Leber stattfindet. Der Befund, daß 15α-Hydroxyoestron als Hauptmetabolit von Oestron in der Nebenniere gebildet wird, läßt vermuten, daß diese Verbindung auch als Oestrogenmetabolit im Urin zu erwarten ist.

Wir danken Herrn Professor O. Wintersteiner, New Brunswick, für wertvolle Hinweise und Herrn Professor J. Fried, Chicago, für die Überlassung von 15 α-Hydroxyœstron. Fräulein Margitta Behm und Herrn O. Haupt sind wir für ihre Mitarbeit sehr zu Dank verpflichtet.

Literatur

1. Breuer, H.: Vitam. u. Horm. 20, 285 (1962).
2. Savard, K., R. I. Dorfman, B. Baggett, L. L. Fielding, L. L. Engel, H. T. McPherson, L. M. Lister, D. S. Johnson, E. C. Hamblen and F. L. Engel: J. clin. Invest. 39, 534 (1960).
3. Smith, E. R., H. Breuer, and H. Schriefers: Biochem. J., 1964 (im Druck).
4. Knuppen, R., M. Behm u. H. Breuer: Hoppe-Seylers Z. physiol. Chem., 1964 (im Druck).
5. — Z. Vitamin-, Hormon- u. Fermentforsch. 12, 355 (1962).
6. —, u. H. Breuer: Biochim. biophys. Acta (Amst.) 58, 147 (1962).
7. — O. Haupt u. H. Breuer: Steroids 3, 123 (1964).
8. Laskin, A. I., P. Grabowich, B. Junta, C. de Lisle Meyers u. J. Fried: J. organ. Chem., 29, 1333 (1964).
9. Loke, K. E., E. J. D. Watson, and G. F. Marrian: Biochim. biophys. Acta 26, 230 (1957).

Aus der endokrinologischen Abteilung des Institutes für Hygiene und Mikrobiologie de
Universität, Homburg/Saar

Konjugation von Steroiden im Nebennieren-, Leber- und Femoralvenenblut

Von

E. KAISER[1] und G. W. OERTEL

Mit 2 Abbildungen

Zahlreiche Untersuchungen beschäftigen sich mit der Sekretion von Steroiden durch die Nebennierenrinde. Bekanntlich übersteigt die Konzentration „freier" Steroide, wie Cortisol *(1—3)*, Corticosteron *(1—3)*, Progesteron *(3)*, Δ^4-Androstendion- (3, 17) *(2—4)* und dessen 11β-Hydroxy-Verbindung *(2—4)* im menschlichen Nebennierenvenenblut die entsprechenden Plasmaspiegel in der Peripherie um ein Vielfaches, was für eine aktive Sekretion der Nebenniere spricht. Dagegen gehen die Ansichten über die Biogenese des mengenmäßig bedeutendsten Plasmasteroids „Dehydroepiandrosteron" bisher noch auseinander. Das liegt daran, daß trotz Nachweises dieses Steroids im Nebennierenvenenblut durch BAULIEU *(5)*, LIEBERMAN u. TEICH *(6)* und andere *(7)* brauchbare Vergleichsuntersuchungen im peripheren Blut fehlen. Auf der anderen Seite ließen sich nach Perfusion peripheren Gewebes beim Hund mit Pregnenolon und 17-Hydroxypregnenolon als den Vorstufen von Dehydroepiandrosteron *(6)* signifikante Mengen von „konjugiertem" Dehydroepiandrosteron *(8)* nachweisen. Eine derartige Umwandlung der aus der Nebennierenrinde stammenden C_{21}-Steroide *(9)* dürfte beim Menschen eine geringere Bedeutung haben *(10)*. Die Befunde BAULIEUs u. Mitarb. *(11)*, die bei einer Patientin mit virilisierendem Nebennierenrindentumor im Nebennierenvenenblut 535 μg Dehydroepiandrosteronsulfat, im peripheren Blut dagegen nur 340 μg der gleichen Verbindung pro 100 ml Plasma fanden, lassen die Sekretion von Dehydroepiandrosteron in Form des Schwefelsäureesters, wie es die gleichen Autoren *(11)* auch für 7-Oxo-dehydroepiandrosteron diskutieren, für möglich erscheinen.

All diese angeschnittenen Fragen und Probleme verlangten nach einer Klärung, worüber folgende Untersuchungen über die Konzentrationen der wichtigsten Steroide und ihrer Metaboliten in freier und gebundener Form im Nebennieren-, peripheren und nicht zuletzt im Lebervenenblut Aufschluß geben sollten, denn die Einflußnahme der Leber auf die Konjugation der Steroide durfte keinesfalls vernachlässigt werden.

Als Ausgangsmaterial standen 322 ml Nebennieren-, 183 ml Leber- und 184 ml Femoralvenenblut zur Verfügung, welches von den Herren Kollegen BETTE, LEPPLA u. WEINHEIMER von der Medizinischen Klinik in Homburg/Saar mittels

---

[1] Neue Anschrift: Frauenklinik der Medizinischen Akademie, Düsseldorf.

Katheter gewonnen und uns freundlicherweise für derartige Untersuchungen überlassen wurde.

Nach Extraktion „freier" Steroide aus dem jeweiligen Plasma durch Chloroform erfolgte die Auftrennung der in der wäßrigen Phase verbliebenen Steroidconjugate durch Chromatographie an einer Säule aus DEAE-Sephadex. Die in den beiden ersten Eluaten anfallenden unpolaren Conjugate und Glucuroniside wurden durch Lösungsmittelverteilung getrennt. Anschließend erfolgte Solvolyse der unpolaren Conjugate aus der organischen Phase und der Sulfate aus dem letzten Säuleneluat, sowie Freisetzung der glucuronosidartig gebundenen Steroide durch β-Glucuronidase. Die Isolierung der aus den Conjugaten freigesetzten Steroide geschah schließlich vermittels Papierchromatographie in geeigneten Systemen. Zur quantitativen Bestimmung diente für Cortisol und Tetrahydrocortisol die Porter-Silber-Reaktion (*12*), für Corticosteron die fluorometrische Bestimmung nach Braunsberg u. James (*13*), für Progesteron die UV-Absorption bei 240 mμ (*14*), für Pregnenolon und Dehydroepiandosteron die Schwefelsäure-Äthanol-Reaktion (*15*), für 17-Hydroxypregnenolon die Vanillin-Phosphorsäure-Reaktion (*16*), desgleichen für 17-Hydroxyprogesteron, sowie für die 17-Ketosteroide die Mikro-Zimmermann-Reaktion (*17*) und für die C_{18}-Steroide die Kober-Reaktion nach Ittrich (*18*).

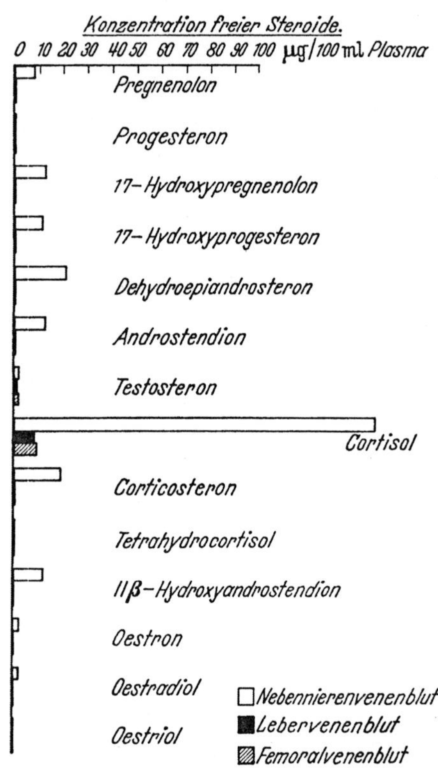

Abb. 1. Konzentration freier Steroide

Diese quantitative Auswertung mittels geeigneter Farbreaktionen läßt erkennen, daß von seiten der Nebennierenrinde von den 7 bestimmten C_{21}-Steroiden: Pregnenolon, Progesteron, 17-Hydroxypregnenolon, 17-Hydroxyprogesteron, Cortisol, Corticosteron und Tetrahydrocortisol, von den 4 C_{19}-Steroiden: Dehydroepiandrosteron, Androstendion, Testosteron und 11β-Hydroxyandrostendion und von den 3 C_{18}-Steroiden: Oestron, Oestradiol und Oestriol außer Tetrahydrocortisol alle aufgeführten Steroide in freier Form ausgeschüttet werden. Durchschnittlich übersteigen die Konzentrationen der einzelnen Steroide im Nebennierenvenenblut die des Leber- und Femoralvenenblutes um das 10—20fache. Im übrigen unterscheidet sich das Leber- und Femoralvenenblut in bezug auf Konzentration an freien Steroiden nur beim Cortisol, das im Lebervenenblut geringfügig vermindert gegenüber der Peripherie vorlag (Abb. 1).

Gleichzeitig erfolgt aber auch eine Ausschüttung von Pregnenolon, 17-Hydroxypregnenolon, Dehydroepiandrosteron, Cortisol, Corticosteron, Oestron, Oestradiol in Form von lipophilen, solvolysierbaren Conjugaten, die nach unserer Ansicht sog. Sulfatidylverbindungen (*19*) darstellen und die erst im weiteren einer Zersetzung zu Steroidsulfaten unterliegen. Unserer Beobachtung nach beträgt die Sulfatfraktion gewöhnlich nicht mehr als 20—50% der gesamten solvolysierbaren

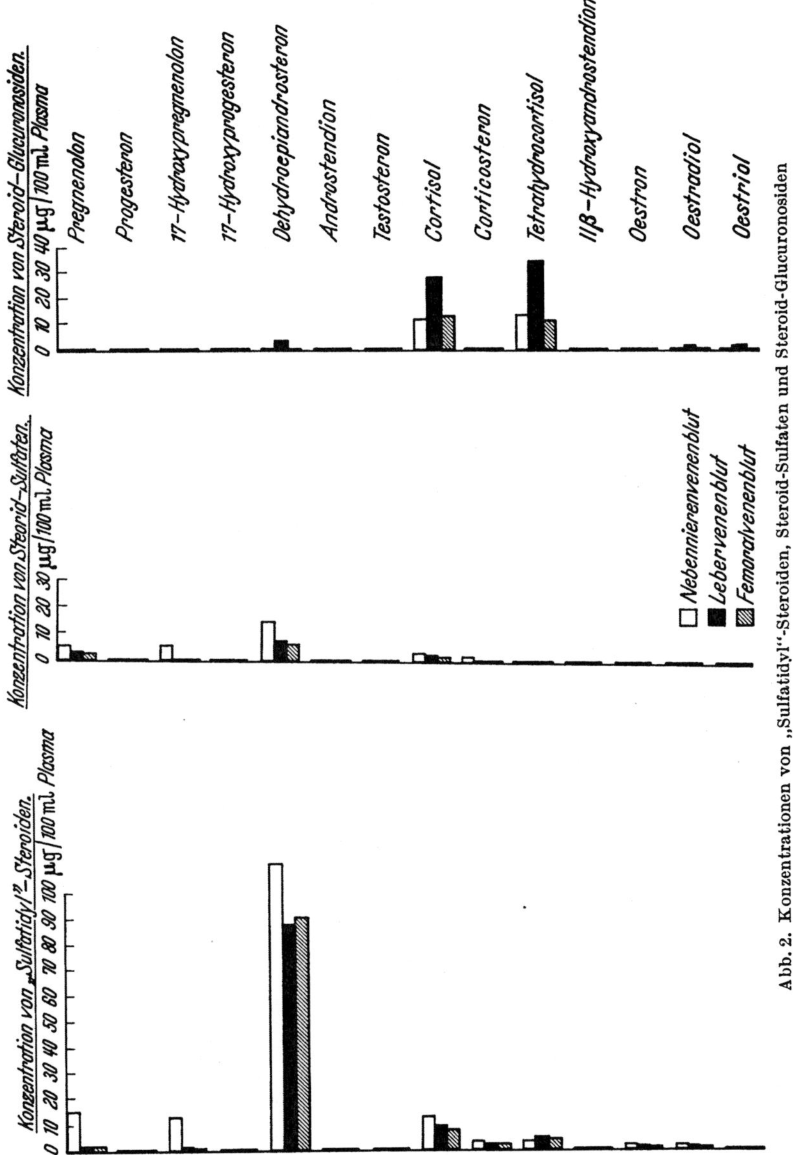

Abb. 2. Konzentrationen von „Sulfatidyl"-Steroiden, Steroid-Sulfaten und Steroid-Glucuronosiden

Conjugate. Auffällig hoch ist die Konzentration derartiger Verbindungen von Dehydroepiandrosteron, Cortisol, Corticosteron und Tetrahydrocortisol im Leber- und Femoralvenenblut.

Die Betrachtung der Konzentrationen an Steroidglucuronosiden in den einzelnen Plasmaproben läßt deutlich eine praktisch ausschließlich in der Leber stattfindende Kopplung einzelner Steroide wie Cortisol und Tetrahydrocortisol mit

Glucuronsäure erkennen (Abb. 2). Derartige Verbindungen von Cortisol und Tetrahydrocortisol konnten auch im Nebennierenvenenblut nachgewiesen werden.

Diese vorliegenden Resultate bestätigen die Beobachtungen anderer Untersucher, wonach Cortisol, Corticosteron, Androstendion in der Nebennierenrinde gebildet und als „freie", d. h. mit Chloroform extrahierbare Steroide ausgeschüttet werden. Da auch deutliche Unterschiede in der Konzentration „freien" Dehydroepiandrosterons im Nebennierenvenenblut und peripheren Blut bestehen, dürfte wohl über die Sekretion dieses freien 17-Oxosteroids seitens der Nebennierenrinde kein Zweifel mehr bestehen. Auch die potentiellen Vorstufen, wie Pregnenolon und 17-Hydroxypregnenolon werden offensichtlich von der Nebennierenrinde ausgeschieden. Infolge ihrer nur Bruchteile betragenden Konzentration im peripheren Plasma kann wohl eine Bildung von Dehydroepiandosteron in der Peripherie aus diesen C_{21}-Steroiden unter physiologischen Bedingungen beim Menschen als bedeutungslos angesehen werden. Eine Sezernierung von Oestron und Oestradiol in „freier" Form durch die Nebennierenrinde läßt sich ebenfalls diskutieren. Obwohl sich die Spiegel der unpolaren, solvolysierbaren Conjugate, der sog. Sulfatidylsteroide, die sich nur bei schonendster Aufarbeitung nachweisen lassen, nicht signifikant außer denen der Δ^5-3β-OH-Steroide und möglicherweise des Cortisols von denen im Leber- und Femoralvenenblut unterscheiden, möchten wir in diesen Fällen doch eine Sekretion derartiger Substanzen neben freien Steroiden durch die Nebennierenrinde befürworten. Unserer Ansicht nach handelt es sich bei diesen Sulfatidylverbindungen um ein Primärprodukt, wogegen die Sulfate im wesentlichen nur sekundärer Natur sind. Dagegen läßt sich ihre Biogenese in der Leber auf Grund eigener anderer Untersuchungen noch nicht sicher von der Hand weisen. Wenn man schließlich die hohe Clearance von Glucuroniden bedenkt, wie sie BONGIOVANNI u. EBERLEIN beschrieben (20), dann sind die geringen Anteile der mit β-Glucuronidase spaltbaren Steroidconjugate gemessen an der Gesamtmenge vorhandener Conjugate nicht verwunderlich. Sicher läßt sich abschließend sagen, daß eine Sekretion konjugierter Δ^5-3β-OH-Steroide seitens der Nebennierenrinde von Bedeutung zu sein scheint.

Literatur

1. REICH, H., D. H. NELSON, and A. ZAFFERONI: J. biol. Chem. 187, 411 (1950).
2. ROMANOFF, E. P., P. HUDSON, and G. PINCUS: J. clin. Endocr. 13, 1546 (1953).
3. SHORT, R. V.: Biochem. Soc. Symp. 18, 59 (1960).
4. LOMBARDO, M. E., C. McMORRIS, and P. B. HUDSON: Endocrinology 65, 426 (1959).
5. BAULIEU, E.-E.: C. R. Acad. Sci. (Paris) 251, 1421 (1960).
6. LIEBERMAN, S., and S. TEICH: J. clin. Endocr. 13, 1140 (1953).
7. REVOL, A., R. MALLEIN et P. GUINET: Ann. Biol. clin. 18, 565 (1960).
8. OERTEL, G. W., and K. B. EIK-NES: Endocrinology 65, 766 (1959).
9. — — Arch. Biochem. 93, 392 (1961).
10. SOLOMON, S., A. C. CARTER, and S. LIEBERMAN: J. biol. Chem. 235, 351 (1960).
11. BAULIEU, E.-E., R. EMILIOZZI u. C. CORPECHOT: Experientia (Basel) 17, 110 (1961).
12. EIK-NES, K. B.: J. clin. Endocr. 17, 502 (1957).
13. BRAUNSBERG, H., and V. H. T. JAMES: J. Endocr. 21, 327 (1960).
14. ZANDER, J., u. H. SIMMER: Klin. Wschr. 32, 529 (1954).
15. OERTEL, G. W., u. K. B. EIK-NES: Analyt. Biochem. 31, 98 (1959).
16. CARSTENSEN, H., G. W. OERTEL, and K. B. EIK-NES: J. biol. Chem. 234, 2510 (1959).
17. OERTEL, G. W., and E. KAISER: Clin. chim. Acta 7, 221 (1962).
18. ITTRICH, G.: Hoppe-Seylers Z. physiol. Chem. 320, 103 (1960).
19. OERTEL, G. W.: Biochem. Z. 334, 431 (1961).
20. BONGIOVANNI, A. M., and W. R. EBERLEIN: J. clin. Endocr. 17, 238 (1957).

Aus der 2. Medizinischen Klinik und Poliklinik der Medizinischen Akademie Düsseldorf
(Direktor: Prof. Dr. K. Oberdisse)

Über den Einfluß der Schilddrüsenfunktion auf Metabolite des Fettstoffwechsels im Serum

Von

K. Jahnke[1], F. A. Gries, H. Bethge und H. Fehlings

Mit 1 Abbildung

In den letzten Jahren wurde berichtet, daß die Plasmakonzentration der nichtveresterten Fettsäuren bei Hyperthyreose erhöht ist (*1, 2*). Diese Fettsäuren werden unter dem Einfluß verschiedener lipolytisch wirksamer Hormone bei der Spaltung der Triglyceride im Fettgewebe freigesetzt. Soweit sie dort nicht wieder rückverestert werden, treten sie in das Blut über. Vor allem bei unzureichendem Glucoseumsatz stellen sie eine wesentliche Quelle des muskulären Energiestoffwechsels dar. Sie werden außerdem in der Leber (und der Niere) rückverestert und in der Leber zu Ketokörper abgebaut.

Auch Glycerin wird bei der lipolytischen Spaltung der Triglyceride freigesetzt. Im Gegensatz zu freien Fettsäuren kann freies Glycerin im Fettgewebe nicht verwertet werden. Es wird daher vollständig an das Blut abgegeben und in der Leber und Niere nach Phosphorylierung metabolisiert. Deutlicher noch als die Konzentration der freien Fettsäuren im Serum reflektiert daher die des freien Glycerins Lipolysevorgänge im Fettgewebe.

Um den Einfluß der Schilddrüsenfunktion auf den Fettstoffwechsel genauer zu analysieren, haben wir bei 33 euthyreoten, 28 hyperthyreoten und 6 hypothyreoten Personen verschiedene Zwischenprodukte des Fettstoffwechsels gleichzeitig unter Grundumsatzbedingungen untersucht, nämlich freie Fettsäuren (FFS) nach Dole (*3*), freies Glycerin (FGlyc) nach Kreutz (*4*), veresterte Fettsäuren (VFS) nach Rosenthal, Pfluke und Callerami (*5*) im Serum, sowie Acetacetat (AcAc) nach Walker (*6*) und Glucose (BZ) nach Grady und Lamar (*7*) im Blut. Die Ergebnisse sind in der Tab. 1 zusammengefaßt.

Bei *unbehandelter Hyperthyreose* mit ausgeprägter Tachykardie, Grundumsatzsteigerung, PBI-Erhöhung und Hypocholesterinämie fanden wir eine erhebliche Vermehrung zirkulierender Metabolite des Fettstoffwechsels.

Sie ist am ausgeprägtesten für freies Glycerin, dessen Konzentration nahezu das 6fache des euthyreoten Wertes erreicht. Die Konzentration der freien Fettsäuren ist um etwa das 2fache erhöht. Sehr deutlich ist auch die Konzentration des Acetacetats, nämlich um das $2^1/_2$fache angestiegen. Lediglich die veresterten Fettsäuren verhalten sich umgekehrt: ihre Konzentration ist auf etwa $^2/_3$ des euthyreoten Wertes erniedrigt.

---

[1] Mit dankenswerter Unterstützung der Deutschen Forschungsgemeinschaft.

Tabelle 1

	Euthyreot	Hyperthyreot unbehandelt	Hyperthyreot in Remission	Hypothyreot
n	33	15	18	6
Pulsfrequenz (pro min)		111 ± 15	89 ± 13	76 ± 5
Grundumsatz (%) .		+72 ± 17	+50 ± 14	—21 ± 6
PBI (γ-%)		11,4 ± 2,8	7,7 ± 2,4	2,8 ± 0,5
Cholesterin (mg-%)	223 ± 46	172 ± 42	172 ± 46	479 ± 74
Blutzucker mg-%) .	84 ± 10	83 ± 7	81 ± 28	76 ± 10
FFS (μmol/l) . . .	394 ± 128	762 ± 198	538 ± 157	540 ± 185
FGlyc (μmol/l) . .	43 ± 16	254 ± 151	116 ± 17	126 ± 46
(FFS:FGlyc) . . .	9,8 ± 3,3	3,7 ± 1,3	4,8 ± 1,9	4,4 ± 2,6
AcAc (γ-%) . . .	433 ± 149	1110 ± 482	426 ± 166	335 ± 106
VFS (mg-%) . . .	421 ± 105	287 ± 85	325 ± 110	848 ± 227

Diese Konzentrationsverschiebungen sind *unter der Behandlung* mit Thyreostatica oder Radiojod rückläufig, wie an Verlaufsuntersuchungen nachzuweisen war. Das zeigen auch die in der Tab. 1 aufgeführten Werte bei Hyperthyreose in Remission. Hier ist die Erhöhung der Konzentration für freies Glycerin noch deutlich, weniger die der freien Fettsäuren. Der Acetacetatspiegel ist wieder völlig normalisiert.

Bei *Hypothyreose* fanden wir nicht etwa eine eindeutige Umkehr dieser Verhältnisse. Im Vergleich zu euthyreoten Personen wurden bei diesen Fällen, die mit ausgeprägter Grundumsatzerniedrigung, erniedrigtem PBI und beträchtlicher Hypercholesterinämie einhergingen, ebenfalls erhöhte Durchschnittskonzentrationen für freie Fettsäuren und freies Glycerin gemessen. Die Einzelwerte streuen aber erheblich. Auch erniedrigte Konzentrationen sahen wir. Der Acetacetatspiegel war sicher nicht erhöht. Eine eindeutige Umkehr zeigten lediglich die veresterten Fettsäuren: ihre Konzentration ist beträchtlich, nahezu um das Doppelte des euthyreoten Wertes erhöht.

Wie soll man sich die beschriebenen Konzentrationsänderungen der Metabolite des Fettstoffwechsels unter dem Einfluß der abnormen Schilddrüsenfunktion erklären?

Am schwierigsten ist das bei der *Hypothyreose*, zumal die Fallzahl gering und die Streuung der Einzelwerte relativ groß ist.

Bisher haben nur Harlan u. Mitarb. (8) über das Verhalten der freien Fettsäuren im Plasma bei Hypothyreose berichtet. Sie fanden eher erniedrigte Werte. Aus in vitro-Versuchen am Fettgewebe von Ratten mit induzierter Hypothyreose, die von Debons und Schwartz (9) sowie Deykin und Vaugham (10) durchgeführt wurden, muß man schließen, daß zur regelrechten Lipolyse nur Fettgewebe unter euthyreoten Bedingungen fähig ist. Wenn erhöhte Konzentrationen der freien Fettsäuren und des freien Glycerins gefunden werden, wird man das am ehesten mit einem verminderten Umsatz erklären können.

Andererseits haben Freinkel (11) und kürzlich Vaugham und Steinberg (12) gezeigt, daß TSH in vitro die Freisetzung von Fettsäuren und Glycerin aus Fettgewebe steigert. Eine gewisse Konzentrationszunahme dieser Metabolite im Serum bei Menschen mit primärem Myxödem und vermehrter TSH-Produktion wäre demnach denkbar. Inwieweit das hier praktisch in Betracht zu ziehen ist, läßt sich aus unseren Untersuchungen nicht entscheiden.

Wesentlich besser lassen sich die Befunde bei *Hyperthyreose* interpretieren:

Die Verschiebungen der Metabolit-Konzentrationen sind offenbar mit der Grundumsatzsteigerung verbunden. Das ließ sich unter dem Einfluß einer kurzfristigen Behandlung mit Lugolscher Lösung zeigen. Innerhalb von 9 Tagen sanken die Grundumsatzsteigerung auf 42% und gleichzeitig die Konzentrationen für freie Fettsäuren auf 50 ± 11%, für freies Glycerin auf 52 ± 27%, für Acetacetat auf 67 ± 30% des Ausgangswertes ab, wobei die Konzentration des Cholesterins um 32 ± 25 % und die der veresterten Fettsäuren um 38 ± 24% anstiegen.

Für die erhebliche Vermehrung der freien Fettsäuren und des freien Glycerins muß man eine gesteigerte Lipolyse im Fettgewebe verantwortlich machen. Das konnte auch in vitro am Fettgewebe von Trijodthyronin-behandelten Ratten nachgewiesen werden (*9, 12*). Solches Fettgewebe zeigt zudem eine Potenzierung der Adrenalin-induzierten Lipolysesteigerung. Trijodthyronin allein hat in vitro am Fettgewebe jedoch keine Wirkung. Es wurde daher angenommen, daß Schilddrüsenhormone nur indirekt zur Lipolysesteigerung führen, indem sie den Abbau des lipolytisch aktiven Adrenalins durch Hemmung der Monoaminooxydase verzögern.

Fett- und Kohlenhydratstoffwechsel unterliegen einem inversen Basis-Regulationsmechanismus. Gesteigerte Lipolyse geht mit vermindertem Glucoseumsatz einher und umgekehrt. Das läßt sich auch bei Hyperthyreose zeigen (Abb. 1).

Innerhalb weniger Minuten nach intravenöser Injektion von 25 g Glucose sieht man einen steilen Abfall der zuvor erheblich gesteigerten Konzentrationen von freien Fettsäuren, freiem Glycerin und Acetacetat, wobei die normalen Nüchternwerte noch unterschritten werden. Wesentlich ist, daß dieser Abfall erst bei und nach erhöhtem Blutzuckerspiegel zustande kommt.

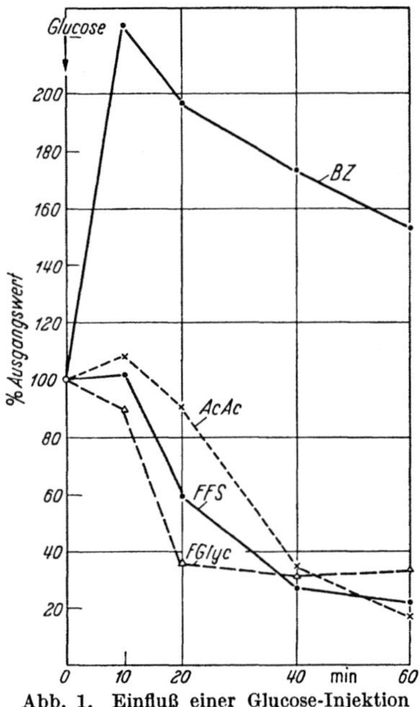

Abb. 1. Einfluß einer Glucose-Injektion (25 g i.v.) auf Metabolite des Fettstoffwechsels im Serum bei Hyperthyreose

Die Annahme, daß Hyperthyreote nur infolge ungenügenden Kohlenhydratangebotes auf gesteigerte Lipolyse eingestellt sind, bleibt aber unbefriedigend.

Hier ist auf Untersuchungen hinzuweisen, die kürzlich aus dem Arbeitskreis von RANDLE (*13*) zusammenfassend publiziert wurden. Danach vermindert ein erhöhtes Angebot freier Fettsäuren und Ketokörper die Insulinwirksamkeit und damit die Glucoseaufnahme in die Zelle. Wahrscheinlich handelt es sich um eine Hemmung des insulinabhängigen Transportmechanismus durch die Zellmembran. Offenbar wird diese Hemmwirkung auch hier erst bei höherem Blutzuckerspiegel und durch Glucoseinjektion stimulierter Insulinsekretion überspielt. Unter diesen Bedingungen wird dann auch der Ketokörperstau im Blut beseitigt.

Wenn man diese Gesichtspunkte berücksichtigt, ergeben sich für den Einfluß der Hyperthyreose auf den Fettstoffwechsel folgende Verhältnisse:

Unter der intensivierten Wirkung lipolytischer Hormone, wahrscheinlich Adrenalin, ist die Freisetzung von Fettsäuren und Glycerin aus dem Fettgewebe bei Hyperthyreose gesteigert. Freie Fettsäuren fließen dem erhöhten Umsatz in der Peripherie zu. Ein beträchtlicher Teil wird von der Leber abgefangen, dort nur in geringem Ausmaß wieder verestert, in weitaus größerem zu Ketokörpern umgebaut. Infolge des erhöhten Angebotes an freien Fettsäuren und Ketokörpern kommt es zu einer Hemmung der Insulinwirkung, damit zu einer Verminderung der Glucose-umsatzrate und sekundär zu einer weiteren Steigerung der Lipolyse.

Die gesteigerte Lipolyse bei Hyperthyreose erklärt sich demnach nicht bloß durch den erhöhten Substratumsatz, sondern auch aus einer Hemmung der insulin-abhängigen Glucoseutilisation im Muskel. Das entspricht den Verhältnissen im Hunger oder bei kohlenhydratarmer Ernährung (*13*).

Damit lassen sich jetzt auch ältere Beobachtungen erklären, die auf eine herab-gesetzte Glucose-Toleranz bei Hyperthyreose, die relative Häufung von Glucosurien bei Hyperthyreose und die Intensivierung des experimentellen Diabetes unter Trijodthyronin hinweisen (*14*).

Literatur

1. Rich, C., E. L. Bierman, and J. L. Schwartz: Plasma nonesterified fatty acids in hyper-thyroid states. J. clin. Invest. **38**, 275 (1959).

2. Marks, B. H., J. Kiem, and A. G. Hills: Effect of hyperthyroidism on fasting nonesteri-fied fatty acid concentration and on its response to glucose-ingestion. Metabolism **9**, 1130 (1960).

3. Dole, V. P.: A relation between nonesterified fatty acids in plasma and the metabolism of glucose. J. clin. Invest. **35**, 150 (1956).

4. Kreutz, F. W.: Enzymatische Glyzerinbestimmung. Klin. Wschr. **40**, 362 (1962).

5. Rosenthal, H. L., M. L. Pfluke, and I. Callerami: The colorimetric estimation of serum fatty acids. Clin. chim. Acta **4**, 329 (1959).

6. Walker, P. G.: A colorimetric method for the estimation of acetoacetate. Biochem. J. **58**, 639 (1954).

7. Grady, H. J., and M. A. Lamar: Glucose determination by automatic chemical analysis. Clin. Chem. **5**, 542 (1959).

8. Harlan, W. R., J. Laszlo, M. D. Bogdonoff, and E. Estes jr.: Alterations in free fatty acid metabolism in endocrine disorders. Part I: Effect of thyroid hormone. J. clin. Endocr. **23**, 33 (1963).

9. Debons, A. F., and J. L. Schwartz: Dependence of the lipolytic action of epinephrine in vitro upon thyroid hormones. J. Lipid Res. **2**, 86 (1961).

10. Deykin, D., and M. Vaugham: Release of NEFA by adipose tissue from rats treated with trijodthyronine or propylthiouracil. J. Lipid Res. **4**, 200 (1963).

11. Freinkel, N.: Extrathyroidal action of pituitary thyrotropin: effects on the carbohydrate-, lipid- and respiratory metabolism of adipose tissue. J. clin. Invest. **40**, 476 (1961).

12. Vaugham, M., and D. Steinberg: Effects of Hormones on lipolysis and Esterification of NEFA during incubation of adipose tissue in vitro. J. Lipid. Res. **4**, 194 (1963).

13. Randle, P. J., P. B. Garland, C. N. Hales, and E. A. Newsholme: The glucose-fatty acid cycle. Lancet **1963 I**, 785

14. Joslin, E. P., H. F. Root, P. White, and A. Marble: The treatment of Diabetes mellitus. 10th Ed. Philadelphia: Lea & Febiger 1959.

Diskussion

K. Oberdisse:

Es ist auffällig, daß sich bei der Untersuchung der Metabolite des Fettstoffwechsels bei der Hypothyreose so wenig übereinstimmende Ergebnisse zeigen, die zudem nicht immer im Gegensatz zu den bei der Hyperthyreose gefundenen Werten stehen, was man eigentlich erwarten sollte. Ich möchte darauf aufmerksam machen, daß dies sehr wohl mit Schwere und vor allen Dingen Dauer der Hypothyreose in Zusammenhang stehen kann, da die Produktion des den Fettstoffwechsel beeinflussenden TSH nur im Beginn des Krankheitsbildes erhöht ist. Später ist dies nicht mehr der Fall, offenbar weil die Aktivität des hypophysären Reglermechanismus zur Ruhe kommt, wenn der Hypophysenvorderlappen durch niedrige Konzentration der Schilddrüsenhormone im Blut selbst geschädigt worden ist.

Aus der Medizinischen Poliklinik der Universität Freiburg/Br.
(Direktor: Prof. Dr. H. Sarre)

24 Stunden-Rhythmik
der Serumkonzentration von PBI[131]

Von

D. P. Mertz

Mit 2 Abbildungen

Ebenso wie bei Tier und Pflanze ist die Tagesperiodik von Funktionskreisen beim Menschen in den meisten Fällen durch einen sinusförmigen Ablauf, dessen Wendepunkte in den frühen Morgen- und Nachmittagsstunden liegen, gekennzeichnet (Wachsmuth 1945). Nach Aschoff (1954, 1955) beruht die biologische Tagesperiodik nicht auf exogenen Faktoren, sondern auf einer endogenen selbsterregten Komponente, die nach Ausschluß von Umweltfaktoren als Eigenfrequenz mit einer bestimmten Periodendauer nachweisbar ist. Pittendrigh (1961) schrieb der Biorhythmik die Eigenschaften eines Systems zusammengekoppelter Oscillatoren zu. Beim Menschen fand Aschoff (1963), daß eine von der Eigenfrequenz der Oscillatoren bestimmte Spontanperiode unter den Bedingungen einer geringen Beleuchtungsstärke zwischen 24,6 und 26 Std dauert. Durch verschiedene physikalisch-chemische und soziologische Zeitgeber (Aschoff 1954) können die Spontanfrequenzen und die Phasenlage in begrenztem Umfang moduliert und auf die Erdumdrehung eingestellt werden. Hierbei ist zu berücksichtigen, daß Niveau und Frequenz der Eigenschwingung bei lichtaktiven Tieren und möglicherweise auch beim Menschen in direktem Zusammenhang mit der Lichtintensität stehen.

Auf endokrinologischem Gebiet sind wir auf Grund der in den letzten 20 Jahren durchgeführten Untersuchungen am besten über die Tagesperiodik der 17-Hydroxycorticosteroide (Bliss et al. 1953; Tyler et al. 1954; Arner et al. 1962) und der 17-Ketosteroide (Pincus 1943; Wooster 1943; Zimmermann und Hofschläger 1953) orientiert. Der Jodhaushalt der Schilddrüse wurde auf ein tagesrhythmisches Verhalten so gut wie überhaupt nicht geprüft. Nach allem, was bis heute über die Tagesperiodik im allgemeinen bekannt ist, dürfte eine solche Tagesrhythmik auch für den endogenen Jodstoffwechsel zu erwarten sein.

Da der Zweiphasen-Radiojodtest (Horst 1952) die Grundlage der meisten Jodstoffwechselstudien in der klinischen Medizin darstellt, interessierte uns die Frage, ob man bei der Auswertung eines Radiojodtestes etwa ebenso tagesrhythmische Schwankungen mit berücksichtigen muß, wie dies beispielsweise für das Elektrokardiogramm (Engelbertz und Hildebrandt 1953) zutrifft.

In orientierenden Untersuchungen an 15 euthyreoten, gesunden Studenten, die über die Versuchsdauer ihre üblichen Lebensgewohnheiten so weit wie möglich beibehielten, prüften wir die Serum-PBI¹³¹-Konzentration. 48 Std nach Gabe einer Spürdosis von trägerfreiem NaJ¹³¹ und dann fortlaufend alle 4 Std bis 24 Uhr sowie um 8 Uhr vormittags des folgenden Versuchstages wurde ambulant (in der Wohnung der Probanden) Venenblut zur Doppelbestimmung abgenommen.

Wie Abb. 1 zeigt, lassen sich bei 11 der 15 Versuchspersonen tagesrhythmische Schwankungen der PBI¹³¹-Serumkonzentration einwandfrei nachweisen, wobei das

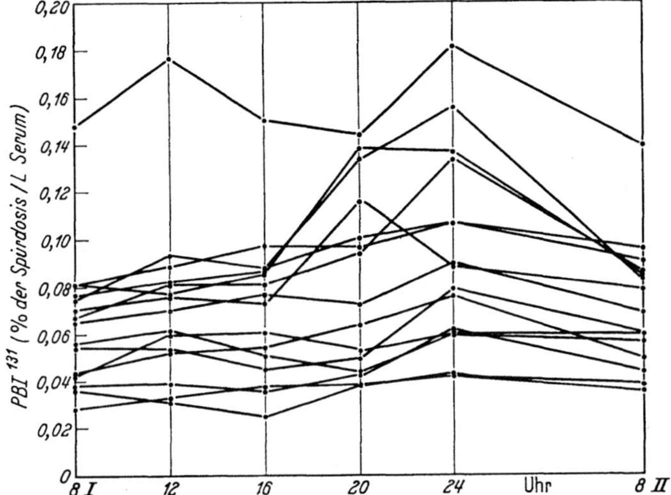

Abb. 1. Individuelles Verhalten tagesrhythmischer Schwankungen der PBI¹³¹-Konzentration im Serum

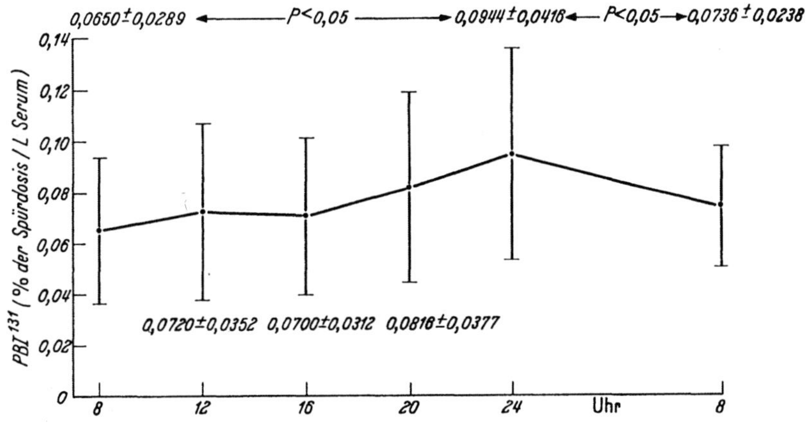

Abb. 2. 24 Stunden-Rhythmik der PBI¹³¹-Konzentration im Serum

Maximum in 9 Fällen um Mitternacht und bei 2 Probanden um 20 Uhr auftritt. Keine Tagesperiodik konnte bei 4 Personen mit sehr niedrigen PBI¹³¹-Werten gefunden werden. Es ist möglich, daß tagesrhythmische Schwankungen hierbei durch analytische Ungenauigkeiten maskiert wurden. Bei 2 weiteren Fällen beobachteten wir einen kleinen Gipfel um 12 Uhr, der jedoch den mitternächtlichen Maximalwert nicht erreichte.

Nach Abb. 2 ist das mitternächtliche Maximum der PBI[131]-Konzentration im Kollektiv signifikant von einem morgendlichen Minimum um 8 Uhr different. Der 24 Uhr-Mittelwert übertrifft die jeweiligen 8 Uhr-Bezugsmittelwerte um 44 bzw. 27%. Da die mittleren PBI[131]-Konzentrationen um 12 und 16 Uhr (mit Ausnahme von 2 Fällen) praktisch vom morgendlichen Wert um 8 Uhr nicht abweichen, ist es für die Durchführung eines Radiojodtestes irrelevant, ob dieser am Vormittag oder in den frühen Nachmittagsstunden begonnen wird.

Da es sich in der Mehrzahl der Versuchspersonen um sog. Nachtarbeiter handelt, bleibt die Frage zu diskutieren, inwieweit diese Besonderheit einen Einfluß auf die beschriebene PBI[131]-Tagescharakteristik ausübt. Nach Aschoff (1963) ist ja bekannt, daß Umweltfaktoren zu empfindlichen Störungen des „zirkadianen Systems" (Halberg 1960) führen können. In diesem Zusammenhang sind ältere Befunde von Bergfeld (1930) erwähnenswert, wonach Lichtmangel an der Ratten-schilddrüse morphologische Veränderungen hervorruft, die für eine Funktions-steigerung sprechen, während sich bei Dauerbelichtung eine normale Schilddrüsen-struktur fand. Eine Antwort auf diese Frage wird man nur, wie Menzel in seiner Monographie (1962) ausführt, bei Simultanbestimmung mehrerer Parameter, die im vorliegenden Fall den Hormonjodstoffwechsel und andere Hormonsysteme um-fassen müßten, gewinnen können. Auf Grund unserer Befunde glauben wir jedoch, daß der Dynamik des Jodstoffwechsels eine ungefähre „Tag-Nacht-Rhythmik" überlagert ist, auch wenn die Ergebnisse einiger Autoren, die zu verschiedenen Tageszeiten die proteingebundene Jodkonzentration im Serum chemisch bestimmt haben, dem zu widersprechen scheinen. Tingley et al. (1958) sowie Schatz und Volpé (1959) stellten keine tagesrhythmischen Schwankungen fest. Bei einer Aus-wertung dieser Befunde muß man indessen zu bedenken geben, daß die Intervalle der Messungen zu groß gewesen sind bzw. zeitlich ungünstig gelegen haben. Tingley et al. (1958) entnahmen zwischen 17 Uhr und 7 Uhr, Schatz und Volpé (1959) zwi-schen 17.30 Uhr und 3 Uhr morgens kein Blut. Zum anderen räumten die Autoren selbst die Möglichkeit ein, daß die angewandte Technik der PBI-Bestimmung (nach Barker et al. 1951) zu wenig empfindlich war, um Tagesschwankungen zu erfassen. Erwähnenswert ist ein Ergebnis von Walfish et al. (1961), wonach die Verschwinderate von etikettiertem L-Thyroxin nach einmaliger Injektion aus dem Serum euthyreoter Personen im Laufe des Tages progressiv zunimmt und um 2 Uhr morgens ein Maximum erreicht. Als Erklärung wurden Veränderungen des Plasmavolumens angenommen. Für die von uns gefundenen tagesrhythmischen Schwankungen der PBI[131]-Konzentration im Serum können Änderungen des Plasma-Volumens wahrscheinlich nicht verantwortlich gemacht werden, da dann infolge eines „Verdünnungseffektes" um Mitternacht nicht ein Maximum, sondern ein Minimum hätte auftreten müssen. Gemessen am Hämatokrit und Hämoglobin-gehalt sowie an der Plasma-Eiweißkonzentration ist das Plasma-Volumen gegen 2.30 Uhr morgens am größten (Renbourn 1947).

Vorerst muß die Frage unbeantwortet bleiben, ob es sich bei der Tagesperiodik der Dynamik des endogenen Jodstoffwechsels um ein zentrales Phänomen handelt und/oder ob periphere Interferenzen mit anderen Faktoren eine Rolle spielen. Kausalzusammenhänge mit Tagesgängen von anderen Komponenten liegen natür-lich immer nahe und sind auf dem Gebiet der Tagesperiodik oft viel zu früh behaup-tet worden. Leider stehen wir hier ganz am Anfang unseres Verständnisses.

Zusammenfassung

Die Frage nach tagesperiodischen Schwankungen des endogenen Jodstoffwechsels ist bisher praktisch nicht bearbeitet worden. In orientierenden Untersuchungen an 15 euthyreoten gesunden jungen Männern fand sich ein mitternächtliches Maximum der Serumkonzentration von PBI[131], das im Kollektiv von einem Minimum um 8 Uhr morgens signifikant unterschiedlich ist. Individuell ergaben sich manche Abweichungen vom Gruppenverlauf. Wahrscheinlich ist der Dynamik des Jodstoffwechsels eine ungefähre Tag-Nacht-Rhythmik überlagert. Über Kausalzusammenhänge läßt sich gegenwärtig nichts aussagen.

Literatur

ARNER, B., P. HEDNER u. T. KARLEFORS: Acta endocr. (Kbh.) **40**, 421 (1962).

ASCHOFF, J.: Naturwissenschaften **41**, 49 (1954); **42**, 569 (1955).

— Dtsch. med. Wschr. **88**, 1930 (1963).

BARKER, S. B., M. J. HUMPHREY, and M. H. SOLEY: J. clin. Invest. **30**, 55 (1951).

BERGFELD, W.: Endokrinologie **6**, 269 (1930).

BLISS, E. L., A. A. SANDBERG, D. H. NELSON and K. EIK-NES: J. clin. Invest. **32**, 818 (1953).

ENGELBERTZ, P., u. G. HILDEBRANDT: Arch. Physik. Ther. **5**, 361 (1953).

HALBERG, FR.: Cold Spr. Harb. Symp. quant. Biol. **25**, 289 (1960).

HORST, W.: Klin. Wschr. **30**, 439 (1952).

MENZEL, W.: Menschliche Tag-Nacht-Rhythmik und Schichtarbeit. Basel und Stuttgart: B. Schwabe & Co. 1962.

PINCUS, G.: J. clin. Endocr. **3**, 195 (1943).

PITTENDRIGH, C. S.: Harvey Lect. **56**, 93 (1961).

RENBOURN, E. T.: J. Hygiene **45**, 455 (1947).

SCHATZ, D. L., R. VOLPÉ: J. clin. Endocr. **19**, 1495 (1959).

TINGLEY, T. O., A. W. MORRIS and S. R. HILL: Clin. Res. Proc. **6**, 134 (1958).

TYLER, F. H., C. MIGEON, A. A. FLORENTIN, and C. T. SAMUELS: J. clin. Endocr. **14**, 774 (1954).

WACHSMUTH, G.: Erde und Mensch. Kreuzlingen-Zürich 1945.

WALFISH, P. G., A. BRITTON, P. H. MELVILLE, and C. EZRIN: J. clin. Endocr. **21**, 582 (1961).

WOOSTER, H.: J. clin. Endocr. **3**, 483 (1943).

ZIMMERMANN, W., u. J. HOFSCHLÄGER: Acta endocr. (Kbh.) **12**, 255 (1953).

Aus der II. Med. Univ.-Klinik Wien
(Vorstand: Prof. Dr. K. Fellinger)

Akropachie — eine seltene Komplikation der Basedowschen Erkrankung

Von

R. Höfer und E. Ogris

1920 untersuchte der Wiener Internist Högler eine größere Anzahl von Fällen mit Trommelschlegelfingern und uhrglasförmig veränderten Nägeln, und es fiel ihm dabei auf, daß diese Erscheinungen nicht nur bei Lungen-, sondern auch bei Herz-, Leber- oder Darmerkrankungen vorkommen können. Er schlug daher als übergeordnete Bezeichnung den Ausdruck *Akropachie* vor, um der verwirrenden Fülle der damals verwendeten Namen für das beschriebene Zustandsbild zu entgehen. 1933 wurde erstmals von Thomas eine Akropachie im Zusammenhang mit einer Schilddrüsenerkrankung beobachtet und deshalb *"thyroid acropachy"* genannt.

Seither wurden unseres Wissens insgesamt 30 weitere Fälle publiziert (s. Tab. 1).

Die unter dem Begriff *"thyroid acropachy"* zusammengefaßten Veränderungen bestehen im einzelnen in der Ausbildung von Trommelschlegelfingern mit Uhrglasnägeln und analogen Veränderungen an den Zehen. Im Bereich des Skeletes lassen sich subperiostale Knochenneubildungen an den Diaphysen der Grund- und etwas seltener auch der Mittelphalangen sowie an den Metacarpal- oder Metatarsalknochen nachweisen. In 2 Fällen wurden ähnliche Skeletveränderungen auch an langen Röhrenknochen beobachtet (Thomas 1933, Gimlette 1960). Schließlich kommt es noch zu charakteristischen Schwellungen der Weichteile über den soeben beschriebenen Knochenveränderungen.

Diese Erscheinungen traten bei allen 30 Fällen und auch bei dem von uns beobachteten zusammen mit einem Exophthalmus auf und mit Ausnahme zweier Fälle von Freeman (1938) bestand auch stets ein Myxoedema circumscriptum praetibiale.

In der Regel tritt die Akropachie erst nach erfolgreicher Therapie der Hyperthyreose auf, gleichgültig ob chirurgisch, medikamentös oder mit Radiojod behandelt worden war. Das Intervall zwischen Beginn der Hyperthyreose und Auftreten der Akropachie variierte zwischen 8 Monaten und 22 Jahren. Die Stoffwechsellage war zum Zeitpunkt des Auftretens der Akropachie in rund einem Drittel der Fälle euthyreot und in den anderen zwei Dritteln hypothyreot. Ausnahmen für dieses sehr verallgemeinernde Schema bilden 5 der bisher veröffentlichten Fälle:

Bei drei Patienten (Gimlette, 1960, Freeman, 1958, Jallut, 1962) traten die Symptome einer Akropachie bei unbehandelten und eindeutig hyperthyreoten Patienten auf. Die beiden weiteren Ausnahmen bildeten die Fälle von Naydar

(1963) und GREENE (1951), bei denen sich anamnestisch keinerlei Hinweise für eine hyperthyreote Phase erheben ließen, obwohl ein beträchtlicher Exophthalmus und ein prätibiales Myxödem bestanden.

Es sei darauf hingewiesen, daß die Akropachie bisher nie bei Kindern, sonst aber in allen Lebensaltern beobachtet wurde. Da von den bisher beobachteten 31 Patienten etwa die Hälfte Männer waren, scheint diese Komplikation der Hyperthyreose bevorzugt bei Männern aufzutreten, da ja die Schilddrüsenüberfunktion selbst eindeutig das weibliche Geschlecht bevorzugt.

1. Kasuistik

In unserem Falle handelte es sich um eine 48jährige Patientin (H.B., Krankengeschichte 0643, II. Med. Univ.-Kl., siehe auch Tab. 1, Fall 31), die seit 1959 an Herzklopfen und leichter Belastungsdyspnoe litt. Im Dezember 1960 kam es im Anschluß an einen grippalen Infekt zum Auftreten eindeutig hyperthyreoter Symptome mit Gewichtsabnahme (6 kg in 3 Monaten), Zunahme des Halsumfanges und Auftreten eines Exophthalmus mit leichten Lidödemen. Einige Wochen vor der Aufnahme im März 1961 war auch ein beidseitiges circumscriptes prätibiales Myxödem aufgetreten.

Die Thyreoidea war diffus vergrößert, Schwirren nicht palpabel. Die Konvergenz der Bulbi war gestört, eine Blickbewegung über die Horizontale nach oben nicht möglich. Für eine pulmonale oder kardiale Erkrankung war klinisch und röntgenologisch kein Anhaltspunkt gegeben.

Der Radiojodplasmatest (FELLINGER et al., 1953) war eindeutig im Sinne einer Hyperthyreose verändert. Als Therapie der Wahl wurde die Radiojodausschaltung angesehen und am 23.3.1961 mit 6 mC durchgeführt. Innerhalb kurzer Zeit kam es zu einer kompletten Remission der hyperthyreoten Symptome. Bei den nächsten Kontrollen fiel die Verschlechterung des Exophthalmus und Abnahme des Sehvermögens auf. Drei Monate nach der Radiojodtherapie bemerkte die Patientin erstmals Weichteilschwellungen über den Fingergrundphalangen und Uhrglasnägel an beiden Händen sowie analoge Veränderungen in nicht so starkem Maße an den Füßen. Radiologisch ließen sich in diesem Bereich eindeutig Veränderungen im Sinne subperiostaler kleiner, teilweise ballonförmiger Knochenappositionen an den Grund- und Mittelphalangen beider Hände nachweisen.

Die Vestibulariserregbarkeit und das Röntgenbild der Sella waren normal. Es bestanden ein Ausfall im Bereich der unteren Gesichtsfeldhälfte am linken Auge sowie eine Einschränkung der Motilität beider Bulbi ohne Doppelbildangabe.

Man entschloß sich wegen des zunehmenden Exophthalmus zur Röntgenbestrahlung der Hypophyse und der Retroorbitalregion. Zusätzlich wurde Thyreoidea sicca gegeben. Nach anfänglicher Besserung auf die Bestrahlung kam es im November 1961 wieder zu einer Sehverschlechterung und neuerdings zum Auftreten hyperthyreoter Symptome. Der Radiojodplasmatest (FELLINGER et al., 1953) war eindeutig im Sinne einer Hyperthyreose verändert, ein Kontroll-GU unter Nembutal (HÖFER und VETTER, 1958) betrug +28%. Aus diesem Grunde wurde abermals eine Dosis von 6 mC Jod-131 verabreicht. Erst danach kam es zu einer sukzessiven Besserung nicht nur der subjektiven hyperthyreoten Beschwerden sondern auch zur Besserung des Augenbefundes: Die Kontrolle der Gesichtsfelder zeigt nur mehr ein kleines Restskotom am linken Auge bei praktisch normalen Außengrenzen.

2. Diskussion

Thomas vertrat die Meinung, daß die "thyroid acropachy" im wesentlichen nichts anderes darstellte als das Auftreten einer Osteoarthropathie nach einer bisher noch nicht beschriebenen Grundkrankheit. Dieser Meinung sind auch DEGOS (1962), ELSON (1961) und JALLUT (1961). Seit GIMLETTE (1960) hingegen werden vielfach die Veränderungen bei der thyreogenen Akropachie von jenen der Osteoarthropathie deutlich abgegrenzt. Wir selbst möchten uns der letztgenannten Meinung anschließen.

15*

Tabelle

Zahl	Autor	Geschlecht	Alter b. d. 1. Unters.	Hyperthyreose einst oder jetzt Beginn	Rezidiv	Exophthalmus	prätib. Myxödem	Weichteilschwellungen	Trommelschlegelfinger	Radiologische Veränderungen
1	THOMAS 1933	m	22	1924		+	+	+	+	+
2	CUSHING 1937	m	28	1931		+	+	0	+	?
3	RENEARSON SACASA 1941	m	43	1937		+	+ 1940	0	+	+
4	GREENE 1951	m	74	1932		+++	+	+	+	+
5		m	?	unklar		+++	?	+	+	0
6	LEVITT 1953	m	?	?		+	+	+	+	+
7	ENGEL 1953	w	41	1949		+++	+	+	+	+
8	INCH ROLLAND 1953	w	36	1951		++	+	0	+	?
9		m	39	1933		+ n. Operat.	+	+	+	?
10	WARTHUN BOSHELL 1957	w	35	1945	1947	+++ n. Operat.	+ 1946	+	+	?
11	McGILL 1957	m	51	?		+	+	0	+	+
12	FREEMAN 1958	m	56	1929		+++	0	0	+	+
13		m	54	1956		+++	0	0	+	?
14	DANFORTH HUMPHREY 1958	m	46	1952		+	+	+	+	+
15	KING 1959	w	45	1954	1954 1955	+	0	+	+	+
16	DIAMOND 1959	m	24	1955	1957 1958	+	+	+	0	+
17	GIMLETTE 1960	m	61	1949		+	+ 1951	+	+	+
18		w	49	1955		+++	+ 1956	+	+	+
19		w	51	1946		+	+ 1946	+	+	0
20		w	58	1950 ohne Exophth.	1957	+ n. Thyreoidektomie	+ 1951	+	0	+
21		w	62	1944 ohne Exophth.		+ n. Thyreoidektomie	+ 1944	+	+	+
22		w	36	1955	1959	+	+ 1955	+	+	+
23	DEGOS 1961	w	34	1959		+	+	+	+	0
24	ELSON 1961	w	53	1958		+	+	+	0	+
25	VERNEY 1962	w	52	1946		+	+	+	+	+
26	JALLUT u. Mitarb. 1962	w	53	1954	1956 1957	+ n. Therapie	+	+	+	+
27		m	38	1952		+	+	+	+	?
28		w	31	1954	1955 1957	+	+	+	+	+
29	NAYDAR 1963	m	49	?		+	+	+	+	+
30	MALKINSON 1963	w	48	1955		+	+	+	+	0
31	HÖFER 1964	w	48	1961	1961	+++	+	+	+	+

vorangegangene Therapie TE = Thyreoidektomie	Schilddr.-Funkt. b. Beginn d. AP	Zeit vom Beginn d. Hyperth. bis Auftr. d. AP	Bemerkungen
TE 1927	Hypo	8 Monate	erstmals ges. im hyperth. Zustand vor d. Ther., 1931 Knochenneubild. an Radius u. Ulnadiaph. Sella normal
TE	Hypo?	8—9 Jahre	erstmals ges. im hyperth. Zustand
TE 1938	Hypo	6 Wochen n. TE	erstmals ges. euth. n. Therapie
KJ	Euth	18 Jahre[1]	erstmals ges. euth., 18 Jahre n. d. Therapie, 1932 v. d. Therapie bereits Trommelschlegelf. vorh., jedoch erst 18 Jahre später schwer
?	Hypo	?	Hyperth. nicht nachgewiesen, Praezipit. AK positiv
?	?	?	
KJ, Propylthiourazil, TE	Hypo	1 Jahr	erstmals ges. euth., 4 Mo n. d. Therapie, Hypophysenbestrahlung
Methylthiourazil	Hypo	9 Monate	erstmals ges. n. d. Operation, Hypophysenbestr. ohne Erfolg
Bestrahlung	Euth.	?	erstmals ges. euth., 19 Jahre n. d. Therapie, Exophth. bald nach Bestr., Prätib. Myxödem nach 19 Jahren
TE	Euth	2 Jahre n. TE	erstmals ges. euth., 2 Jahre nach d. Therapie
Thiourea, TE	?	2 Jahre n. TE	erstmals ges. hyperth. vor d. Therapie, Exophth. folgte
TE 1929, Bestrahlung	Hypo	28 Jahre	erstmals ges. euth., 28 Jahre n. Therapie
J 131	Hyper	—	im Stadium einer floriden Hyperthyreose!
J 131	Euth	3 Jahre	erstmals ges. euth., 5 Jahre n. d. Therapie
Tapazol, J 131	Euth	2 Jahre n. J 131	Biopsie im Bereich d. Veränderungen d. Akropachie durchgeführt
J 131, TE 1952, J 131	Euth	3 Jahre	Sella normal
TE 1950	Hypo	6 Jahre n. TE	auch dist. Enden von Radius, Ulna, Tibia und Fibula betroffen, Sella normal
TE 1956	Hypo	2 Jahre n. TE	Auch Zehen betroffen, Praezipit. AK negativ
TE 1946	Euth	12 Jahre n. TE	
TE 1951, J 131	Euth	8 Jahre n. TE	keine Trommelschlegelfinger wie bei DIAMOND
TE 1944	Hypo	13 Jahre?	TRC-Test negativ
TE 1955, J 131	Hyper	—	im Stadium der Hyperthyreose!
J 131	Hypo	2 Jahre	
J 131	Hypo	2 Jahre	beide I. Metacarp. u. Grundphalangen am stärksten betroffen
Thiourazil, TE 1951	Euth	10 Jahre n. TE	Thoraxröntgen und Sella normal, hoher AK-Titer im TRC-Test!
Methimazol, J 131, J 131	Hypo	4 Jahre	Schmerzen im Bereich der Veränderungen durch d. Akropachie
Methylthiourazil TE	Euth	2 Jahre	isolierte akromegaloide Veränderung einer Großzehe, Sella an d. oberen Grenze d. Norm
J 131	Hyper	1 Jahr	aufgetreten während eines Rezidivs bei einer Gravidität, Bestimmung des STH ergab normale Werte
Substitution mit Thyreoidea sicca	Hypo		Histologisch verifizierte Struma lymphom. Hashimoto ohne vorausgeg. Hyperthyreose!
J 131	Hypo	3 Jahre	Prätibiales Myxödem bereits vor d. Therapie
J 131, J 131	Euth	6 Monate n. J. 131	hoher AK-Titer im TRC-Test

[1] MCGILL berichtet, daß später autoptisch ein eosinophiles Adenom des HVL gefunden wurde.

Die klassische Osteoarthropathie ist gekennzeichnet durch das Auftreten von Trommelschlegelfingern und Uhrglasnägeln, sowie analogen Veränderungen an den Zehen. An Skeletveränderungen beobachtet man subperiostale Knochenneubildungen im Bereich der Diaphysen der Phalangen, an den langen Röhrenknochen der Extremitäten, sowie, wenn auch seltener, Veränderungen an weiteren Skeletteilen wie an Rippen, Beckenknochen, Claviculae, Processus transversi der Wirbelsäule und der Maxilla. Darüber hinaus kommt es in etwa einem Drittel der Fälle zu Gelenksveränderungen an Hand-, Fuß-, Ellbogen- und Kniegelenken von Art einer Periarthritis, manchmal auch zu einer Synovitis. In schweren Fällen können diese Gelenksveränderungen zu Arrosionen der Gelenksflächen führen. Weichteilschwellungen werden vorwiegend in den distalen Anteilen der Unterarme und der Unterschenkel beobachtet.

An Verschiedenheiten zwischen der Osteoarthropathie und der thyreogenen Akropachie lassen sich also anführen:

1. Bei der Akropachie sind von den Skeletveränderungen nahezu ausschließlich die Diaphysen von Metacarpal- bzw. Metatarsalknochen sowie die Grundphalangen von Händen und Füßen befallen, während bei der Osteoarthropathie nahezu das ganze Skelet verändert sein kann.

2. Die Weichteilschwellungen liegen bei der Akropachie ausschließlich über den Knochenveränderungen.

3. kommt es bei der Akropachie niemals zu Gelenksveränderungen, eventuelle Bewegungseinschränkungen gehen ausschließlich auf die Behinderung durch Weichteilschwellungen zurück.

4. Schließlich wird auch noch auf röntgenologische Verschiedenheiten der Knochenveränderungen hingewiesen. So beschreiben Lionel und King (1959), sowie Verney (1962), daß bei der Akropachie die Knochenneubildungen senkrecht zur Längsachse des Röhrenknochens stünden und einen rauhen und unregelmäßigen Eindruck machten ("bubbled surface" nach Verney). Bei der Osteoarthropathie hingegen sind die Knochenbälkchen, wie auch histologisch verifiziert, parallel zur Längsachse des Knochens in mehreren dünnen Lagen angeordnet, womit die Veränderungen regelmäßig und leicht gewellt erscheinen.

5. Die Trommelschlegelfinger sowie die Uhrglasnägel sind bei beiden Erkrankungen identisch.

Wir glauben also im Hinblick auf die oben angeführten Argumente, daß sich Osteoarthropathie und thyreogene Akropachie nicht nur durch die Grundkrankheiten, sondern auch durch die Charakteristika der Veränderungen unterscheiden lassen.

Ebenso wie die Genese der Osteoarthropathie ist bisher auch die der Akropachie nicht befriedigend geklärt. Es wurde die Einwirkung hypophysärer Faktoren auf das Bindegewebe vermutet und die thyreogene Akropachie in das thyreohypophysäre Syndrom nach Lamberg (1954) eingeordnet. Argumente für diese Meinung brachte Freeman (1958), der bei 2 Fällen mit "thyroid acropachy" autoptisch einmal eine Hyperplasie der eosinophilen Zellen des HVL und einmal ein eosinophiles Adenom fand. Auch McGill (1957) fand bei der Obduktion eines Falles von Greene ein eosinophiles Adenom des Hypophysenvorderlappens.

Als weiteres Argument für die hypophysäre Genese der Akropachie führt Freeman noch das Zusammentreffen von Schilddrüsenfunktionsstörung mit Exophthalmus und prätibialem Myxödem mit der Akropachie an. Gilliland und Strudwick (1959) fanden im Serum zweier Patienten mit thyreogener Akropachie keine erhöhten TSH-Werte. Diesem Bericht ist allerdings wohl kaum allzuviel Gewicht beizulegen, da die angewendete Methode und die Interpretation ihrer Ergebnisse problematisch sind.

JALLUT (1962) konnte in einem Fall neben den typischen Symptomen einer Akropachie eine isolierte akromegaloide Veränderung an einer Großzehe zeigen und schloß auf eine gemeinsame, zumindest aber verwandte Ursache. Dagegen spricht aber, daß die Akromegalie der Akropachie nur in ihren Anfangsstadien ähnelt, und die Bestimmung des Wachstumshormons bei einem seiner Patienten normale Werte ergab.

THOMAS glaubte die Ursache für das Auftreten einer Akropachie in einer Veränderung der Zirkulationsgrößen im Bereich der Acren zu finden, wie sie beim Übergang einer Hyperthyreose in Eu-, bzw. Hypothyreose auftreten. Gegen diese Theorie spricht die Tatsache, daß Intervalle bis zu 22 Jahren zwischen dem Ausbruch der Hyperthyreose und dem Erscheinen der Akropachie beobachtet wurden, sowie auch, daß die Akropachie auch bei eindeutig hyperthyreoter Stoffwechsellage auftreten kann.

Die eigenartige Lokalisation sowie die Beschränkung der Veränderungen auf einen so kleinen Bereich des Körpers machen es allerdings wahrscheinlich, daß zusätzlich auch noch lokale Faktoren wirksam sind: eine Überlegung, die in gleicher Weise für das prätibiale Myxödem und nicht zuletzt ja auch für den Exophthalmus gilt.

Eine neue Erklärungsmöglichkeit, die vielleicht sowohl einer humoralen Genese als auch einer lokalen Prädisposition gerecht wird, ergibt sich aus eigenen Beobachtungen.

Bei unserer Patientin mit thyreogener Akropachie war der TRC-Test auf zirkulierende Schilddrüsenautoimmunkörper mit dem sehr hohen Titer von 1:2,5 Mill. positiv. Bei genauer Durchsicht der Literatur konnten wir nun finden, daß bei den Fällen von ELSON und VERNEY ebenfalls im TRC-Test auffallend hohe Titer (1:250000, bzw. 1:2,5 Mill.) gefunden wurden. Bei dem von NAYDAR veröffentlichten Fall wurde eine Struma lymphomatosa diagnostiziert, beim zweiten, von GREENE veröffentlichten Fall können wir mit großer Wahrscheinlichkeit eine solche vermuten.

Diese auffallende Beobachtung veranlaßte uns dazu, alle jene Patienten nachzuuntersuchen, bei denen eine andere seltene und circumscripte Komplikation einer Hyperthyreose bestand: das prätibiale Myxödem. Von insgesamt 9 solcher Patienten, bei denen der Beginn des prätibialen Myxödems bis zu 6 Jahren zurücklag, wurde bei 8 ein positiver Ausfall des TRC-Testes gefunden. Die Titerstufen waren: dreimal 1:250, zweimal 1:2500, zweimal 1:25000 und einmal 1:2,5 Mill. Es ist dies zweifellos ein auffallendes Ergebnis, wenn man bedenkt, daß ROITT und DONIACH (1960) bei Hyperthyreosen nur in 63% der Fälle positive Ergebnisse fanden, wobei nur in einigen wenigen Fällen die Titer über 1:25000 lagen. Noch auffallender ist, daß auch schon eben diese Autoren 1960 zu folgenden Betrachtungen kommen: sie konnten feststellen, daß, obwohl die Struma lymphomatosa Hashimoto bei Frauen 17mal häufiger vorkommt, als bei Männern, die Incidenz von lymphadenoiden Veränderungen und letzten Endes Myxödemen für beide Geschlechter gleich ist. Darüber hinaus wurden bei Patienten mit progressivem Exophthalmus in außergewöhnlich hohem Prozentsatz Schilddrüsenautoimmunkörper von hohem Titer gefunden. Eine auffallende Parallelität, wenn man bedenkt, daß bei der "thyroid acropachy" ebenfalls Männer und Frauen im gleichen Maße betroffen sind.

Wir möchten deshalb als weitere Theorie der Entstehung der "thyroid acropachy" und wohl auch des prätibialen Myxödems die Möglichkeit einer autoimmunologischen Genese zur Diskussion stellen, wie dies Roitt und Doniach in gewissem Sinn für den malignen Exophthalmus bereits getan haben.

Literatur

Bamberger, E.: Z. klin. Med. 18, 193 (1890).
Camp, J. D., and R. L. Scanlan: Radiology 50, 581 (1948).
Cushing, E. H.: Int. Clin. 2, 200 (1937).
Danforth, W. H., and H. A. Humphrey: J. clin. Endocr. 18, 1302 (1958).
Degos, R., J. Delort et R. Touraine: Bull. Soc. franç. Derm. Syph. 68, 155 (1961).
Diamond, M. Th.: Ann. intern. Med. 50, 206 (1959).
Ebstein, E.: Derm. Wschr. 68, 113 (1919).
Elson, G. D.: Amer. J. Rœntgenol. 85, 114 (1961).
Engel, F. L.: J. clin. Endocr. 13, 1132 (1953).
Fellinger, K., E. Mannheimer u. H. Vetter: Wien. Z. inn. Med. 34, 359 (1953).
Flavell, G.: Lancet 1956 I, 260
Freeman, A. G.: Lancet 1958 II, 57.
Fried, B. M.: Arch. int. Med. 72, 565 (1943).
Gilliland, I. C., and J. I. Strudwick: Brit. med. J. 1956 I, 378.
Gimlette, T. M. D.: Lancet 1960 I, 22.
Ginsburg, J.: Quart. J. Med. 27, 335 (1958).
—, and J. B. Brown: Lancet 1961 II, 1274.
Greene, R.: Proc. roy. Soc. Med. 44, 159 (1951).
Hall, G. H.: Lancet 1959 I, 750.
Höfer, R., u. H. Vetter: Wien. klin. Wschr. 70/10, 169 (1958).
Högler, F.: Wien. Arch. inn. Med. 1, 35 (1920).
Holling, H. E., and R. S. Brodey: J. Amer. med. Ass. 178, 977 (1961).
— — and H. C. Boland: Lancet 1961 II, 1269.
Inch, R. S., and C. F. Rolland: Lancet 1953 II, 1239.
Jallut, O., M. P. Koenig u. A. Labhart: Schweiz. med. Wschr. 92, 225 (1962).
King, L. R.: J. clin. Endocr. 19, 1323 (1959).
Lamberg, B. A.: Acta med. scand. 148, 225 (1954).
Levitt, T.: Lancet 1953 II, 1359.
Malkinson, F. D.: Arch. Derm. 88, 303 (1963).
McGill, D. A.: Guy's Hosp. Rep. 106, 47 (1957).
Mendlowitz, M.: Medicine 21, 269 (1942).
Naydar, A.: J. clin. Endocr. 23, 215 (1963).
Roitt, I. M., D. Doniach: Brit. med. Bull. 16/2, 152 (1960).
Rynearson, E. H., and C. F. Sacasa: Proc. Mayo Clin. 16, 353 (1941).
Thomas, H. M.: Arch. int. Med. 51, 571 (1933).
Verney, G. I.: Brit. J. Radiol. 35, 644 (1962).
Warthun, T. A., and B. R. Boshell: Arch. int. Med. 100, 319 (1957).

Aus der Medizinischen Fakultätsabteilung des allgemeinen Krankenhauses „Dr. O. Novosel“,
Universität Zagreb, Jugoslawien (Vorstand: Prof. Dr. E. Hauptmann)

Ergebnisse cytochemischer Untersuchungen an Schilddrüsenpunktaten[1]

Von

Z. Škrabalo[2]

Die einzelnen Phasen in der Funktion einer endokrinen Drüse kann man histologisch verfolgen. Das cytologische Studium der Schilddrüse mittels Aspirationsmaterial ist jedoch eine erst seit kurzer Zeit bekannte Untersuchungsmethode (Söderström 1952; Škrabalo 1960, 1961; Črepinko 1960; Myren 1962).

Einen weiteren Fortschritt des Studiums an der einzelnen Schilddrüsenzelle stellt die Cytochemie dar. Die Cytochemie vermag gewisse Feinstrukturen der Zelle sichtbar zu machen, einen Einblick in die Biochemie der feineren Elemente der Zelle zu gewähren und biochemische Abläufe unter normalen und unter pathologischen Verhältnissen zu erfassen. (Wachstein 1949, 1955; Friederici 1955; Kaplow 1955; Kenny 1957; Hertl 1959; Lambers 1962, 1963; Bauer-Sič 1963.)

1. Arbeitsmethode

a) Die Aspirationspunktion

Bei der Aspirationspunktion an Schilddrüsenkranken waren wir bestrebt, Einsicht in die cytologischen Veränderungen an den Schilddrüsenzellen bei Hyper- und Hypothyreosen sowie beim euthyreotischen Kropf zu bekommen. Die Punktion wurde mit einer Nadel von 0,2—0,5 mm Durchmesser, ohne Anaesthesie mit Aspiration nach der Y-Technik ausgeführt. Das erhaltene Material wurde auf Objektträger ausgestrichen und für cytologische Untersuchungen nach der Methode Pappenheim (May-Grünwald-Giemsa) gefärbt. Dabei wurde nicht eine einzige Komplikation festgestellt.

b) Cytochemische Untersuchungen

Für cytochemische Untersuchungen sind bisher folgende Reaktionen zum Nachweis chemisch definierter Zellbestandteile angewandt worden:

a) Nachweis von Kohlenhydraten: Perjodsäure-Schiff-Methode (Hotchkiss 1948; Wachstein 1949; Gärtner 1961);

---

[1] Die Untersuchungen waren dank der verständnisvollen Unterstützung durch Herrn Prof. Dr. Jores und Herrn Prof. Dr. Nowakowski ermöglicht worden während meines Aufenthaltes als Humboldt-Stipendiat in Hamburg (II. Med. Universitätsklinik und Poliklinik) 1962/63.

[2] Infolge unverschuldet späten Eingangs der Vortragsmeldung konnte die Abhandlung nicht mit in das Programm aufgenommen, soll sie hier aber dennoch gedruckt werden.

b) Nachweis von Fetten: Sudanschwarz B-Methode und Sudan-III-Färbung (Sheehan u. Storey 1947; Romeis 1948; Kay 1941);

c) Nachweis von proteingebundenen Sulfhydrilgruppen: Ferroferricyanid-Methode (Chèvremont 1943; Campanacci 1955)

d) Nachweis von Ribonucleinsäure und Desoxyribonucleinsäure: Methylgrün-Pyronin-Methode (Perry 1956)

e) Nachweis von Peroxydase: Benzidinmethoden (Sato 1928; Graham-Knoll nach Undritz 1952)

f) Nachweis von alkalischer Phosphatase: Azofarbstoff-Methode (Kaplow 1955; Merker 1960)

g) Nachweis von saurer Phosphatase: Azofarbstoff-Methode (Gössner 1958; Löffler 1962)

Wir werden in dieser Mitteilung nur über die Ergebnisse mit der cytochemischen Perjodsäure-Schiff-Methode für den Nachweis von Kohlenhydraten berichten.

c) Perjodsäure-Schiff-Methode — Reaktionsverlauf

a) 10 min fixieren in Methanol und abspülen mit aqua bidest.

b) 25 min einstellen in 0,5%ige Perjodsäurelösung, abspülen mit aqua bidest.

c) 25 min einstellen in Schiffsches Reagens (1 g acridinfreies Pararosanilin in 30 ml 1 n-Salzsäure unter Schütteln lösen, 1 g Kaliumpyrosulfat in 170 ml aqua bidest. gelöst zufügen; 24 Std im Dunkel stehen lassen; 0,6 g Aktivkohle zusetzen, schütteln und filtrieren).

d) 3 mal je 5 min einstellen in frisches schwefeldioxydhaltiges Wasser (300 ml Leitungswasser, 15 ml 1 n-Salzsäure, 15 ml 10%iges Natriumpyrosulfit).

e) 10 min spülen in fließendem Wasser.

f) 8 min gegenfärben mit Hämalaun nach Mayer.

g) 20 min spülen in fließendem Wasser.

Kontrolle: Fixierte Ausstriche 3 Std bei 37° C einstellen in 0,1%ige Diastaselösung (0,1 Diastase hochgereinigt in 100 ml 0,9%iger Kochsalzlösung) und anschließend wie oben weiterbehandeln.

2. Material und Resultate

1962 und 1963 sind an der II. Med. Universitätsklinik und Poliklinik in Hamburg insgesamt 174 Schilddrüsenpunktionen durchgeführt worden. Von diesen Punktaten wurden 150 mit der Perjodsäure-Schiff-Methode für den Nachweis von Kohlenhydraten untersucht. Bei allen Patienten liegen die Ergebnisse der klinischen und Laboruntersuchungen vor (Grundumsatz, Radiojodstudium, Cholesterin, Elektrokardiogramm, Röntgenuntersuchung, cytologische Beurteilung des Aspirationspunktates der Schilddrüse usw.). Bei 57 Patienten ist auch eine histologische Untersuchung des Strumagewebes nach der Operation durchgeführt worden.

In den Ausstrichen von normalem Schilddrüsengewebe kann man folgende cytologische Elemente bemerken:

Thyreocyt oder Schilddrüsenparenchymzelle, welche sich cytologisch in zwei Formen manifestieren kann: inaktiver Thyreocyt und aktiver Thyreocyt. Ersterer ist eine kleine Zelle, dem Lymphocyten ähnlich, der einen runden Kern mit dichtem Chromatin und spärliches, scharf begrenztes blaues Cytoplasma hat. Der aktive Thyreocyt hat entweder einen größeren Kern oder reichliches Cytoplasma. Sein Kern ist rund oder oval mit lockerem Chromatin und kann scharf begrenzte Nucleolen enthalten. Das Cytoplasma, das oft reichlich und unscharf begrenzt ist, kann basophil oder acidophil sein.

Phagocyt. Eine Zelle mit kleinem, oft pyknotischem, an den Zellrand gedrückten Kern und reichlichem, schaumigem Cytoplasma, welches grünes oder schwarzes Pigment in verschiedenen Mengen enthält.

Die Schilddrüsenriesenzelle ist ein Konglomerat von radiär geordneten Thyreocytenkernen, deren Cytoplasma miteinander konfluiert. Wenn solch eine Riesenzelle Kolloid enthält, wird sie Mikrofollikel genannt.

Kolloid ist in den Ausstrichen nach PAPPENHEIM (MAY-GRÜNWALD-GIEMSA) violett bis dunkelblau gefärbt und kann in verschiedenen Mengen vorkommen. In ihm sind oft Thyreocyten und Phagocyten eingeschlossen (ŠKRABALO 1960, 1961).

Unsere Fragestellung war: Welche Veränderungen sind in Thyreocyten, Phagocyten und im Kolloid durch die Färbung nach der Perjodsäure-Schiff-Methode feststellbar, d. h. welche Zellbestandteile enthalten Kohlenhydrate oder kohlenhydratähnliche Substanzen?

In ihrer Beschreibung der Veränderungen in cytologischen Punktionen bei Hyperthyreose, Hypothyreose und euthyreotischer diffuser und knotiger Strumen haben einige Autoren (ŠKRABALO 1960, 1961, 1963; ČREPINKO et al. 1960) schon vor einigen Jahren auf die Tatsache hingewiesen, daß für das cytologische Bild der Hyperthyreose die „hyperaktiven Thyreocyten" charakteristisch sind. Die Art und die chemische Beschaffenheit der Granulation in diesen hyperaktiven Thyreocyten konnte nicht durch die morphologische Färbungsmethode nach PAPPENHEIM (MAY-GRÜNWALD-GIEMSA) geklärt werden. Da sie aber in gesteigerten Mengen bei den medikamentös nicht behandelten Hyperthyreosen vorhanden sind, können sie als morphologische Äußerungsform der Hyperaktivität betrachtet werden.

3. Ergebnisse

Die Ergebnisse unserer Untersuchungen der cytologischen Aspirationspunktionen mit der Perjodsäure-Schiff-Methode sind in der Tab. 1 dargestellt und lassen sich wie folgt zusammenfassen:

1. Die Thyreocyten unterscheiden sich nach dieser Methode in der Farbe des Cytoplasmas, je nach den verschiedenen funktionalen Zuständen und Erkrankungen der Schilddrüse, wie auch nach den Granulationen dieses Cytoplasmas. So ist die Farbe des Cytoplasmas ausgesprochen blaß bei Hypothyreosen, von mittlerer Intensität bei euthyreotischen Strumen (gleichgültig ob sie diffus oder knotig sind) und von stärkstem Rot bei Hyperthyreosen und beim Schilddrüsencarcinom. Bei euthyreotischen Strumen findet man im Cytoplasma der Thyreocyten winzige feine Granulationen, die bei Hyperthyreosen grob und gruppiert sind, bei Hypothyreosen und beim Carcinom ebenfalls vorkommen — doch seltener sind.

2. Die Phagocyten haben bei der Anwendung der Perjodsäure-Schiff-Methode die gleichen morphologischen Charakteristiken bei allen funktionalen Zuständen und Erkrankungen (s. die Tab. 1). Sie unterscheiden sich einzig durch eine größere oder kleinere Anzahl.

3. Die Farbabstufung des mit Perjodsäure-Schiff-Methode gefärbten Kolloids variiert: von der ganz blassen bei Hypothyreosen über die hellrote bei einigen euthyreotischen diffusen Strumen bis zur roten bei anderen euthyreotischen diffusen und knotigen Strumen, Hyperthyreosen und dem Krebs.

Tabelle 1

Diagnose	Thyreocyt	Phagocyt	Kolloid	Zahl der Fälle
Euthyreotische diffuse Struma	Zahl der Zellen: + + Der Kern: ungefärbt Das Cytoplasma: hellrot Die Granulation: ∅ oder PAS + Material in feingranulärer Form	Zahl der Zellen: + Der Kern: ungefärbt Das Cytoplasma: ungefärbt oder hellrot Die Granulation: sehr wenig bis sehr viel winziges oder größeres grobscholliges PAS + Material	diffuse hellrote oder rote Färbung, einige Kolloid-Partikel sind intensiv rot gefärbt	42
Euthyreotische knotige Struma	Zahl der Zellen: + Der Kern: ungefärbt Das Cytoplasma: hellrot Die Granulation: ∅ oder PAS + Material in feingranulärer Form	Zahl der Zellen: + + + Der Kern: ⎫ Das Cytoplasma: ⎬ wie oben Die Granulation: ⎭	diffuse hellrote oder rote Färbung, einige Kolloid-Partikel sind intensiv rot gefärbt	70
Hyperthyreose	Zahl der Zellen: + + + Der Kern: ungefärbt Das Cytoplasma: rot Die Granulation: grobscholliges PAS + Material in Gruppen	Zahl der Zellen: ∅ oder + Der Kern: ⎫ Das Cytoplasma: ⎬ wie oben Die Granulation: ⎭	diffuse rote Färbung	28
Hypothyreose	Zahl der Zellen: + Der Kern: ungefärbt Das Cytoplasma: ganz hellrot Die Granulation: ∅ oder seltene Gruppen von grobscholligem PAS + Material	Zahl der Zellen: ∅ oder + Der Kern: ⎫ Das Cytoplasma: ⎬ wie oben Die Granulation: ⎭	diffuse ganz hellrote Färbung	2
Schilddrüsen-Carcinom	Zahl der Zellen: + + + Der Kern: ungefärbt Das Cytoplasma: rot Die Granulation: ∅ oder seltene Gruppen von grobscholligem PAS + Material	Zahl der Zellen: ∅ oder + Der Kern: ⎫ Das Cytoplasma: ⎬ wie oben Die Granulation: ⎭	diffuse rote Färbung, häufig sind intensivrot gefärbte Partikel von Kolloid vorhanden	8

4. Bei den Versuchen mit der Anwendung des Diastase-Kontroll-Testes hat man feststellen können, daß die Perjodsäure-Schiff-Färbung nach der Anwendung der Diastase eine weit geringere Intensität aufgewiesen hat, als das der Fall bei den Präparaten ohne Diastasetest war. Doch die Granulationen in den hyperaktiven Thyreocyten waren nach der Diastase-Wirkung nicht lösbar.

Mit allen Einschränkungen einer vorläufigen Mitteilung könnte man evtl. die Hypothese aufstellen, daß das Perjodsäure-Schiff (PAS) + und in Granulationen gruppierte Material morphologische Äußerungsform hyperaktiver Zustände bedeutet und Häufungen von Schilddrüsenhormonen in Thyreocyten in diesen Zuständen darstellen könnte.

Demnach würde die vorläufige Antwort auf die zu Anfang gestellte Frage lauten: In Thyreocyten, Phagocyten und im Kolloid bestehen bestimmte Unterschiede in verschiedenen funktionellen Zuständen und pathologischen Veränderungen, und diese sind durch die Perjodsäure-Schiff-Färbungsmethode feststellbar. PAS + Substanzen sind in ungleichen Mengen im Cytoplasma der Thyreocyten, Phagocyten und des Kolloids enthalten, oft aber findet man dieses

PAS + Material auch in feinen Granulationen des Thyreocytenplasmas, in groberen Granulationen des Phagocytenplasmas, als besonders charakteristische Häufungen aber in grobkörnigen Gruppen des Thyreocytenplasmas bei Hyperthyreosen.

Nach all den Auslegungen also wäre die Perjodsäure-Schiff-Färbungs-Methode ein weiterer Beweis, daß bei hyperaktiven Zuständen in Schilddrüsenzellen morphologische Änderungen nachweisbar sind. Änderungen, die vielleicht durch das Konzentrieren von Schilddrüsenhormonen bedingt sind und cytochemisch sich in der Form von Granulationen PAS + Materials im Cytoplasma der Thyreocyten manifestieren.

Literatur

BAUER-SIČ, P.: Zbl. Vet.-Med. 10, 365 (1963).

CAMPANACCI, L., F. PIERANGOLI e S. TURA: G. Clin. med. 36, 572 (1955).

CHÈVREMONT, M., et J. FREDERIC: Arch. Biol. (Liège) 54, 589 (1943).

ČREPINKO, I., Z. ŠKRABALO, and E. HAUPTMANN: Advance Abstracts of Short Communications, First International Congress of Endocrinology, Periodica, Copenhagen. 1159 (1960).

FRIEDERICI, L.: Klin. Wschr. 33, 1020 (1955).

GÄRTNER, J., and A. NORDEN: Acta med. scand. 169, 289 (1961).

GÖSSNER, W.: Histochemie 1, 48 (1958).

HERTL, M.: Folia haemat. (Lpz.) 3, 269 (1959).

HOTCHKISS, R. B.: Arch. Biochem. 16, 131 (1948).

KAPLOW, L. S.: Blood 10, 1023 (1955).

KAY, W., u. R. WHITEHEAD: J. Path. Bact. 53, 279 (1941).

KENNY, J., u. W. C. MOLONEY: Blood 2, 295 (1957).

LAMBERS, K., u. P. BAUER-SIČ: Dtsch. med. Wschr. 87, 1913 (1962); 9. Freiburger Symposium über Zyto- und Histochemie in der Hämatologie. S. 254. Berlin-Göttingen-Heidelberg: Springer 1963.

LÖFFLER, H., u. W. BERGHOFF: Klin. Wschr. 40, 363 (1962).

MERKER, H., u. L. HELLMEYER: Dtsch. med. Wschr. 86, 253 (1960).

MYREN, J., u. E. SIVERTSSEN: Acta endocr. 39, 431 (1962).

PERRY, S., u. J. REYNOLDS: Blood 11, 1132 (1956).

ROMEIS, B.: Mikroskopische Technik. München: Leibnitz 1948.

SATO, K.: Kongr. Zbl. Ges. inn. Med. 51, 775 (1928).

SHEEHAN, H. L., u. G. W. STOREY: J. Path. Bact. 59, 336 (1947).

ŠKRABALO, Z., I. ČREPINKO, Z. GRGIČ i E. HAUPTMANN: Liječn Vjesn. 83, 1035 (1961).

— —, and E. HAUPTMANN: Advance Abstracts of Short Communications, First International Congress of Endocrinology, Periodica, Copenhagen, 1157 (1960).

— — — Zdr. Novine 16, 154 (1963).

SÖDERSTRÖM, N.: Acta med. scand. 144, 237 (1952).; Acta Soc. Med. upsalien 63, 53 (1958).

UNDRITZ, E.: Hämatologische Tafeln. Sandoz 1952.

WACHSTEIN, M.: Blood 4, 54 (1949).

Aus der 2. Med. Klinik und Poliklinik der Med. Akademie Düsseldorf
(Direktor: Prof. Dr. K. OBERDISSE)

Tierexperimentelle Untersuchungen zum Vorkommen von Jod in extrathyreoidalen Geweben

Von

H. A. VON SCHWEINITZ und M. ECKSTEIN

Mit 1 Abbildung

Jod ist im Serum des menschlichen und des tierischen Organismus als anorganisches Jodid, als freies organisches Jod und als sog. Hormonjod nachweisbar. Das Hormonjod im weiteren Sinne (das PBI) faßt Thyroxin, Trijodthyronin, Jodtyrosine und andere organische Jodverbindungen zusammen [4, *8*, *9*]. Die chemische Bestimmung des Jod allein gibt keinen Aufschluß über die Menge und Verteilung von Jodid und Schilddrüsenhormonen. Eine Abgrenzung jodhaltiger Eiweißfraktionen voneinander ist durch die Chromatographie möglich. Die Anwendung von Jod¹³¹ oder Jod¹³¹-markierten Schilddrüsenhormonen erlaubt darüber hinaus eine Beurteilung der Verteilung derselben sowie des Jodumsatzes im gesamten Organismus (*4, 5*).

Nach älteren Untersuchungen wurde bei einigen Tierarten eine bevorzugte Speicherung von Jod und Jodverbindungen im Hypothalamus und in der Hypophyse, vorwiegend im Hinterlappen, festgestellt (*1, 2, 3, 6, 7, 10, 11, 12*). Der Grund ist nicht ohne weiteres ersichtlich und wird von einigen Autoren (*2, 7, 11, 12*) darin gesehen, daß in diesem Bereich der primäre Angriffspunkt des Thyroxins in dem Reglermechanismus Hypothalamus-Schilddrüse liege.

In diesem Zusammenhang interessierten uns folgende Fragen:

1. In welcher Form und in welcher Verteilung sind markiertes Jod und jodhaltige Verbindungen in extrathyreoidalen Geweben überhaupt nachweisbar?

2. Inwieweit ist die extrathyreoidale Jodaufnahme durch experimentelle Maßnahmen, die das Gleichgewicht des Hypophysen-Schilddrüsensystem stören, zu beeinflussen?

1. Methodik

Nach Injektion von 1 mC/kg Körpergewicht beim Kaninchen, bzw. 0,25 mC/ 100 g Ratte trägerfreien Jod¹³¹ bzw. Jod¹³¹-markierten Thyroxins (spez. Aktivität 2 μC/μg) und Trijodthyronins (spez. Aktivität 25 μC/μg) erfolgte die Tötung der Tiere zu verschiedenen Zeitpunkten (1—72 Std). Die Organe wurden homogenisiert. Das Homogenat wurde einerseits mit Butanol extrahiert, andererseits mit Trypsin hydrolysiert und aufsteigend eindimensional auf Papierstreifen in einem System von Butanol-Dioxan-Ammoniumhydroxyd chromatographiert (*4, 5*). In jeder Versuchsreihe liefen als Kontrollsubstanzen markiertes Jod, Thyroxin und Trijodthyronin. Sie wurden zur Identifizierung außerdem nichtmarkiert auf den Startpunkt der Organhomogenate aufgetragen. Ihre Identität wurde angenommen bei Übereinstimmung der Laufstrecke des Isotops mit der im Spraytest nachgewiesenen Substanz. Die Radioaktivität der einzelnen Bande wurde mit dem Bohrlochkristall bestimmt. Daneben wurden noch Autoradiogramme der Gewebe nach der sog. Strippingmethode hergestellt, um die Lokalisation des Jod^{131} in den verschiedenen Gewebsarealen festzustellen.

Der Einfluß folgender experimenteller Maßnahmen auf die extrathyreoidale Jodverteilung wurde untersucht: Einserseits Hypophysektomie und Thyreoidektomie, andererseits die Behandlung mit Thiouracil und TSH — allein und nach Ausschaltung von Hypophyse und Schilddrüse. Versuchstiere waren Ratten und Kaninchen unter gleichartigen Ausgangsbedingungen.

2. Ergebnisse

Tab. 1 (I. Jod^{131})

4 Std nach intravenöser Injektion von Jod^{131} ist dies in allen untersuchten Organen nachzuweisen. Bevorzugt sind beim Kaninchen neben exkretorischen Drüsen (insbesondere Speicheldrüse) die Hypophysenanteile. Diese Aktivitäten liegen jedoch unter der des Serums. Bei der Ratte sind in Hypophyse und besonders Speicheldrüse nur geringe Mengen von Jod^{131} nachweisbar. Die in HHL und Hypothalamus bevorzugt nachzuweisende Anreicherung von Jod wird durch Thyreodektomie gesteigert, durch TSH-Behandlung gemindert. Demgegenüber ist nach Hypophysektomie nur eine minimale Zunahme der Aktivität im Hypothalamus im Vergleich zum Normaltier festzustellen. Erfolgt eine TSH-Behandlung nach vorhergehender Thyreoidektomie, so ist diese ohne Einfluß auf die Jodaufnahme in den untersuchten Geweben.

Nach der chromatographischen Untersuchung des Serums fällt das Serumjodid innerhalb der ersten 4 Std nach Injektion von Jod^{131} steil, dann langsamer ab. Gleichzeitig kommt es zu einem Anstieg von Thyroxin und Trijodthyronin, hierbei ist der Anstieg von Trijodthyronin geringer. Entsprechende Verhältnisse finden sich in allen untersuchten Geweben, jedoch ist der zeitliche Ablauf gegenüber dem Serum insgesamt verzögert.

4 Std nach Injektion von Jod^{131} findet sich in den Organen noch fast ausschließlich Jodid, Hormonjod nur geringgradig in Organen mit vermehrter Jodidaufnahme. Nach Thyreoidektomie ist sowohl im Serum wie in sämtlichen Organen nur Jodid nachweisbar, vermutlich wegen der fehlenden Hormonjodbildung durch die Schilddrüse. Ein entsprechendes Verhalten ist nach Thiouracilbehandlung zu beobachten. Die Bildung des Hormonjods wird durch TSH-Behandlung gesteigert, jedoch findet sich dieses nur im Serum, die Organe zeigen keine meßbare Zunahme des Hormonjod. Nach Hypophysektomie ist ebenfalls in den Organen kein Hormonjod nachweisbar, im Serum weniger als beim Normaltier. Bemerkenswert ist hierbei, daß unter den verschiedenen Versuchsbedingungen kein prinzipieller Unterschied in der Verteilung von Jodid und Hormonjod in den Geweben bei Ratten und Kaninchen besteht, obwohl beim Kaninchen hohe Jodkonzentrationen in Hypothalamus und Hypophyse nachweisbar sind.

Tabelle 1. *Bestimmung der Radioaktivität nach i.v.-Injektion von J^{131} (I) und Radiothyroxin (II) bei der Ratte (a) und beim Kaninchen (b) in verschiedenen Organen nach Chromatographie: Anteile an Jod (J), Thyroxin (T_4) und Trijodthyronin (T_3) von der Gesamtaktivität nach Injektion von J^{131}*

I. J^{131}	%	Se-rum	Leber	Hypophyse				Gehirn				Speichel-drüse	
				Vorder-lappen		Hinter-lappen		Hypo-thalamus		Frontal-lappen			
		a+b	a+b	a	b	a	b	a	b	a	b	a	b
		1,0	0,5	0,25	0,5	0,2	0,9	0,2	0,4	0,1	0,1	0,2	0,6
Normaltier	J	80	85	100	90	100	90	100	95	100	100	100	85
	T_4	15	15		10		10		5				15
	T_3	5											
		1,0	0,7	0,3	0,6	0,3	*1,0*	0,2	0,6	0,2	0,2	0,3	0,8
Thyreoidektomie	J	100	100	100	100	100	100	100	100	100	100	100	100
	T_4												
	T_3												
		1,0	0,7	—	—	—	—	0,3	0,5	0,1	0,1	0,2	0,6
Hypophysekto-mie	J	90	100					100	100	100	100	100	100
	T_4	10											
	T_3												
		1,0	0,8	0,2	0,4	0,2	0,5	0,1	0,3	0,1	0,1	0,1	0,3
TSH-Behand-lung	J	55	75	100	95	100	95	100	100	100	100	100	80
	T_4	35	25		5		5						20
	T_3	10											

II. Thyroxin[131]

Normaltier		1,0	0,6	0,3	0,3	0,25	*3,0*	0,9	0,8	0,1	0,1	0,2	0,7
Thyreoidektomie		1,0	0,8	0,3	0,3	0,4	*5,0*	0,25	*1,3*	0,1	0,1	0,4	0,8
Hypophysektomie		1,0	0,7					0,2	0,6	0,2	0,2	0,2	0,8
THS + Thyreoidektomie		1,0	0,7	0,2	0,2	0,3	*4,0*	0,2	0,9	0,2	0,2	0,2	0,6

Tab. 1 (II. Thyroxin[131])

4 Std nach Injektion von Radiothyroxin haben wir im HHL eine wesentlich höhere Aktivität gemessen als im Serum. Dies gilt jedoch nur für das Kaninchen, nicht für die Ratte. Geringer, nicht über dem Serum liegend, ist die Aktivität im Hypothalamus. Nach Thyreoidektomie kommt es zu einer stärkeren Aufnahme des Thyroxins im HHL. Ein ähnliches Verhalten ist nach Behandlung mit Thiouracil zu beobachten. Bei TSH-Behandlung wird die Aufnahme des Thyroxins gemindert. Erfolgt nach Thyreoidektomie eine TSH-Behandlung, so liegt die Anreicherung des Thyroxins in der Hypophyse beim Kaninchen nach wie vor deutlich über der des Serums. Nach Hypophysektomie kommt es zu einer geringeren Thyroxinspeicherung im Hypothalamus. Die übrigen Organe zeigen durch die angeführten Maßnahmen keine wesentliche Änderung der Thyroxinaufnahme.

Nach der Chromatographie handelt es sich bei der nachgewiesenen Aktivität fast ausschließlich um Thyroxin in sämtlichen Organen. Lediglich im Serum ist 4 Std nach Injektion von Radiothyroxin geringgradig Jodid nachweisbar.

Die Aufnahme des Thyroxins (Abb. 1) durch die verschiedenen Organe ist zeitabhängig. Im Hypophysenhinterlappen kommt es bis zur vierten Stunde post injectionem zu einem steilen Anstieg der Aktivität, dann zu einem kontinuierlichen Abfall. In der Leber wird die maximale Konzentration schon in der ersten Stunde nach der Injektion erreicht. Demgegenüber nimmt die Aktivität im Plasma kontinuierlich ab, im Gehirn und den übrigen Organen mäßig zu.

Versuche mit Trijodthyronin zeigen ein ähnliches Verhalten, jedoch ist hier die Aufnahme wie der Abfall der Aktivität in den Organen gegenüber dem Thyroxin wesentlich beschleunigt.

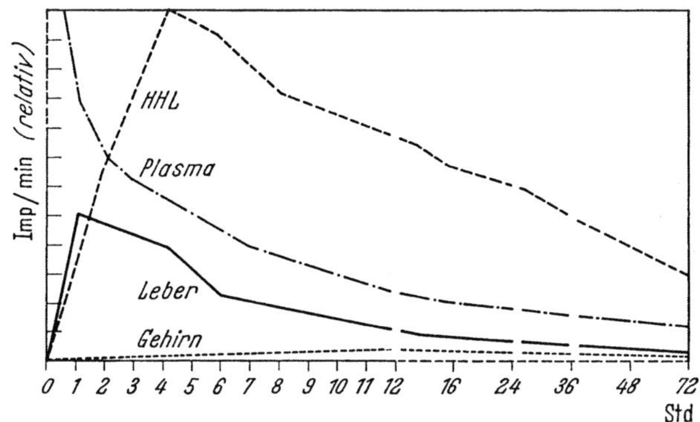

Abb. 1. Radioaktivität nach i.v.-Injektion von Radiothyroxin beim Kaninchen. ——— Leber, - - - - HHL, — · — · — Plasma · · · · · · Gehirn

In Gewebsautoradiogrammen des Hypothalamus, in denen eine gesteigerte Aufnahme von Thyroxin gefunden worden ist, liegt die Aktivität vorwiegend im Bereich des nucleus paraventricularis, des nucleus supraopticus und am Ansatz des Hypophysenstiels. Diese Verteilung wird grundsätzlich nicht durch Thyreoidektomie oder Hypophysektomie beeinflußt.

Zusammenfassung

Zusammenfassend läßt sich sagen:

1. Jodid, Thyroxin und Trijodthyronin werden nach intravenöser Injektion von allen Organen aufgenommen, jedoch in unterschiedlicher Konzentration. Die Anreicherung in den Organen variiert zwar bei einzelnen Tierarten, jedoch stehen zumeist Leber und exkretorische Drüsen im Vordergrund, beim Kaninchen (nach anderen Autoren auch beim Hund (1)) HHL und Hypothalamus.

2. Thyreoidektomie und Hypophysektomie bewirken eine gesteigerte Jodaufnahme in HHL und Hypothalamus.

3. TSH-Behandlung hebt den Effekt der Hypophysektomie hinsichtlich der Jodaufnahme im extrathyreoidalen Gewebe wieder auf. Demgegenüber ist kein nennenswerter Einfluß der TSH-Behandlung nach Thyreoidektomie festzustellen. Somit scheint eine unmittelbare Wirksamkeit des TSH hinsichtlich des Jodstoffwechsels in extrathyreoidalen Geweben nicht wahrscheinlich.

4. Das Verhalten von Jod und Hormonjod scheint in extrathyreoidalen Geweben ausschließlich von der Schilddrüsenfunktion abhängig zu sein, die ihrerseits wieder von der Höhe des Blutjodspiegels beeinflußt wird. In vivo markierte organische Jodverbindungen nach Jod[131]-Gaben werden extrathyreoidal nur bei intakter Schilddrüse nachweisbar.

5. Sicher besteht eine gewisse Affinität des Jods und des Hormonjods zu den anatomisch und phylogenetisch als Einheit anzusehenden Hypophysen-Hypothalamussystem. Unserer Meinung nach treffen mehrere Faktoren zusammen, die für die bevorzugte Aufnahme von Jod und jodhaltigen Verbindungen im HHL und Hypothalamus von Bedeutung sein können. Unter anderem findet sich eine Lockerung der Liquor-Gehirnschranke in diesem Bereich. Weiterhin ist auch gegenüber H³-markierten Aminosäuren ein andersartiges Verhalten als in anderen Gehirnteilen festzustellen. Möglicherweise handelt es sich jedoch lediglich um eine erhöhte Speicherung von Jod in diesen Geweben. Hierfür könnten auch die Untersuchungen anderer Autoren (12) sprechen. Sie stellten nach Hypophysenstieldurchtrennung, die also die Funktion der Hypophyse weitgehend ausschaltet, eine unveränderte Speicherung von Thyroxin fest. Nach unseren Untersuchungen läßt sich kein Beweis dafür finden, daß der Hypothalamus ein direkter oder alleiniger Angriffspunkt des Hormonjods im Reglermechanismus des Hypophysen-Schilddrüsen-Systems ist.

Literatur

1. Courrier, R.: Contribution a l'endocronologie de la Thyreoid. Acta endocr. (KbH) 7, 57 (1951).
2. Ford, D. H., and J. Gross: The Localization of J¹³¹-labelled Triiodothyronin and Thyroxine in the Pituitary and Brain of the Male Guinea Pig. Endocrinology 63, 549 (1958).
3. — S. Kantounis, and R. Lawrence: The Localization of Jod¹³¹-labelled Triiodothyronine in the Pituitary and Brain of Normal and Thyroidectomized Male Rats. Endocrinology 64, 997 (1959).
4. Klein, E.: Der endogene Jodhaushalt des Menschen und seine Störungen. Stuttgart: G. Thieme 1960.
5. — Der normale und pathologische Umsatz von Schilddrüsenhormonen in der Körperperipherie. Klin. Wschr. 40, 3 (1961).
6. Jensen, J. M., and D. E. Clark: Localization of radioactive 1-Thyroxine in neurohypophysis. J. Lab. clin. Med. 38, 663 (1951).
7. Jentzer, A.: Relation between Hypophysis and the Thyroid Gland. Experiments with radioactive Jodine. Acta endocr. (Kbh.) 12, 264 (1953).
8. Pitt-Rivers, R.: The Chemistry of Thyroid Diseases. Springfield, Illinois, USA: Ch. C. Thomas 1960.
9. —, and J. R. Tata: The Thyroid Hormones. London: Pergamon Press 1959.
10. Schittenhelm, A., u. B. Eisler: Über die Verteilung des Jods im Zentralnervensystem nach Zufuhr von Schilddrüsenstoffen. Über die Verteilung des Jods im Zentralnervensystem bei Mensch und Tier. Z. ges. exp. Med. 86, 275, 290 (1933).
11. Sturm, A., u. W. Wernitz: Hormonjod im Gehirn. Klin. Wschr. 93 (1956).
12. Taurog, A., G. W. Harris, W. Tong, and J. L. Chaikoff: The Uptake of Jod¹³¹-labelled Thyroxine and Triiodothyronine by the Neurohypophysis. Endocrinology 59, 34 (1956).

Aus der 2. Med. Klinik und Poliklinik der Med. Akademie Düsseldorf
(Direktor: Prof. Dr. K. Oberdisse)

Zur Pathogenese der endokrinen Ophthalmopathie

Von

F. A. Horster und E. Klein[1]

Mit 1 Abbildung

Die endokrine Ophthalmopathie mit ihren Kardinalsymptomen Exophthalmus, Lidödem und Augenmuskellähmung kann als benigne oder durch örtliche Komplikationen maligne gewordene Form auftreten bzw. verlaufen. Beide können mit oder ohne Schilddrüsenüberfunktion einhergehen, so daß man eine hyperthyreote und eine euthyreote endokrine Ophthalmopathie unterscheidet [7, *10, 12*].

Nähere Auskunft über die Frequenz der Erkrankung gibt die Tab. 1. Unter 5890 Patienten mit Schilddrüsenfunktionsstörungen befanden sich 385 mit einer endokrinen Ophthalmopathie, von denen etwa zwei Drittel hyperthyreot und ein Drittel euthyreot waren; insgesamt 299 Frauen und 86 Männer. Im gleichen Zeitraum wurden 382 Hyperthyreosen ohne Ophthalmopathie diagnostiziert. Es fällt auf, daß endokrine Ophthalmopathien und Hyperthyreosen ohne Augenbeteiligung unter allen Schilddrüsenkranken zu gleichen Teilen vertreten waren: jeweils 6,5%.

Im folgenden interessiert zunächst das Lebensalter, in dem sich Hyperthyreosen und endokrine Ophthalmopathien am häufigsten manifestieren. Ausgewertet wurden nur die bei den Frauen ermittelten Daten, da die Zahl der Männer für eine statistische Aussage zu gering erschien. Abb. 1 zeigt, daß die Hyperthyreosen ohne Augensymptome von der Pubertät bis zum Klimakterium zunehmen, die Mehrzahl der Hyperthyreosen aber erst jenseits des 45. Lebensjahres auftritt. Endokrine Ophthalmopathien dagegen manifestieren sich überwiegend vor dem 55. Lebensjahr; ihre euthyreoten Verlaufsformen sind im höheren Alter sogar ausgesprochen selten anzutreffen. Man kann dieser Abbildung auch entnehmen,

Tabelle 1. *Anteil der Hyperthyreosen und endokrinen Ophthalmopathien unter 5890 Patienten mit Schilddrüsenfunktionsstörungen*

Krankheit	Anzahl der Fälle			♀:♂	% von 5890
	Gesamt	♀	♂		
Hyperthyreote endokrine Ophalmopathien	253	203	50	4,0:1	4,3
Euthyreote endokrine Ophthalmopathien	132	96	36	2,6:1	2,2
Total	385	299	86	3,4:1	6,5
Hyperthyreosen ohne endokrine Ophthalmopathie	382	329	54	6,1:1	6,5

[1] Mit dankenswerter Unterstützung der deutschen Forschungsgemeinschaft. — Herrn Prof. Dr. H. W. Bansi, Hamburg, zum 65. Geburtstag.

16*

daß die dem Klimakterium entsprechenden Jahre bevorzugtes Manifestationsalter für diese Krankheiten sind.

Neben dem Klimakterium gibt es noch andere endokrinologisch bedeutsame Faktoren, die für das Auftreten einer endokrinen Ophthalmopathie pathogenetisch bedeutsam sein könnten; sie sind in der Tab. 2 zusammengestellt. Außerdem sind die Anzahl der Fälle, deren prozentualer Anteil und die Schilddrüsenfunktion vor Anftreten der endokrinen Ophthalmopathie angegeben. Es bekamen 12 Frauen

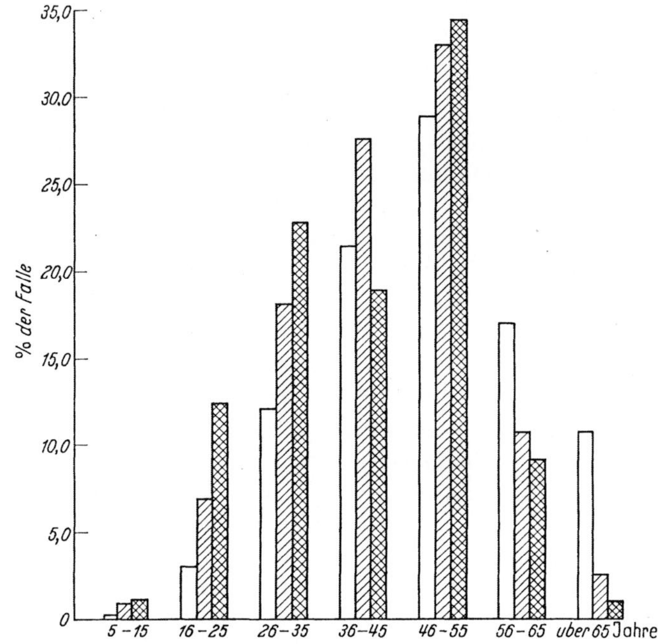

Abb. 1. Vergleich des Manifestationsalters von Hyperthyreosen und endokrinen Ophthalmopathien (628 Patientinnen). ☐ Hperthyreosen ohne endokrine Ophthalmopathie (n = 329), ▨ Hyperthyreote endokrine Ophthalmopathien (n = 203), ▩ Euthyreote endokrine Ophthalmopathien (n = 96)

Tabelle 2. *Zur Pathogenese der endokrinen Ophthalmopathie bei 299 Patientinnen*

Pathogenetischer Faktor	Anzahl	%	Schilddrüsenfunktion vor Beginn der e.O.		
			normal (ohne Struma)	Blande Struma	Hyperthyreose (mit u. ohne Struma)
Pubertät	3	1,0	3	—	—
Partus	12	4,0	4	8	—
Abortus.	3	1,0	—	3	—
Totalop. (Praeklimakterium) . . .	9	3,0	7	2	—
Klimakterium-Beginn.	63	21,0	7	45	11
Strumektomie	30	10,0	—	19	11
^{131}J-Therapie.	1	0,3	—	—	1
Therapie mit antithyreoidalen Substanzen ohne HVL-Bremsung . .	39	13,0	—	—	39
Nicht indizierte Therapie mit antithyreoidalen Substanzen	6	2,0	—	6	—
Spontaner Beginn	133	44,7	111	22	—
Gesamt	299	100,0	132	105	62

im Anschluß an eine Geburt, 3 nach einer Fehlgeburt und 9 nach einer gynäkologischen Totaloperation eine endokrine Ophthalmopathie. Oft lag bereits eine Struma vor, ehe das pathogenetisch vermutlich ausschlaggebende Ereignis eintrat. Dies gilt besonders für die Fälle, bei denen die endokrine Ophthalmopathie mit dem Klimakterium begann. Bei zehn Prozent unserer Patientinnen trat sie im Anschluß an eine Strumektomie auf und überraschend oft auch während einer Therapie mit antithyreoidalen Substanzen, wenn diese nicht mit einer Bremsung der HVL-Tätigkeit durch Schilddrüsenhormone verbunden war. In sechs Fällen war eine blande Struma fehlerhafterweise mit antithyreoidalen Substanzen behandelt und dabei eine endokrine Ophthalmopathie entstanden. Unter den bei „spontaner Beginn" vermerkten Fällen befinden sich viele im Alter von 40—60 Jahren, ohne daß anamnestisch verwertbare endokrine Ereignisse zu eruieren waren.

Eine Beurteilung der angeführten pathogenetischen Faktoren läßt erkennen, daß nicht ausschließlich Störungen zwischen Schilddrüse und Hypophyse eine endokrine Ophthalmopathie auszulösen scheinen, obgleich der bei 85% aller endokrinen Ophthalmopathien beschleunigte intrathyreoidale Jodumsatz auf eine thyreotrope Stimulierung hinweist [10]. Allein die Tatsache, daß mehr Hyperthyreosen ohne als mit einer Ophthalmopathie einhergehen zeigt, daß Störungen der TSH-Sekretion und des homöostatischen Reglermechanismus nicht die wesentliche Ursache für das Auftreten einer endokrinen Ophthalmopathie sein können.

Bestimmt man im Serum von Patienten, die an einer unbehandelten endokrinen Ophthalmopathie leiden, den Gehalt an Thyreotropin (TSH) und an Exophthalmus produzierendem Faktor (EPF), so findet man für TSH Werte, die eher im unteren Normalbereich liegen [4, 8, 11, 13, 14], während der EPF stets nachzuweisen ist [5, 9]. Der Exophthalmusfaktor findet sich nie im Serum von gesunden Menschen, ist aber in allen bisher untersuchten Hypophysen enthalten [1, 3, 6]. Es ist unbekannt, welche Umstände den EPF veranlassen, die Hypophyse zu verlassen und speziell im orbitalen und periorbitalen Gewebe, selten auch in anderen Bindegewebsbezirken pathologische Veränderungen herbeizuführen [2, 16].

Unsere Beobachtungen könnten dafür sprechen, daß nicht nur zwischen den glandotropen Hormonen und ihren zugehörigen peripheren Drüsen, sondern auch unter den HVL-Hormonen selbst ein gewisser Reglermechanismus existiert, der die Partialfunktionen des Hypophysenvorderlappens steuert. Der EPF wäre dann z. B. — etwa in Analogie zu den Jodfehlverwertungen — als ein Precursor glandotroper Hormone aufzufassen, der nur dann in die Peripherie gelangt und dort wirkt, wenn die physiologische Sekretion der glandotropen Hormone gestört wird. Solche Störfaktoren sind u. a. operative oder medikamentöse Eingriffe in das Endokrinium und plötzliche Änderungen der HVL-Sekretion, etwa bei der Hyperthyreose zu Beginn des Klimakteriums, bei Partus oder Abortus. Im Tierexperiment konnten wir zeigen, daß eine Injektion von Gonadotropinen das Auftreten eines experimentellen Exophthalmus begünstigt oder verstärkt. Auch der bei etwa 5% aller Cushing-Patienten zu beobachtende Exophthalmus könnte dadurch erklärt werden, daß eine Störung der glandotropen Hormonsekretion die EPF-ausschüttung veranlaßt: Bei diesen Cushing-Fällen ist der EPF im Serum positiv [15].

Zusammenfassend ergibt sich die Vermutung, daß die endokrine Ophthalmopathie Ausdruck einer mehrseitigen hypophysären Funktionsstörung ist.

Literatur

1. Bottari, P. M., and B. T. Donovan: J. Physiol. (Lond.) **140**, 36 P (1958).
2. Brunish, R., and B. Sorensen: Acta endocr. (Kbh.) **44**, 606 (1963).
3. Condliffe, P. G.: In: Thyrotropin. Springfield (USA): C. C. Thomas Publisher 1963.
4. D'Angelo, S. A.: J. clin. Endocr. **23**, 229 (1963).
5. der Kinderen, P. J., M. Houstra-Lanz, and F. Schwarz: J. clin. Endocr. **20**, 712 (1960).
6. Dobyns, B. M., A. Wright, and L. Wilson: J. clin. Endocr. **21**, 648 (1961).
7. Horst, W., u. K. Ullerich: Hypophysen-Schilddrüsenerkrankungen und endokrine Ophthalmopathie. Stuttgart: F. Enke-Verlag 1958.
8. Horster, F. A.: Vortrag 70. Tg. Dtsch. Gesellsch. Inn. Med., Wiesbaden 1964.
9. —, u. E. Klein: In: Verhandlungen 10. Symp. Dtsch. Gesellsch. Endokrinologie, Heidelberg: Springer 1964.
10. Klein, E., H. Zimmermann u. H. Lins: Endokrinologie **39**, 44 (1960).
11. El Kabir, D. J.: In: Advances in Thyroid Research. London: Pergamon Press 1961.
12. Lamberg, B. A.: In: Fortschritte der Schilddrüsenforschung. Stuttgart: Thieme Verlag 1962.
13. McKenzie, J. M.: Physiol. Rev. **40**, 398 (1960).
14. Pimstone, B. L., R. Hoffenberg, and E. Black: J. clin. Endocr. **23**, 336 (1963).
15. Schwarz, F., P. J. der Kinderen u. M. Houstra-Lanz: J. clin. Endocr. **22**, 718 (1962).
16. Wegelius, O.: Acta endocr. (Kbh.) **33**, 453 (1960).

Aus der endokrinologischen Abteilung des Institutes für Hygiene und Mikrobiologie der
Universität Homburg/Saar und dem Dept. Gynécol. Obstétr., Hôpital Civil,
Université de Strasbourg, France

Biosynthese von Steroiden bei Stein-Leventhal-Syndrom

Von

G. W. OERTEL und R. GANDAR

Mit 2 Abbildungen

Das Vorkommen von Androgenen im menschlichen Ovar ist durch die eingehenden Untersuchungen verschiedener Arbeitskreise, wie der von ZANDER und von SIMMER, hinlänglich bewiesen. Wenngleich neben den klassischen Androgenen Androstendion und Testosteron auch potentielle Vorstufen, z. B. Dehydroepiandrosteron, nachzuweisen waren, so blieb die Konzentration dieser Steroide doch beinahe unter der Erfassungsgrenze. Demgegenüber fanden MAHESH und GREENBLATT bei Stein-Leventhal-Syndrom relativ hohe Konzentrationen von 17-Hydroxypregnenolon und Dehydroepiandrosteron im Ovarialgewebe. Des weiteren zeigten in vitro-Versuche von AXELROD und GOLDZIEHER, daß bei der Bebrütung von Ovarialschnitten bei Stein-Leventhal-Syndrom mit ^{14}C-markiertem Pregnenolon eine signifikante Umwandlung des Substrats in 17-Hydroxypregnenolon und Dehydroepiandrosteron erfolgt. Da bei den letztgenannten Experimenten eine Kontrollbebrütung normalen Ovarialgewebes mit Pregnenolon nicht beschrieben wurde, ist eine quantitative Aussage über die offenbar veränderte Biosynthese von Steroiden bei Stein-Leventhal-Syndrom nicht möglich. Aus diesem Grunde unternahmen wir folgende Versuche (Abb. 1). Ovarialschnitte von drei histologisch gesicherten Stein-Leventhal-Ovarien sowie von zwei normalen Ovarien wurden in Krebs-Ringer-Phosphatpuffer von pH 7,2 mit 0,6—2,5 μg 7α-^{3}H-Pregnenolon als Substrat, 4 μMol ATP, 2 μMol DPNH und 2 μMol TPNH als Coferment und 10 μMol Magnesiumchlorid als Cofaktor pro 100 mg Gewebe 2 Std bei 37° C an der Luft bebrütet. Anschließend extrahierte man die Inkubate dreimal mit je 2 Vol. eiskalten Chloroforms zur Entfernung der freien Steroide, reinigte die vereinigten Extrakte durch Waschen mit 0,5 Vol. 0,1 n-Natronlauge und zweimal je 1 Vol. Wasser und dampfte die über Natriumsulfat getrocknete Lösung unter schonenden Bedingungen zur Trockne ein. Die präextrahierten Inkubate wurden sodann mit 5 Vol. Äthanol-Aceton (1:1 v/v) 24 Std bei Zimmertemperatur behandelt. Nach Absaugen ausgefällten Eiweißes erfolgte die Chromatographie der im Filtrat enthaltenen gesamten Steroidkonjugate an einer Säule (10 × 1 cm) aus aktiviertem und vorbehandeltem DEAE-Sephadex A-50. Enthielten die ersten mit Chloroform-Methanol-Wasser (1:9:2 v/v) und 90% Methanol eluierten Fraktionen früheren Versuchen zufolge die lipophilen und solvolysierbaren Steroid-

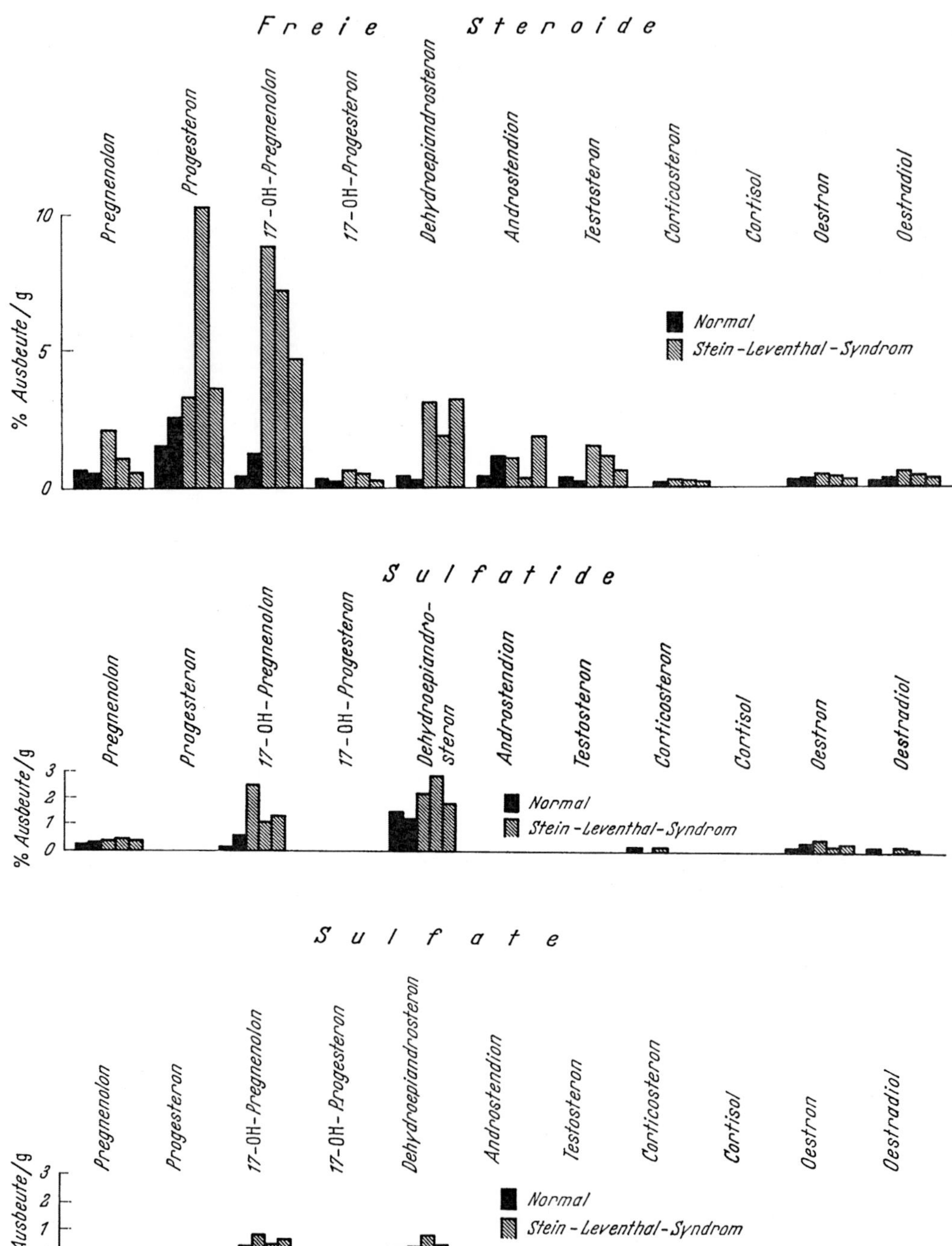

Abb. 1. Methode der Bebrütung von Ovarialgewebe mit Pregnenolon. Ovarialschnitte in Krebs-Ringer-Phosphat-puffer, pH 7,2; *Substrat:* 0,6—2,5 µg 7 α-³H-Pregnenolon; *Coenzyme:* 4 µM ATP, 2 µ MDPNH, 2 µM TPNH-*Cofaktor:* 10 µM Magnesiumchlorid; *Dauer:* 2 Std bei 37° C an der Luft

konjugate, die von uns als Sulfatidylsteroide angesehen werden, sowie gegebenenfalls vorhandene Glucuronoside, so befanden sich die Steroidsulfate ausschließlich in der letzten, mit saurer Ammoniumsulfatlösung gewonnenen Fraktion. Die in den ersten beiden Eluaten vorkommenden lipophilen Steroidkonjugate wurden durch eine zusätzliche Lösungsmittelverteilung zwischen Äthylacetat und 5% Natriumbicarbonat gereinigt und sodann ebenso wie die Fraktion der Steroidsulfate einer Spaltung durch Solvolyse unterworfen. Die Auftrennung der freigesetzten Steroide erfolgte durch mehrfache Papierchromatographie in bewährten

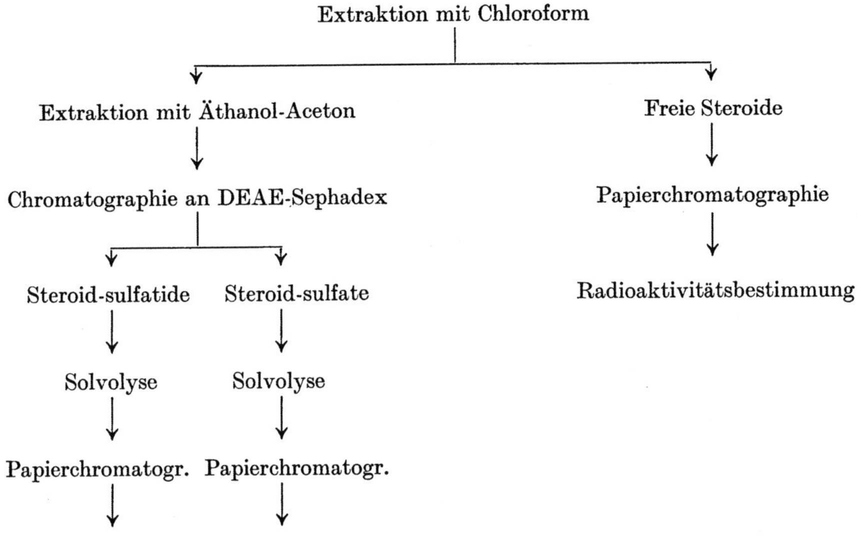

Abb. 2. Ergebnisse der Bebrütung von Ovarialgewebe mit Pregnenolon

Lösungsmittelsystemen, die quantitative Auswertung der isolierten Steroidfraktionen aber durch Messung der Radioaktivität im Packard Tri-Carb Szintillationszähler. Für die einwandfreie Charakterisierung der einzelnen Steroide sorgte die Verdünnung anfallenden Materials mit entsprechenden authentischen, jedoch nichtmarkierten Verbindungen und eine Reinigung bis zur konstanten spezifischen Radioaktivität. Die in der Abb. 2 zusammengestellten Versuchsergebnisse, die in prozentualer Umwandlungsrate pro g Frischgewebe, bezogen auf die Radioaktivität eingesetzten Substrats angegeben sind, lassen erkennen, daß die Biosynthese von Steroiden aus Pregnenolon bei normalen und Stein-Leventhal-Ovarien durchaus verschieden ist. So zeigt sich z. B. bei der Auswertung der freien Steroidfraktion eine deutlich gesteigerte Umwandlung des Substrats in 17-Hydroxypregnenolon und Dehydroepiandrosteron wie auch in Testosteron. Im zweiten Falle kam es auch zu einer vermehrten Bildung von Progesteron. Während bei allen Bebrütungsversuchen eine geringe Menge von Corticosteron sowie von Oestron und Oestradiol nachzuweisen war, konnte Cortisol nicht identifiziert werden. Die Konzentration der hier aufgeführten Steroide blieb bei normalem und Stein-Leventhal-Ovarialgewebe praktisch unverändert. Ähnliche Unterschiede in der Ausbeute einzelner Steroide erhielt man bei dem Vergleich der Biosynthese von konjugierten Steroiden im normalen bzw. Stein-Leventhal-Ovar. Die Konzentration an lipophilen und solvolysierbaren sog. Sulfatidylsteroiden war im Stein-Leventhal-Gewebe deutlich erhöht, wenigstens was die Δ^5-3β-Hydroxysteroide 17-Hydroxypregnenolon und Dehydroepiandrosteron angeht. Ähnliche Ergebnisse wurden bei der quantitativen Analyse der aus der Sulfatfraktion gewonnenen Steroide beobachtet. Auch hier lag die Konzentration der isolierten Δ^5-3β-Hydroxysteroide 17-Hydroxypregnenolon und Dehydroepiandrosteron in den Inkubaten von Stein-Leventhal-Ovarien über den im Normalgewebe festgestellten Werten.

Betrachtet man die hier genannten Ergebnisse, so fällt eine merkliche Steigerung der Umwandlungsrate von Pregnenolon in 17-Hydroxypregnenolon und Dehydroepiandrosteron bei Stein-Leventhal-Syndrom besonders ins Auge. Des weiteren wird Testosteron in erheblich vermehrtem Umfange gebildet, wogegen die Biosynthese von Oestrogenen kaum beeinflußt erscheint. Die Biosynthese von Androgenen verläuft bekanntlich auf zwei Wegen, von denen der eine von Pregnenolon über Progesteron und 17-Hydroxyprogesteron, der andere aber über 17-Hydroxypregnenolon und Dehydroepiandrosteron zu dem Androgen Androstendion führt. Letztere Verbindung aber steht bekanntlich mit Testosteron in einem reversiblen Redoxgleichgewicht.

Da in unseren Versuchen die Konzentration der aus Pregnenolon gebildeten Δ^5-3β-Hydroxysteroide vergleichsweise erhöht erscheint, nehmen wir an, daß die Biosynthese der offensichtlich vermehrten Androgene vornehmlich über diese Zwischenstufen verläuft. Eine derartige Ansicht steht in Einklang mit den Befunden von Mahesh und Greenblatt. Inwieweit ein direkter Abbau von Cholesterin zu Dehydroepiandrosteron bei der Biogenese von Testosteron beteiligt ist, bleibt abzuwarten. Daß die Bildung von Androgenen im Ovar auch auf dem klassischen Wege, d. h. über Progesteron und 17-Hydroxyprogesteron erfolgen kann, haben jüngste Versuche mit einem andersartigen Ovarialtumor gezeigt, wo statt der Δ^5-3β-Hydroxysteroide die Zwischenstufen mit der Δ^4-3-Ketogruppe bevorzugt entstanden.

Zusammenfassend darf festgestellt werden, daß bei der Bebrütung von Ovarialgewebe normaler Personen und verschiedener, an Stein-Leventhal-Syndrom leidender Patienten mit Tritium-markiertem Pregnenolon deutliche Unterschiede in der Umwandlung des Substrates zu erkennen sind. Gegenüber der Norm wurde bei Stein-Leventhal-Syndrom eine bemerkenswerte Zunahme von 17-Hydroxypregnenolon und Dehydroepiandrosteron, sowohl in der Fraktion freier Steroide wie auch in der lipophiler, solvolysierbarer Konjugate und Sulfate. Des weiteren ließ sich eine gesteigerte Bildung von Testosteron nachweisen, so daß die Annahme einer vermehrten Androgenproduktion über Zwischenstufen der Δ^5-3β-Hydroxysteroide berechtigt sein sollte.

Literatur

Axelrod, L. R., and J. W. Goldzieher: J. clin. Endocr. **22**, 431 (1962).
Mahesh, V. B., and R. B. Greenblatt: J. clin. Endocr. **22**, 441 (1962).
Simmer, H., u. H. E. Voss: Klin. Wschr. **38**, 819 (1960).
Zànder, J.: Klin. Wschr. **35**, 1101 (1957).
— J. biol. Chim. **232**, 117 (1958.)

Aus der Universitäts-Hautklinik Hamburg-Eppendorf
Direktor: Prof. Dr. Dr. J. Kimmig

Acanthosis nigricans benigna und Zusammenhang mit primärem Hypogonadismus

Von

C. Schirren

Über die Ätiologie und Pathogenese der Acanthosis nigricans bestehen verschiedene Auffassungen. Alle Autoren sind sich allerdings über eine Differenzierung der A. n. in eine benigne und eine maligne Form und in eine sog. Pseudo-Acanthosis nigricans einig. Die Pseudo-Acanthosis nigricans ist als ein Teilsymptom bei übermäßiger Gewichtszunahme und Adipositas anzusehen; sie bildet sich mit der Gewichtsabnahme zurück und steht mit speziellen Grundkrankheiten nicht in Zusammenhang. Demgegenüber wird für die benigne und maligne Acanthosis nigricans seit den Arbeiten von Ollendorff-Curth die Ansicht vertreten, daß „Sexualhormone die A. n. benigna und Krebsstoffe die A. n. maligna aktivieren". Der echte Zusammenhang zwischen einer endokrinen Störung und dem Auftreten einer A. n. benigna wird in der älteren Literatur nur aufgrund von Spekulationen hergestellt; in der neueren Literatur fehlen die entsprechenden endokrinologischen Spezialuntersuchungen, um den Verdacht erhärten zu können. Es erscheint daher gerechtfertigt, eine eigene Beobachtung vorzutragen, bei welcher der Verlauf über mehrere Jahre zu verfolgen war und bei der aufgrund dieser Verlaufsbeobachtung die Zusammenhangsfrage einer Acanthosis nigricans benigna mit einer endokrinen Störung gesichert zu sein scheint.

S. G., geb. 4.9.1910. Prot. Nr. 275/61 und 13383/63.

Familienanamnese. o.B.

Eigene Anamnese. 1934 Geburt einer gesunden Tochter; 1939—1945 Wehrmacht und 1945—1949 Gefangenschaft in Rußland. 1941 Minenverletzung der BWS. 1948 Mumpsorchitis bds. 1956 beginnend Auftreten einer Braunschwarzverfärbung der Genital-, Axillar- und Halsregion; später traten an diesen Stellen zusätzlich polypöse Wucherungen auf. Verschiedentlich Röntgenbestrahlungen und Lokalmaßnahmen ohne jeden Erfolg. In den letzten Jahren zunehmend Impotentia coeundi.

Befund. 54jähriger Mann mit mäßiger Adipositas und ausgedehnten sichtbaren Hautveränderungen im Gesicht, am Hals, in den Achseln und der Ano-Genital-Region bei allgemeiner Hyperpigmentierung der Haut. In diesen Bereichen findet sich eine braunschwarze Verfärbung der Haut mit starker Vergröberung der oberflächlichen Hautfelderung und starker Faltenbildung. Daneben zahlreiche linsen- bis erbsengroße, rundliche, weiche, braunschwarz verfärbte halbkugelige, z. T. gestielte Tumoren mit rauher Oberfläche.

Genitale. Penisgröße an der unteren Normgrenze, Vorhaut reponibel; Hoden bds. an der unteren Normgröße, von weicher Konsistenz. Nebenhoden bds. verdickt und verhärtet. Ductus deferens unauffällig. Prostata flach, Sulcus kaum ausgeprägt, keine Infiltrate. Sehr dicke, elephantiastische Scrotalhaut. — Geschlechtschromatinbefund: negativ. — Eingehende klinische Untersuchungen zum Ausschluß eines Neoplasmas seit 5 Jahren stets negativ (vgl. Schirren 1963 und 1964).

Tabelle 1. *Übersicht der speziellen Untersuchungsbefunde bei Acanthosis nigricans benigna*

	1961	1964
BSG	6/18	11/25
Grundumsatz.	+16%	+13%
Gesamt-Cholesterin	300 mg-%	315 mg-%
Spermiogramm	Aspermie	Oligospermie
Fructose.	1940 γ/ml	1100 γ/ml
17-Ketosteroide.	8,73 mg/24 Std	7,7 mg/24 Std
	5,72 mg/24 Std	6,26 mg/24 Std
	9,0 mg/24 Std	
Gonadotropine	192 ME/24 Std	192 ME/24 Std
	384 ME/24 Std	384 ME/24 Std

Diskussion

Bei der vorliegenden Beobachtung erkrankte ein bisher gesunder Mann im Jahre 1948 an einer doppelseitigen Mumpsorchitis, in deren Folge sich ein primärer hypergonadotroper Hypogonadismus mit den bekannten Symptomen entwickelte. 8 Jahre später traten die ersten Zeichen einer Acanthosis nigricans auf, die bis zum gegenwärtigen Zeitpunkt als sog. *benigne* Form angesehen werden muß. Die Manifestation der hormonalen Störung ist also vorausgegangen, so daß sich der Zusammenhang mit dem Endokrinium aufdrängt. Wenn überhaupt etwas als beweisend anerkannt werden kann, dann ist es der Verlauf dieser Erkrankung, der sich in dieser Form nicht in der Literatur findet. Unter kritischer Würdigung aller Momente glaube ich, daß man den Zusammenhang als zweifelsfrei ansprechen darf, sofern man Verlaufsbeobachtungen überhaupt anerkennen will.

Es ist nach eigener Auffassung unerheblich, wenn von den Gegnern der „endokrinen Theorie" ins Feld geführt wird, daß zu viele endokrine Einflüsse angeschuldigt werden, ohne daß umgekehrt diese Störungen stets mit einer Acanthosis nigricans vergesellschaftet auftreten. Ähnlichen Gegebenheiten begegnet man in der Medizin immer wieder. Hier tritt der Individualfaktor Gottrons in Aktion. Für die vorliegende Beobachtung ist es besonders wichtig, daß *vor* Auftreten der Acanthosis nigricans endokrine Faktoren wirksam waren, die auf die Haut eingewirkt haben und den entscheidenden Anstoß für das Auftreten der A.n. gegeben haben. Gerade das nicht unmittelbare Auftreten der A. n. nach der Mumpsorchitis muß als besonders beweiskräftiges Argument *für* den Zusammenhang mit einer hormonalen Störung angesehen werden.

Zusammenfassung

An einer eigenen Beobachtung von Acanthosis nigricans bei einem 53 jährigen Mann kann der Zusammenhang der Erkrankung in diesem Falle mit einer echten endokrinen Störung — einem primären, postinfektiösen, hypergonadotropen Hypogonadismus — in Zusammenhang gebracht werden.

Literatur

Schirren, C.: Die Beteiligung der Haut bei endokrinen Störungen unter besonderer Berücksichtigung der Acanthosis nigricans. Internist 4, 501—509 (1963).
— Untersuchungen und Beobachtungen bei Acanthosis nigricans. Arch. klin. exp. Derm. (1964) (im Druck).

Diskussion

W. W. Kühnau (Bingen):

Könnte der Patient damals auch eine unbemerkte Mumps-Meningitis mit einer Schädigung des Hypothalamus absolviert haben ?

C. Schirren:

Theoretisch sind die von Ihnen gemeinten Erkrankungen im Bereich des Hypothalamus durchaus denkbar. Da bei der von mir vorgetragenen Beobachtung aber klare endokrinologische Befunde vorliegen, möchte ich ihnen den Vorrang vor Spekulationen geben.

G. Bettendorf (Hamburg):

Sind Fälle bekannt, bei denen die Acanthosis mit einer Ovarialinsuffizienz einhergeht ?

C. Schirren:

Aus der Literatur sind mir Zusammenhänge zwischen Acanthosis nigricans und Oophoritis nach Mumps nicht bekannt. Dagegen habe ich einige eigene Beobachtungen, bei denen andere endokrine Erkrankungen eine wesentliche Rolle beim weiblichen Geschlecht zu spielen scheinen. Über eine Beobachtung wurde gemeinsam mit Herrn Nowakowski vor Jahren auf einer Hamburger Dermatologen-Tagung berichtet; es handelte sich um eine Frau mit einer offenbar hypophysär bedingten Amenorrhoe bei Morbus Cushing.

Aus dem Anatomischen Institut der Universität Bonn
(Direktor: Prof. Dr. E. Tonutti)

Hypophysäre Hemmwirkung auf die basale Follikelentwicklung im Eierstock?[1]

Von

S. Fetzer und J. Hillebrecht

Mit 2 Abbildungen

Bei Kernmessungen an der Membrana granulosa der Eierstockfollikel der Ratte haben wir gefunden, daß zwischen kleinen, mittelgroßen und großen Follikeln sprunghafte Änderungen des Kernvolumens auftreten (Abb. 1). Bei den kleinen Follikeln mit einem Durchmesser von 100 μ und mehrreihiger Granulosa ohne Höhlenbildung liegt das Häufigkeitsmaximum der Verteilungskurve der Kernvolumina bei 50 μ^3. Die mittelgroßen Follikel haben einen Durchmesser von 200—360 μ und zeigen den Beginn einer Höhlenbildung. Hier ist das Häufigkeitsmaximum der Kernvolumina auf 90—95 μ^3 verschoben.

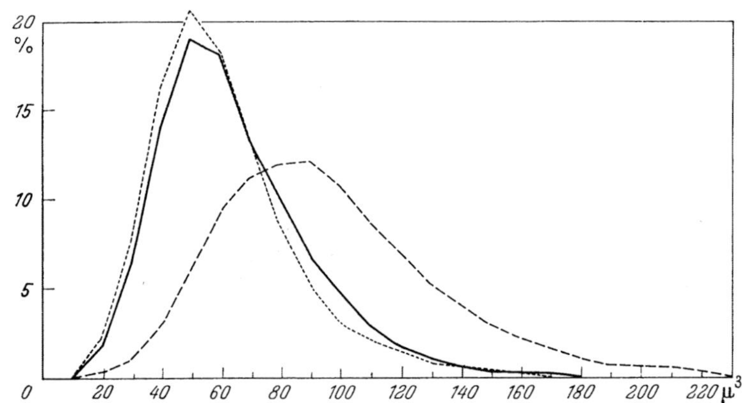

Abb. 1. Häufigkeitsverteilung der Kernvolumina der Granulosazellen kleiner, mittelgroßer und großer Follikel bei Normaltieren. Jede Kurve aus den Meßwerten aller Follikel der betreffenden Größenordnung berechnet. Kleine Follikel: ausgezogene Linie; mittelgroße Follikel: gestrichelt; große Follikel: punktiert

Follikel dieser Größenordnung finden sich auch noch bei langhypophysektomierten Tieren. Sie gehören also zu dem autonomen, von der Hypophyse unabhängigen Bestand, der die Basis für die cyclischen Reifung einer Art-abhängigen Anzahl von Follikeln bildet.

Bei den großen, reifen Follikeln mit einem Durchmesser von 600—800 μ geht das Häufigkeitsmaximum auf das bei den kleinen Follikeln gemessene Volumen, also auf 50 μ^3, zurück.

---

[1] Mit Unterstützung der Deutschen Forschungsgemeinschaft.

Nach unserer Auffassung dient die Volumenverdoppelung der Granulosakerne bei den mittelgroßen Follikeln zur Bereitstellung von Kernsubstanz, um die beschleunigte Weiterentwicklung zum großen, sprungreifen Follikel im Falle der Stimulierung zu erleichtern. Diese Anschauung, daß die Kernvergrößerung eine Wachstumsreserve darstellt, wird gestützt durch die Untersuchungen von LANE und DAVIS über das Verhalten des Mitose-Index: Bei 100 μ großen Follikeln beträgt er 11, bei 300 μ großen erreicht er den Wert 24, um dann stetig zu fallen und bei 600 μ großen Follikeln den Wert 13 und schließlich bei 700μ großen den Wert 8 zu erreichen.

Betrachtet man das Verhalten der „Gesamtvolumina" der Einzeltiere (d. h. den Meßwert von je 100 Zellkernen), so liegen im Falle der kleinen Follikel alle Werte mit Ausnahme eines „Ausreißers" sehr eng zusammen (Abb. 2). Bei den mittelgroßen und großen Follikeln der Normaltiere streuen die Werte in einem weiteren Bereich. Wir führen dieses unruhige Bild auf die intensivierten Wachstums- und Reifungsvorgänge bei beiden Follikelgruppen zurück.

Wir haben weiterhin bei 12 langhypophysektomierten Ratten, bei denen die Operation zwischen 34 und 98 Tagen zurücklag, das Kernvolumen des Follikelepithels bestimmt. Dabei ergab sich ein überraschendes Ergebnis: Die Meßwerte der kleinen Follikel waren nicht mehr wie bei den Normaltieren in einem engen Bereich massiert, sondern lagen weit auseinandergezogen und waren z. T. sogar vergrößert (Abb. 2). Eliminiert man den „Ausreißer"-Wert bei den kleinen Follikeln der Normaltiere aufgrund des Chauvenetschen Kriteriums, da seine Abweichung vom gereinigten Mittelwert das 21fache von σ_x beträgt, so ist die Differenz der Mittelwerte zwischen normalen und hypophysenlosen Ratten signifikant (P-Wert wenig größer als 0,001).

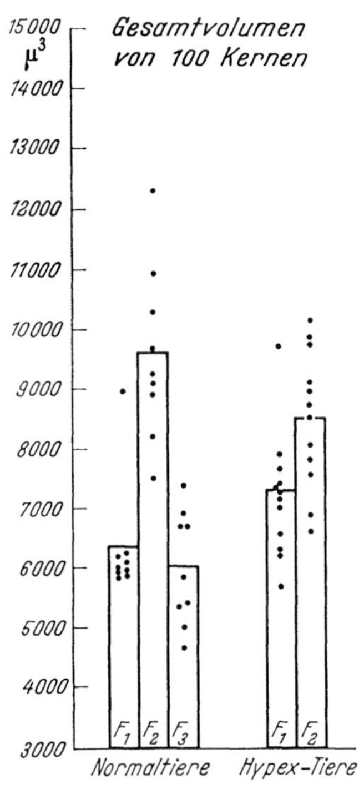

Abb. 2. „Gesamtvolumen" für je 100 Zellkerne von Granulosazellen je Follikel. F 1: kleine Follikel; F 2: mittelgroße Follikel; F 3: große Follikel; Hypex: Hypophysektomie

In weiteren Untersuchungen an hypophysenlosen Ratten konnte dieses Ergebnis bestätigt werden. Auch hier hatte also ein Teil der kleinen Follikel auf die Entfernung der Hypophyse mit einem Impuls zur Kernvergrößerung des Follikelepithels reagiert.

PAESI u. a. hatten bei infantilen Ratten festgestellt, daß nach Entfernung der Hypophyse die Zahl der kleinsten Follikel mit einem Durchmesser von 23 bis 32 μ stark vermehrt und die Follikel mit einem Durchmesser von 43—142 μ vermindert werden. Nach seiner Deutung hemmt die Hypophyse die Entwicklung der Eier zu Primärfollikeln, so daß der Ausfall der Hypophyse zu einer starken Vermehrung der Primärfollikel führt.

Unsere Ergebnisse zeigen, daß dieser regulative Bremseffekt der Hypophyse sich auch auf größere Follikel mit einem Durchmesser von 100μ erstreckt.

Welcher Faktor für dieses Hemmphänomen der Follikelreifung verantwortlich ist, ist ungeklärt. Zu diskutieren ist

1. die Existenz eines besonderen hypophysären Faktors,
2. die etwaige Hemmwirkung eines bekannten Hypophysenhormons,
3. der Einfluß der Ovarialhormone.

Bei den vorhin erwähnten hypophysenlosen Tieren fand sich das Phänomen der Enthemmung auch bei Tieren, die 6 Tage mit Choriongonadotropin behandelt worden waren. Es ist daher unwahrscheinlich, daß die Hemmwirkung durch die ICSH-Aktivität oder durch Oestrogene ausgeübt wird.

Literatur

Burkl, W., u. G. Kellner: Acta anat. (Basel) **23**, 49 (1955).

Koller, S.: Statistische Auswertungsmethoden in Biochemisches Taschenbuch, S. 1206. Herausgeg. v. H. M. Rauen, Berlin-Göttingen-Heidelberg: Springer-Verlag 1956.

Lane, C. E., and F. R. Davis: Anat. Rec. **73**, 429 (1939).

Paesi, F. J. A.: Acta endocr. (Kbh.) **3**, 89 (1949).

Wissenschaftliche Tabellen: J. R. GEIGY A.G., Basel, Ausgabe 1955, S. 43.

Swezy, O.: Ovogenesis and its Relation to the Hypophysis. Lancaster, Pennsylvania: The Science Press 1933 (zitiert nach Paesi).

Aus dem Hauptlaboratorium der Schering AG Berlin (Direktor: Dr. RASPÉ) und aus der Universitäts-Frauenklinik Göttingen (Direktor: Prof. Dr. H. KIRCHHOFF)

Effekte von Zeitfaktor, Hypophysektomie und abgestuften Mengen Prolactin auf die Deciduomareaktion der weißen Ratte

Von

I. v. BERSWORDT-WALLRABE, R. v. BERSWORDT-WALLRABE und U. HERLYN

Mit 2 Abbildungen

Die Deciduomareaktion an der pseudograviden, weißen Ratte ist zu Untersuchungen über die hormonale Kontrolle der Implantation geeignet.

An insgesamt 60 intakten Ratten wurde nachgewiesen, daß die Deciduomareaktion sowohl bei der durch Prolactin pseudograviden, als auch bei der durch Portioreiz pseudograviden Ratte quantitativ gleichwertig ist.

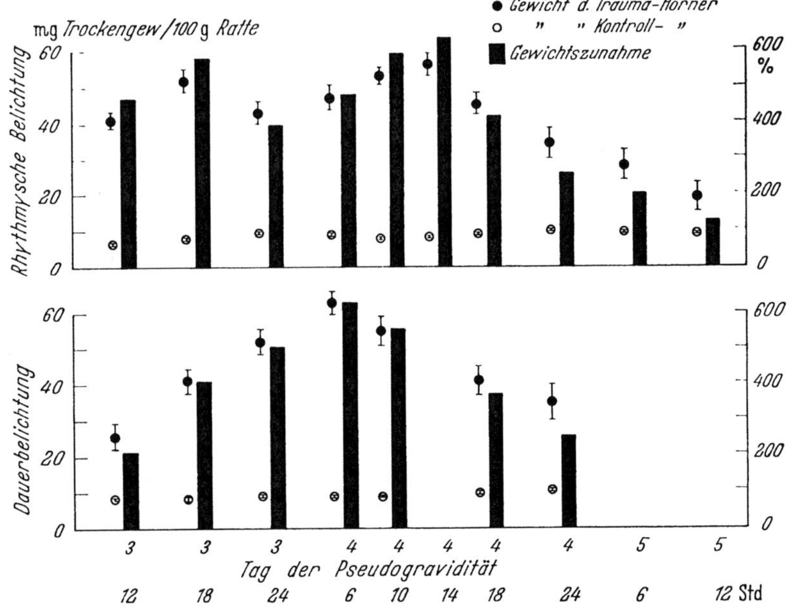

Abb. 1. Maximum der uterinen Sensibilität bei rhythmisch belichteten und dauerbelichteten intakten Inzuchtratten unter Prolaktinbehandlung (1 mg/Tag)

Die uterine Sensibilität gegenüber dem Endometriumtrauma ist am intakten Tier auf Tag 3 und 4 der Pseudogravidität beschränkt. Dauerbehandlung mit Prolactin vermag die Sensibilität des Uterus nicht über Tag 4 hinaus zu verlängern.

Dauerbelichtung hat eine Verschiebung der maximalen Sensibilität des Uterus bei
gleicher Versuchsanordnung auf den frühen Tag 4 der Pseudogravidität zur Folge;
andere Faktoren, vermutlich hypophysären Ursprungs, scheinen demnach an der
luteotrophen Wirkung beteiligt zu sein (Abb. 1).

An insgesamt 198 Ratten wurde gezeigt, daß am intakten Tier steigende Dosen
von Prolactin bei einem Schwellenwert von 0,125 mg/Tag eine nahezu gleichwertige
maximale deciduale Reaktion hervorrufen. Am hypophysektomierten Tier kommt

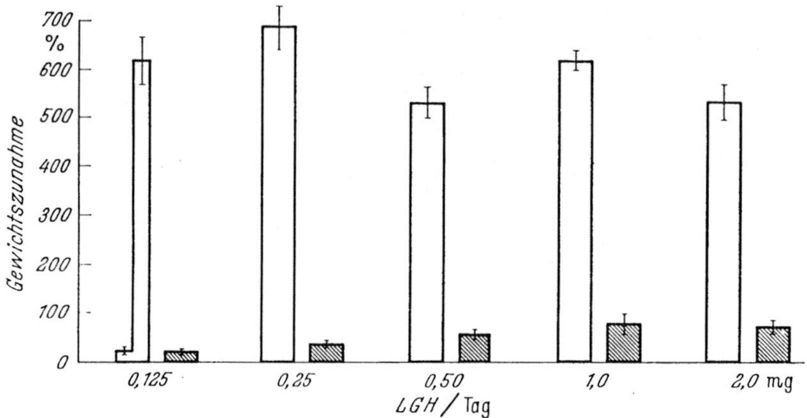

Abb. 2. Deciduoma-Reaktion bei steigender Dosierung von Prolaktin in intakten = ☐ und in
hypophysektomierten = ▨ Ratten

es lediglich zu einer geringen decidualen Zellreaktion, wenn Prolaction substituiert
wird. Steigende Dosen haben dabei einen geringen linearen Anstieg der Reaktion
zur Folge (Abb. 2).

Die Beobachtungen, daß

1. Prolactin die uterine Sensibilität nicht über den Tag 4 hinaus zu verlängern
vermag,

2. Lichteinflüsse die maximale Sensibilität Prolactin-behandelter Ratten zeitlich verschieben,

3. hypophysektomierte Tiere bei Prolactin-Behandlung nur eine geringe
Deciduoma-Bildung zeigen,
führen zu der Schlußfolgerung, daß es verfrüht erscheint, das lactogene Hormon
mit dem luteotrophen Hormon zu identifizieren. Andere hypophysäre Faktoren
scheinen beteiligt zu sein.

Aus der Universitäts-Frauenklinik Hamburg-Eppendorf
(Komm. Dir.: Priv.-Doz. Dr. G. UHLMANN)
Endokrinologische Abteilung (Priv.-Doz. Dr. G. BETTENDORF)

Ovulationsauslösung durch hypophysäres Human-Gonadotropin bei Fehlen der gonadotropen Hypophysenfunktionen

Von

G. BETTENDORF und M. BRECKWOLDT

Mit 2 Abbildungen

Die klinische Wirksamkeit des hypophysären Humangonadotropin (HHG), das wir seit einigen Jahren nach einem eigenen Verfahren aus menschlichen Hypophysen isolieren, haben wir bei zahlreichen Patientinnen mit einer hypogonadotropen bzw. normogonadotropen Ovarialinsuffizient überprüft. Von insgesamt 128 Patientinnen mit einer primären oder mit einer sekundären Amenorrhoe besaßen 37% eine nicht meßbare Gonadotropinausscheidung, 49% zeigten normale Werte und bei 14% war die Ausscheidung erhöht. Im Vergleich zu den Gonadotropinwerten wurden ebenfalls die Gesamtoestrogene bestimmt. In allen drei Patientengruppen schwankt die Oestrogenausscheidung um einen Mittelwert, der bei der hypogonadotropen Form bei 24 μg, bei der normogonadotropen Gruppe bei 34 μg und in der hypergonadotropen Gruppe bei 23 μg liegt. Es ergibt sich hieraus, daß von der Größe der Oestrogenausscheidung im Einzelfall keine Rückschlüsse auf die Gonadotropin-Ausscheidung und damit auf die Ursache der vorliegenden Störung möglich ist.

Bereits bei den ersten Untersuchungen mit HHG stellte sich heraus, daß eine Stimulierung der Ovarialfunktion möglich war, die über das physiologische Ausmaß hinausging. Dies zeigte sich sowohl in der starken Steigerung der Oestrogen- und Pregnandiolausscheidung als auch in den häufig beobachteten multiplen Ovulationen. Insgesamt wurden bisher 44 Behandlungen durchgeführt. In 36 Fällen war es möglich, eine Ovulation zu induzieren, und lediglich in 8 Fällen trat keine biphasische Reaktion im Anschluß an die Behandlung auf. Die anfänglich sehr hoch gewählte Dosierung haben wir bald reduziert, und wir können jetzt genaue Angaben über die erforderlichen Aktivitätsmengen machen.

Nach den bisherigen Ergebnissen läßt sich sagen, daß die Verabreichung von HHG in einer gesamten Dosierung von 4000—5000 HMG-Einheiten, verteilt auf 10 bis 14 Tage, in allen Fällen zu einer biphasischen Ovarialfunktion führt (Abb. 1). Dies gilt auch für 6 Behandlungen bei hypophysektomierten Patientinnen, die in der hypogonadotropen Gruppe enthalten sind. Im Hinblick auf die FSH-Aktivität ist eine Gesamtdosierung von 150—200 Äquivalenten im Vergleich zum FSH-NIH-S1 erforderlich. Beim LH lassen sich die Werte noch nicht genau festlegen. Für die Induktion einer Ovulation waren jedoch bei den normogonadotropen Formen der Amenorrhoe insgesamt 1—2 Äquivalente im Vergleich zur benutzten Bezugssubstanz ausreichend.

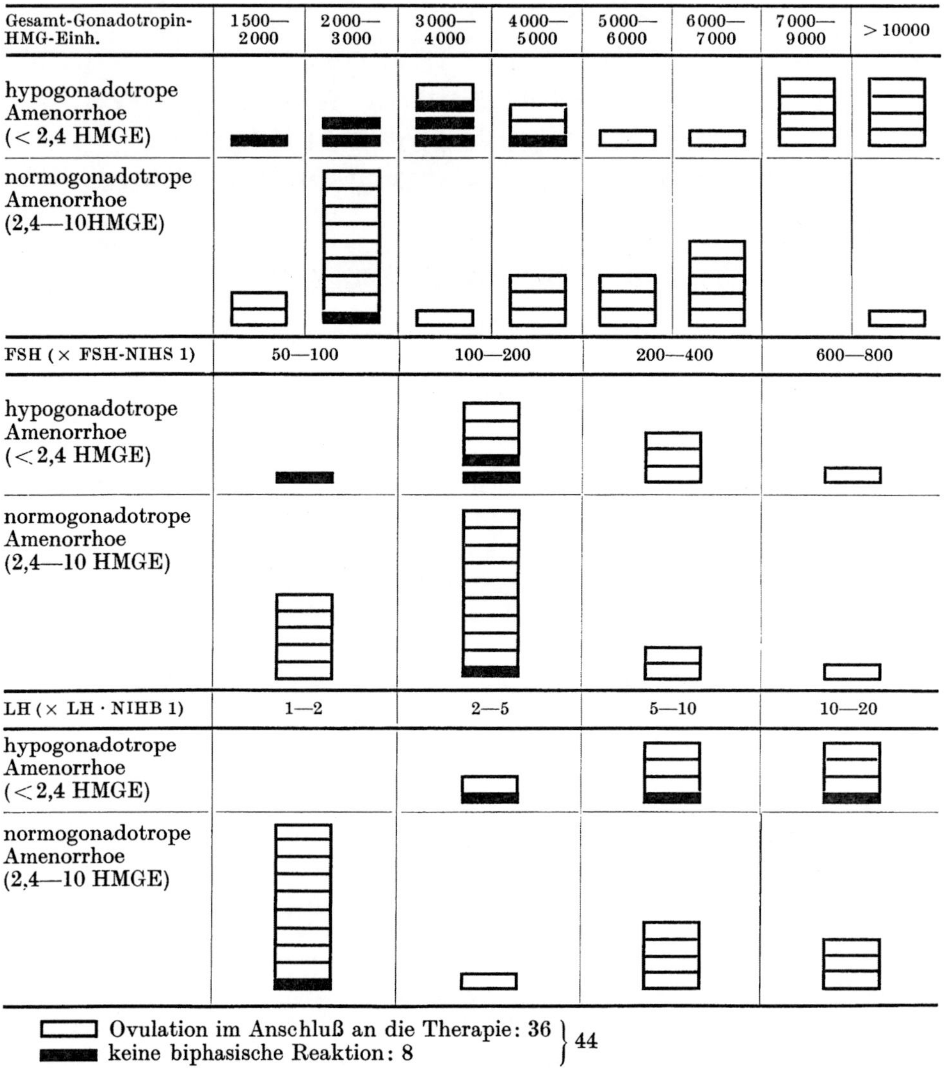

Gesamt-Gonadotropin-HMG-Einh.	1500—2000	2000—3000	3000—4000	4000—5000	5000—6000	6000—7000	7000—9000	> 10000
hypogonadotrope Amenorrhoe (< 2,4 HMGE)								
normogonadotrope Amenorrhoe (2,4—10HMGE)								

FSH (× FSH-NIHS 1)	50—100	100—200	200—400	600—800
hypogonadotrope Amenorrhoe (< 2,4 HMGE)				
normogonadotrope Amenorrhoe (2,4—10 HMGE)				

LH (× LH · NIHB 1)	1—2	2—5	5—10	10—20
hypogonadotrope Amenorrhoe (< 2,4 HMGE)				
normogonadotrope Amenorrhoe (2,4—10 HMGE)				

☐ Ovulation im Anschluß an die Therapie: 36 ⎱ 44
■ keine biphasische Reaktion: 8 ⎰

Abb. 1. Klinische Wirkung des HHG in Abhängigkeit von der Gonadotropinausscheidung und der verabreichten HHG-Dosierung

Vergleicht man die Wirkung des HHG bei der primären Amenorrhoe mit der bei der sekundären Amenorrhoe sowie bei hypophysektomierten Patientinnen, so ergeben sich die gleichen Verhältnisse wie in der ersten Tabelle. Die ovarielle Reaktion ist nicht abhängig von der hormonellen Ausgangssituation und nicht von der Dauer der vorliegenden Störung (Abb. 2).

Bei einem Teil der 8 negativ verlaufenden Behandlungscyclen haben wir das HHG jeden zweiten bzw. jeden dritten Tag verabreicht. Gleichzeitig wurde eine

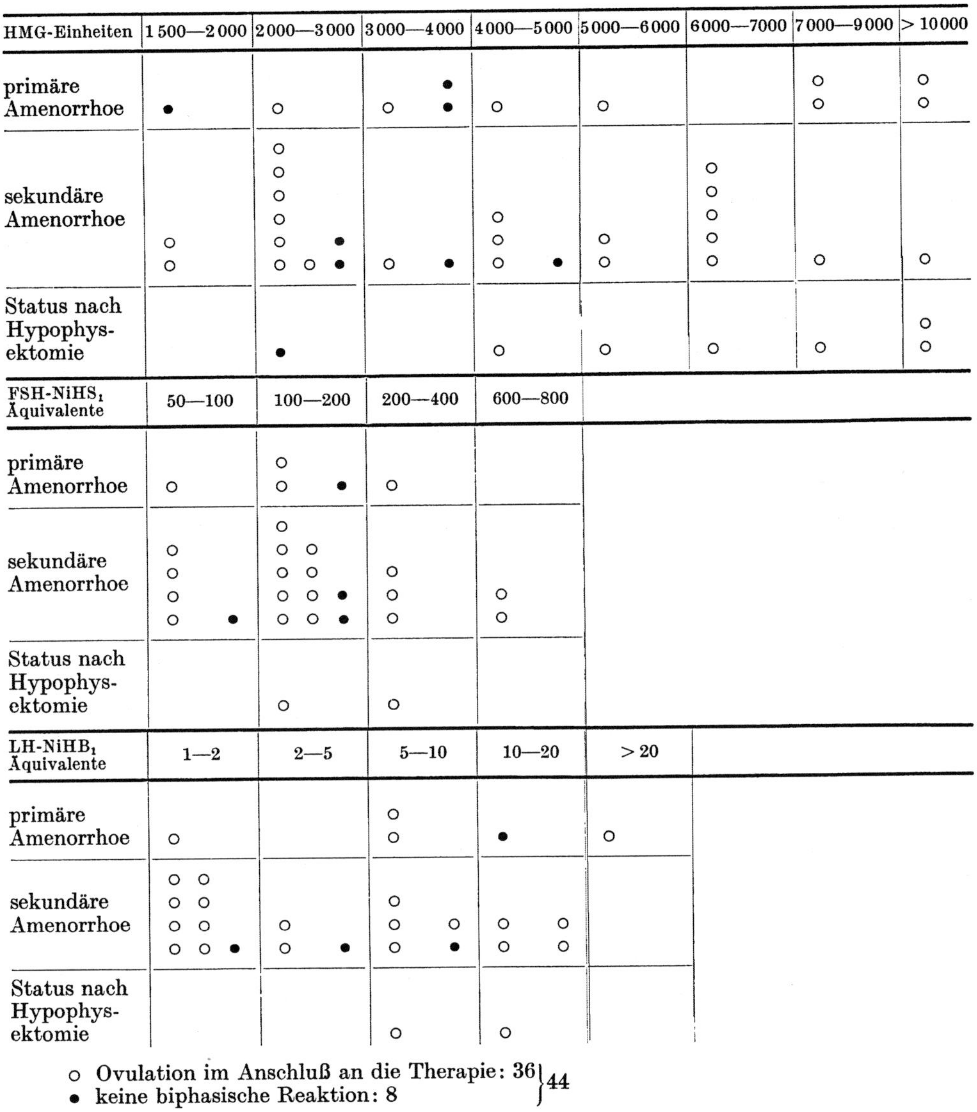

HMG-Einheiten	1 500—2 000	2 000—3 000	3 000—4 000	4 000—5 000	5 000—6 000	6 000—7 000	7 000—9 000	> 10 000
primäre Amenorrhoe	●	○	○ ● ●	○	○		○ ○	○ ○
sekundäre Amenorrhoe	○ ○	○ ○ ○ ○ ○ ○ ○ ● ●	○ ●	○ ○ ○ ●	○ ○	○ ○ ○ ○ ○	○	○
Status nach Hypophysektomie		●		○	○	○	○	○ ○

FSH-NiHS$_1$ Äquivalente	50—100	100—200	200—400	600—800				
primäre Amenorrhoe	○	○ ○ ●	○					
sekundäre Amenorrhoe	○ ○ ○ ○	○ ○ ○ ○ ○ ○ ○ ○ ○ ● ●	○ ○ ○	○ ○				
Status nach Hypophysektomie		○	○					

LH-NiHB$_1$ Äquivalente	1—2	2—5	5—10	10—20	> 20			
primäre Amenorrhoe	○		○ ○	●	○			
sekundäre Amenorrhoe	○ ○ ○ ○ ○ ○ ○ ○ ●	○ ●	○ ○ ○ ○ ●	○ ○ ○ ○				
Status nach Hypophysektomie			○	○				

○ Ovulation im Anschluß an die Therapie: 36 }
● keine biphasische Reaktion: 8 } 44

Abb. 2. Klinische Wirkung des HHG bei verschiedenen Amenorrhoe-Formen in Abhängigkeit von der verabreichten Gesamtgonadotropin,- FSH-, bzw. LH-Aktivität

Dosis gegeben, die unter 400 HMG-Einheiten lag. Eine Stoßbehandlung mit hohen Dosen HHG über kurze Zeit war ebenfalls nicht wirkungsvoll. Auf der anderen Seite ist es möglich, durch die Verabreichung von 400 HMG-Einheiten HHG über 10—14 Tage bei allen Patientinnen eine Ovulation zu induzieren, diese Menge entspricht einer FSH-Aktivität von 15—20 FSH/NIH S 1 Äquivalenten.

Beim Vergleich der Wirkungen des HHG mit verschiedenen HMG-Präparaten ergab sich, daß die klinische Wirkung der verschiedenen Substanzen von der Höhe der follikelstimulierenden Aktivität und nicht von der gonadotropen Gesamtaktivität abhängig ist.

Der eigentliche Beweis für eine induzierte Ovulation mit der gleichzeitigen Freigabe eines befruchtungsfähigen Eies wird letzten Endes nur durch das Eintreten einer Schwangerschaft erbracht. Bisher haben wir zwei Schwangerschaften beobachtet, die mit Sicherheit auf eine durch das HHG hervorgerufene Ovulation zurückzuführen sind. In einem Fall handelt es sich um eine 27 jährige Patientin, die nach dem Eintreten der Menarche mit 15 Jahren nur für etwa 1 Jahr unregelmäßig menstruiert war. Alle zur Verfügung stehenden therapeutischen Maßnahmen versagten. Bei der wiederholten Behandlung mit HHG trat immer eine Ovulation mit einer nachfolgenden echten Menstruation auf. Im 7. Behandlungscyclus mit HHG und HCG kam es zur Konzeption. Es besteht jetzt eine Schwangerschaft der 20. Woche[1].

Bei einer anderen Patientin bestanden bis zum 21. Lebensjahr normale Menstruationen, dann kam es zu einer sekundären Amenorrhoe. 5 Jahre später konnte als Ursache hierfür ein Hypophysentumor festgestellt werden. Bei der Operation wurde ein chromophobes Adenom des Hypophysenvorderlappens entfernt. Im Anschluß an die Exstirpation des Tumors fanden sich erniedrigte 17-Keto-Steroid- und 17-Hydroxy-corticosteroid-Werte. Es erfolgte eine Substitution mit Corticoiden und Schilddrüsenpräparaten. Mehrfache Behandlungsversuche, die wegen eines dringenden Kinderwunsches durchgeführt wurden, waren erfolglos. Die erstmalige Behandlung mit HHG in steigender Dosierung hatte eine erhebliche Anregung der Ovarialtätigkeit zur Folge. Sie machte sich zunächst in einer starken Steigerung der Oestrogenausscheidung bemerkbar. Am 2. Tag nach HCG-Verabreichung kam es zu einem Sprung der Basaltemperatur und im Anschluß hieran zu einer erhöhten Pregnandiolausscheidung. Am 39. Tag nach dem Basaltemperatursprung, d. h. nach der möglichen Konzeption, war der Schwangerschaftstest positiv. Bis zur 24. Woche entsprach die Ausscheidung der Oestrogene und des Pregnandiol etwa den Werten, die wir auch sonst in der Frühgravidität sehen. Die Ausscheidung der 17-Keto-Steroide lag jetzt zwischen 10 und 20 mg pro 24 Std und die 17-Hydroxy-Corticoid-Ausscheidung um 5 mg. Bis zur 20. Woche war die Größenzunahme des Uterus normal, zeigte ab dann jedoch kein weiteres Wachstum. In der 26. Woche post conceptionen fiel die Oestrogenausscheidung stark ab, und wenige Tage später war der Pregnosticontest zum erstenmal negativ. Zum gleichen Zeitpunkt setzten leichte Schmierblutungen ein, und die Röntgenaufnahmen ließen einen abgestorbenen Feten erkennen, der deutlich kleiner war, als nach der Schwangerschaftsdauer zu erwarten war. Da bisher noch keine Ausstoßung erfolgt ist, können noch keine weiteren Angaben gemacht werden[2]. Die Frage muß offen bleiben, ob die fehlende oder zumindest stark herabgesetzte Hypophysenfunktion für den Verlauf verantwortlich zu machen ist, oder ob eine andere Ursache in Frage kommt.

Die Ergebnisse zeigen, daß es möglich ist, eine fehlende gonadotrope Funktion durch die Verabreichung entsprechender menschlicher Gonadotropin-Präparate vollwertig zu ersetzen. Damit können wir auch bei den Formen der Ovarialinsuffizienz, für die es bisher keine wirksame Behandlung gab, eine erfolgreiche Therapie treiben. Gleichzeitig werden uns die Untersuchungen mit den Gonadotropinen bessere Einblicke in die physiologische Regulierung der Ovarialfunktion geben.

---

[1] *Anmerkung bei der Korrektur:* Die Pat. steht jetzt wenige Tage vor der Entbindung.

[2] Die Pat. hat inzwischen 2 weibliche Feten ausgestoßen und bei vier weiteren Pat. trat eine Gravidität ein.

Die Untersuchungen wurden mit Unterstützung der *Deutschen Forschungsgemeinschaft* durchgeführt, für die Überlassung der Testsubstanzen danken wir der Endocrinology Study Section des NIH Bethesda/USA.

Literatur

BETTENDORF, G., M. APOSTOLAKIS, and K. D. VOIGT: Proceedings. III. World Congr., Gynec. Obstet. 1, 76 (1961).
BETTENDORF, G.: Habil.-Schrift Hamburg 1961.
— Geburtsh. u. Frauenheilk. 22, 928—931 (1962).
— M. APOSTOLAKIS, and K. D. VOIGT: Acta endocr. (Kbh.) 41, 1, 13 (1962).
— Int. J. Fertil. 8, 799 (1963).
— Int. J. Fertil. 9, 351 (1964).
—, u. M. BRECKWOLDT: Arch. Gynäk. 199, 423 (1964).
DICZFALUSY, E., E. JOHANNISSON, K. G. TILLINGER, and G. BETTENDORF: Acta endocr. (Kbh.) Suppl. 90, 35 (1964).
HELLER, C. G. F., H. LAWSON, and E. L. SEVERINGHAUS: Amer. J. Physiol. 121, 364 (1938).
PARLOW, A. F.: Human Pituitary Gonadotrop (Hrsg. A. ALBERT) p. 300. Springfield: Ch. C. Thomas 1961.
STEELMAN, S. L., and F. M. POHLEY: Endocrinology 53, 604 (1953).

Diskussion

J. HAMMERSTEIN (Berlin):

Die hohen Oestrogen-Ausscheidungswerte bei Ihren Amenorrhoe-Patienten haben mich überrascht. Könnten Sie hierzu Stellung nehmen? Unsere eigenen Befunde bei der Amenorrhoe liegen stets unter 30 γ/die. — Haben Sie bei der hypophysektomierten Patientin einen Metopirontest durchgeführt? Damit ließen sich Anhaltspunkte dafür gewinnen, ob die Hypophyse bei der Operation vollständig entfernt worden ist oder nicht, was für die Beurteilung Ihres Falles von Wichtigkeit zu sein scheint.

G. BETTENDORF:

1. Eine Erklärung für die teilweise auffallend hohen Oestrogenwerte bei den Patientinnen mit einer nicht meßbaren Gonadotropinausscheidung kann zur Zeit noch nicht gegeben werden. Die Oestrogene wurden als Gesamtoestrogene nach der Methode von ITTRICH bestimmt.

2. Bei der hypophysektomierten Patientin, die gravide wurde, haben wir keinen Metopiron-Test durchgeführt. Mit Sicherheit ist die Hypophysenfunktion jedoch über lange Zeit vor der Hypophysektomie gestört gewesen und durch die Operation zusätzlich beeinflußt worden. Eine Gonadotropinausscheidung war vor Eintreten der Schwangerschaft nicht nachweisbar, und die Ausscheidung der Ovarialsteroide war erniedrigt.

Aus der Frauenklinik der Medizinischen Akademie Düsseldorf
(Direktor: Prof. Dr. med. R. ELERT)

Über den Einfluß von Choriongonadotropin (HCG)-Antiserum auf die biologische Aktivität verschiedener Gonadotropine

Von

H. SCHMIDT-ELMENDORFF und R. BUCHHOLZ

Mit 2 Abbildungen

Untersuchungen von WIDE (1962) zeigten, daß ein im Kaninchen gewonnenes HCG-Antiserum nicht nur spezifisch mit HCG reagiert, sondern auch in Form einer sog. "cross-reaction" mit menschlichem hypophysärem luteinisierendem Hormon (LH).

Im folgenden möchten wir über unsere Ergebnisse vom Einfluß dieses HCG-Antiserum auf die biologische follikelstimulierende (FSH)-, luteinisierende (LH)- und gesamtgonadotrope Aktivität verschiedener Gonadotropine berichten.

Nachstehende Gonadotropinpräparate wurden untersucht:

1. das menschliche Choriongonadotropin — HCG,
    a) der internationale Standard für HCG,
    b) Pregnyl[1].
2. Das menschliche Menopausegonadotropin — HMG,
    a) HMG - 24,
    b) HMG - Res. P. 3605[1],
    c) Pergonal[2].
3. Das Stutenserumgonadotropin — PMS-G.
    a) der internationale Standard für PMS-G.
    b) Gestyl[1].

In allen quantitativen Untersuchungen wurde der internationale Standard verwandt, während die qualitativen Vergleiche mit den entsprechenden Handelspräparaten vorgenommen wurden.

Die folgenden biologischen Nachweismethoden wurden angewandt:

1. der Augmentation-Test bei Ratten;
2. der Prostata-Test bei hypophysektomierten Ratten;
3. der Prostata-Test bei intakten Ratten;
4. der Ratten-Uterus-Test;
5. der Ratten-Ovar-Test;
6. der Maus-Uterus-Test.

---

[1] Organon, Oss (Holland).
[2] Istituto Sereno, Rom (Italien).

Während Test 1, der Augmentation-Test, als spezifisch für FSH-Aktivität gilt, ist der Prostata-Test bei hypophysektomierten Ratten ein hochspezifischer Indicator für LH-Aktivität. Test 3—6: Der Prostata-Test bei intakten Ratten, der Ratten-Uterus-, der Ratten-Ovar- und der Maus-Uterus-Test sind weder für FSH, noch für LH-Aktivität spezifisch, sondern sie reagieren auf Gonadotropine, die sowohl FSH als auch LH-Aktivität enthalten. Sie können deshalb als Nachweis für gesamtgonadotrope Aktivität dienen.

Nun zu den Ergebnissen:

Untersuchungen des Einflusses von HCG-Antiserum auf die biologische Wirkung von HCG haben ergeben, daß 125 μg Antiserum imstande sind, die Wirkung von HCG auf den Maus-Uterus-Test bis zu einer Dosis von 2,5 IE HCG pro Tier vollständig zu inaktivieren, d. h. eine IE HCG + 50 μg Antiserum besitzen keine Wirkung mehr auf den Maus-Uterus-Test.

Weiterhin wurde der Einfluß von HCG-Antiserum auf die biologische Wirkung von HMG-24 untersucht. Jede Maus erhielt 25 μg Antiserum und zusätzlich, je nach Gruppe, 0,5 bis 2,5 E HMG-24. Es zeigte sich, daß 25 μg Antiserum imstande sind, die biologische Wirkung von bis zu 1,25 E HMG-24 zu inaktivieren.

Als der vorige Versuch mit Pergonal wiederholt wurde, vermochten 25 μg Antiserum ungefähr 0,75 E Pergonal zu inaktivieren.

Von besonderem Interesse erscheint die Wirkung von HCG-Antiserum auf PMS-G, welches ja tierischen Ursprungs ist. Der Zusatz von 1 mg Antiserum pro Maus bewirkte, daß die Wirkung von 0,25 IE PMS-G praktisch aufgehoben wurde. Die Dosenwirkungskurve für 0,5 und 1,0 IE PMS-G + HCG-Antiserum lag signifikant niedriger als für PMS-G allein. Bemerkenswert ist jedoch die relative Unempfindlichkeit von PMS-G gegen HCG-Antiserum. So sind über 4 mg Antiserum notwendig, um die biologische Aktivität von einer IE PMS-G auf den Maus-Uterus-Test zu unterdrücken. Die entsprechende Menge beträgt für eine IE HCG ungefähr 50 μg, für eine E HMG-24 20 μg und für eine E Pergonal etwa 30 μg Antiserum.

Ein Vergleich der Inaktivierbarkeit dieser Gonadotropine durch HCG-Antiserum mit ihrer biologischen LH-Aktivität zeigt, daß zwischen der immunologischen und der biologischen LH-Aktivität von HCG, HMG und Pergonal ein angedeuteter Zusammenhang besteht. Dieser wird jedoch zwischen den menschlichen Gonadotropinen und dem tierischen PMS-G vermißt.

Gonadotripon 1 Einheit	Antiserum μg	Biologische LH-Aktivität-HMG-24
HCG	ungefähr 50	1,7
HMG-24	ungefähr 20	1,0
Pergonal	ungefähr 30	1,1
PMS-G	ungefähr 4000	0,62

Das erste Bild zeigt die biologische Wirkung von HMG mit und ohne HCG-Antiserum auf das Ratten-Ovar. Die ausgezogenen Linien stellen die Dosenwirkungskurve von HMG allein dar, die unterbrochenen die Kurven von HMG, welches mit HCG-Antiserum vorbehandelt wurde, und zwar wurden pro 1 E HMG 50 μg Antiserum verwandt.

Während die Dosenwirkungskurven von Ratten-Uterus- und -Ovar-Test anzeigen, daß vorbehandeltes HMG seine Wirkung auf diese Organe verloren hat, zeigt der Verlauf der Kurven des FSH spezifischen Augmentation-Test, daß HMG auch nach Behandlung mit dem Antiserum noch seine ursprüngliche FSH-Aktivität besitzt. Als der vorhergehende Versuch noch einmal mit Pergonal wiederholt wurde, zeigte sich, daß auch dieses Menopause-Gonadotropin nach Zusatz von HCG-Antiserum seine ursprüngliche FSH-Aktivität beibehält, während seine Wirkung auf die Oestrogenproduktion und die Ovar-Gewichtszunahme und damit Follikelstimulation verlorengeht.

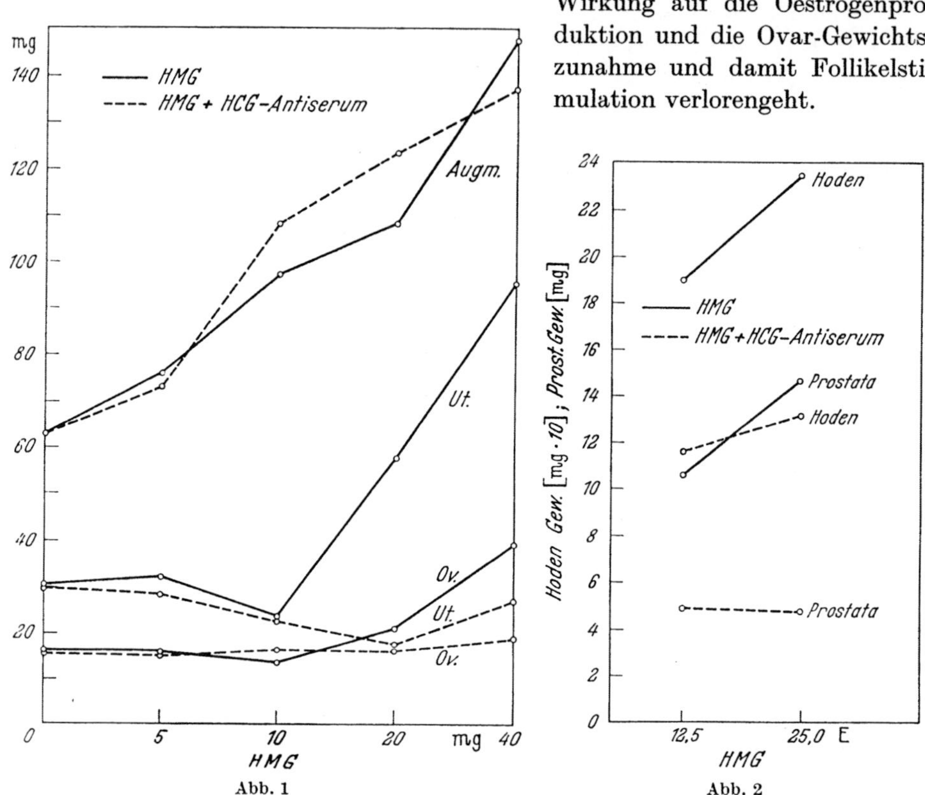

Abb. 1. Augmentation-Reaktion (+ 50 IE HCG), Uterusgewicht, Ovargewicht

Abb. 2. Hoden- und Prostata-Gewicht der hypophysektomierten Ratte

Das zweite Bild zeigt fernerhin, daß mit HCG-Antiserum behandeltes HMG keine LH-Aktivität mehr besitzt. Es zeigt keine Wirkung mehr auf die Prostata-Gewichtszunahme und damit die Androgenproduktion der hypophysektomierten Ratte.

Der vorhergehende Versuch ist noch einmal mit HMG und Pergonal bei der intakten männlichen Ratte wiederholt worden. Wieder zeigte es sich, daß HCG-Antiserum die gesamtgonadotrope Wirkung, in diesem Fall in Prostata- und Samenblasen-Gewichtszunahme ausgedrückt, zu inaktivieren vermag.

Zusammenfassung

1. HMG verliert nach Einwirkung von HCG-Antiserum seine LH-spezifische Wirkung in den für LH-spezifischen Rest am ventralen Prostatalappen der hypo-

physektomierten infantilen Ratte sowie seine Wirksamkeit in den gesamtgonado-
tropen Tests. Das mit HCG-Antiserum inaktivierte HMG zeigt keinen Einfluß
mehr auf die Oestrogen- und die Androgenproduktion sowie auf die Ovar-
Gewichtsvergrößerung und damit die Follikelstimulation.

2. Untersuchungen des HMG mit und ohne HCG-Antiserum auf den Aug-
mentations-Test dagegen zeigen, daß die Gegenwart oder das Fehlen der inakti-
vierten Komponente des HMG keinen Einfluß auf den Test haben. Da der Aug-
mentations-Test in beiden Fällen gleich reagiert, wird angenommen, daß nur die
LH-Aktivität inaktiviert wird, während die FSH-Komponente erhalten bleibt.

Literatur

WIDE, L.: An immunological method for the assay of human chorionic gonadotrophin. Acta
endocr. (Kbh.) Suppl. **70** (1962).

Aus der Frauenklinik der Medizinischen Akademie Düsseldorf
(Direktor: Prof. Dr. med. R. ELERT)

Über den Einfluß verschiedener Kombinationen von HMG und HCG auf die Ovarfunktion der Ratte

Von

R. BUCHHOLZ und H. SCHMIDT-ELMENDORFF

Mit 2 Abbildungen

Untersuchungen von E. DICZFALUSY, E. JOHANNISSON, K. G. TILLINGER und G. BETTENDORF haben gezeigt, daß nicht alle Präparate des menschlichen hypophysären Gonadotropins oder HMG geeignet sind, in Verbindung mit HCG eine Ovulation beim Menschen auszulösen.

Im folgenden soll untersucht werden, in welcher Weise die FSH-Aktivität von HMG-Präparaten mit der LH- bzw. der LTH-Aktivität von HCG kombiniert werden muß, um eine optimale Wirkung auf das Ovar auszuüben.

Nachstehend sind die getesteten Präparate aufgeführt:

1. menschliches Menopause-Gonadotropin — HMG — Res. P. 3605[1];

2. HMG + HCG-Antiserum;

3. menschliches Menopause-Gonadotropin-Pergonal[2];

4. Pergonal + HCG-Antiserum;

5. menschliches Choriongonadotropin-HCG.

Präparat 2 unterscheidet sich von Präparat 1 dadurch, daß es durch Zusatz von HCG-Antiserum vorbehandelt wurde. Wie aus dem vorhergehenden Vortrag hervorgeht, wird dadurch die LH-Komponente inaktiviert, während die FSH-Komponente erhalten bleibt.

Was das Pergonal anbetrifft, so entspricht in unseren Untersuchungen eine Einheit Pergonal einer Einheit HMG-24, bezogen auf die gesamtgonadotrope Aktivität, wie sie im Maus-Uterus-Test nachgewiesen wird. Vergleichende Untersuchungen der biologischen Aktivität von HMG und Pergonal haben ergeben, daß beim Pergonal das Verhältnis von FSH zu LH mehr zugunsten des FSH liegt, und zwar im Verhältnis von ungefähr 6:1.

Pergonal + HCG-Antiserum ist wie HMG + HCG-Antiserum eine Präparation, die nur noch FSH-Aktivität aufweist.

Das Präparat Nr. 5, das HCG, besitzt im Vergleich zu den beiden Menopause-Gonadotropinen eine geringere FSH-Aktivität, dagegen vorwiegend luteinisierende sowie luteotrope Eigenschaften.

---

[1] Organon, Oss (Holland).
[2] Istituto Sereno, Rom (Italien).

Die Planung der Experimente war wie folgt:

Die Versuchstiere wurden in 6 Gruppen aufgeteilt. Die erste und die letzte Gruppe wurden mit 100% des HMG bzw. des HCG injiziert. Die übrigen Gruppen erhielten 80% HMG + 20% HCG, 60% HMG + 40% HCG usw. Gleichzeitig wurde ein Parallelversuch unternommen, bei welchem HMG + HCG-Antiserum mit HCG kombiniert wurde. Jeder Versuch wurde dreimal wiederholt und die Mittelwerte der Ergebnisse zu einer Kurve vereinigt. Jeder Punkt auf den Kurven entspricht dem Mittelwert von mindestens 12 Tieren.

Das erste Bild zeigt, daß eine Kombination von HMG und HCG einen deutlichen Synergismus auf die Ovar-Gewichtszunahme und damit die Follikelstimulation ergibt, insbesondere, wenn 40% HMG mit 60% HCG kombiniert werden. Das entspricht einer Kombination von 20 E HMG mit 30 IE HCG. Mit anderen Worten: 20 E HMG + 30 IE HCG haben eine bessere Wirkung auf das Ovar als z. B. 10 E HMG + 40 IE HCG oder 40 E HMG + 10 IE HCG.

Im Falle der Kombination von HMG + HCG-Antiserum mit HCG zeigt sich der Gipfel des Synergismus durch Verlust der LH-Aktivität von HMG bei 20% HMG + HCG-Antiserum und 80% HCG.

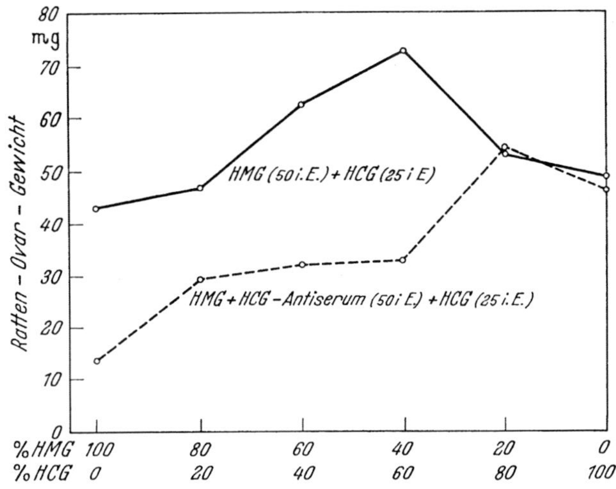

Abb. 1. Ovargewichtszunahme nach Kombination von HMG (mit und ohne HCG-Antiserum) und HCG

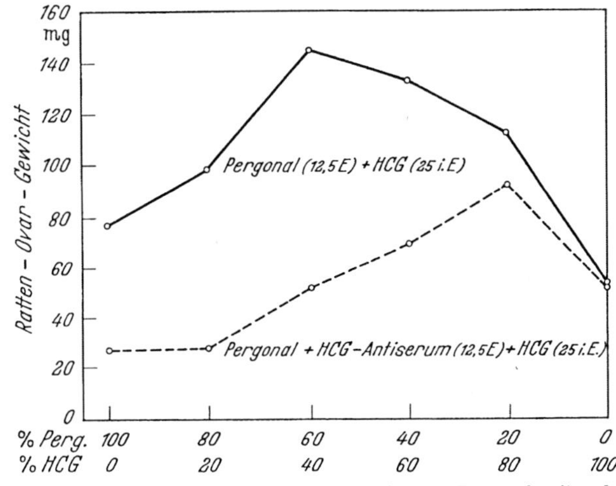

Abb. 2. Ovargewichtszunahme nach Kombination von Pergonal (mit und ohne HCG-Antiserum) und HCG

Wiederholt man den vorhergehenden Versuch mit Pergonal und HCG, so ergeben sich die beiden Kurven auf dem zweiten Bild.

Hier ist die optimale Kombination bei 60% Pergonal und 40% HCG erreicht. Das entspricht einer Mischung von 7,5 E Pergonal und 10 IE HCG. Obwohl im Vergleich zu 20 E HMG + 30 IE HCG im vorigen Versuch bei dieser Kombination viel weniger HMG E Pergonal verwandt wurden, nämlich 7,5 E und entsprechend weniger IE HCG, so ist die Stimulation des Ovars hier trotzdem erheblich größer. Die Ursache dafür dürfte in dem relativ höheren FSH-Gehalt des Pergonals liegen.

Nach Verlust seiner LH-Aktivität zeigt Pergonal, ähnlich wie HMG, ein Kombinationsoptimum bei 20% Pergonal + HCG-Antiserum und 80% HCG.

Bemerkenswert ist auch hier, daß 100% Pergonal + HCG-Antiserum keine Ovar-Gewichtsvergrößerung und damit Follikelstimulation hervorzurufen vermögen, obwohl, wie im vorigen Vortrag nachgewiesen werden konnte, die FSH-Aktivität des Pergonal nach Vorbehandlung mit HCG-Antiserum erhalten geblieben ist. Es bedarf also neben der FSH-Aktivität noch des Zusatzes von LH- bzw. LTH-Aktivität, um eine optimale Gewichtsvergrößerung und damit Follikelstimulation hervorzurufen.

Es ergibt sich nach diesen Resultaten die Frage, ob die FSH-Aktivität für sich allein eine biologische Bedeutung hat, da sie ja nur im Zusammenspiel mit LH- bzw. LTH-Aktivität ihre volle Wirkung auszuüben vermag.

Zusammenfassung

Nur eine bestimmte Kombination von HMG und HCG ist imstande, eine optimale Reaktion auf die Vergrößerung des Ratten-Ovars auszuüben. Dieses Mischungsverhältnis von HMG und HCG verschiebt sich nach Verlust der LH-Aktivität des HMG zu Lasten des HCG.

Literatur

Diczfalusy, E., E. Johannisson, K. G. Tillinger, and G. Bettendorf: Comparison of the clinical and steroid metabolic effect of human pituitary- and urinary gonadotrophins in amenorrhoeic women. Acta endocr. (Kbh.) Suppl. **90** (1964).

Diskussion

P. J. Czygan (Hamburg):

Mit welchen Präparationen von HCG wurden die Kaninchen immunisiert, war es eine sehr reine Ausgangssubstanz, die praktisch keine Anti-FSH-Wirkung auslöste? Aus Ihren Ergebnissen könnte man schließen, daß man durch die Reaktion von Anti-HCG mit dem LH-Anteil einer Substanz reines FSH gewinnen kann.

R. Buchholz:

Das HCG-Antiserum wurde uns freundlicherweise von N. V. Organon, Oss (Holland), zur Verfügung gestellt. Es ist dasselbe HCG-Antiserum, welches auch zur Herstellung des Pregnosticon-Tests verwandt wird. Das Antiserum wurde mit hochgereinigtem HCG gewonnen.

G. Bettendorf (Hamburg):

1. Ich halte es für verfrüht, bei den gonadotropen Substanzen von reinem FSH bzw. reinem LH zu sprechen. Da die Antikörper mit Substanzen hergestellt wurden, die ebenfalls nicht als rein zu bezeichnen sind, muß offen bleiben, ob die LH-Wirkung durch den Zusatz dieser Antikörper völlig blockiert wird.

2. Zur Frage der Prolactin- bzw. LTH-Wirkung möchte ich nur darauf hinweisen, daß im HHG praktisch kein Prolactin nachweisbar ist und trotzdem die Corpus-luteum-Funktion durch die Verabreichung des HHG aufrechterhalten werden kann. Dies trifft auch für hypophysektomierte Patientinnen zu, wo eine endogene Prolactinwirkung nicht vorhanden ist.

3. Beim Vergleich verschiedener HMG-Präparate mit der Wirkung des HHG konnten wir zeigen, daß die FSH-Aktivität der verschiedenen Präparate entscheidend ist. Auf Grund der sog. gonadotropen Gesamtaktivität, gemessen im Maus-Uterus-Test, können keine Voraussagen über die klinische Wirksamkeit gemacht werden. Die Verabreichung von etwa 400 HMG-Einheiten HHG über 10—14 Tage führt immer zu einer Ovulation. Im Vergleich hierzu ist die Verabreichung von 400 HMG-Einheiten eines HMG-Präparates für die gleiche Zeit völlig

wirkungslos gewesen. Wenn man die Dosis des HMG-Präparates jedoch so weit erhöht, daß die gleiche Menge an FSH-Aktivität verabreicht wird, so resultiert gleichfalls eine positive Wirkung auf die Ovarialfunktion.

R. Buchholz:

Wie im Prostata-Test bei hypophysektomierten infantilen Ratten festgestellt werden konnte, besitzt HMG + HCG-Antiserum bis zu einer Dosierung von 40 E HMG keinen Einfluß mehr auf diesen LH-spezifischen Test, während die Wirkung auf den FSH-spezifischen Augmentationstest erhalten bleibt.

R. v. Bersworth-Wallrabe (Berlin):

Haben Sie die FSH-Aktivität in intakten Ratten bei gleichzeitiger Applikation massiver Dosen HCG bzw. LH gemessen oder in hypophysektomierten Tieren, die keine LH-Komponente enthielten?

R. Buchholz:

Wir haben das HMG, welches mit HCG-Antiserum behandelt worden war, im Augmentationstest bei intakten Ratten gemessen. Dieser Test beruht auf dem Prinzip, daß hohe Dosen LH- bzw. LTH-Aktivität dem Versuchstier gleichzeitig mit dem zu testenden Präparat zugeführt werden. Man nimmt an, daß so die im zu testenden Präparat enthaltene LH-Aktivität keinen Einfluß mehr auf den Test ausübt.

Die Zusammenhänge der Nebennierenrinde- und Ovarialfunktion in bezug auf die Sterilität

Von

G. Silló-Seidl

Nach der Literatur sind die neutralen 17-Ketosteroid-Werte bei den virilen Frauen erhöht. Zander hatte nachgewiesen, daß 60% der virilen Frauen kinderlos sind. Ich möchte diese Frage von einer anderen Seite betrachten: Wieviele kinderlose Frauen weisen übermäßige Körperbehaarung auf? In diesem Zusammenhang läßt sich noch eine weitere Frage stellen:

Welche Zusammenhänge bestehen zwischen Nebennierenrinde- und der Ovarialfunktion?

Ich habe bei 100 Frauen, ohne Auswahl, die 17-Kerosteroide im Urin ermittelt. Bei diesen Frauen wurde besonders auf die Körperbehaarung geachtet, und ich teilte die Frauen in Normalbehaarte und Überbehaarte ein. Die familiäre Überbehaarung oder eine Überbehaarung durch öfteres Rasieren bestimmter Körperteile (bei Ballerinen, Schauspielerinnen und eitlen Frauen), wurde außer Acht gelassen. Die Ovarialfunktion wurde durch prämenstruelle Strichcurettage, vaginale Abstriche, funktionale Cervixdiagnostik und Messungen der Morgentemperaturen kontrolliert. Als ovarialgestörte Frauen galten diejenigen, die in erster Linie nicht ovulierten und solche, bei denen eine Endometriose vorhanden war. Nach dieser Einteilung wurde durchschnittlich folgende 17-Ketosteroid-Ausscheidung gefunden (Tab. 1).

Tabelle 1

	Ovarialfunktion			
	gestörte		normale	
	Ketosteroidwerte in mg/Tag		Ketosteroidwerte in mg/Tag	
Überbehaarte	19,52	17	18,37	24
Normalbehaarte	17,53	30	15,96	29
Durchschnittlich	18,03	47	17,05	53

Anhand dieser Tabelle ist tatsächlich nachzuweisen, daß überbehaarte Frauen höhere Ketosteroid-Werte ausscheiden als die Normalbehaarten. Gleichzeitig zeigten die Zahlen, daß die niedrigste Ausscheidung in der normalbehaarten, ovulierenden Gruppe war. Sogar normalbehaarte und ovarialgestörte Frauen zeigten erhöhte Ketosteroid-Werte. Auffallend hoch war die Zahl überbehaarter Patientinnen (41) bei den sterilen Frauen.

Wenn man innerhalb der oben beschriebenen Einteilung *die* Fälle, die Ketosteroide normal und *die*, welche erhöht ausscheiden, unterteilt (Tab. 2), läßt sich nur in der Gruppe ovarialgestörter Frauen zwischen Überbehaarten und Normalbehaarten innerhalb der normalausscheidenden Patienten ein beachtenswerter Unterschied nachweisen (13,47:10,99). Die Grenze der normalen und erhöhten Ausscheidung legte ich nach TAYLOR (15 mg/Tag Wert) fest.

Tabelle 2

	Ovarialfunktion						
	gestörte				normale		
	Ketosteroidwerte in mg/Tag				Ketosteroidwerte in mg/Tag		
	normal		erhöhte		normal		erhöhte	
Überbehaarte . .	13,47	6	21,91	11	11,80	8	21,65	16
Normalbehaarte. .	10,99	12	21,89	18	11,87	16	21,00	13

Bemerkenswert an dieser Tabelle ist die hohe Zahl der Patienten (29:29) die erhöhten Ketosteroid-Wert aufweisen.

Wenn man diese Tabellen betrachtet und die Ausscheidungen mit der jeweiligen Vorgeschichte und der später gestellten Diagnose vergleicht, dann sollte man der Ermittlung der 17-Ketosteroid-Ausscheidung in der Sterilitätsdiagnostik einen größeren Platz einräumen, als es bisher der Fall war. In Verbindung mit der aufgezählten Untersuchungs-Systematik, läßt der Nachweis der neutralen 17-Ketosteroid-Ausscheidung die Zusammenhänge der Nebennieren-Funktion und Ovarial-Funktion für praktische Zwecke sich in groben Zügen folgendermaßen zusammenstellen.

Falls Anovulation mit Überbehaarung und hohem Ketosteroid-Wert vergesellschaftet ist, wird die Sterilitätsursache in einer Nebennierenrindenfunktionsstörung gesucht und die Behandlung mit Glucocorticoide vorgeschlagen. War Überbehaarung bei normaler Androgen-Ausscheidung vorhanden, ist die Ursache der Ovarial-Störung in den Eierstöcken selbst zu suchen. Meist weisen diese Fälle ein Arrhenoblastom auf. Sowohl bei diesen Fällen, wie auch bei Fällen von Stein-Leventhal-Syndrom, ist eine chirurgische Therapie angezeigt. Die normalen oder erniedrigten Ketosteroid-Werte mit normaler Körperbehaarung und Ovarialstörung, deuten auf eine zentrale Genese hin; für eine Diagnose sind weitere Untersuchungen notwendig. Von 13 Frauen mit normaler Ovarialfunktion und normaler Körperbehaarung, aber erhöhter Ketosteroidausscheidung konnte ich bei 5, von 16 mit normaler Ketosteroidausscheidung nur bei einer keine Ursache für die Sterilität feststellen.

Hier muß eine Frage gestellt und offen gelassen werden:

Inwieweit spielt die Nebennierenrindenhyperfunktion trotz nachgewiesener Ovulation für die Sterilität eine Rolle?

Meine nächste Aufgabe wird die Nachprüfung dieser Frage sein. Zusammenfassend kann man sagen, daß die Ermittlung der 17-Ketosteroidwerte beim Suchen nach der Ursache der Kinderlosigkeit eine nicht zu unterschätzende Hilfe bietet.

Aus diesem Grund erscheint die Bestimmung als Routineuntersuchung empfehlenswert.

Aus dem Hauptlaboratorium der Schering AG., Berlin

Untersuchungen zur Ovulationshemmung an der Ratte mit Kombinationen von Steroiden verschiedener Stoffklassen

Von

F. Neumann und A. Domenico

Mit 1 Abbildung

Es ist seit langem bekannt, daß es mit verschiedenen Steroidhormonen sowohl bei der Frau als auch beim Versuchstier gelingt, die Ovulation zu unterdrücken. Da jedoch Androgene und Oestrogene (bei alleiniger Gabe) wegen ihrer starken Nebenwirkungen bzw. schlechten Verträglichkeit in der Klinik für diese Indikation ausschieden und Progesteron nur bei Gabe extrem hoher Dosen wirksam war (Pincus, 1956 u. a.), fand diese Beobachtung lange Zeit keine Beachtung. Erst 1954 fand man unter Derivaten des 19-Nor-testosterons (Zaffaroni 1954, Sandoval et al. 1955) oral stark wirksame Gestagene, die sich außer durch eine relativ geringe androgene Wirkung vor allem durch eine starke zentrale Hemmwirkung auszeichneten (Herz et al. 1954, Pincus et al. 1956 u. a.). In der Klinik haben sich Substanzen dieser Stoffklasse in Verbindung mit einem Oestrogen zur hormonalen Konzeptionsverhütung bewährt (Rock et al. 1956, Garcia et al. 1958, Pincus et al. 1959, Peeters et al. 1960 u. a.). Zur tierexperimentellen Testung der antiovulatorischen Aktivität von Steroidhormonen werden meistens Kaninchen, seltener Ratten oder andere Tiere benutzt.

Wir arbeiten mit geschlechtsreifen weiblichen Ratten, die regelmäßige 4- bzw. 5 tägige Cyclen aufweisen müssen, um in den Versuch zu gelangen. (Der Cyclus wird vor Beginn des Versuchs kontrolliert und Tiere mit unregelmäßigen Cyclen scheiden aus.) Am Ende eines Cyclus, im sog. Metöstrus, wird mit der Verabfolgung der Substanzen begonnen, insgesamt wird entsprechend der Cyclusdauer über 4 Tage behandelt. Es ist entscheidend, daß die Verabfolgung der Testsubstanzen am Ende eines Cyclus beginnt. Wird den Tieren die Testsubstanz erst ein oder zwei Tage vor der zu erwartenden Spontanovulation verabfolgt (die Ratte ovuliert gegen Morgen in der Nacht nach dem Prooestrus), so sind zur Verhinderung der Ovulation 100 bis 1000 mal höhere Dosen notwendig. Nach 4 tägiger Versuchsdauer werden die Tiere einseitig kastriert, die Tuben abpräpariert und aus ihnen Quetschpräparate angefertigt (Eindeckungsmedium ist physiologische Kochsalzlösung). Unter dem Mikroskop kann dann leicht festgestellt werden, ob eine Ovulation stattgefunden hat oder unterblieben ist. Am 5. Tage werden die Tiere getötet und die zweite Tube in gleicher Weise aufgearbeitet.

Die Abb. 1 zeigt rechts eine Rattentube mit Eizellen in der stark erweiterten Ampulle. Die Eizellen werden durch Granulosa-Zellen, die bei der Ovulation mit ausgestoßen werden, wie von einem Hof umgeben. Links im Bild sehen Sie eine Rattentube bei gehemmter Ovulation. Die Auswertung geschieht nach dem Alles-oder-Nichts-Prinzip, d. h. nur dann gilt die Ovulation als gehemmt, wenn keine Eizelle in der Tube festgestellt wird. Zur Genauigkeit des Testes ist zu sagen, daß unbehandelte Kontrolltiere in 10% aller Fälle versagen, d. h. nicht am Morgen nach dem Prooestrus ovulieren. Die Annahme, daß jeder zweite Cyclus bei der Ratte anovulatorisch verläuft, ist somit sicher falsch.

Wir prüften unter der Behandlung mit Steroidhormonen verschiedener Stoffklassen fortlaufend das Verhalten des Cyclus und konnten folgende Zusammenhänge zwischen Cyclus und gehemmter Ovulation feststellen.

Bei Dosen von gestagenen, androgenen und auch anabolen Steroiden, die etwa das 3fache der 50%igen Ovulationshemmungsdosis überschreiten, ist der Cyclus vollständig unterdrückt, die Tiere bleiben im Dioestrus. Werden jedoch Dosen verabfolgt, die bei etwa 50% der Versuchstiere die Ovulation unterdrücken, so ist mitunter auch bei den Tieren der Cyclus normal, die nicht ovulieren.

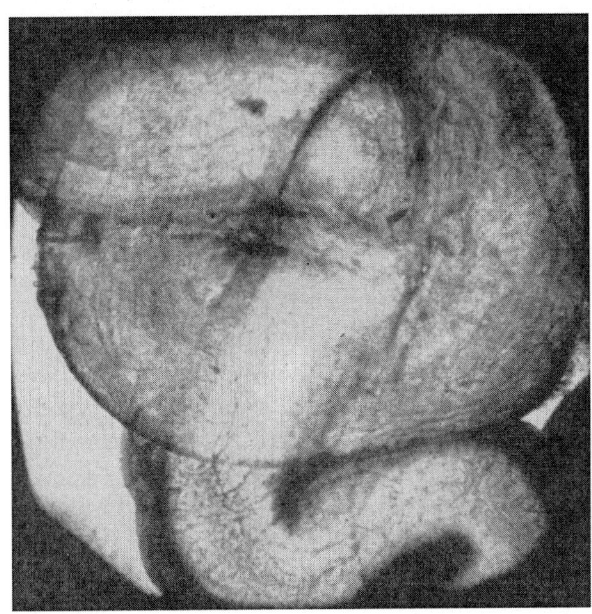

a

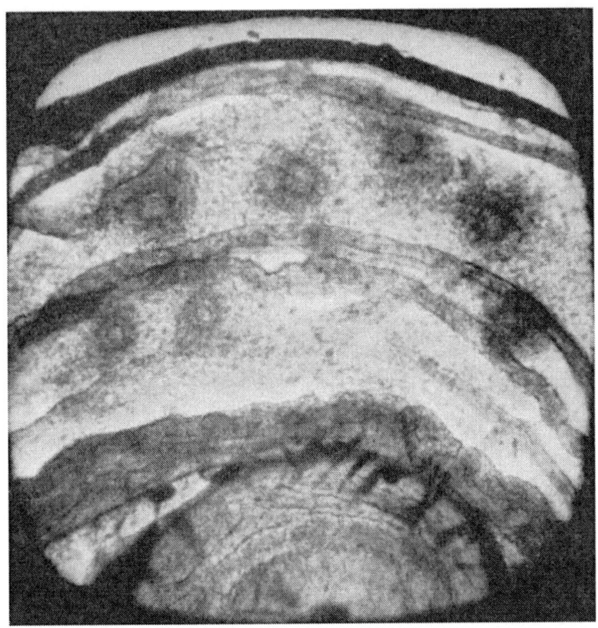

b

Abb. 1 a u. b. Quetschpräparate von Rattentuben (nativ) a gehemmte Ovulation; b stattgefundene Ovulation, von Granulosazellen umgebene Eizellen in der erweiterten Ampulle

18*

Höhere Oestrogendosen führen zu einem Daueroestrus bei den behandelten Tieren. Bei etwa 50%igen Wirkungsdosen verhalten sich Oestrogene ähnlich wie Verbindungen der vorher genannten Stoffklassen.

Wir haben nun versucht, folgende Fragen zu klären:

1. Welche der beiden Komponenten eines Oestrogen-Gestagen-Gemisches ist bei wechselnden Dosenverhältnissen wirksam?

2. In welcher Weise beeinflussen sich die geprüften Verbindungen bei simultaner Gabe?

3. Inwieweit kann die oestrogene oder gestagene Komponente durch ein anabol wirksames Steroid ersetzt werden?

Für diese Untersuchungen haben wir folgende Verbindungen benutzt:

17 α-Äthinyl-19-nor-testosteronacetat (Gestagen)

17 α-Äthinyl-oestradiol (Oestrogen)

1-Methyl-Δ^1-androsten-17 β-ol-3-on-17-acetat $\Big\}$ Anabolika
1 β-17 α-Dimethyl-Δ^2-androsten-17 β-ol[1]

Jede dieser vier aufgeführten Verbindungen ist bei alleiniger Gabe in der Lage, die Ovulation zu verhindern.

Aus Tab. 1 wird ersichtlich, daß z. B. 0,03 mg Äthinyloestradiol etwa genauso stark wirksam sind wie 3 mg Äthinyl-19-nor-testosteronacetat. Beide Steroide unterscheiden sich in ihrer Wirkungsstärke also wie etwa 100:1. (Dabei verlaufen

Tabelle 1. *Ovulationshemmung (geprüft an weiblichen Ratten bei oraler Gabe über 4 Tage)*

| Dosis (mg/Tier/Tag); pro Dosis 10 bis 30 Tiere | | | | % der Tiere mit gehemmter Ovulation |
17 α-Äthinyl-19-nor-testosteronacetat	17 α-Äthinyl-östradiol	1-Methyl-Δ^1-androsten-17 β-ol-3-on-17 β-acetat	1 β,17 α-Dimethyl-Δ^2-androsten-17 β-ol	
10,0	—	—	—	85
3,0	—	—	—	60
1,0	—	—	—	8
—	1,0	—	—	100
—	0,3	—	—	95
—	0,1	—	—	65
—	0,03	—	—	55
—	0,01	—	—	28
—	—	10,0	—	80
—	—	3,0	—	25
—	—	—	1,0	100
—	—	—	0,3	50
—	—	—	0,1	20
3,0	0,01	—	—	47 (?)
—	0,01	10,0	—	45 (?)
5,0	—	5,0	—	80
1,5	—	1,5	—	35
5,0	—	—	0,5	100
1,7	—	—	0,17	40
0,5	—	—	0,05	20
5,0	0,01	5,0	—	80
1,7	0,03	—	0,17	77

[1] Die Substanz 1 β-17 α-Dimethyl-Δ^2-androsten-17 β-ol wurde von Herrn Dr. Müller, Schering AG. synthetisiert.

die Dosiswirkungskurven nicht ganz parallel.) Im entsprechenden Handelspräparat, das diese beiden Wirkungskomponenten enthält, liegen bei einem Dosenverhältnis von 80:1 (80 Teile des Gestagens, 1 Teil des Oestrogens) ganz ähnliche Verhältnisse vor, d. h. bei diesem Dosenverhältnis sind beide Steroide an der Ovulationshemmung beteiligt — wenn unsere Rattenversuche auf den Menschen übertragbar sind. Bei einem Dosenverhältnis von 1:3 ist ausschließlich die oestrogene und bei einem Dosenverhältnis von 1000:1 ausschließlich die gestagene Komponente wirksam. (Voraussetzung dafür sind natürlich submaximale Wirkungsdosen.) Zwischen diesen beiden Bereichen 1000:1—3:1 sind beide Komponenten an der Ovulationshemmung beteiligt.

Nun zur Frage, inwieweit sich ein Oestrogen und Gestagen bei simultaner Gabe in ihrer Wirkung beeinflussen. Es liegt ganz sicher keine Potenzierung der Wirkung vor, denn bei gleichzeitiger Gabe einer allein etwa 50% wirksamen Gestagendosis und einer oestrogenen Schwellendosis ist der Prozentsatz an Tieren mit gehemmter Ovulation nicht höher als nach der Applikation des Gestagens allein.

Es sei darauf hingewiesen, daß diese Befunde im Widerspruch zu Untersuchungen an ovariektomierten Affen stehen. Mit hohen Dosen von Progesteron konnte dabei die Gonadotropinausscheidung (Maß der Hypophysenbremsung) im Harn nicht herabgesetzt werden. Erst die Zugabe einer allein unwirksamen Oestradioldosis bewirkte dann eine Herabsetzung der Gonadotropinausscheidung (SALHANIK et al. 1952).

Auch anabol-wirksame Steroide können die Ovulation verhindern, und zwar in Dosen, die (an der Ratte geprüft) nicht oder nur wenig androgen wirksam sind. 1-Methyl-Δ^1-androsten-17β-ol-3-on-17-acetat ist nur wenig schwächer wirksam als Äthinyl-19-nor-testosteronacetat. Die zweite von uns geprüfte anabol wirksame Substanz 1β-17α-Dimethyl-Δ^2-androsten-17β-ol ist sogar etwa 10mal stärker ovulationshemmend als Äthinyl-19-nor-testosteronacetat (s. Tab. 1).

Bei gleichzeitiger Gabe eines Gestagens und Anabolikums zeigt sich in jedem Falle eine Addition der Wirkungen.

Bei gleichzeitiger Gabe einer mittelgradig bis submaximal wirksamen Dosis des Gestagens oder Anabolikums und unterschwelligen Oestrogendosis ist letztere ohne Einfluß auf die Wirkung der zwei übrigen Komponenten.

Schließlich konnten wir auch eine Addition der Wirkung bei gleichzeitiger Gabe einzeln nur schwach wirksamer Dosen eines Gestagens, Oestrogens und Anabolikums feststellen.

Zusammenfassung

Es gelingt bei der Ratte durch Oestrogene, Gestagene und auch anabol wirksame Steroide die Ovulation zu unterdrücken. 17α-Äthinyl-oestradiol und 17α-Äthinyl-19-nor-testosteronacetat unterscheiden sich dabei in ihrer Wirkungsstärke etwa wie 100:1. Im entsprechenden Handelspräparat beträgt das Gestagen-Oestrogen-Verhältnis 80:1, d. h. bei diesem Dosenverhältnis dürften beide Steroide an der Ovulationshemmung beteiligt sein. Bei simultaner Verabfolgung beider Substanzen kommt es vermutlich zu einer Addition der Wirkung. Das gleiche gilt für die simultane Gabe eines Anabolikums und Gestagens und schließlich auch für die gleichzeitige Gabe von Substanzen aller drei geprüften Stoffklassen.

Fräulein G. BÖHNISCH möchten wir für die Ausführung der Versuche danken.

Literatur

Garcia, C. R., G. Pincus, and J. Rock: Effects of Three 19-Nor-Steroids on Human Ovulation and Menstruation. Amer. J. Obstet. Gynec. **75**, 82 (1958).

Herz, R., W. Tullner, and E. Raffelt: Progestional Activity of Orally Administered 17 α-Ethinyl-19-Nortestosterone. Endocrinology **54**, 228 (1954).

Peeters, F., M. van Roy, and H. Oeyen: Ovulationsunterdrückung durch Progestagene. Geburtsh. u. Frauenheilk. **20**, 1306 (1960).

Pincus, G., M. C. Chang, E. S. E. Hafez, M. X. Zarrow, and A. Merrill: Effects of Certain 19-nor-Steroids on Reproductive Processes in Animals. Science **124**, 890 (1956).

— C. R. Garcia, J. Hock, M. Paniagua, A. Pendleton, F. Laraque, R. Nicolas, R. Borno, and V. Pean: Effectiveness of an Oral Contraceptive. Science **130**, 81 (1959).

Rock, J., G. Pincus, and C. R. Garcia: Effects of Certain 19-Nor-Steroids on the Normal Human Menstrual Cycle. Science **124**, 891 (1956).

Salhanik, H. A., F. L. Hisaw, and M. X. Zarrow: The action of Estrogene and Progesterone on the Gonadotropine Content of the Petuitary of the Monkey. J. clin. Endocr. **12**, 310 (1952).

Sandoval, A., G. H. Thoma, C. Djerassi, G. Rosenkranz, and F. Sondheimer: LXI. Synthesis of 19-Nor desoxycorticosterone, a Potent Mineralocorticoid Hormone. J.Amer. chem. Soc. **77**, 148 (1955).

Zaffaroni, A., H. J. Ringold, G. Rosenkranz, F. Sondheimer, G. H. Thomas, and C. Djerassi: Steroids LXIII Synthesis of Δ^4-19-Nor-Pregnene-11 β,17 α,21-Triol-3,20 Dione (19-Norhydrocortisone) and Related 19-Nor-Adrenal-Hormones. Amer. chem. Soc. **76**, 6210 (1954).

N. V. Organon, Oss (Holland)

Klinische und endokrinologische Beobachtungen mit einer neuen Kombination von Lynestrenol und Mestranol

Von

G. L. IJZERMAN und H. KOPERA

Die Unterlagen für diesen Bericht verdanken wir einer Anzahl Untersucher in der Klinik, die so freunlich waren, uns ihre Ergebnisse zur weiteren Verarbeitung und zur Kompilation bekannt zu machen und freizugeben. Die Daten beziehen sich ausschließlich auf die cyclische Verwendung einer Kombination von 2,5 mg Lynestrenol und 75 γ Mestranol in einer Tablette. Es wurde immer eine Tablette pro die (meistens am Abend) verabreicht, und zwar vom 5. bis zum 25. Tage des Cyclus.

Die Abb. 1 zeigt eine Tabelle, der zunächst Umfang und Zusammensetzung des Beobachtungsgutes zu entnehmen sind. Insgesamt sind 854 Frauen während 4159 Cyclen behandelt worden, unter ihnen 330 Frauen (38%) während mehr als 6 Cyclen, das sind 2488 Cyclen oder 60% der gesamten Zahl. Aus dieser Tabelle kann man weiter ersehen, daß auch der Prozentsatz der Frauen mit erwiesener hoher Fertilität in der Serie verhältnismäßig groß war.

In derselben Abbildung werden die Daten aufgezeigt, aus denen man schließen darf, daß auch mit dieser niedrigen Dosierung von Lynestrenol und Mestranol eine zuverlässige Unterdrückung der Fertilität mittels Hemmung der Ovulation gewährleistet ist. Die Zuverlässigkeit geht aus der Tatsache hervor, daß bei den 854 Frauen während 4159 Cyclen keine einzige Schwangerschaft auftrat. Bei 60 Frauen wurden Pregnandiolausscheidungsbestimmungen vorgenommen. Daß in keinem Fall ein cyclusgerechter Anstieg der Ausscheidung festgestellt werden konnte, ist ein starker Hinweis dafür, daß keine Ovulation stattfand. Die Unterdrückung der Ovulation wurde bei 16 Frauen während einer Laparotomie einwandfrei nachgewiesen.

Tabelle 1

	Frauen		Cyclen	
	Anzahl	%	Anzahl	%
Gesamtes Beobachtungsgut.	854		4159	
Während 6 oder mehr Cyclen behandelt . . .	330	38	2488	60
Mit bekannter hoher Fertilität	207	24	961	23
Ovulationshemmung wahrscheinlich (Pregnandiolausscheidung nicht cyclusgerecht erhöht)	> 60			
Ovulationshemmung erwiesen (Laparotomie) .	16			
Schwangerschaften	0	0		

Tabelle 2

	Frauen		Cyclen	
	Anzahl	%	Anzahl	%
Zwischenblutungen (Breakthrough Bleeding and Spotting)	71 von 630	11,2	119 von 3228	3,7
Ausbleiben der Abbruchblutung (Pseudo-Amenorrhoe)			68 von 2844	2,4
Nausea (einschl. leichter Übelkeit)	41 von 831	5		

Der Eintritt der Abbruchblutungen erfolgte mit großer Regelmäßigkeit am 3. oder 4. Tage nach Einnahme der letzten Tablette, also am 27. oder 28. Tage des artifiziellen Cyclus. Nur in seltenen Fällen ist die Latenszeit vor dem Eintreten der menstruationsähnlichen Blutung kürzer oder länger. Die Tatsache, daß die Abbruchblutungen auch bei dieser niedrigen Dosierung nicht eher erfolgten als mit den gleichen Wirkstoffen in höherer Dosierung, worüber wir früher berichtet haben, war eigentlich einigermaßen unerwartet. Die Menge des Blutverlustes ist jedoch, wie erwartet, im Durchschnitt etwas größer als bei höheren Einzeldosen, obwohl immer noch geringer als bei einer normalen Menstruation. Die in der 2. Tabelle aufgeführten Nebenwirkungen zeigen gleichfalls alle die zu erwartenden Verschiebungen gegenüber der Medikation höherer Dosen und auch Zwischenblutungen jeglicher Stärke wurden hier etwas häufiger gesehen, nämlich bei 11,2% der Frauen oder in 3,7% der Cyclen, was für diese Behandlungsart immer noch einen sehr niedrigen Prozentsatz darstellt. Die in der angelsächsischen Literatur oft als "silent menstruation" gedeutete Verschiebung der Abbruchblutung über den Zeitpunkt hinaus, der aus Sicherheitsgründen als Anfang für einen neuen Tablettencyclus empfohlen wird, möchten wir lieber als Pseudoamenorrhoe bezeichnen. Diese kommt bei der niedrigen Dosierung der untersuchten Kombination außerordentlich selten vor, und zwar in nur 2,4% der kontrollierten Cyclen. Auch die Häufigkeit der bei manchen derartigen Präparaten sehr störenden Nausea, wurde mit der neuen Kombination infolge der niedrigen Konzentration anscheinend weitgehend zurückgedrängt und bei unserem Beobachtungsgut in nicht mehr als 5% der Fälle beobachtet. Hierbei ist noch zu bemerken, daß auch die geringste Übelkeit mitgezählt wurde. Alle andere von Ovulationshemmern bekannte Nebenwirkungen traten, wenn überhaupt, nur vereinzelt in Erscheinung.

Abgesehen von der für die Anwendung derartiger Präparate selbstverständlich wichtigen Wirtschaftlichkeit, bietet also die niedrige Konzentration der neuen Kombination auch wesentliche medizinische Vorteile.

Literatur

Anberg, A.: Persönliche Mitteilung, wird demnächst ausführlich veröffentlicht.
Borglin, N. E.: Vorgetragen auf dem 2nd European Meeting of the International Fertility Association, Brüssel, 2.—3. März 1963. (Im Druck: Int. J. Fertil.).
Bouillaud, E. L. P.: Thèse de Bordeaux, 1963.
Brehm, H.: Persönliche Mitteilung.

FERIN, J.: Persönliche Mitteilung.

—, and R. DEMOL: Vorgetragen auf dem 2nd European Meeting of the International Fertility Association, Brüssel, 2.—3. März 1963. (Im Druck: Int. J. Fertil.).

FORTUIN-BLITZ, A. F.: Persönliche Mitteilung.

GERBER, E.: Vorgetragen auf der 3. Bayerisch/Österreichisch/Schweizerischen Gynäkologen-Tagung, Luzern, 2.—4. Oktober 1963.

GRANAAT, D.: Persönliche Mitteilung.

HAUSER, G. A.: Vorgetragen auf der 3. Bayerisch/Österreichisch/Schweizerischen Gynäkologen-Tagung, Luzern, 2.—4. Oktober 1963.

KOPERA, H., M. N. G. DUKES, and G. L. IJZERMAN: Vorgetragen auf dem 2nd European Meeting of the International Fertility Association, Brüssel, 2.—3. März 1963. (Im Druck: Int. J. Fertil.).

LINTHORST, G.: Persönliche Mitteilung, wird demnächst ausführlich veröffentlicht.

OVERBEEK, G. A., Z. MADJEREK, and J. DE VISSER: Acta endocr. (Kbh.) 41, 351—370 (1962).

SZONTÁGH, F. E.: Im Druck im Zbl. Gynäk.

TAUSK, M.: Ned. T. Geneesk. 107, 1976—1980 (1963).

TURPEINEN, K.: Persönliche Mitteilung.

WIBAUT, F. P.: Persönliche Mitteilung.

WINTER, M. S. DE, C. M. SIEGMANN, and S. A. SZPILFOGEL: Chem. and Ind. 1960, 905.

Der Einfluß der Hypervitaminose A auf die Fertilität des Rattenbockes

Von

U. Laschet (Landeck/Pfalz), W. Hohlweg und W. Scheithauer

Mit 1 Abbildung

Die Verabfolgung von hohen Dosen Vitamin A an trächtige Rattenweibchen führt zu Mißbildungen der Feten, wie die Forschergruppe von Giroud und Martinet in zahlreichen Arbeiten nachweisen konnte. Daneben beschrieb Cohlan auch eine Verminderung der Fertilität der Rattenweibchen, wenn Vitamin A vom 2.–16. Tag post coitum täglich in Dosen von 35 000 iE zugeführt wurde. Gebauer hingegen konnte diese Ergebnisse von Cohlan nicht bestätigen. Daß es nach Verabfolgung hoher Vitamin A-Dosen bei Rattenweibchen tatsächlich zu einer Fertilitätsverminderung kommt, konnte ich in Zusammenarbeit mit Hohlweg und Weise feststellen.

Uns beschäftigte nunmehr die Frage, ob die Hypervitaminose A auch zu einer Fertilitätsstörung beim Rattenbock führt. v. Drigalski und Laubmann untersuchten im Jahre 1933 erstmals die Wirkung hoher Vitamin A-Gaben auf die Hodenstruktur. Sie verabfolgten einen hochgereinigten Vitamin A-haltigen Leberextrakt längstens über 14 Tage per Schlundsonde und fanden bei den Böcken nach dieser Behandlungszeit deutliche degenerative Veränderungen der Hoden. Die Struktur der Hoden wird von ihnen als unregelmäßig beschrieben; teilweise seien die Hodenkanälchen völlig nekrotisch und allgemein atrophisch gewesen. Sie fanden fast keine ausgereiften Spermien sondern nur die Vorstufen, darunter auch Riesenzellen. Weslaw, Wroński, Wróblewski und Wróblewski fanden bei ähnlichen Untersuchungen 1938 keine Schädigung der Spermiogenese. Sie verabfolgten täglich etwa 20 000 iE Vitamin A per Schlundsonde über einen Gesamtzeitraum von 52–55 Tagen. Da es sich bei den von v. Drigalski und auch von Weslaw u. Mitarb. angewandten Präparate um Leberextrakte handelte, die neben Vitamin A auch größere Mengen Vitamin D und Ballaststoffe enthielten, überprüften wir diese Ergebnisse mit synthetischem Vitamin A-Palmitat. Unsere Voruntersucher hatten über die Fertilität der Böcke keine Auskunft gegeben. Wir bezogen Fertilitätsuntersuchungen in unsere Anordnung mit ein.

20 erwachsene Rattenböcke des Rehbrücker Inzuchtstammes erhielten täglich 60 000 iE Vitamin A-Palmitat in öliger Lösung per Schlundsonde verabreicht. Die Versuchsdauer betrug wie bei v. Drigalski 14 Tage. In dreitägigen Abständen wurden je 3 Tiere getötet und zunächst die Hoden histologisch untersucht. Unsere erhobenen Befunde bestätigen die Ergebnisse von v. Drigalski (vgl. Abbildung 1). Erst nach etwa 840 000 iE beginnt das Hodengefüge sich zu verändern. Während die meisten Tubuli noch funktionstüchtig sind, liegen in verschiedenen

Lumina Spermiocyten, die sich sonst am Rand der Hodenkanälchen befinden. Nach insgesamt 1 000 000 iE Vitamin A arten die Hodenveränderungen in eine

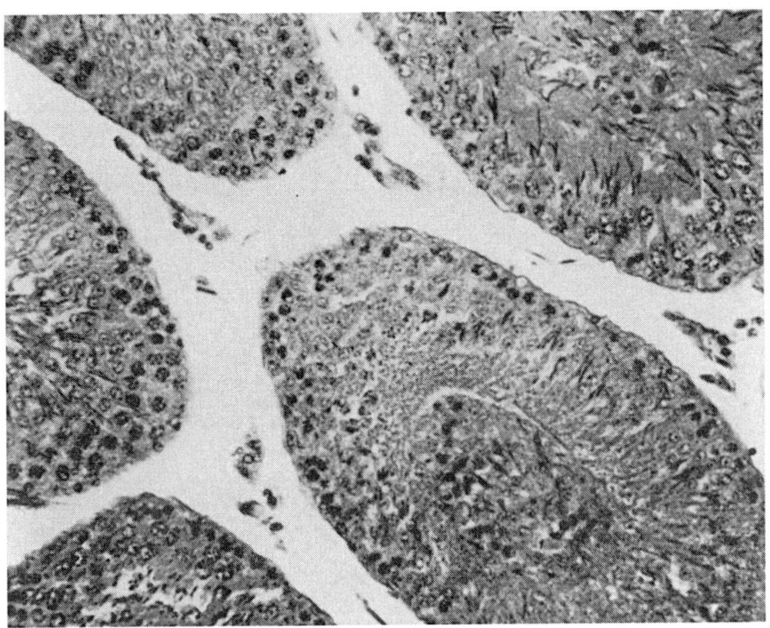

Abb. 1a. Hoden nach 840 000 iE Vitamin A-Palmitat oral

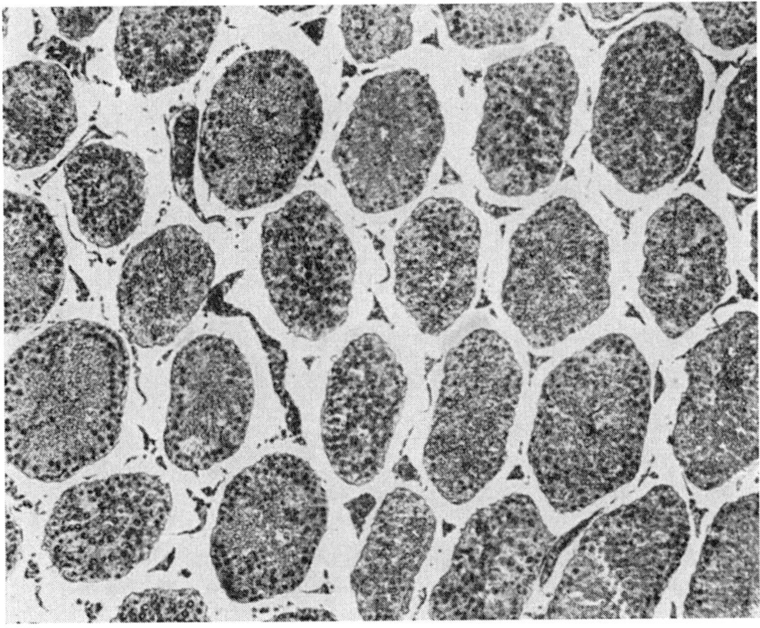

Abb. 1b. Hoden nach 1 020 000 iE Vitamin A-Palmitat oral

regelrechte Hodendestruktion aus. Die Samenkanälchen besitzen keine Lumina mehr. Spermien, Spermatiden und Präspermatiden sind nicht mehr zu finden. Das

Chromatin der Kerne verdichtet sich am Rande; es entstehen Siegelringzellen ähnlich wie nach hohen Oestrogendosen. In einigen dieser Riesenzellen sind die

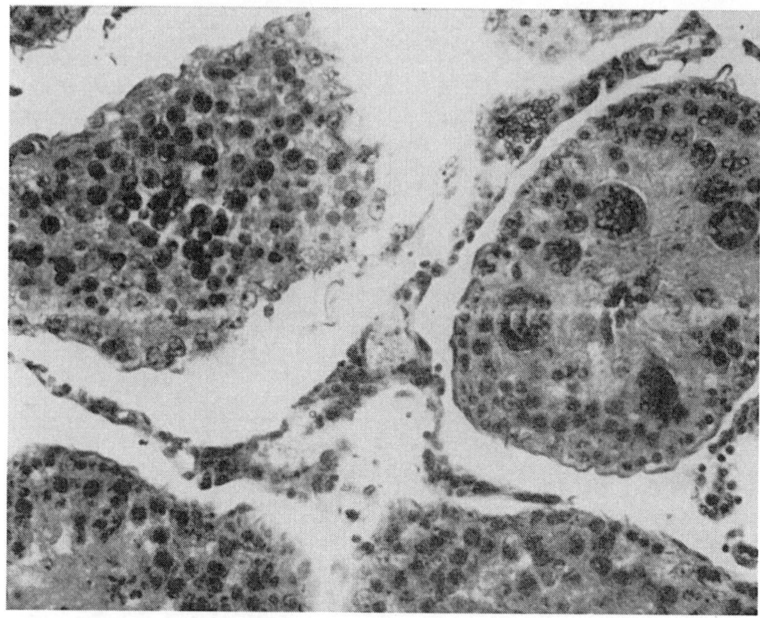

Abb. 1c. Mehrkernige Riesenzellen im Hoden nach 1 020 000 iE Vitamin A-Palmitat oral

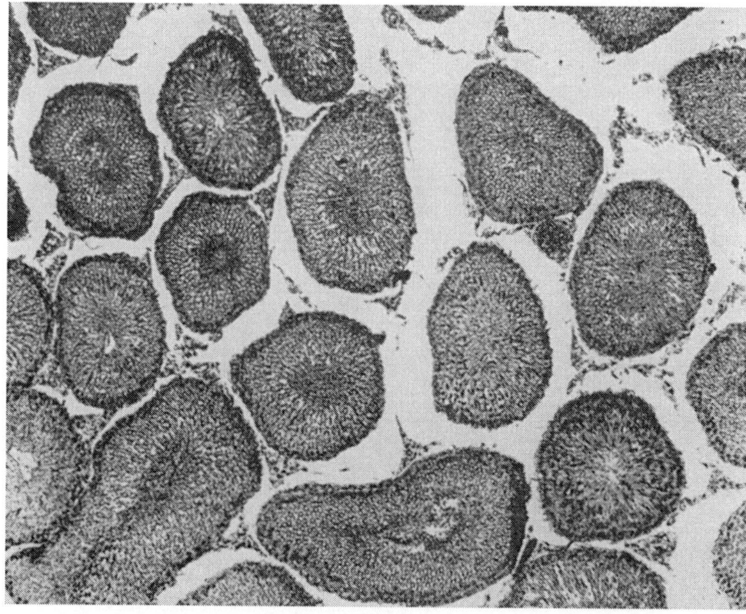

Abb. 1d. Hoden 30 Tage nach der letzten Vitamin A-Gabe von insgesamt 1 020 000 iE

Kerne an der Peripherie perlschnurartig aufgereiht. Zum Teil sind die Zellen in die Spermiogenesereihe nicht mehr einzuordnen. 30 Tage nach Beendigung der Vitamin A-Zufuhr hat sich das histologische Hodenbild wieder normalisiert.

Zur Untersuchung der Fertilität wurden die Tiere einer zweiten Versuchsgruppe nach Verabfolgung von insgesamt 1020000 iE Vitamin A in insgesamt 5 sich jeweils über 6 Tage erstreckenden Fertilitätsprüfungen untersucht. Jeder Rattenbock wurde am Tag nach der letzten Vitamin A-Verabfolgung mit jeweils 5 fertilen Rattenweibchen 6 Tage lang in einem gemeinsamen Käfig gehalten. Dieser ersten Testserie folgten 4 weitere, so daß die Fertilität der Böcke über einen Zeitraum von insgesamt 30 Tage hin untersucht wurden. Als Kontrollen dienten 15 Böcke, die 0,2 ml Rinderklauenöl täglich oral erhielten. Bei den zur Prüfung verwendeten weiblichen Tieren wurden am 14. Tag nach der Entfernung des Bockes die Zahl der Fruchtanlagen ermittelt und die Feten auf externe Mißbildungen untersucht.

Die ausgeprägten Hodendestruktionen führten bei den Böcken zum Erlöschen der Fertilität. Die ersten beiden Paarungsversuche verliefen völlig negativ, d. h. die Rattenweibchen der Serien 1 und 2, die bis zum 12. Tag nach der letzten Vitamin A-Gabe mit den Böcken zusammen waren, wurden nicht trächtig. Erst zwischen dem 13. und 18. Tag nach Absetzen der Vitamin A-Zufuhr wurden von insgesamt 65 untersuchten Weibchen 9 trächtig. Diese wiesen eine normale Anzahl äußerlich normal gestalteter Feten auf. Nach dem 18. Tag nach Beendigung der Vitamin A-Zufuhr kommt es mit fortschreitender Regeneration der Hodenstruktur auch zur Regeneration der Fertilität. Von 65 weiblichen Tieren der 4. und 5. Serie wurde je 16 Tiere trächtig. Von den gleichzeitig untersuchten Öl-behandelten Kontrollen wurden von 75 Weibchen 26 trächtig. Es liegt die Vermutung nahe, daß eine Subfertilität bei einem Teil der Böcke erhalten bleibt.

Fassen wir die Befunde kurz zusammen, so können wir feststellen, daß eine Überdosis von Vitamin A beim Rattenbock zu schweren Schäden der Spermiogenese führt, die eine völlige Hemmung der Fertilität nach sich ziehen. Die Schädigung der Spermiogenese ist völlig, die Hemmung der Fertilität weitgehend reversibel.

Literatur

COHLAN, S. Q.: Amer. J. Dis. Child. **86**, 348 (1953).
DRIGALSKI, W. v., u. W. LAUBMANN: Klin. Wschr. **12**, 1171 (1933).
GEBAUER, H.: Pharmazie **9**, 684 (1954).
GIROUD, A., et M. MARTINET: Arch. franc. Pédiat. **12**, 1936 (1955).
— — C. R. Soc. Biol. (Paris) **149**, 1088 (1955).
— — Arch. Anat. micr. Morph. exp. **45**, 77 (1956).
— — C. R. Soc. Biol. (Paris) **153**, 201 (1959).
— — Arch. franc. Pédiat. **16**, 1 (1959).
— — C. R. Soc. Biol. (Paris) **154**, 1353 (1960).
LASCHET, U., W. HOHLWEG u. W. WEISE: Int. Z. Vitamin-Forsch. **30**, 77 (1959).
— — — Int. J. Fertil. **5**, 317 (1960).
WESLAW, W., B. WROŃSKI, A. WRÓBLEWSKI u. B. WRÓBLEWSKI: Klin. Wschr. **17**, 777, 879 (1938).

Aus dem Zoophysiologischen Institut der Universität Uppsala (Direktor: Prof. Dr. P. E
Lindahl) und aus der Hautklinik der Medizinischen Akademie Düsseldorf
(Direktor: Prof. Dr. Dr. A. Greither)

Tierexperimentelle Beobachtungen über einen hormonal gesteuerten männlichen Sexualcyclus

Von

O. Hornstein

Das Problem, über das hier kurz berichtet werden soll, läßt sich in die Frage
zusammenfassen: Gibt es auch bei männlichen Säugetieren einen Cyclus von
Sexualfunktionen, der zum Cyclus des weiblichen Oestrus in biologische Parallele
gesetzt werden kann? Die Frage erscheint schon deshalb berechtigt, weil im
Hypophysenvorderlappen *bei beiden Geschlechtern* identische, d. h. geschlechts-
unspezifische gonadotrope Hormone gebildet werden.

Wenn wir von der allgemein geläufigen, an bestimmte Jahreszeiten gebundenen
Brunstperiode der meisten wildlebenden Säugetiere absehen und unser Augenmerk
nur auf domestizierte Arten richten, so war bis vor wenigen Jahren über ein
männliches Analogon zum cyclischen Oestrus weiblicher Säuger nichts Sicheres
bekannt. Erst in den letzten Jahren sind bei männlichen Kaninchen sowie bei
Zuchtbullen periodische, statistisch gesicherte Rhythmen verschiedener Sexualfunktionen nachgewiesen worden (Kihlström). So bestehen beim Kaninchen
cyclische Schwankungen des Ejaculat-Volumens, der Zahl und Konzentration der
Spermien, des Vorkommens einer gelatinösen Substanz im Ejaculat, ferner cyclische Variationen der sexuellen Erregbarkeit und der morgendlichen Körpertemperatur. Die Ursache dieser cyclischen Vorgänge vermutete man in einer
hormonalen Regulation, konnte sie bislang aber nicht beweisen.

1963 hatten wir Gelegenheit, am Zoophysiologischen Institut Uppsala gemeinsam mit Kihlström, dem Entdecker der meisten dieser cyclischen Phänomene,
weitere Tierexperimente zur Frage eines männlichen Sexualcyclus durchzuführen.
Unser Versuchsmaterial, geschlechtsreife männliche Kaninchen, wurde dabei über
längere Zeit *täglich* den folgenden gleichbleibenden Bedingungen unterworfen:

1. Jeweils morgens wurden mittels eines angefeuchteten Glasstäbchens Zell-
abstriche von den vorderen $^2/_3$ der Urethra entnommen und mikroskopisch untersucht.

2. Anschließend wurde die rectale Körpertemperatur mit einem elektrischen
Thermometer gemessen.

3. Anschließend wurde mittels einer künstlichen Vagina Ejaculat abgenommen und die Menge bestimmt.

Angaben über Einzelheiten der Methodik befinden sich z. Z. in den Acta Endocrinologica im Druck.

Pro Tier wurden täglich 100 Epithelzellen der Urethra ausgezählt und cytologisch typisiert, wobei sich 5 Zelltypen unterscheiden ließen: Typ 1 und 2 als unreife, großkernige Zellen von verschiedener Dimension, Typ 3 als weiter ausgereifte, aber noch unverhornte Zelle, Typ 4 als verhornte Zelle mit pyknotischem Kern, Typ 5 als gleichfalls verhornte, aber großkernig gebliebene Zelle.

Sämtliche Befunde wurden erst nach Beendigung der Versuchsserie verglichen und nach dem von KERMACK und McKENDRICK angegebenen Verfahren statistisch ausgewertet. Es kommt dabei auf den Nachweis einer vorhandenen oder fehlenden Übereinstimmung von empirisch gefundenen Meßwerten mit theoretisch auf Zufallsverteilung berechneten Vergleichswerten an, und zwar mit Hilfe der χ^2-Methode. Auf diese Weise wurden nicht nur die Schwankungen der urethralen Epitheldesquamation, der Körpertemperatur und des Ejaculatvolumens isoliert auf cyclisches Verhalten geprüft, sondern diese biologischen Kriterien wurden auch auf etwa bestehende synchrone Beziehungen untereinander untersucht.

Als Ergebnis dieser Untersuchungen fand sich ein eindeutiger, *statistisch gesicherter Cyclus sowohl bezüglich der Exfoliation des Urethralepithels, als auch des Temperaturverlaufs und des Volumens der Ejaculate.* Die durchschnittliche, in Tagen angegebene Cyclusdauer der einzelnen Kriterien ist für jedes Versuchstier getrennt in Tab. 1 angegeben:

Tabelle 1. *Durchschnittliche Dauer der gefundenen Cyclen in Tagen*
Untersuchte Characteristica

Kaninchen Nr.	Zelltyp 1+2	Zelltyp 3	Zelltyp 5	Temperatur	Ejaculat-Volumen
191	4,2	4,6	5,5	4,7	4,1
190	4,0	3,6	4,5	4,7	3,5
252	3,2	3,3	3,8	4,0	3,9
198	4,8	5,0	5,5	3,7	4,3
154	4,6	3,7	8,5	4,4	3,9
155	4,7	4,1	8,0	4,4	3,9
156	4,2	3,6	3,8	3,5	4,3
157	4,0	4,5	—	4,0	4,0
158	5,0	4,0	4,0	4,7	3,9

Bei 4 Zelltypen (Typ 1, 2, 3, 5) konnte ein Rhythmus der Exfoliation gefunden werden, dessen mittlere Dauer — bezogen auf alle 9 Versuchstiere — 4,2 (Typ 1 und 2) bzw. 3,9 (Typ 3) bzw. 5,1 (Typ 5) Tagen entspricht. Die Verlängerung der mittleren Cyclusdauer beim Zelltyp 5 erklärt sich durch die statistische Interferenz einzelner Cyclen von verdoppelter Dauer (sog. Multiple des Grundcyclus, z. B. bei Versuchstier 154 und 155). Nur die verhornten, kernpyknotischen Zellen des Typus 4 zeigten keine regelmäßige Periodizität ihrer täglichen prozentualen Schwankungen.

Für das *Samenvolumen* bestätigten sich die schon früher gefundenen cyclischen Schwankungen mit Intervallen zwischen den Minima von knapp 4 Tagen (interpolierte Werte aller Versuchstiere). Die Abnahme der Samenmenge war dabei statistisch häufiger mit einer Abnahme von Zelltyp 3 und einer Zunahme von Zelltyp 4 korreliert als mit dem umgekehrten Verhalten.

Schließlich zeigte auch die *Temperaturkurve* einen cyclischen Verlauf mit einer durchschnittlichen Periodenlänge von annähernd $4^{1}/_{2}$ Tagen. Jedoch ließ sich hier keine statistisch gesicherte Beziehung zum exfoliativen Cyclus der in der Tab. 1 aufgeführten Zelltypen feststellen.

In einer anschließenden mehrwöchigen Versuchsreihe haben wir nun die gleichen cytologischen Untersuchungen und Temperaturmessungen *nach Kastration* der Versuchstiere wiederholt. Dabei zeigte sich, daß spätestens vom 8. Tag an *der zuvor regelmäßige Exfoliationscyclus zusammenbricht* und außerdem eine charakteristische Verschiebung in der durchschnittlichen prozentualen Häufigkeit der desquamierten Epithelzellen auftritt (Tab. 2). Die großkernigen jugendlichen Zellen (Typ 1 und 2) und die großkernigen verhornten Zellen (Typ 5) nehmen erheblich zu, die Zellen des Typus 3 nehmen stark ab. Die Unterschiede sind statistisch hochsignifikant. Nur Zelltyp 4, der bereits beim intakten Kaninchen keine cyclischen Schwankungen der Desquamation erkennen ließ, bleibt unbeeinflußt.

Tabelle 2. *Vergleich zwischen der prozentualen Häufigkeit der verschiedenen exfoliierten Zelltypen vor und nach Kastration*

Zelltyp	Vor Kastration Mittelwerte	Nach Kastration Mittelwerte	Wahrscheinlichkeit der Differenz
1	$1,18 \pm 0,14$	$6,04 \pm 0,68$	$< 0,0005$
2	$4,45 \pm 0,13$	$10,15 \pm 0,85$	$< 0,0005$
3	$44,68 \pm 3,25$	$14,33 \pm 1,49$	$< 0,0005$
4	$46,89 \pm 4,14$	$46,95 \pm 2,20$	—
5	$1,36 \pm 0,48$	$23,36 \pm 1,74$	$< 0,0005$

Entgegen unserer ursprünglichen Erwartung blieb dagegen der Temperaturcyclus auch nach der Kastration erhalten. Schon bei den intakten Kaninchenböcken hatte sich keine sichere Beziehung zwischen dem Exfoliationscyclus der Zelltypen 1 bis 3 und 5 und dem Temperaturcyclus herstellen lassen. Lediglich zwischen der Zunahme des (keiner cyclischen Desquamation unterworfenen) Zelltyps 4 und dem Abfall der Temperatur war bei den Normaltieren ein statistischer Zusammenhang gefunden worden.

Welche Schlußfolgerungen können aus diesen Befunden gezogen werden?

1. Bei geschlechtsreifen männlichen Kaninchen besteht eine cyclisch sich ändernde Exfoliation des Urethralepithels, die ein morphologisches Äquivalent zur cyclischen Abschilferung des Vaginalepithels darstellt und mit dem Oestrus des domestizierten weiblichen Kaninchens biologisch vergleichbar ist.

2. Dieser Exfoliationscyclus weist Beziehungen zu cyclischen Schwankungen der täglichen Ejaculatmenge auf. Er unterliegt dem hormonalen Einfluß der Testes, da er nach Kastration verschwindet.

3. Ein außerdem beobachteter Temperaturcyclus ist von der Hodenfunktion unabhängig bzw. durch Kastration nicht aufhebbar.

4. Der geschilderte Exfoliationsrhythmus ist *Ausdruck eines männlichen Sexualcyclus*, dessen Existenz bislang für das Kaninchen und den Stier bewiesen worden ist. Es gibt also — jedenfalls bei diesen Tieren — eine Rhythmik der Sexualfunktionen bei *beiden* Geschlechtern. Ob in dieser Feststellung eine auch für andere Species gültige biologische Gesetzmäßigkeit liegt, ist vorläufig noch unbekannt.

Diskussion

Jöchle (München):

Tägliches Absamen-Lassen von Bullen wie Kaninchenböcken über längere Zeit bedingt eine Überlastung und Erschöpfung der Spermienproduktion; diese Erfahrung läßt vermuten, daß es sich bei den nachgewiesenen periodischen Veränderungen in der Ejaculatqualität um einen artifiziell induzierten Erholungsrhythmus handeln könnte.

O. HORNSTEIN:

Die Bestimmung des Ejaculat-Volumens wurde ebenso wie die cytologische Untersuchung mit voller Absicht täglich durchgeführt, da bei größeren Abständen cyclische Schwankungen nicht zu erfassen gewesen wären. Nur zweimal wöchentlich Ejaculat abzunehmen, wie Sie vorschlagen, wäre bei einem 4tägigen Sexualcyclus völlig unergiebig. Im übrigen müßte, wenn Ihr Einwand einer erschöpfungsbedingten Volumenverminderung richtig wäre, die Samenmenge allmählich immer geringer werden. Das ist aber nicht der Fall, die cyclische Schwankung bleibt in der ursprünglichen Amplitude erhalten. Ich möchte nochmals betonen, daß die hier untersuchten Kriterien nur Teilphänomene eines männlichen Sexualcyclus sind, die sich in ihrer Rhythmik gut in das Ensemble der bereits gefundenen cyclischen Phänomene einfügen.

E. TONUTTI (Bonn):

Gibt es im Rahmen der sog. cyclischen Veränderungen eine Periode der Infertilität ?

O. HORNSTEIN:

Untersuchungen über wechselnde Fertilität der Kaninchenböcke haben wir nicht unternommen, sie wären angesichts der kurzen Cyclusdauer wahrscheinlich mit Unsicherheitsfaktoren belastet. Die Zahl und Konzentration der Spermien ist aber schon früher von Herrn KIHLSTRÖM untersucht worden, wobei sich statistisch gesicherte Unterschiede zwischen Minima und Maxima zeigten. Ob die ersteren eine Subfertilität bedingen, erscheint möglich. Bei Bullen sind von KIHLSTRÖM cyclische Schwankungen der Fertilität nachgewiesen worden, korreliert mit cyclischem Absinken der Spermienzahl. Auch hierbei wurde täglich Ejaculat mittels einer künstlichen Vagina entnommen.

Herrn JÖCHLES Einwand, daß diese Belastung unphysiologisch sei, mag für die Praxis der künstlichen Besamung in den Zuchtanstalten zutreffen; es kam aber zunächst nur auf den Nachweis einer cyclischen Variation an, und diese hat sich auch nach mehrmonatiger Dauer des Versuches bestätigt. Das Verfahren hat aber auch eine bedeutsame praktische Seite:

Wird Ejaculat nur während der cyclischen Gipfelperioden gewonnen, so steigt die Befruchtungsrate erheblich. Übrigens hat sich gezeigt, daß der gefundene Rhythmus bei alten Bullen zurückgeht.

Aus der Universitäts-Frauenklinik Mainz

Der Einfluß hoher Kochsalzgaben auf die Aldosteron- und Elektrolytausscheidung bei gesunden Schwangeren

Von

G. Stark

Mit 2 Abbildungen

Die Ätiologie und die Pathophysiologie der Schwangerschaftstoxikosen in dem letzten Schwangerschaftsdrittel ist noch nicht eindeutig geklärt. Die charakteristische Symptomentrias dieses Krankheitsbildes zeigt neben einer Hypertonie, einer Proteinurie, eine ganz erhebliche Wasserretention, so daß es zur Ausbildung von Ödemen kommt.

Es ist heute bekannt, daß an der Entstehung hydropischer Zustandsbilder eine tubuläre Retention von Natrium entscheidend beteiligt ist, die ihrerseits auf eine hier nicht näher zu diskutierende Weise eine Retention von Wasser nach sich zieht. Als eine Ursache für die Natriumretention ist heute die gesteigerte Sekretion des Natrium-retinierenden Hormons — Aldosteron — durch die Nebennierenrinde anzusehen, so daß bei diesen Fällen mit Ödemen eine über die Norm hinausgehende Ausscheidung im Urin nachzuweisen ist.

Bei der Untersuchung solcher Gestosen haben wir, ähnlich wie andere Untersucher, zeigen können, daß der Mittelwert der Aldosteronausscheidung mit etwa 25 γ/Tag signifikant kleiner ist, als der Mittelwert bei der normalen Schwangerschaft, der bei uns bei 80 γ/Tag lag. Auch die Einzelwerte der Gestosen lagen im allgemeinen niedriger, als sie bei der normalen Gravidität gefunden werden. Am eindrucksvollsten waren unsere Befunde bei der Eklampsie, die mit typischen Krampfanfällen einhergeht. Bei dieser Form der Schwangerschaftstoxikose fanden wir teilweise Werte, die sich im Bereich nichtschwangerer Frauen bewegten, und zwar zwischen 5—10 γ/Tag.

Die Ursache dieser außerordentlich niedrigen Aldosteronausscheidung bei den Gestosen ist noch nicht klar. Sicher dürfte sein, daß es bei diesen Krankheitsbildern zu einer vermehrten Natriumretention kommt. So konnte man bei den Schwangerschaftstoxikosen mit Hilfe von radioaktiv markierten Substanzen einen vergrößerten Natriumraum sowie erhöhte Werte des austauschbaren Natriums nachweisen. Besonders deutlich waren die Unterschiede bei den Eklampsien und Präeklampsien mit starker Ödemneigung.

Als mögliche Ursache der niedrigen Aldosteronausscheidung bei den Gestosen wird von Bartholomey und Onnis sowie Antoni eine Unterfunktion der Nebennierenrinde angenommen, da sie bei an Eklampsie verstorbenen Patientinnen in der Nebennierenrinde eine histochemisch deutlich verminderte Aktivität gefunden haben.

Andererseits ist bekannt, daß die Steuerung des Aldosteron-Stoffwechsels weitgehendst von dem jeweiligen Natrium-Bestand des Körpers abhängen kann.

Zur Klärung der Frage, ob möglicherweise die starke Natriumretention bei den Gestosen die Ursache für die verminderte Aldosteronausscheidung im Urin ist, haben wir bei gesunden Schwangeren am Ende der Gravidität diese mit hohen Natrium-Gaben in Form von Kochsalz belastet und die Ausscheidung des Aldosterons und der Elektrolyte nachgeprüft.

Zur Untersuchung kamen 11 gesunde Schwangere im 7. bis 10. Schwangerschaftsmonat, die während des Versuchs eine Standarddiät erhielten. Der Urin wurde quantitativ gesammelt und das Aldosteron nach der von NEHER und WETTSTEIN angegebenen Methode bestimmt. Um Störungen der Resorption im Magen-Darm-Kanal zu vermeiden, wurde das Kochsalz intravenös als 1,8%ige Lösung in 1000 ml entsprechend = 18 g Kochsalz oder 308 mval/Tag infundiert. Alle Infusionen wurden von den Patientinnen ohne Reaktion vertragen. In einer Synopse wurden alle Einzelwerte und Befunde bei einer Patientin jeweils graphisch dargestellt. Jeder Versuch bestand aus einer

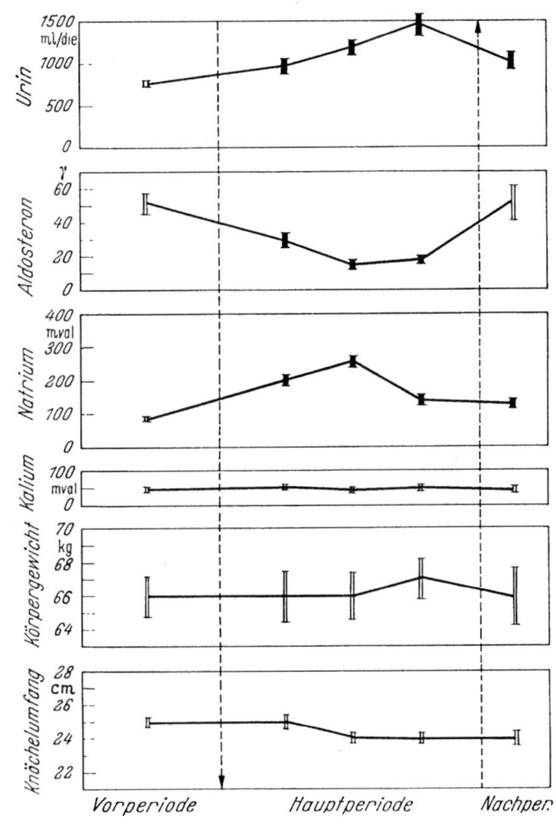

Abb. 1. Der Einfluß hoher NaCl-Gaben über mehrere Tage bei gesunden Schwangeren auf die Aldosteron- und Elektrolytausscheidung (Mittelwerte). Schwarze Säulen: Signifikante Veränderungen zur Vorperiode

5tägigen Vor-, einer 10tägigen Versuchs- und einer 2tägigen Nachperiode. Neben dem Natrium und Kalium sowie der Urin- und Aldosteronausscheidung wurden zusätzlich auch noch das Gewicht, der Blutdruck und als Ausdruck einer Ödembildung der Knöchelumfang kontrolliert.

Das eindrucksvollste Ergebnis war die Tatsache, daß sich in jedem Fall unserer untersuchten Patientinnen eine deutlich verminderte Ausscheidung des Aldosterons des in der Vorperiode erhöhten Aldosterons unter der Kochsalzbelastung nachweisen ließ. Nach dem Absetzen der Kochsalz-Infusionen kehrten die Werte wieder zu denen der Vorperiode zurück.

Abbildung 1 gibt die Mittelwerte unserer Versuche wieder, wobei wir die Versuchsperiode in 3 Abschnitte unterteilt haben. Besonders eindrucksvoll war der Aldosteron-Abfall zwischen dem 4. und 6. sowie 7. und 10. Tag der Versuchsperiode, wobei die schwarzen Säulen eine signifikante Veränderung gegenüber der Vorperiode bedeuten. Das fortlaufend registrierte Körpergewicht zeigte in Einzelfällen eine Zunahme, in anderen Fällen jedoch bleibt dieses unverändert, so daß die Mittelwerte keine signifikanten Veränderungen aufweisen. Das gleiche gilt für den Knöchelumfang und den hier nicht aufgezeichneten Blutdruck. Es ist verständlich, daß infolge der erhöhten Natrium- und Wasserzufuhr die Ausscheidungsmittelwerte von Natrium und Wasser in der Versuchsperiode gegenüber der Vorperiode signifikant erhöht sind, während Kalium unverändert bleibt.

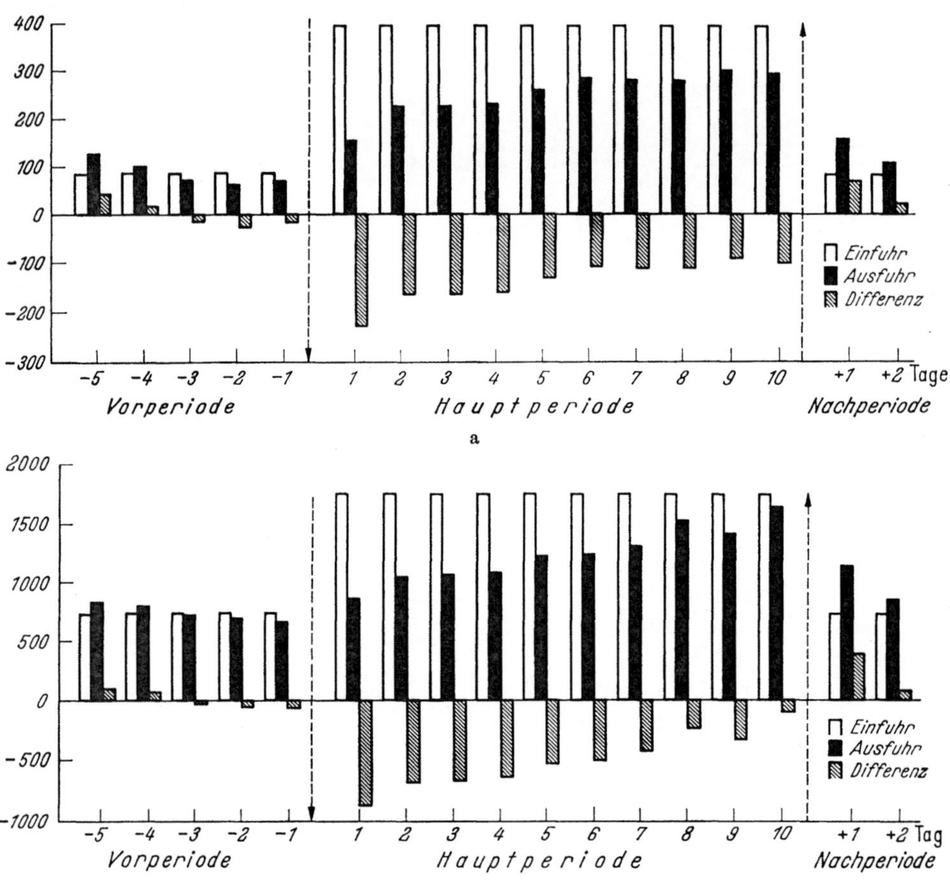

Abb. 2 a u. b. Das Verhalten der Natrium- und Wasser-Retention bei gesunden schwangeren Frauen nach hohen NaCl-Gaben. a Natrium-Retention (Mittelwerte in mval/die); b Wasser-Retention (Mittelwerte in ml/die)

Doch trotz der signifikant erhöhten Wasser- und Natriumausscheidung in der Versuchsperiode ließ sich bei der Gegenüberstellung der Ein- und Ausfuhr — der Bilanzierung — doch eine deutliche Retention von Natrium (s. Abb. 2a) und Wasser (s. Abb. 2b) feststellen, die besonders eindrucksvoll in den ersten Tagen der Kochsalzbelastung ist, während sie sich am Ende der Hauptperiode langsam auszugleichen scheint.

Unsere Befunde stimmen damit mit denen von Ehrlich u. Mitarb. überein, die bei Untersuchung an 3 gesunden Schwangeren am Ende der Zeit nach einer zusätzlichen Kochsalzbelastung von 6—12 g auch eine deutliche Senkung der Aldosteronausscheidung feststellten.

Bei Vorliegen dieser Befunde fragt man sich, inwieweit die Hormone der Nebennierenrinde überhaupt ursächlich für die Entstehung der Präeklampsie und Eklampsie anzusprechen sind. Bekannt ist, daß es trotz einer Adrenal-Ektomie zu den Erscheinungsbildern einer Eklampsie kommen kann. Daß eine Natrium-Überladung des Organismus primär für die Ausbildung einer Gestose verantwortlich zu machen ist, ist aufgrund unserer Untersuchungen sowie der von ROBINSON u. a. als sehr unwahrscheinlich anzusehen, da diese Autoren bei gesunden Schwangeren und selbst bei leichten Präeklampsien nach einer kochsalzreichen Diät bzw. zusätzlichen Kochsalzbelastung keine Verschlechterung des klinischen Zustandsbildes gesehen haben. Auch die zusätzliche Gabe von Doca und Aldosteron zu den Kochsalzgaben führte bei gesunden Schwangeren nicht zu den Erscheinungen einer Toxikose bzw. zu einer deutlichen Veränderung des Elektrolytstoffwechsels.

Da mit der Clearance-Methode bei den Gestosen eine veränderte Funktion der Niere nachgewiesen wurde, muß dieses Organ hierbei offenbar eine besondere Rolle spielen. Infolge einer verminderten Nierendurchblutung kommt es hier zu einer Abnahme des Glomerulus-Filtrates und Herabsetzung der Natriumausscheidung mit nachfolgender glomerulotubulärer Imbalanz. Bioptische Untersuchungen der Niere, die bei den Gestosen eine Verdickung der Kapselmembran, gelegentlich auch Gefäßsklerosen zeigen, erklären die pathologisch veränderte Funktion der Niere.

Wir glauben deshalb, daß es vorwiegend aufgrund der veränderten Nierenfunktion bei den Schwangerschaftstoxikosen zu einer vermehrten Natriumretention kommt, und daß der daraus resultierende erhöhte Salzbestand des Organismus für die niedrige Aldosteron-Ausscheidung bei den Gestosen verantwortlich zu machen ist.

Ähnlich wie bei unseren Kochsalz-Belastungsversuchen besteht bei diesen Krankheitsbildern bereits ein so hoher Natriumbestand im Organismus, daß nicht noch zusätzlich eine vermehrte Bildung des Natrium-retinierenden Hormons — Aldosteron — zur Natriumretention erforderlich ist.

Aus der Medizinischen Universitäts-Poliklinik Bonn
(Direktor: Prof. Dr. F. Tiemann)

Das Verhalten der β-Glucuronidase-Aktivität in Leber und Niere im Verlauf der Schwangerschaft

Von

R. E. Dohrmann und B. Louven

Mit 1 Abbildung

Nach den Untersuchungen von Fishman ist kein Zweifel daran zu erheben, daß Beziehungen zwischen der β-Glucuronidase (β-Gluc.) und dem Oestrogenstoffwechsel bestehen. So zeigt sich im letzten Drittel der Schwangerschaft ein deutlicher Anstieg der Serum-β-Gluc., die nach dem Partus innerhalb von 5 Tagen wieder auf den Normalwert abfällt (5). Unter dem Einfluß von Oestrogenen vermindert sich demgegenüber die Fermentaktivität im Puerperium nur geringfügig und bleibt während dieser Hormonsubstitution auf erhöhten Werten (5). Anhand von tierexperimentellen Untersuchungen konnte Fishman zeigen, daß es bei ovariektomierten Mäusen unter dem Einfluß von Oestrogenderivaten zu einem starken Anstieg der β-Gluc.-Aktivität des Uterus kommt, während sich die Fermentkonzentrationen von Leber, Milz und Nieren kaum veränderten (3, 4).

In der vorliegenden Arbeit wurde das Verhalten der β-Gluc.-Aktivität von Leber- und Nierengewebe im Verlauf der Schwangerschaft und nach dem Partus untersucht. Im weiteren sollte zur Frage Stellung genommen werden, ob sich während dieser Zeitabschnitte Hemm- oder Aktivierungsfaktoren in den Organen nachweisen lassen, die eine evtl. Aktivitätsänderung bedingen können.

1. Methodik

Als Versuchstiere dienten Mäuse des gleichen Stammes (BLH) mit einem durchschnittlichen Gewicht um 25 g. Bei einer Schwangerschaftsdauer von 21 Tagen dienten für die Fermentbestimmung Tiere des 5., 10., 15. und 20. Schwangerschaftstags sowie 5 und 10 Tage post partum. Es wurden jeweils 3 Tiere der gleichen Schwangerschafts- und Puerperalphase in einem Versuch zusammengefaßt. Desgleichen wurden die Fermentaktivitäten von Leber und Nieren der Jungtiere des 5. und 10. Lebenstages bestimmt.

Die Fermentbestimmung und Ermittlung des Hemm- bzw. Aktivierungsfaktors erfolgte nach früher angegebenen Methoden (1, 2).

Die Aktivitäten in der Tabelle und Abbildung sind angegeben in „γ-freigesetztes Phenolphthalein/g Gewebe $\times$ Std". Die Hemmung bzw. Aktivierung ist ausgedrückt in „prozentualer Abweichung der gemessenen von der errechneten Aktivität".

2. Ergebnisse

Die β-Gluc.-Aktivität der Leber steigt von einem Ausgangswert mit 3166 $\pm$ 158 γ/g Gewebe $\times$ Std von Beginn der Schwangerschaft ab kontinuierlich an und erreicht ein Maximum gegen Ende derselben am 20. Schwangerschaftstag mit

6242 ± 193 γ/g Gewebe × Std (Abb. 1). Bei einer Kontrollgruppe nicht schwangerer weiblicher Mäuse des gleichen Stammes wurde zum selben Zeitpunkt nur eine Aktivität der Leber-β-Gluc. von 3151 ± 183 γ/g Gewebe × Std gemessen (s. Tab. 1). Damit dürfte bewiesen sein, daß der Fermentanstieg nicht durch andere Faktoren (Nahrung, Umweltsbedingungen usw.), sondern durch die Gravidität bedingt ist.

Die β-Gluc.-Aktivität der Nieren stieg während der Schwangerschaft von einem Ausgangswert mit 1107 ± 28 γ/g Gewebe × Std gleichfalls stetig an (Abb. 1).

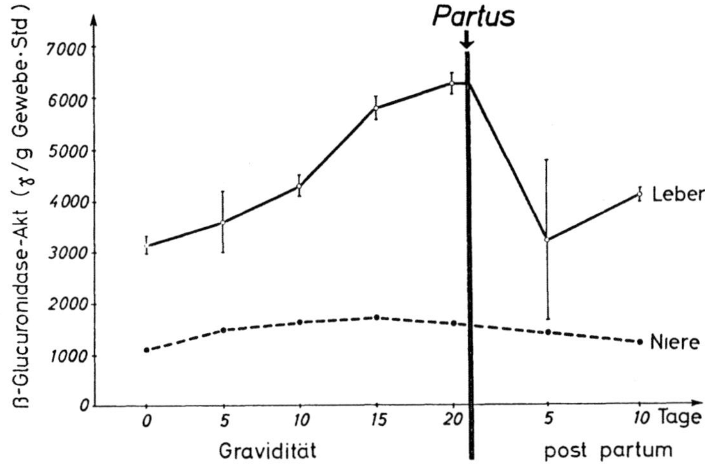

Abb. 1. Verhalten der β-Glucuronidase-Aktivität von Leber- und Nierengewebe der Maus während der Schwangerschaft und post partum

Das Maximum wurde hierbei aber schon am 15. Schwangerschaftstag mit 1674 ± 85 γ/g Gewebe × Std erreicht. Am Ende der Schwangerschaft war die Aktivität auf 1590 ± 146 γ/g Gewebe × Std wieder etwas abgefallen. Der Kontrollwert nicht schwangerer Mäuse lag zum gleichen Zeitpunkt mit 1272 ± 114 γ/g Gewebe × Std deutlich niedriger und entsprach etwa dem Ausgangswert (s. Tab. 1).

Tabelle 1

Die Zahlen stellen den Mittelwert von jeweils drei einzelnen Versuchsergebnissen dar

Zeitphase der Gravidität resp. des Puerperiums	Fermentaktivität (γ/g Gewebe × Std)		Hemm- resp. Aktivierungsfaktor (proz. Abweichung der gemess. v. errechneter Aktiv.)	
	Leber	Niere	Leber	Niere
Leerwert	3166 ± 158	1107 ± 28	+ 2	− 3
5. Tag	3590 ± 618	1469 ± 80	+ 2,3	+ 0,3
10. Tag	4295 ± 193	1621 ± 95	− 0,3	+ 2,8
15. Tag	5787 ± 260	1674 ± 85	+ 3,1	+ 3,1
20. Tag	6242 ± 193	1590 ± 146	− 6	− 2,7
Leerwert	3151 ± 183	1272 ± 114	− 4,1	− 4,1
5. Tag p. p. . . .	3204 ± 1750	1431 ± 296	0	− 1,7
Tochtergen. . . .	3283 ± 807	2249 ± 705	− 6	− 2,5
10. Tag p. p. . . .	4124 ± 103	1170 ± 57	+ 3,2	− 5,5
Tochtergen. . . .	4113 ± 523	1997 ± 93	− 2,5	− 5,7

Bereits 5 Tage nach dem Partus war die Fermentkonzentration der Leber mit 3204 ± 1750 γ/g Gewebe $\times$ Std auf den Ausgangswert zurückgegangen, während die Normalisierung der Nieren-β-Gluc. erst 10 Tage nach dem Partus erreicht war (Abb. 1).

Bei der Tochtergeneration entsprach die Leber-β-Gluc. der mütterlichen Fermentkonzentration. Demgegenüber lag die β-Gluc.-Aktivität der Nieren bei der Tochtergeneration deutlich höher als bei den Muttertieren. Die entsprechenden Fermentaktivitäten sind in der Tabelle wiedergegeben.

Ein Hemm- bzw. Aktivierungsfaktor für β-Gluc. konnte in der Leber und Niere während der Schwangerschaftsperioden und der Puerperalzeit nicht ermittelt werden (s. Tab. 1). Die durchschnittlichen prozentualen Abweichungen der gemessenen von den errechneten Aktivitäten betrugen bei der Vorlage des Reinfermentes in einer Konzentration um 10000 γ/100 ml $\times$ Std für die Leber zwischen $+ 3,2\%$ und $- 6\%$, für die Nieren zwischen $+ 3,1\%$ und $- 5,7\%$. Die Abweichungen innerhalb der einzelnen Versuchsgruppen waren darüber hinaus nicht einheitlich im Sinne einer Hemmung oder Aktivierung.

3. Besprechung

Unter der Vorstellung eines Einflusses von β-Gluc. auf die Glucuronidierungsvorgänge folgerte Fishman anhand seiner Befunde, daß die Oestrogene am Ort ihrer Wirkung durch die fermentative Kopplung an Glucuronsäure für ihre biologische Funktion im Uterusgewebe nutzbar werden (4). Diese Hypothese dürfte bei den heutigen Kenntnissen über den Hormonstoffwechsel und die Entgiftungsvorgänge im Rahmen der Glucuronidierung kaum noch aufrechterhalten werden können.

Entgegen den Befunden von Fishman u. Mitarb. zeigen die vorliegenden Ergebnisse, daß die β-Gluc. während der Schwangerschaft sowohl in der Leber als auch im Nierengewebe ansteigt. Die Zunahme der Fermentaktivität beträgt in der Leber gegen Ende der Gravidität etwa 100%, während der Anstieg in der Niere mit 50% über dem Ausgangswert schon am Ende des dritten Viertels der Schwangerschaft erreicht wird. Dieser Anstieg der β-Gluc.-Akt. ist nicht auf aktivierende oder hemmungsbeeinflussende Faktoren zurückzuführen, sondern wahrscheinlich als Folge einer Zunahme von Fermenteiweiß aufzufassen. Dies könnte aus der Zunahme der Fermentaktivität, bezogen auf den Eiweißgehalt des Homogenates, gefolgert werden, was bei den vorliegenden Versuchen auch gemessen wurde.

Nachdem Fishman unter dem alleinigen Einfluß von Oestrogenen nur einen β-Gluc.-Anstieg im Uterusgewebe feststellte, wird die nachgewiesene Zunahme der Fementaktivität in Leber und Nieren während der Schwangerschaft wahrscheinlich durch die hierbei erhöhte Konzentration an Gestagenen und möglicherweise auch Steroiden bedingt sein. In Analogie zu anderen Befunden, die ein gleichsinniges Verhalten zwischen β-Gluc.-Aktivität und Glucuronidierungsrate erkennen lassen, können wir bei der bis heute unbekannten biologischen Funktion dieses Fermentes den beschriebenen Vorgang lediglich als hormonale Enzyminduktion deuten.

Zusammenfassung

Während der Gravidität kommt es sowohl in der Leber als auch in der Niere zu einem Anstieg der β-Glucuronidase-Aktivität. Dies ist wahrscheinlich auf eine

Zunahme des Fermenteiweißes zurückzuführen und kann nicht als Folge von aktivierenden Faktoren auf das Ferment gedeutet werden. Nach dem Partus erreicht die Leber nach etwa 5 Tagen und die Niere nach 10 Tagen wieder ihre normale Fermentaktivität.

Literatur

1. DOHRMANN, R. E.: Dtsch. Arch. klin. Med. **206**, 322—333 (1960).
2. —, u. H. J. UHLES: Klin. Wschr. **11**, 527—531 (1963).
3. FISHMAN, W. H., and L. W. FISHMAN: J. biol. Chem. **152**, 487—488 (1944).
4. — J. biol. Chem. **169**, 7—15 (1947).
5. — L. D. ODELL, J. E. GILL, and R. A. CHRISTENSEN: Amer. J. Obstet. Gynec. **59**, 414 (1950).

Aus dem Pathologischen Institut der Universität Würzburg

Die DNS-Synthese im Hypophysenvorderlappen der Ratte bei funktioneller Belastung[1]

Von

G. Dhom und E. Stöcker

Mit 1 Abbildung

Im Hypophysenvorderlappen (HVL) sind regressive Zellveränderungen auch in lebensfrisch fixierten Organen ein regelmäßiger Befund. Man muß also annehmen, daß ständig ein gewisser Zellwechsel stattfindet. Sein Umfang ist bis jetzt aber unbekannt. Der Zellersatz soll nach Romeis bevorzugt über Amitosen erfolgen. Auf der anderen Seite sind Mitosen, speziell im Vorderlappengewebe von Versuchstieren ein geläufiger Befund. Schließlich führt die Stimulation des Vorderlappens durch Eingriffe in das endokrine System auf die Dauer zu einer echten Hyperplasie bestimmter Zellpopulationen bis zur Geschwultsbildung.

Um zu genaueren quantitativen Vorstellungen über die DNS-Neubildung im HVL unter Ruhebedingungen und bei funktioneller Belastung zu kommen, haben wir die H^3-Thymidininkorporation in die Zellkerne des Vorderlappens von Ratten autoradiographisch untersucht.

1. Methodik

Männliche, 185 bis 190 g schwere Albinoratten (BD II) erhielten 500 μC H^3-Thymidin i.p. (spez. Akt. 6,0 C/mMol, Schwarz Bio Research New York). Eine Gruppe diente als Kontrolle, 2 weitere Gruppen waren 14 Tage vor dem Eingriff thyreoidektomiert, bzw. kastriert worden. Tötung der Tiere jeweils 1 Std nach Applikation von H^3-Thymidin zu identischer Tageszeit. Fixation der Hypophyse für 48 Std in neutralem Formalin, Paraffineinbettung. Waschen der entparaffinierten Schnitte in hochkonzentrierten Lösungen von inaktivem Vorläufer, um das nicht in die neugebildete DNS-inkorporierte, d. h. nur an die Gewebestruktur adsorbierte H^3-Thymidin auszutauschen. Autoradiographie mit stripping-film (Kodak, A.R.10). Exposition der Autoradiogramme 8—14 Tage. Nachfärbung mit Haematoxylin-Eosin. Zur Differenzierung der Vorderlappenzellen wurde ein Teil der Präparate PAS-vorgefärbt und im Anschluß an die Exposition nach der Methode von Siperstein mit Hämatoxylin-Methanilgelb nachgefärbt. Bestimmt wurden der H^3-Index, d. h. der Prozentsatz der innerhalb der Verfügbarkeitszeit von H^3-Thymidin DNS-synthetisierender Kerne, und der Mitoseindex, d. h. der Prozentsatz der Mitosen. Dabei wurde stets die Gesamtfläche des horizontal

---

[1] Mit Unterstützung durch die Deutsche Forschungsgemeinschaft und das Bundesministerium für Wissenschaftliche Forschung.

geschnittenen Vorderlappens ausgewertet, pro Tier zwischen 45000 und 70000 Parenchymzellen. An den PAS-Hämatoxylin-Methanilgelb gefärbten Präparaten wurden pro Tier 100 markierte Parenchymzellen typenmäßig differenziert.

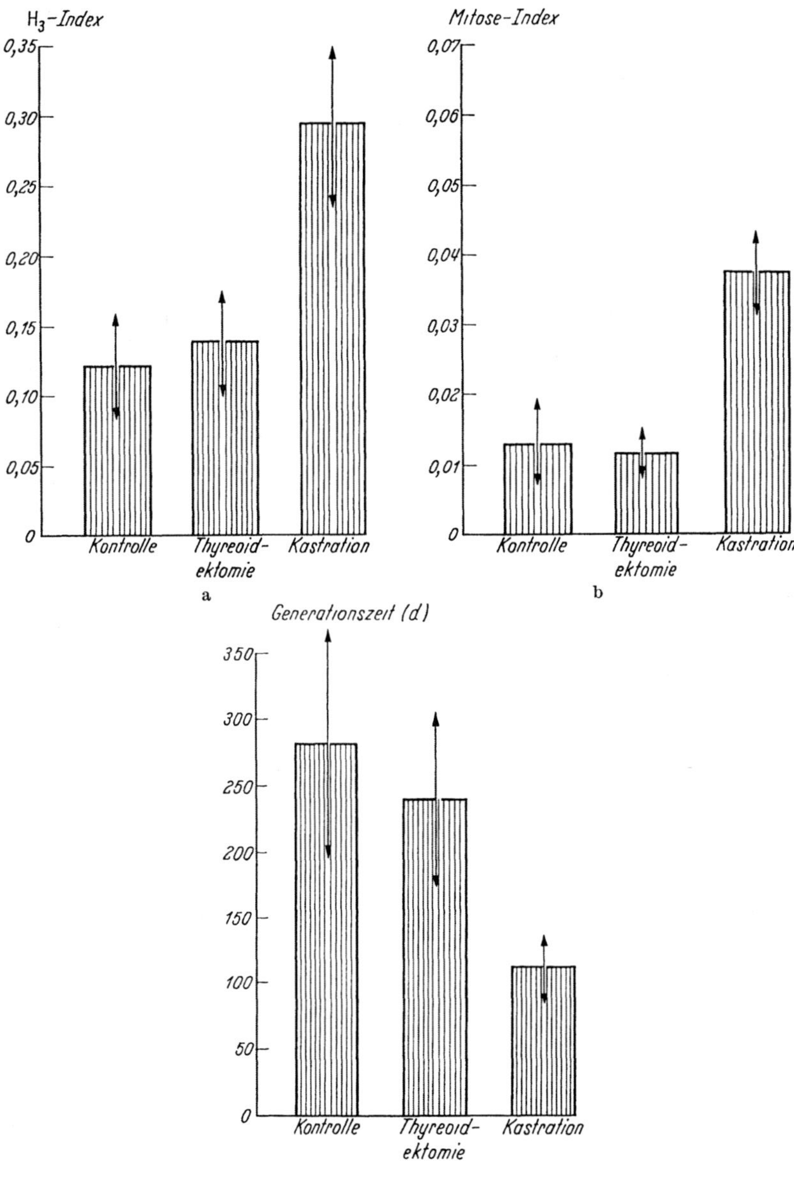

Abb. 1a—c. H³-Index, Mitose-Index und Generationszeit (in Tagen) der Parenchymzellen des Hypophysenvorderlappens der Ratte

2. Ergebnisse

Unter Ruhebedingungen findet sich im HVL ein niedriger H³-Index, im Mittel sind 12 von 10000 Parenchymkernen markiert mit einem relativ großen Schwankungsbereich (Abb. 1a u. b). Der Quotient H³-Index/Mitose-Index weist dabei

Werte auf, die mit denen anderer Gewebe durchaus vergleichbar sind (Schultze und Oehlert 1960, Noltenius, Schellhas und Oehlert 1963, Stöcker und Altmann 1964). 14 Tage nach Thyreoidektomie zeigen H^3-Index und Mitose-Index keinen signifikanten Anstieg. Nach Kastration dagegen steigt die Zahl der DNS-synthetisierenden Zellkerne etwa um den Faktor 2,5 an und in gleichem Sinne nimmt der Mitoseindex zu, der Quotient H^3-Index/Mitose-Index bleibt somit angenähert im gleichen Niveau. Dies ist ein gewichtiges Argument dafür, daß die Zellneubildung ganz überwiegend über Mitosen laufen muß. Amitosen können, wenn überhaupt, nur eine ganz untergeordnete Rolle spielen. Aus den gegebenen Werten kann man die Dauer des Generationscyclus der Vorderlappenzellen errechnen (Abb. 1c). Für die Kontrollen beträgt die mittlere Lebensdauer 281 Tage. Der HVL der Ratten gehört demnach unter „Ruhebedingungen" zu den langsam proliferierenden Geweben und ist größenordnungsmäßig etwa dem Leberparenchym an die Seite zu stellen (vgl. Stöcker und Altmann 1964). Im Gegensatz zu Nakamura et al. ist aber seine DNS-Synthese keineswegs gleich 0 zu setzen. Nach Kastration sinkt — entsprechend der größeren Zellneubildungsrate — die Dauer des Generationscyclus in unserer Versuchsreihe im Mittel auf 110 Tage ab. Die Berechnung dieser Generationszeit — ausgehend von einem steady state der Zellproliferation — bezieht sich auf das Gesamtkollektiv der Vorderlappenzellen. Man wird freilich damit rechnen müssen, daß bei längeren Zeitintervallen zwischen Operation und Tötung bestimmte Teilkollektive der Vorderlappenzellen in eine mehr exponentielle Proliferationsphase eintreten, die zur Hyperplasie führt.

Es erhebt sich daher die Frage nach der Verteilung der radioaktiv markierten Elemente auf die einzelnen Zelltypen des HVL. An den mit PAS vorgefärbten und nach der Methode von Siperstein (1963) mit Hämatoxylin-Methanilgelb nachgefärbten Autoradiogrammen kommt man zu einer guten Differenzierung der verschiedenen Vorderlappenzellen und stellt fest, daß grundsätzlich in allen Formen H^3-markierte Kerne auftreten können (Demonstration). Dabei bilden die sog. Chromophoben zweifellos einen Sammeltopf aus kleinen, undifferenzierten „Stammzellen" und größeren, aktiven, degranulierten Formen. In beiden Typen, auch in den degranulierten frühen Thyreoidektomiezellen kann eine DNS-Synthese beobachtet werden. Betrachten wir die prozentuale Verteilung der markierten Zellen auf die verschiedenen Typen (Tab. 1), so ergibt sich folgendes: Bei den Kontrollen gehören mehr als 80% der H^3-markierten Elemente zu den sog. chromophoben Zellen. Das entspricht zweifellos den Vorstellungen, die man sich

Tabelle 1. *Prozentuale Verteilung H^3-Thymidin-markierter Vorderlappenzellen bei Ruhe und funktioneller Belastung*

	Chromophobe	Gonadotrope	Thyreotrope	Acidophile
Kontrollen	85 % M = 82,7% 80,4%	5,3% M = 4,2% 3,2%	3,5% M = 5,0% 6,5%	6,2% M = 8,0% 9,9%
Thyreoidektomie 14 Tage	74,7% M = 72,4% 70,2%	3,1% M = 3,0% 2,9%	21,0% M = 20,1% 19,2%	1,2% M = 4,4% 7,7%
Kastration 14 Tage	60,8% M = 59,5% 58,3%	30,4% M = 30,7% 31,0%	5,9% M = 6,8% 7,7%	2,9% M = 2,9% 3,0%

auch bisher schon theoretisch machen konnte, daß nämlich der Zellersatz überwiegend von den undifferenzierten Stammzellen ausgeht, die hier einen wesentlichen Teil der DNS-synthetisierenden Elemente ausmachen. Ein Rest verteilt sich auf jeweils einige Prozent gonadotroper, thyreotroper (PURVES und GRIESBACH 1955, 1956) und acidophiler Zellen. Nach Thyreoidektomie steigt der Prozentsatz degranulierter Thyreotroper, nach Kastration der Anteil hyperplastischer Gonadotroper deutlich an. 14 Tage nach Thyreoidektomie werden rund 20% von überwiegend degranulierten Thyreotropen, nach Kastration rund 30% von hyperplastischen Gonadotropen gestellt. Entsprechend sinkt der Anteil der chromophoben Elemente ab. Die Zellneubildung geht also unter funktioneller Belastung gezielt auf einen Teil der spezifisch differenzierten Elemente über, ein neuerlicher Beweis für die funktionelle Spezialisierung dieser Zellformen in der Rattenhypophyse. Dabei ist interessant, daß 14 Tage nach Thyreoidektomie zwar die Gesamtzahl DNS-synthetisierender Elemente nicht signifikant erhöht ist und trotzdem eine wesentliche Steigerung des Anteils markierter Thyreotroper gegenüber den Kontrollen beobachtet werden kann.

Das im Zeitraum von 2 Wochen quantitativ unterschiedliche Ansprechen auf die beiden gewählten Eingriffe kann aufgrund dieser ersten autoradiographischen Befunde noch nicht geklärt werden. An sich übertrifft die Zahl der gonadotropen Zellen im HVL der männlichen Ratte die Summe der thyreotropen Elemente. Nach eigenen Zählungen beträgt die Relation rund 10:7. Dieser Befund kann aber nicht allein den signifikant größeren Anstieg der markierten Kerne nach Kastration befriedigend erklären. Man muß vielmehr damit rechnen, daß der Stimulus zur DNS-Neubildung nach Thyreoidektomie und nach Kastration zeitlich verschiedene Gipfel aufweist. In weiteren Untersuchungen soll daher versucht werden, die Zahl DNS-synthetisierender Kerne in Abhängigkeit von der Dauer der verschiedenen funktionellen Belastungen genauer zu erfassen.

Literatur

NAKAMURA, R. M., D. S. MIYADA, and D. L. MOYER: Nature (Lond.) **199**, 707 (1963).
NOLTENIUS, H., H. SCHELLHAS u. W. OEHLERT: Naturwissenschaften **51**, 15 (1963).
PURVES, H. D., and W. E. GRIESBACH: Endocrinology **56**, 374 (1955).
— — J. Endocr. **13**, 365 (1956).
ROMEIS, B.: Hypophyse. Handbuch mikrosk. Anatomie VI, 3. Teil. Berlin 1940.
SCHULTZE, B., and W. OEHLERT: Science **131**, 737 (1960).
SIPERSTEIN, E. R.: J. cell. Biol. **17**, 521—546 (1963).
STÖCKER, E., u. H.-W. ALTMANN: Naturwissenschaften **51**, 15 (1964).

Zusammenfassung

Autoradiographische Untersuchungen des Hypophysenvorderlappens der Ratte nach H³-Thymidin-Inkorporation zeigen einen niedrigen H³-Index bei Kontrollen. 14 Tage nach Thyreoidektomie steigt er nicht signifikant an, 14 Tage nach Kastration ist er dagegen deutlich erhöht. Die DNS-Neubildung findet bei den Kontrollen überwiegend in den chromophoben Zellen statt, nach Thyreoidektomie wächst der Anteil der Thyreotropen, nach Kastration der Anteil der Gonadotropen an der DNS-Synthese.

Wir danken Fräulein H. BREIDER, Fräulein H. ARLET und Fräulein H. NITZSCHE für ihre technische Hilfe bei der Durchführung der Untersuchungen.

Diskussion

R. Elert (Düsseldorf):

Obwohl die Rattenhypophyse kein geeignetes Objekt für das Studium der Schwangerschaftsveränderungen ist, erlaube ich mir die Anfrage, ob Sie auch Hypophysen trächtiger Ratten mit der von Ihnen beschriebenen Methode untersucht haben.

G. Dhom:

Wir haben bisher nur mit männlichen Tieren gearbeitet, da bei ihnen die Entwicklung der gonadotropen Elemente reichlicher und konstanter ist, als beim Weibchen, bei dem die gonadotropen Zellen mit dem Cyclus stark variieren.

Aus der Medizinischen Universitäts-Poliklinik, Bonn
(Direktor: Prof. Dr. F. Tiemann)

Der Einfluß von anabolen Steroiden auf Blutgerinnungsfaktoren

Von

H.-L. Krüskemper und R. Klesper

Mit 1 Abbildung

Die Untersuchung des Einflusses von anabolen Steroiden auf Gerinnungsfaktoren im Plasma wurde aus zwei Erwägungen heraus vorgenommen, einmal weil die meisten Gerinnungsfaktoren spezifische Proteide sind, deren Reaktion auf anabole Steroide von Bedeutung sein kann, zum andern da aus typischen Verschiebungen im Aktivitätsmuster der Gerinnungsfaktoren auf gewisse Funktionsstörungen der Leber geschlossen werden kann.

Versuchspersonen waren 26 gesunde Männer, 25—40 Jahre alt. 7 erhielten ein Placebo; die übrigen Vp. wurden 3 Wochen lang mit 1-Methyl-androst-1-en-17 β-ol-3-on behandelt [Gruppe 1 (10 Vp.): 30 mg 1-Methyl-androst-1-en-17 β-ol-3-on-17 β-acetat pro Tag, oral; Gruppe 2 (9 Vp.): 50 mg 1-Methyl-androst-1-en-17 β-ol-3-on pro Tag, oral]. Diese Verbindung erschien besonders interessant, da es sich um das erste oral wirksame anabole Steroid handelt, welches in 17-Position nicht alkyliert ist. Vor Versuchsbeginn und in wöchentlichen Abständen wurden die Aktivitäten folgender Faktoren im Plasma bestimmt: Prothrombin, Faktoren V und VII nach Biggs u. McFarlane (*1*), Faktor X nach Bachmann et al. (*2*), Thrombinzeit und Progressiv-Antithrombin (III) nach Quick (*3*) und das Thrombelastogramm nach Hartert (*4*).

Tabelle 1. *Das Verhalten des Prothrombins und der Faktoren V und X im Plasma gesunder Männer unter oraler Verabfolgung von 1-Methyl-androst-1-en-17 β-ol-3-on bzw. dessen 17 β-Acetats. Angegeben sind die Abweichungen vom Wert vor Beginn der Steroidmedikation in Prozent.*

Placebo: $n = 7$

	Prothrombin			Faktor V			Faktor X		
	1	2	3	1	2	3	1	2	3
		Wochen			Wochen			Wochen	
M . . .	+1,0	—0,6	—1,0	—1,0	+1,0	+2,0	—4,0	—2,9	—3,6
σM . . .	1,2	1,6	1,5	4,0	2,0	2,4	2,8	3,0	1,0

1-Methyl-androst-1-en-17 β-ol-3-on-17 β-acetat
(30 mg p.d.) $n = 10$

	Prothrombin			Faktor V			Faktor X		
M . . .	+1,5	+5,1	+5,5	+13,6	+14,3	+16,1	—4,9	—2,3	+1,9
σM . . .	1,6	2,6	3,0	4,9	5,2	5,1	2,7	3,6	3,7

1-Methyl-androst-1-en-17 β-ol-3-on
(50 mg p.d.) $n = 9$

	Prothrombin			Faktor V			Faktor X		
M . . .	+8,6	+7,6	+8,0	+28,0	+35,5	+32,2	+6,7	+6,0	+7,1
σM . . .	3,2	2,4	2,0	14,3	9,8	9,6	5,1	5,5	6,3

Die wichtigsten Ergebnisse sind in Tab. 1 zusammengefaßt. Während Thrombinzeit, Thrombelastogramm und Faktor VII keine Veränderungen zeigten, stiegen die Werte für die Faktoren V und X, das Prothrombin und das Progressiv-Antithrombin an. Dieser Anstieg konnte für die Gruppe 2 (50 mg 1-Methyl-androst-1-en-17β-ol-3-on) statistisch gesichert werden. Am Beispiel des Faktors V (Abb. 1) kann demonstriert werden, daß auch eine Dosisabhängigkeit der Wirkung des anabolen Steroids besteht.

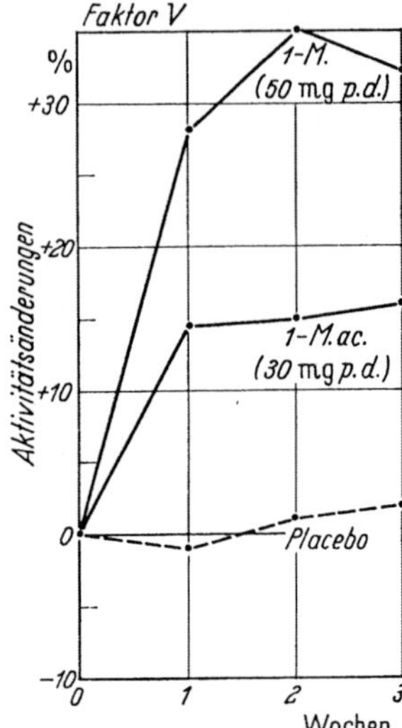

Abb. 1. Dosisabhängigkeit des Anstieges von Faktor *V* im Plasma gesunder Männer unter Behandlung mit 1-Methyl-androst-1-en-17β-ol-3-on   (1-M.) bzw. dessen 17β-Acetates 1-M. ac.)

Im ganzen sind die Verschiebungen während der Gabe des anabolen Steroids zwar eindeutig nachweisbar, jedoch nicht von solchem Ausmaß, daß (bei gesunden Versuchspersonen) Änderungen der Gerinnungszeit resultierten.

Eine befriedigende pathophysiologische Einordnung dieser Befunde ist derzeit noch nicht möglich. Zu diskutieren sind zwei Hypothesen:

1. Die Zunahme der Gerinnungsfaktoren ist Ausdruck der anabolen Wirkung des Steroids, d. h. einer Stimulation der Synthese einzelner Faktoren.

2. Die Zunahme der Faktoren ist Zeichen einer intrahepatischen Cholostase.

Mit dieser zweiten Möglichkeit ist durchaus zu rechnen, da 1-Methyl-androst-1-en-17β-ol-3-on zu einer gesteigerten BSP-Retention führen kann, und da gerade die kombinierte Zunahme der Faktoren V, X und des Prothrombins ein typischer Befund bei Cholostase ist (5).

Die erstgenannte Hypothese wird z. Z. geprüft. Ausgehend von der Tatsache, daß nach Verabfolgung natürlicher Androgene (z. B. von Testosteron bzw. seiner 17β-Ester) bislang keine Störung der Ausscheidungsfunktion der Leber nachgewiesen werden konnte, wurde orientierend untersucht, ob Unterschiede im Einfluß von 17α-Methyltestosteron und Testosteron-17β-propionat auf Gerinnungsfaktoren bestehen. Dabei zeigte sich, daß 17α-Methyltestosteron ähnliche Veränderungen wie 1-Methyl-androst-1-en-17β-ol-3-on hervorruft, während Testosteron-17β-propionat keine sichere Wirkung zu entfalten scheint; definitive Aussagen sind jedoch noch nicht möglich.

Wenn es sich bei der hier beschriebenen Zunahme von Gerinnungsfaktoren im Plasma um den Ausdruck einer steroidbedingten Leberfunktionsstörung handelt, so hat man mit der Gerinnungsanalyse ein Diagnostikum zur Hand, welches anderen Methoden (z. B. der Fermentdiagnostik und dem BSP-Test) an Empfindlichkeit weit überlegen zu sein scheint.

Literatur

1. Biggs, R., and G. McFarlane: Human Blood Coagulation and its Disorders; 2nd ed. Oxford 1957.

2. Bachmann, F., F. Duckert, M. Geiger, P. Baer, and F. Koller: Thrombos. Diathes. haemorrh. (Stuttg.) **1**, 169 (1957).

3. Quick, A. J.: Hemorrhagic Disease. Philadelphia 1957.

4. Hartert, H.: Naunyn-Schmiedebergs Arch. exp. Path. Pharmak. **222**, 154 (1954).

5. Broicher, H., H. Egli u. K. Kesseler: Dtsch. med. Wschr. **1960**, 350.

Aus den Wissenschaftlichen Laboratorien der Farbwerke Hoechst AG.

Veränderungen der mechanischen Eigenschaften des Bindegewebes unter dem Einfluß von Hormonen

Von

G. VOGEL

Mit 2 Abbildungen

In früheren Untersuchungen [1, 2, 3] konnten wir zeigen, daß bereits eine einmalige Injektion von Hormonen die Reißfestigkeit des Binde- und Stützgewebes beeinflußt.

Als einfachstes und empfindlichstes Modell dient uns die Prüfung der Reißfestigkeit der distalen Femurepiphysenfuge bei Ratten. In dieser Versuchsanordnung wird das im Hüftgelenk exartikulierte Hinterbein einer jungen Ratte am Femurkopf in einer Schlinge aufgehängt. Am Sprunggelenk wird mittels eines Hakens mit konstanter Geschwindigkeit gezogen und an einer Drehfederwaage die Reißfestigkeit gemessen.

Aufgrund unserer Kenntnisse über die Eiweiß-katabole Wirkung der Corticosteroide würde man erwarten, daß die Reißfestigkeit des Bindegewebes unter dem Einfluß dieser Hormone abnehmen würde. Überraschenderweise steigt nach einer einmaligen Injektion von Hydrocortisonacetat oder Prednisolonacetat in Kristallsuspension die Reißfestigkeit schon nach wenigen Stunden an und erreicht nach 24 bzw. 48 Std ein Maximum, um dann wieder langsam abzufallen (Abb. 1).

Von verschiedenen Corticosteroiden lassen sich Dosenwirkungskurven der Reißfestigkeitszunahme aufstellen. Wie auch aus der Relation der antiinflammatorischen Wirkung bekannt, ist Prednisolonacetat etwa 4—5mal stärker als Hydrocortisonacetat. Desoxycorticosteronacetat ist schwächer wirksam.

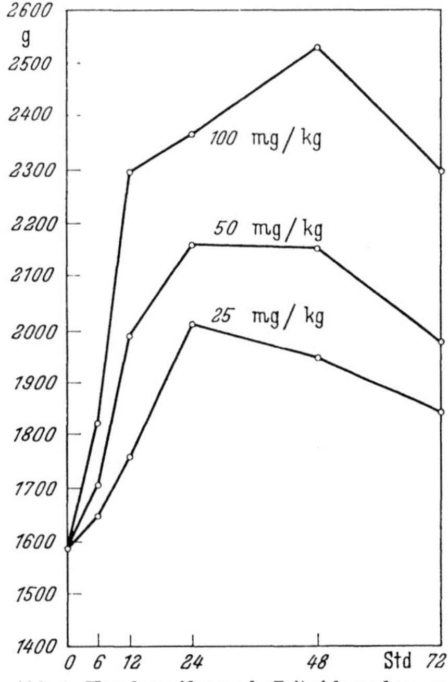

Abb. 1. Knochenreißversuch, Zeitwirkungskurven nach einmaliger Injektion von Hydrocortisonacetat

Im Kurzversuch erniedrigen Schilddrüsenhormone die Reißfestigkeit der Femurepiphysenfuge. l-Trijodthyronin ist etwa 3mal wirksamer als l-Thyroxin.

Wenn man Corticoide und Schilddrüsenhormone gleichzeitig injiziert, so tritt ein Antagonismus auf. Die durch 10 mg/kg Prednisolonacetat gesteigerte Reißfestigkeit wird durch gleichzeitige Gabe von 1 mg/kg l-Trijodthyronin fast wieder auf den Wert der unbehandelten Tiere reduziert. Progesteron setzt ebenfalls die Belastbarkeit der Femurepiphysenfuge herab und vermindert die durch gleichzeitige Gabe von Prednisolonacetat hervorgerufene Steigerung. Die Exstirpation der Nebennieren erniedrigt, die Hypophysektomie und Thyreoidektomie erhöhen die Reißfestigkeit der Epiphysenfugen.

Die an diesem Modell erhobenen Befunde ließen sich qualitativ und quantitativ an der isolierten Schwanzsehne der Ratte reproduzieren, deren Struktur wesentlich

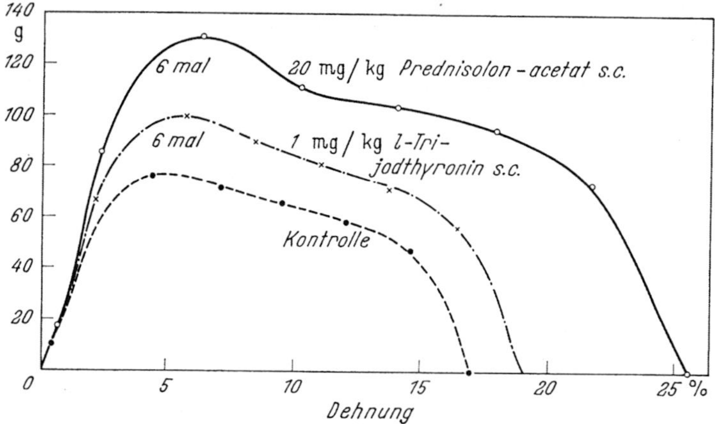

Abb. 2. Kraft-Dehnungs-Kurven an Sehnenfäden des Rattenschwanzes. Einspannlänge 2 cm, Abzugsgeschwindigkeit 1 cm/min

einfacher ist. Auch an diesem Modell lassen sich Dosenwirkungskurven und Zeit-Wirkungskurven der Änderung der Reißfestigkeit aufstellen. An Sehnenfäden, deren Durchmesser unter dem Präparationsmikroskop bestimmt wurde, konnten wir mittels des Instron [R]-Gerätes Kraftdehnungsdiagramme aufnehmen.

Über einen bestimmten Bereich entspricht das Kraftdehnungsdiagramm dem Hookeschen Gesetz: Die Kraft ist der Dehnung proportional. Mit weiterer Dehnung durchläuft die Spannung ein Maximum, um vor dem eigentlichen Reißen langsam abzufallen. Nach Vorbehandlung mit Corticosteroiden steigt das Kraftdehnungsdiagramm steiler an, d. h. die Faser ist härter geworden. Das Maximum der Spannung liegt höher und der Abriß erfolgt erst bei höherer Dehnung. Während Schilddrüsenhormone bei einmaliger Gabe die Reißfestigkeit erniedrigen, steigern sie bei mehrmaliger Gabe die Belastbarkeit, so daß das Kraftdehnungsdiagramm nach 6maliger Injektion von l-Trijodthyronin zwischen dem des Prednisolonacetats und dem der Kontrollen liegt (Abb. 2).

Werden die Sehnenfäden einer bestimmten Zugbelastung ausgesetzt und in der bis dahin erreichten Dehnung festgehalten, so fällt die gemessene Kraft erst schnell, dann immer langsamer ab. Nach 5 min ist annähernd ein Endwert erreicht. Dieser Wert ist nach Vorbehandlungen mit Corticosteroiden wesentlich höher als bei den Kontrollen, mit anderen Worten, die Relaxation ist nach Corticoiden erniedrigt. Die Relaxation der Kollagenfasern ist nach einmaliger l-Trijodthyronin-Behandlung erhöht, nach mehrmaliger erniedrigt.

Man muß aus diesen Untersuchungen den Schluß ziehen, daß Corticosteroide in niedrigen und mittleren Dosen am wachsenden Organismus entgegen der Erwartung eine Verfestigung der bindegewebigen Struktur, d. h. eine Erhöhung der Reißfestigkeit der Epiphysenfugen und Sehnenfäden bewirken. Die Festigkeitszunahme tritt überraschend schnell ein.

Während wir in anderen Untersuchungen [3] zeigen konnten, daß die Bruchfestigkeit des Diaphysenschaftes unter Corticosteroiden in Abhängigkeit von der Dosis von einer Erhöhung in eine Erniedrigung umschlägt, war unter Schilddrüsenhormonen die Zeitwirkungskurve der Reißfestigkeit zweiphasisch, d. h. es trat erst eine Erniedrigung, später eine Erhöhung der Belastbarkeit auf.

Literatur

1. THER, L., H. SCHRAMM u. G. VOGEL: Acta endocr. (Kbh.) **42**, 29 (1963).
2. VOGEL, G., u. L. THER: Naunyn-Schmiedebergs Arch. exp. Path. Pharm. **246**, 72 (1963).
3. — — Verh. dtsch. Ges. Path. **47**, 167 (1963).

Diskussion

E. TONUTTI (Bonn):

Wurden die Untersuchungen auch an kollagenen Fasern durchgeführt, bei deren *Entstehung* der Organismus bereits unter Cortisoneinfluß stand?

G. VOGEL:

Nein. Die hier vorgetragenen Befunde beziehen sich auf Versuche an jungen Ratten im Gewicht zwischen 100 und 200 g. Die Änderungen des mechanisch-physikalischen Verhaltens des Binde- und Stützgewebes unter Hormonen sind bei jungen Tieren ausgeprägter als an älteren.